PRINCIPLES OF BIOLOGY

VOLUME 2:
EVOLUTION, DIVERSITY, AND THE BIOLOGY OF ORGANISMS

Dr. Snigdha Singh and
Dr. Shubham Sewariya

ALLIED
HEALTH
PRESS

Principles of Biology
Volume 2: Evolution, Diversity, and the Biology of Organisms

Authors: **Dr. Snigdha Singh and Dr. Shubham Sewariya**

Allied Health Press
4164 Lakeshore Road
Burlington ON L7L 1A4
Canada
www.alliedhealthpress.com
Email: orders@arclereducation.com

ISBN: 978-1-77956-851-9

ABOUT THE AUTHORS

Dr. Snigdha Singh is an accomplished researcher and educator in medicinal chemistry and nanoscience, currently serving as Assistant Professor in the Department of Chemistry at the University of Delhi and Deputy Director of the Institute of Nano Medical Sciences, University of Delhi. With a distinguished academic trajectory and extensive international research exposure, she has established herself as a leading young scientist in the field of chemical and biomedical research.

She earned her Ph.D. in Medicinal Chemistry from the University of Delhi (2021), under the supervision of Prof. Ramesh Chandra, where her thesis focused on the synthesis and development of novel heterocycles and their biological evaluation. Prior to that, she completed her M.Tech. in Chemical Synthesis and Process Technologies (CSPT) at the University of Delhi with first division (2016) and her B.Sc. (Hons) in Chemistry from Hindu College, University of Delhi (2013), graduating with top merit distinctions. Her academic excellence was recognized early with a merit fellowship and multiple university-level awards.

Dr. Singh's professional journey reflects a unique blend of teaching, research, and leadership roles. Beyond her faculty appointment, she has served as Resident Tutor at Ambedkar-Ganguly Students House for Women, and has held prestigious international positions, including Postdoctoral Associate at Rensselaer Polytechnic Institute (USA), Guest Scientist at Humboldt University (Germany), and Visiting Scientist at Lobachevsky University (Russia). She was also selected as an Indo-UK Newton Bhabha Scholar at the University of Oxford and an Indo-Italy Exchange Scholar at the University of Siena, further expanding her global research collaborations. Her research contributions span drug design, nanostructures for biomedical applications, and innovative drug delivery systems. She has published extensively in high-impact journals such as ACS Sustainable Chemistry & Engineering, RSC Advances, Bioorganic & Medicinal Chemistry, and Inorganic Chemistry Communications, alongside several book chapters and authored textbooks. She is also a co-inventor on an international patent related to hydroxyethylamine-based piperazine compounds (2022). In addition, she has mentored Ph.D. candidates and supervised more than 27 graduate and postgraduate students.

Dr. Singh's work has been recognized through numerous honors, including the Indo Asian Young Women Researcher Award (2024), Young Researcher in Medicinal Chemistry Award (2024), and the Teacher Excellence Award (2023). She has received multiple international travel fellowships and was ranked 19th All-India in the UGC-CSIR NET Examination (2017).

An active member of leading scientific societies such as the Royal Society of Chemistry, American Chemical Society, and Indian Science Congress Association, Dr. Singh also serves as a reviewer for several international journals. Her career reflects a deep commitment to advancing chemical sciences while fostering global academic exchange and mentoring the next generation of researchers.

Dr. Shubham Sewariya is a Research Scientist at the University of Delhi, India, with expertise in medicinal chemistry and drug discovery. His research focuses on developing novel anticancer drug leads by combining synthetic chemistry, biological insights, and computational approaches. He holds an Integrated BS–MS degree in Chemistry from the Indian Institute of Science Education and Research (IISER), Thiruvananthapuram, and has completed his Ph.D. jointly at University of Delhi and the University of Lancashire, UK. His doctoral work centers on the discovery and development of secondary metabolites, including noscapinoids and coumarins, as potential anticancer agents. Dr. Sewariya has received several prestigious fellowships and awards, including the Commonwealth Split-site PhD Scholarship (2023–24) funded by the UK Foreign, Commonwealth & Development Office, the CSIR Senior Research Fellowship, and the INSPIRE SHE Fellowship from Department of Science and Technology, Govt of India. He has also presented his work at international platforms such as the American Chemical Society (ACS) conferences in San Francisco and Chicago. With more than a dozen peer-reviewed publications, his contributions span synthetic chemistry, nanomedicine, and biophysical studies, particularly in designing intelligent nanocarriers and molecular probes for cancer therapy. Beyond the laboratory, Dr. Sewariya is deeply committed to advancing translational research that bridges science and medicine for the betterment of human health.

Brief Table of Contents

Getting Started with This Book

Chapter Openers

A list of Chapter Outline is presented providing a brief Introduction to what will be covered in the chapter.

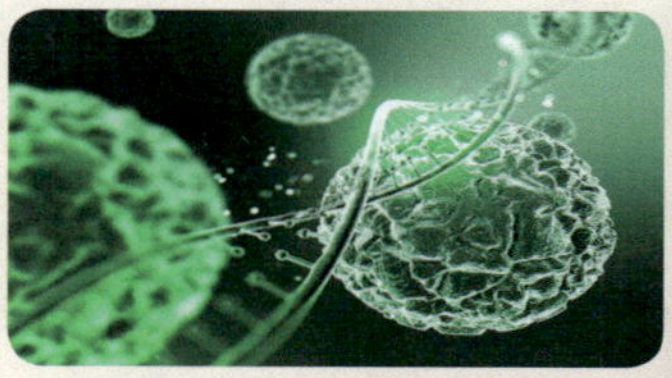

LEARNING OBJECTIVES

After studying this chapter, you will be able to:

1. Discuss the scientific method in biology;
2. Explain the cell theory;
3. Discuss the genetics and heredity;
4. Evolution and natural selection;
5. Focus on ecology and ecosystems;
6. Learn about human biology;
7. Evaluate the plant biology;
8. Explain the microbiology;
9. Focus on current trends in biology.

Introduction

Often referred to as the science of life, biology is a broad field of study aimed at comprehending the several kinds of life on Earth, from tiny creatures to sophisticated multicellular entities. Comprising several subfields, including molecular biology, ecology, and evolutionary biology, each of which advances our knowledge of living systems, this field clarifies the mechanisms controlling life events, including development, reproduction, and interaction with the surroundings, which is the central goal of biology. Moreover, biology fills in the void between the natural sciences and medicine by offering a necessary understanding of health, illness, and the interactions between organisms and their environments. Dealing with urgent global issues such as climate change, biodiversity loss, and new diseases depends on an awareness of these complex biological interactions, therefore highlighting the critical part biology plays in modern civilization. Fundamentally, biology explores and tests ideas about the living world using the scientific method. Through methodical observation and experimentation,

1.1. The Scientific Method in Biology

Learning Outcome

1. Describe the steps of the scientific method;
2. Focus on hypothesis formation;
3. Learn about experimental design.

Learning Outcomes

The beginning of each section of every chapter contains a set of Learning Outcomes that inform students of concepts they should understand. These outcomes help students understand what they should be able to do if they have mastered the material in that section.

Test Your Knowledge

Throughout each chapter, Test Your Knowledge cues you to note which topics have been covered and to self-test your understanding of the material presented.

Test Your Knowledge

1. Explain why correlation does not imply causation in scientific experiments.
2. How does the scientific method help reduce the influence of personal opinions in research?

Chapter Summary

The Summary Section appear at the end of each chapter and summarize main concepts covered throughout the chapter. These points provide key concepts in an engaging and meaningful way.

Summary

The investigation of biological principles highlights the importance of a comprehensive view of life by reflecting the intricate interactions that occur within living systems. The complexity of biology goes beyond individual species or systems, as this essay highlights; it also includes the interactions that mold ecosystems and affect biodiversity. Researchers can conduct in-depth analyses of metabolomic data using sophisticated technologies like MetaboAnalyst, which offer insights into the molecular mechanisms underlying these interactions. As demonstrated by the developments in MetaboAnalyst, for example, the integration of diverse omics data promotes a thorough understanding of metabolic pathways and helps identify important biological markers necessary for research on development and disease (Chong et al., 2018). Additionally, the phenomenon of ecological invasions is a key component of biological research and a vital field of study. The introduction of non-native species into unfamiliar habitats has significant ramifications and calls for multidisciplinary management strategies. Examining the origins and effects of these biological incursions, invasion science has become a crucial field that highlights the need to comprehend ecological dynamics and community interactions. As interest in biological invasions increases, new findings are emerging that put conventional ecological theories and practices to the test. The importance of incorporating various scientific viewpoints is shown by this developing discipline, which also highlights how invasive species shape ecosystems and advances our knowledge of ecological resilience (David et al., 2013). When taken as a whole, these discoveries from invasion science and metabolomics highlight the value of multidisciplinary biology research. Fostering cross-disciplinary collaboration and utilizing technological breakthroughs to improve our comprehension of intricate biological interactions are crucial as we look to the future of biological research. In addition to deepening our understanding of the complexities of life, this method lays the groundwork for tackling urgent environmental issues. Future studies can ensure the continued relevance of biology in a world that is constantly changing by acknowledging the interconnection of biological systems and opening the door to creative solutions that support sustainability and biodiversity conservation. We might imagine a thorough comprehension of life that serves both academia and society at large through ongoing investigation and integration of scientific knowledge.

Knowledge Assessment

The chapter includes material designed to review and reinforce each content. It includes Critical Thinking Questions to help test of your understanding of information presented in the chapter and Short answer Questions to review your understanding of the key concepts presented in the chapter.

Critical Thinking Questions

1. **Who is known as the "Father of Modern Biology"?**
   a. Aristotle
   b. Charles Darwin
   c. Gregor Mendel
   d. Louis Pasteur
2. **What is the basic unit of life?**
   a. Atom
   b. Molecule
   c. Cell
   d. Organism
3. **Which branch of biology studies the interaction of organisms with their environment?**
   a. Genetics
   b. Ecology
   c. Physiology
   d. Botany
4. **Which macromolecule is the main source of energy for living organisms?**
   a. Proteins
   b. Lipids
   c. Carbohydrates
   d. Nucleic acids
5. **DNA is primarily responsible for which function in living organisms?**
   a. Energy production
   b. Protein synthesis
   c. Storing genetic information
   d. Regulating body temperature

Short Answer Questions

1. What is biology, and why is it important to study biology?
2. What are the three main domains of life, and how are they different from one another?
3. Define homeostasis and give an example of how it works in living organisms.
4. What is the cell theory, and why is it fundamental to biology?
5. How do autotrophs and heterotrophs differ in the way they obtain energy?
6. What is the scientific method, and how is it applied in biological research?
7. Discuss how biology helps address global challenges such as climate change, disease, and food security.
8. Explain the levels of biological organization from the smallest unit to the largest. Provide examples at each level.
9. Describe the relationship between structure and function in biology, giving specific examples from cells or organs.
10. How has the discovery of genetics shaped our understanding of inheritance and variation among organisms?

Supplementary Materials

Key Terms: A list of key terms is given for quick reference

KEY TERMS FROM THIS CHAPTER

Annotation enrichment analysis	Biological structure	Cell
Cell theory	Cellular structure	Ecology
Ecosystems	Genetics	Human biology
Microbiology	Natural selection	Organism
Plant biology		

Further Resources: Recommendations for additional readings, websites, and multimedia resources.

References

1. Argelaguet, R., Velten, B., Arnol, D., Dietrich, S., Zenz, T., Marioni, J. C., Buettner, F., et al. (2018). Multi-omics factor analysis—A framework for unsupervised integration of multi-omics data sets. *Molecular Systems Biology, 14*. https://doi.org/10.15252/msb.20178124.
2. Bacher, S., Blackburn, T. M., Essl, F., & Genovesi, P. (2017). *Socio-Economic Impact Classification of Alien Taxa (SEICAT)*. https://core.ac.uk/download/286567949.pdf (accessed on 8th March 2025).
3. Barnett, R., Brace, S., Cucchi, T., & Dobney, K. (2013). *Divergent Evolutionary Processes Associated with Colonization of Offshore Islands*. https://core.ac.uk/download/20664364.pdf (accessed on 8th March 2025).

Detailed Table of Contents

CHAPTER 5

Fungi and Plant Diversity 143

CHAPTER 6

Animal Diversity and Body Plans 179

CHAPTER 7

Plant Structure and Function 215

CHAPTER 8

Animal Structure and Function 253

CHAPTER 9

The Nervous and Endocrine Systems 291

LIST OF FIGURES

PREFACE

The investigation of living systems is an incredibly diverse set of forms, functions, and interrelations that make up our world. From the genetic modifications that propel adaptation to the intricate cell assemblies that lead to tissues and organs, the biological sciences offer a mode of understanding how life arises, develops, and exists. Here, we redirect our attention to the processes that produce diversity at all levels of organization: the genetic processes underlying evolutionary change, the origin of life on Earth, and the numerous lineages that inhabit our planet today.

In the study of life on Earth, it is clear that a single perspective cannot possibly encompass all of its complexity. Organisms range from the easily understood prokaryotic cells to the most complex multicellular animals and plants, each category maintaining characteristics that reflect its evolutionary history. Likewise, the relationships between species—ranging from predator-prey to symbiotic relationships—signal the dynamic webs that infuse ecosystems. By tracing both the functional blueprints and the evolutionary pathways of life, we build a more cohesive understanding of biology, linking molecules to ecosystems.

This book, *Principles of Biology, Volume 2*, aims to lead readers through the key themes of evolution, diversity, and ecological interactions. It provides both a solid foundation and up-to-date information, drawing on the most recent research. In a rational sequence of twelve chapters, it aims to provide a consistent step-by-step progression from the molecular basis of evolution to the complex dynamics of communities and ecosystems.

Chapter 1 introduces evolution, defining terms such as heritable variation and selective processes. It explains how genetic variation arises, the historical development of evolutionary theory, and evidence of common descent. Using observations of patterns in the fossil record, comparative anatomy, and molecular biology, Chapter 1 provides a basis for understanding how species evolve.

Chapter 2 outlines the process of evolution, including the roles of mutation, genetic drift, gene flow, and natural selection. It explains allele frequency change in populations, population size, the influence on genetic variation, and how different modes of selection operate on phenotypic characters. Examples of wild populations and model organisms illustrate each process in action.

Chapter 3 addresses the origin of life and the early evolution of life on Earth. Chapter 3 addresses speculations on the prebiotic synthesis of organic compounds, the emergence of self-replicating systems, and the evolution of RNA genetic systems into DNA and proteins. Milestone events, such as the development of protocells and the formation of metabolic networks, are discussed in relation to the early conditions of the Earth's environment.

Chapter 4 provides an overview of prokaryotes and protists, summarizing the variety of bacterial and archaeal lineages as well as the key groups of single-celled eukaryotes. Chapter 4 highlights their ecological roles, from cycling nutrients to symbiotic relationships, and their structural and metabolic specializations, permitting them to exist in extreme conditions.

Chapter 5 covers fungal and plant diversity, beginning with the evolutionary history of fungi and their unique life cycles. It then addresses the major clades of green plants, from algae to seed vascular plants, emphasizing key innovations such as the development of vascular tissues, seeds, and flowers. Chapter 5 also mentions the ecological role of fungi in decay and nutrient cycling.

Chapter 6 addresses animal diversity and body plans, explaining the principal animal phyla and the features that distinguish them. Following the origin of the most significant features—segmentation, coelomic cavities, bilateral symmetry, and how developmental pathways yield an incredible array of anatomical structures.

Chapter 7 covers plant structure and function, outlining the cellular structure of roots, stems, leaves, and reproductive organs. The chapter explores how water and nutrients are distributed, how photosynthesis occurs at the cellular level, and how plants coordinate environmental cues to regulate growth and development.

Chapter 8 provides an overview of the body structure and function of animals, starting with a review of tissues and organ systems. It provides close-ups of the circulatory, respiratory, digestive, and musculoskeletal systems to detail how each maintains physiological balance and responds to internal and external stresses.

Chapter 9 covers the endocrine and nervous systems, describing how animals sense stimuli, evaluate information, and coordinate body responses. It compares fast electrical signaling to slower chemical hormone channels, showing how these systems are interwoven to regulate behavior, growth, and homeostasis.

Chapter 10 describes reproduction and development from gametogenesis through embryogenesis to birth or germination. It highlights the diversity of sexual and asexual reproductive modes, as well as the molecular signals that guide cell differentiation, tissue pattern formation, and organogenesis.

Chapter 11 deals with population and community ecology, introducing key concepts such as population growth models, species interactions, and ecological community structure. It addresses competition, predation, parasitism, and mutualism and how these interactions influence the distribution and abundance of species.

Chapter 12 concludes with a discussion of conservation biology and ecosystems, summarizing the energy and matter cycles within ecosystems and their associated functions. It addresses the threats posed by non-native species, global warming, and habitat loss and outlines ways to sustain diversity and restore ecological integrity.

This book is targeted at advanced undergraduates, graduate students, and professionals who desire a complete but concise treatment of evolutionary and ecological biology. It will serve as a primary textbook in evolution, diversity, and ecology courses, as well as a reference for scientists seeking concise summaries of basic principles. Whether as a class reader or a personal study supplement, this book aims to provide readers with the conceptual tools and factual data necessary to address contemporary issues in the life sciences.

INTRODUCTION TO EVOLUTION IN BIOLOGY

1

LEARNING OBJECTIVES

After studying this chapter, you will be able to:

1. Explore the theories of evolution;
2. Explain evidence supporting evolution;
3. Identify the evolutionary processes;
4. Define the role of genetics in evolution;
5. Illustrate the evolutionary psychology;
6. List the applications of evolutionary theory.

Introduction

Evolution is the basic concept upon which the field of biological science revolves. It is essentially the accumulation of population alteration across generations caused by processes such as natural selection, mutation, genetic drift, and gene flow. The processes generate a dynamic interaction between organisms and the environment that leads to adaptation that ensures enhanced survival and reproduction. By defining evolution as a gradual process, biologists can then account for the variation in life seen today. Crucially, such a definition does not imply mere changes to physical traits; it comprises the intricate interaction and interrelationship that make up the biosphere on grand chronological scales. Additionally, as the organism adapts to suit the environment, it generates offspring capable of inheriting positive adaptations, detailing how evolution is linked with heredity At the heart of the definition of evolution is the distinction between replicators and interactors, a concept which is useful in explaining processes of evolutionary change. Replicators, such as genes, possess information specifying development and heredity. Interactors, such as entire organisms and their environment, are the things that act upon one another significantly with the replicators. Reydon and Scholz acknowledge the significance of knowing these types in organizational population studies and suggest that organizations are able to change through the capacity to select traits that enable them to thrive in competitive settings. Therefore, the definition of evolution in such terms not only enhances our knowledge of biological inheritance but also accentuates the importance of those interactions that organize evolutionary dynamics in complex systems, both biological and organizational (Aldrich et al., 2014). Modern synthesis of evolutionary theory accounts for both genetic and epigenetic processes, broadening the definition of evolution to include a variety of systems of inheritance that make up biological diversity.

Figure 1.1. Evolution in biology.

Source: https://external-content.duckduckgo.com/iu/?u=https%3A%2F%2Fstudy.com%2Fcimages%2Fcourse-image%2Fevolution-101-intro-to-evolution_174761_large.jpg&f=1&nofb=1&ipt=7ed49a2d40289263c4f9cffa4deac2cde4e5209f93397fa1a9457114a63a0043.

While genetic inheritance is the primary process, epigenetics is also responsible for ensuring the expression of traits as well as their transmission, demarcating an organism's ability to respond to environmental change. Identification of the genetic information-epigenetic interaction enables scientists to understand how evolution processes are more effectively carried out. The results obtained by this approach underscore the complex nature of heredity and its role in evolution processes, showing how systems interact to produce varying forms of life (Griffiths et al., 2017). These are necessary to account for in order to obtain a synthesized understanding of evolution, paving the way for ongoing study of its mechanisms and uses within biological science (Figure 1.1).

Explanation of Biological Evolution

Biology evolution can only be understood by studying the concepts that serve to create this active process. Evolution is caused by various means, which are natural selection, mutation, gene flow, and genetic drift. Natural selection, which has been explained by Charles Darwin, states that living organisms that are well adapted to their surroundings are likely to survive and propagate, and thus transfer beneficial traits to their offspring. In addition, mutations create genetic variation, which is needed for evolution since they supply new characteristics that natural selection has something to work on. Gene flow occurs when populations breed and bring new genes into a population, and genetic drift are non-directional changes in allele frequencies that can influence small populations considerably. Therefore, learning about these mechanisms helps in understanding the ways species evolve and transform over time. In biological anthropology, human evolutionary history is studied within the context of general biological variations observed in species. This area of scholarly inquiry explores the evolution of both human and non-human primates in order to determine the adaptive traits that have emerged due to evolutionary forces. Through comparison of anthropological data and genetic data, scientists are able to look back at our shared ancestors and see how the various versions of hominids, such as Australopithecus and early Homo, developed and adapted to the circumstances of their environment. This comparison reveals the progression of biological changes that characterize physical and behavioral features among human populations. Not only does this discussion highlight the importance of resilience to adversity in the environment, but it also provides us with an awareness of unique characteristics of humans, validating that our evolutionary history is linked with numerous other species (Alfonso-Durruty et al., 2024). Biological evolution is not limited to land-based life; it encompasses a wide range of types of life, such as bacteria and viruses, with significant ecological implications. For instance, research on the evolution of baculoviruses mentions how the host insect and the pathogens have co-evolved in a complicated way and this impacts both the populations. These dynamics can be vital to explain in applications in biological control, and they prove the utilitarian worth of studying evolutionary processes. With scientists attempting to improve viral strains for effective pest management, knowledge of particular molecular mechanisms becomes important. This work not only enlightens us on viral evolution but also helps in engineering new biocontrol agents, and it shows us that evolution is the backbone of manipulating the dynamics of ecosystems and human activities to manage them (Chitneni et al., 2008).

Historical Context of Evolutionary Theory

The evolutionary theoretical historical background is marked by the convergence of philosophical thought and scientific inquiry that has grown exponentially through the ages. Ancient philosophers like Aristotle initially laid the groundwork by defining a teleological point from which to see life, which implied that living things possess built-in goals and purposes expressed in their forms and functions. Aristotle's ideas on causality and substance remain pertinent to this day, as they complicate simplistic understandings of evolution that disregard the facts of form and accident. His perspective provides a dense background upon which to gauge contemporary evolutionary theory, especially within the backdrop of recent disputes regarding the role of teleology in biology, as by Tabaczek et al. (2014). This philosophical basis has established an interactive relationship among the various scientific disciplines in trying to determine the character of life, continuity, and change. In the scientific revolution era, philosophers such as Lamarck and Darwin began a revolution of the existing paradigms of natural science. Lamarck's initial ideas of inheritance and adaptation laid the groundwork for Darwin's revolutionary ideas in On the Origin of Species. Darwin established natural selection as the driving force behind evolution, replacing the fixed species position. Darwin's work offered an empirical foundation but also created philosophical questions about purpose and causality in biotic contexts that pointed towards more systematic familiarity with the richness of life. The reconciliation of Darwinian evolution with Aristotelian motifs has been contentious, such as recent arguments over the interplay between random forces and directed intent, suggesting that historical accounts should inform contemporary evolutionary thought as reported by Kosolosky et al. (2011). Historical background for evolutionary theory has expanded in the past few decades into a wide array of eclectic interdisciplinary methodologies that test and further develop today's models. Contemporary philosophy of biology resolves the tension between historic and a-historic functional explanation approaches, identifying a newly emerging conflict that transcends the scope of traditional evolutionary biology. Scholars have tried to reconcile these opposing paths, forming pluralistic frameworks that respect the pluralism inherent in biological function and context-relevance. These models oppose the teleology vs. mechanism dichotomy in the understanding of evolutionary processes, rather presenting a relation-based understanding of biological traits. Such debates surrounding these concerns emphasize that the understanding of evolutionary theory cannot be grounded on descriptions of the past or scientific facts but must take into consideration philosophical knowledge addressing the complexity and nuance of life's mechanisms as advocated for by Kosolosky et al. (2011) (Figure 1.2).

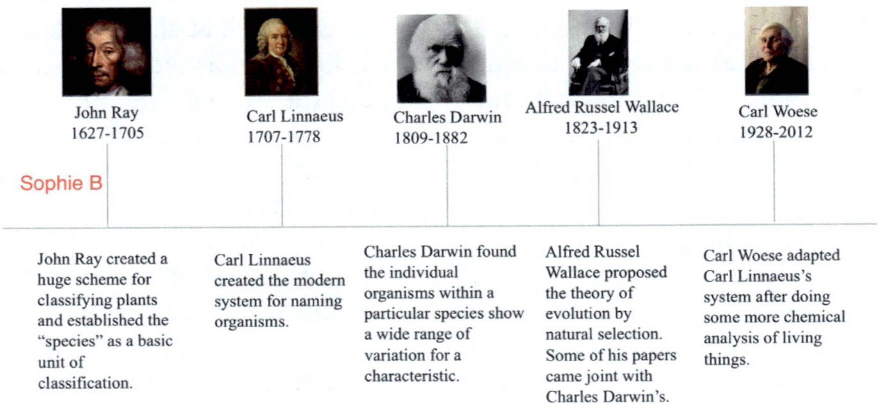

Figure 1.2. Theory of evolution timeline.

Source: https://external-content.duckduckgo.com/iu/?u=https%3A%2F%2Fpbs.twimg.com%2Fmedia%2FEW2Zb2nXYAAs391.jpg%3Alarge&f=1&nofb=1&ipt=73c5b6d8717f3828017b85b71f10bcaaed8a3bb4f4ab6ce2e263c349ad5d8734.

Importance of Evolution in Biology

Evolution is the unifying concept of biology as it provides a general explanation for the nature of the variation of life on Earth. Evolution explains the variation in species over time because of natural selection, genetic drift, and gene flow. This information is relevant to most of the topics, such as ecology, genetics, and conservation biology. For instance, an understanding of evolutionary relationships among species allows biologists to make educated decisions in the conservation of species since the conservation of genetic diversity is crucial in ensuring the survival of species in an ever-changing world. In addition, it offers explanations of the ways in which organisms evolve to live and survive, advocating for evolution not merely as a theoretical model, but as a fundamental scientific principle by which all aspects of biological inquiry can be governed. On a practical level, the relevance of evolution also extends to agriculture and medicine, with its ubiquity found in everyday life. The history of domesticated plants and animals demonstrates the changed impact of selective breeding that has enabled human beings to domesticate efficient and long-lasting food producers. Crop wild relative research, for instance, as detailed by Warschefsky (2018), demonstrates how an understanding of evolutionary processes can be applied in crop improvement programs. Such a union of evolutionary science with agriculture serves to emphasize the central role evolution has in addressing global food security. Similarly, within medicine, the evolution of pathogens necessitates re-adaptation of therapeutic strategies on an adaptive scale; identification of disease agent evolutionary processes can potentially guide more effective public health measures. Evolution is central to such areas as molecular entomology, the co-evolution of insects and the environment. As indicated by UiTM et al. (2017), the study of insect physiology, development, and disease transmission shows that evolutionary mechanisms control such interactions. Insects, being vectors to some of the most life-changing diseases, have evolved some mechanisms through which they can interact with hosts as well as pathogens. These evolutionary mechanisms must be understood to come up with effective disease control measures. With biologists revealing the molecular mechanisms for such interactions, they advance public health interventions and show how evolutionary concepts underpin various fields of biological research and highlight the utility of evolutionary theory in solving complex biological questions.

1.1. Theories of Evolution

Learning Outcome

1. Darwin's theory of natural selection;
2. Modern synthesis and genetic variation;
3. Other evolutionary mechanisms (e.g., genetic drift, gene flow).

Evolutionary theories form the foundation of current biological sciences, describing the complex mechanisms through which species become diverse and change over time. Charles Darwin's theory of natural selection is one of the turning points, stating that better-fitting organisms will survive and reproduce more. It is a process that not only explains the emergence of new traits but also the movement of change within populations. But evolution is not caused by natural selection alone; genetic drift and gene flow are also significant factors. The interplay among these forces demonstrates the intricacies of evolution, emphasizing the dynamic relationship between organisms and the environments around them as they face challenges and opportunities for survival and reproduction. One basic contemporary extrapolation of evolutionary theory is the theory of autopoiesis, built by Humberto Maturana and Francisco J. Varela. Autopoiesis emphasizes the self-organizing nature of living organisms, with organisms possessing the inherent capacity to achieve their own existence by creating elements vital to their survival. Evolution can be seen from an autopoietic viewpoint as a process involving homeostasis on the cellular level, along with rising organizational complexity. Additionally, Maturana's metaphor of natural drift links the evolutionary trajectory of species to intricate interactions with structures within the environment. This perspective not only enriches our understanding of evolutionary events but also bridges the divide between biological theory and cognitive science, illustrating how organisms evolve not only by adaptation but by a dynamic in constant interaction with the environment (Podgórski et al., 2010). Evolution in biological theories philosophically parallels an ongoing dialogue with broader scientific paradigms. Early on, the philosophy of biology was operating within the limits of logical positivism, focusing mainly on the methodologies that were emerging in the physical sciences. But then this paradigm was turned entirely upside down in the 1960s and 1970s, leading to more critical examination and a more sophisticated look into evolutionary biology by philosophers. Consequently, philosophical debates in the present day concerning evolution take into account not just the mechanisms of change but also implications for thinking about such entities as species and classification. Through this continued unfolding of the philosophy of biology, it draws from many sources of enlightenment, giving a greater comprehension of evolutionary mechanisms and supporting the ground for contemporary biological theories (Byron et al., 2006).

1.1.1. Darwin's Theory of Natural Selection

Natural selection theory, which was established by Charles Darwin, is a basic premise in evolutionary biology. The principle in the theory is that the variations in a species are significant in survival; individuals who have desired characteristics have a higher likelihood of surviving and passing them on to the next generation. Over generations, this leads to populations evolving at their basic level to accommodate their genetic makeup. By observing diversity in life, Darwin observed that the environment plays a crucial role in determining what is beneficial in characteristics. These findings demonstrate the effect of environmental factors on informing evolutionary paths, confirming natural selection as not a random process but an adaptation to specific forces in the environment. Darwin's findings transcended survival; Darwin observed the intricate relationship between environment and organisms. His work was how traits that seemed insignificant or detrimental could have strong advantages in certain circumstances. For example, such adaptations are coloration, body size, and behavior, which generally aid an organism in accessing food, evading predators, or locating a mate. Besides that, Darwin's experiments on such creatures as worms proved the complicated interdependence of systems and that minor creatures have a significant role in grand ecological processes such as soil health and landscape formation (Henchman et al., 2018). This is to prove that natural selection is within a system of interactions outside the organism and therefore proves how life is interconnected. Mathematically, the mechanisms underlying Darwin's natural selection have similarly been mathematically better formulated to describe adaptive evolution in populations. In particular, gradual Darwinian evolution models define how some traits can evolve as a response to certain pressures, e.g., intraspecific competition for resources. They suggest evolutionary change can lead to predicted distributions of traits within populations, typically in Gaussian distribution patterns under conditions of equilibrium (Calsina et al., 2013). These theoretical models support Darwin's reasoning, proposing a mathematical theory of the passing on of adaptive characteristics from generation to generation. With such models becoming more complex in the hands of scientists, they validate the empirical grounds on which Darwin's observations are founded, attesting that the principles of natural selection can be applied in contemporary biological science and respecting evolution (Figure 1.3).

Figure 1.3. Darwin's theory of natural selection in biology.

Source: https://cdn.britannica.com/06/180006-138-F7CE1D29/Charles-Darwin-Overview-life-focus-evolution-work.jpg?w=800&h=450&c=crop.

1.1.2. Modern Synthesis and Genetic Variation

The new synthesis is the most significant advance in evolutionary biology, the combination of Darwinian natural selection and genetics to supply a unified view of evolution. The theoretical framework not only emphasizes the role of genetic variation in species but also supplies the mechanism whereby environmental pressures lead to adaptation. With the inclusion of Mendelian inheritance, genetic drift, and gene flow, the new synthesis offers an integrated explanation of how evolution occurs in the long term. Researchers have shown that such an addition is unavoidable in the explanation of the dynamic interactions of organisms with their environments, alongside processes sustaining biological diversity (Gill et al., 2012). This blending firmly shifted evolutionary theory paradigms from being descriptive accounts to good theoretical frameworks that are testable and empirically falsifiable. Genetic variation is an important component in current synthesis which is the raw material on which evolutionary forces are operating. Population variation due to mutation, recombination, and gene flow provides organisms with their own characteristics, some of which are advantageous for survival under varying environmental conditions. Genetic diversity is crucial to the adaptability of species since it allows populations to respond to selective pressures and changes in their environment. The awareness that evolution operates on existing genetic variation has profound implications for conservation biology, selection breeding, and medicine and casts light not just on the value of preserving genetic variation but also on the importance of ethical sensitivity in manipulating it (Gill et al., 2012). As scientists unlock the mystery of genetic variation, they redraw the frontiers of evolutionary biology and sharpen our comprehension of life's complex web. The implications of the modern synthesis extend far beyond first-year biology, affecting numerous multidisciplinary disciplines that incorporate evolutionary principles. The literature in recent years shows that shifting perceptions on genetic inheritance have permeated across varied disciplines from anthropology to psychology and even economics, testifying to the adaptability of evolutionary theory across fields that are capable of breaking down complex structures in nature and society (Hanisch & Eirdosh, 2020). For instance, the application of evolutionary principles to describe social behaviors has produced enormous information on how human societies and cultures develop. This interdisciplinary overlap looks to the new synthesis foundational stance as the pivot of biological knowledge with more comprehensive scientific inquiry, an indicator not only of the revolution in biology but of a paradigm shift which reconfigures methodology and creates new avenues of investigation and application across a range of disciplinary fields of scholarship (Figure 1.4).

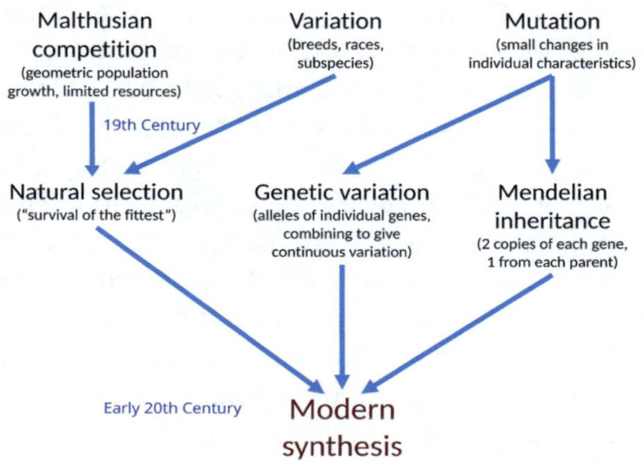

Figure 1.4. Modern synthesis and genetic variation.

Source: https://upload.wikimedia.org/wikipedia/commons/thumb/9/97/Modern_Synthesis.svg/1200px-Modern_Synthesis.svg.png.

1.1.3. Other Evolutionary Mechanisms (e.g., Genetic Drift, Gene Flow)

It is crucial in evolutionary biology to identify mechanisms aside from natural selection that cause evolutionary change. Genetic drift and gene flow are significant processes. Genetic drift refers to a situation when there is a random change in allele frequency within a population, which is more common in small populations where random events can have disproportionate effects. For instance, it may be possible that an arbitrary event in the form of a natural disaster kills off a large proportion of the population, hence altering the gene pool irrespective of desirable traits. This unpredictability of genetic drift provides the potential for such massive divergence among populations over time, showing how evolutionary phenomena can occur irrespective of adaptation and therefore become an important component of the grand narrative of evolution. Gene flow is equally an important mechanism and refers to the movement of genetic material from one population to another via migration. As people from a population mate with others from another population, they contribute new alleles to the gene pool, and this can significantly add to the genetic diversity of a population. This new genetic mutation can increase the capacity of a population to evolve within environments that fluctuate and reverse the impacts of genetic drift. For example, in fragmented habitats under environmental pressure, gene flow contains precious genetic variation that increases resilience (Stewart, 2010). Gene flow is therefore a unifying factor among populations, not only providing the passage of genetic material but also being an evolutionary force that unites divergent lineages within the enormous tapestry of biological evolution. The interaction between gene flow and genetic drift in a dynamic fashion guarantees that evolutionary processes occur, especially in the context of environmental change. They exist together in a delicate balance, sculpting the evolutionary lineages of species. In the majority of situations, for example, in the scenario of the Quaternary period, gene flow and genetic drift have been attributed as the causes of climate change, which has led to changes in the structure of ecological communities as well as evolution or extinction of various species (Stewart, 2010). Since the populations establish new pressures in the ecosystem, genetic variation through gene flow can improve adaptive traits, and genetic drift can eliminate non-adaptive alleles in small relictual populations. These interactions illustrate the intricate routes along which evolution moves, and so scientists are left to continue researching the fine details with which these mechanisms contribute towards ongoing evolution on our planet (Figure 1.5).

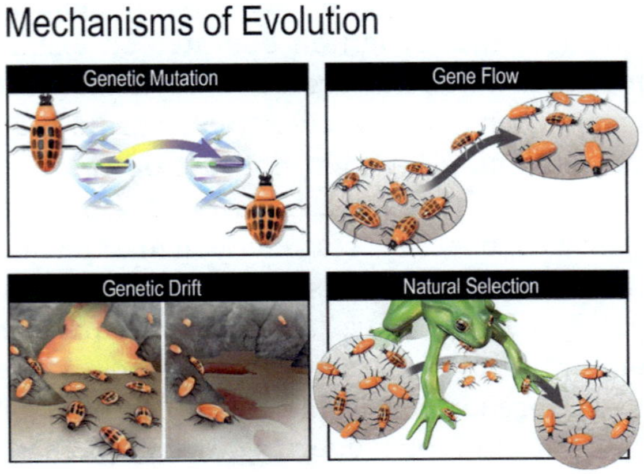

Figure 1.5. Mechanisms of evolution.

Source: https://www.visiblebody.com/hubfs/learn/bio/assets/evolution/mechanisms-of-evolution.jpg.

Test Your Knowledge

1. What is Darwin's theory of natural selection?
2. What are the modern synthesis and genetic variation?

1.2. Evidence Supporting Evolution

Learning Outcome

1. Fossil records and transitional forms;
2. Comparative anatomy and homologous structures;
3. Molecular biology and genetic evidence.

The proof of evolution is vast, spanning from information from a wide range of areas of science, including genetics, paleontology, and comparative anatomy. The most convincing type of evidence may well be the fossil record, which documents the slow creation of seeds over geologic time. Not only does the fossil record tell us about life forms with transitional characteristics, but also about the dying of life forms and the coming into being of new forms. For instance, the transition from fish to tetrapods is best exemplified by fossils such as Tiktaalik with both aquatic and terrestrial characteristics. These fossil records support the theory of descent with modification, where the modern living organisms are descended from previous lineages, validating the theory that all forms of life on Earth share common ancestry. In combination with paleontological evidence, molecular biology offers direct evidence for evolutionary processes from gene studies. Molecular examination of DNA sequences revealed remarkable homologies between distantly related species as evidence of common descent at the molecular level. Similarity in genetics between humans and other primates, for instance, is evidence of great genetic similarity as evidence of a common ancestry. In addition, the concept of homologous structures or anatomical features shared by a common ancestor but for different purposes is evidence for evolution. The human, whale, and bat's forelimbs, for example, are operationally different but have the same bone structure, showing how highly evolved species are interrelated when it comes to evolution. Together, these molecular and anatomical data provide the backing for the hypothesis that evolutionary change is an inherent process of life diversification. Lastly, natural selection theory forms the cornerstone of imagining evolution, citing environmental pressures as a way species are shaped over time. Observational studies of species within their environments repeatedly discover adaptations for greater survival and reproductive benefit. This can be demonstrated in the well-known example of the peppered moth, which experienced color change during the Industrial Revolution in relation to environmental change. The black moths adapted to better survive in soot-dominated, polluted areas, demonstrating the rate of evolutionary ability of natural selection. These are instances of the theory of evolution demonstrating adaptation to environmental change in the here and now. Also, more recent advances in evolutionary biology and genomics are continually reaffirming these overarching principles in hammering home the realization that evolution is not fixed but rather a moving process in adapting itself to our ever-changing tapestry of life on this planet.

1.2.1. Fossil Records and Transitional Forms

The fossil record is a rich store of evidence for the theory of evolution, particularly through the study of transitional forms. Transitional fossils are a demonstration of the gradual transition through geological time that fills the gaps in the anatomy and function among various species. Fossils of Archaeopteryx, for example, illustrate the evolutionary link between birds and dinosaurs with elements of both. These transitional forms not only confirm the theory of common descent but also provide us with concrete evidence of evolutionary processes, such as speciation and adaptation. That they are present in the fossil record is an indication that evolution is not a process that happens instantly, but a gradual one marked by innumerable minor alterations, hence confirming the evolutionary theory's predictions. In addition to the exhibition of anatomical changes, fossil records provide data on the environmental pressures that have influenced the evolutionary lineages of organisms. For instance, a study of South Chinese fossilized leaves demonstrates how, over millions of years, monsoonal climates have significantly shaped ancient ecosystems. These fossils display adaptations equivalent to current tropical monsoon environments, indicating the way species adapt to their surrounding in the course of time (Aleksandrova et al., 2017). This connection between climate and evolutionary adaptation indicates that species never evolve independently; rather, they interact with their surroundings, necessitating a synoptic grasp of ecological and environmental processes when dealing with the study of evolutionary biology. The fossil record is thus a dynamic archive, imparting significance to the intricate relationship of organisms and their surroundings through time. The intricate information gained from fossils can also enhance our understanding of the family relationships between various species. New discoveries, such as the skull of *Oriensmilus liupanensis*, have produced key morphological information that allows us to re-evaluate phylogenetic relations in felids and their ancestors. These studies explain the process of evolution to contemporary cats and make them connected to other carnivorous mammals (Guan et al., 2020). The ability to analyze damaged and complete fossils allows scientists to obtain reliable evidence regarding lineage divergence and adaptation. This two-way appreciation of transitional forms deepens our appreciation of the diversity and evolution of life and teaches us that the fossil record is not merely an history but also a vital aid in unraveling the secrets of biological ancestry and evolutionary transition (Figure 1.6).

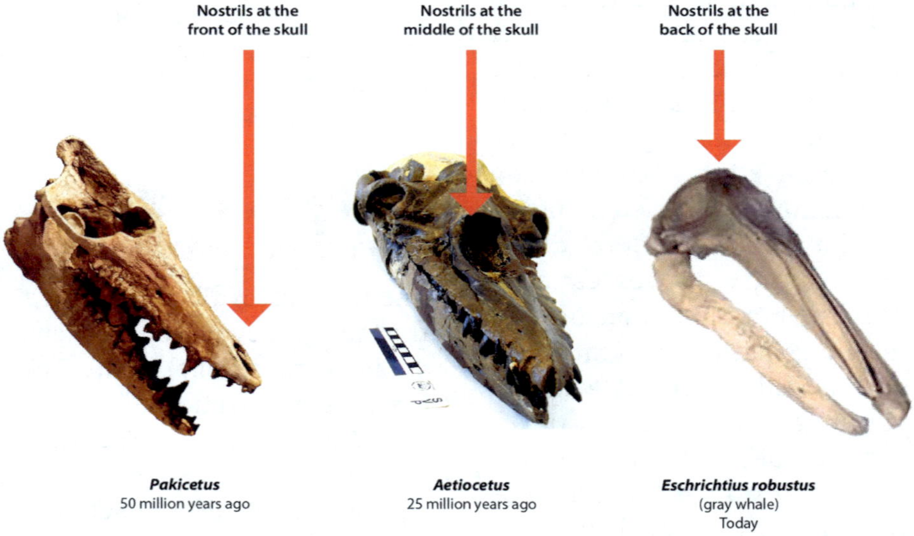

Figure 1.6. Transitional features.

Source: https://evolution.berkeley.edu/wp-content/uploads/2021/10/whale_nostril_transitional_forms_04.png.

1.2.2. Comparative Anatomy and Homologous Structures

It is worth studying comparative anatomy as comparative anatomy explains the similarity of various species in terms of anatomical structures. Comparative anatomy is the study of the comparison between the similarity and the dissimilarity of organism structure with a specific focus on the homologous structures, those structures of the anatomy with a common origin but which possess similar but not similar functions in current species. This process not only displays evolutionary lines but also the way in which species develop and adjust to their environment through time. For instance, the mammalian, avian, and reptilian front limbs display similar bone structures that reflect that they share a common evolutionary heritage. Such comparisons of anatomy give deep insights into the past lives of varied forms of life, supporting the theory of evolution by demonstrating how species alter characteristics of their ancestors due to forces of selection. Homologous structures are also essential evidence when conducting evolutionary studies, as they demonstrate the dynamic nature of anatomical development over time. Homologous structures can have different shapes and functions, depending on the environmental pressures that a species has had to endure. For instance, forelimbs of human, whale, and bat establish homology as the levels of their respective forelimbs are the same, but are used to carry out entirely different functions such as grasping, swimming, and flying. This is an example of how adaptable organisms are and how the hidden genetic forces underpinning evolutionary adaptation are manifested. Furthermore, understanding homologous structures enables scientists to use them in constructing the phylogenetic trees that give a picture of how the species evolved from their common ancestor.

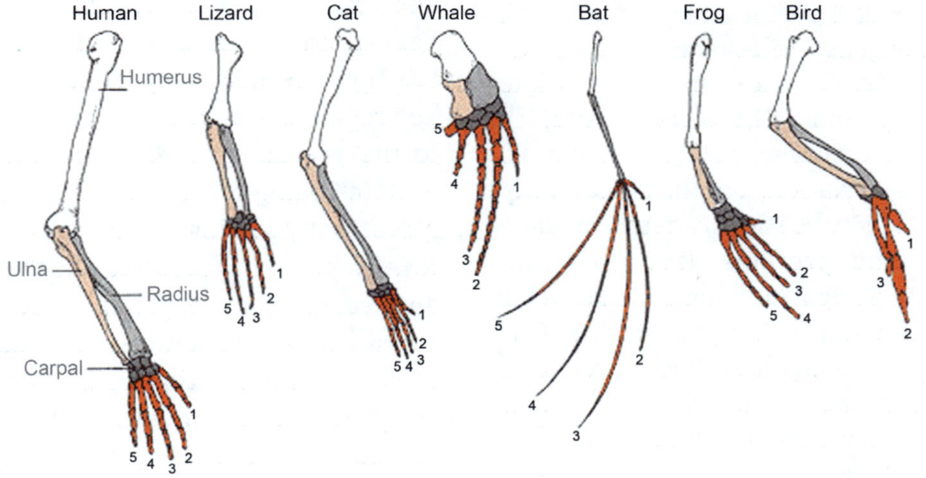

Figure 1.7. Comparative anatomy and homologous structures in biology.

Source: https://open.lib.umn.edu/app/uploads/sites/206/2017/08/c2fig4-e1578701010294.png.

Comparative anatomy thus not only enhances our understanding of anatomical diversity but also enhances our understanding of evolutionary processes. The practice of homologizing based on science follows the acquisition of information about homologous structures by identifying and classifying anatomical homologies between species. Homologization is a process more than the naming of structures, yet it is an active, rule-governed process relative to the attentive observation of the corresponding anatomical structures of interest and their evolutionary history. As noted by the literature, homologizing practice combines knowledge of the concept of homology with evidence in nature and accordingly raises the extent of biological diversity and evolution known. By such homologizing exercises, scientists can explain the sense and meaning of homology, forming an appreciation of natural kinds and the processes for discerning biological classification (Kendig et al., 2016). Comparative anatomy and exercise of homologous structures is thus a critical aspect of evolutionary biology, which reveals the intricacy of life diversification (Barrientos et al., 2016) (Figure 1.7).

1.2.3. Molecular Biology and Genetic Evidence

Molecular biology has greatly improved our ability to understand evolutionary phenomena by explaining the genetic mechanisms of various phenomena in life. Advances in methods like DNA sequencing and RNA were pivotal in helping scientists discover the molecular basis of heredity and variation. For instance, early discoveries that RNA was largely an intermediate between DNA and proteins developed into discoveries of RNA's central functions in cellular biology and evolutionary biology in general. This paradigm shift, as documented, has the function of highlighting the degree to which molecular biology is not only concerned with the research of genetic sequences but also with the complicated systems of regulation which govern multicellular organism development and function. Through these molecular configurations, researchers can track the ancestry and evolutionary changes of species and know much more about life's diversity and interrelatedness (Mattick & Amaral, 2022). The neo-Darwinian synthesis, a foundation of contemporary evolutionary theory, discovered the importance of micro-mutations in evolution. But new information emerging from molecular biology is now starting to contradict this view. Recent research on plant evolution demonstrates that multiple QTLs have significant effects on phenotypic traits, which is contrary to the common belief that small-effect mutations are responsible for most evolutionary change.

Research has demonstrated that the QTLs have a major contribution in generating variation within species and generating interspecific and intraspecific variation. This discovery heralds a significant interface between evolutionary biology and functional genetics, because the investigation of such large-effect loci illuminates the genetic foundation of adaptational characters and complexity to the dominant dogmas of gradual evolutionary adaptation (David et al., 2003). Also, the research concerning non-protein-coding RNA and its role in the evolutionary process has been a critical area of research in molecular biology. The discovery of such RNA forms as key regulators of cellular processes is contrary to the prevailing perceptions that considered protein-coding genes as the focal point of genetic research. By putting the functions of non-coding RNA on paper in processes like the development and regulation of genes, researchers have embarked on redefining our understanding of genetic information complexity. This active story evokes an integrated form of learning evolution by combining molecular information with ecological and environmental contexts. Consistent with the continued growth of the field, the capacity of researchers to apply sophisticated molecular methods indicates the necessity for more research, especially in nature, to explain the complex dynamics behind evolutionary change (Mattick & Amaral, 2022).

Test Your Knowledge

1. What are the fossil records and transitional forms?
2. What is comparative anatomy?

1.3. Evolutionary Processes

Learning Outcome

1. Speciation and the formation of new species;
2. Adaptation and survival in changing environments;
3. Coevolution and interactions between species.

Evolutionary mechanisms include the processes by which species change, diversify, and transform through generations. The main processes include essential principles such as the natural selection process, the process of genetic drift, and gene flow. Natural selection is a theory of evolutionary biology. According to this theory, individuals who possess favorable traits are better allowed to live longer and reproduce, thus passing on their advantageous traits to subsequent generations. This adaptation process is triggered by environmental pressures that shape the genetic variation of the populations. As has been shown in many studies, including ecological relationship analyses, the interaction between evolutionary and environmental processes can yield relevant data about species behavior and traits. For instance, the awareness of adaptation of different traits can enlighten us on general demographic patterns in populations while they become adapted to shifts in ecological conditions (Burger et al, 2024). Another important evolutionary process is genetic drift, which highlights the impact of chance in generating genetic variance. Unlike natural selection, an environment-based non-random process, genetic drift is a random process that most substantially functions in small populations. Under the bottleneck or founder effect, alleles can randomly increase or decrease in frequency and give rise to large deviations from the original population genetic models. Fixation or loss of characters will occur because of random changes and will determine the direction of evolution and formation of species. One such example is in *Pristionchus nematode* research, where various species were created under varying ecological conditions to demonstrate how random genetic mutations can lead to species diversity in certain environments (Herrmann et al., 2006). Gene flow or transfer of genetic information among populations is another factor that should be taken into account when describing evolutionary mechanisms. Gene flow can create genetic diversity between populations that is in opposition to the effect of genetic drift and permits the introduction of new alleles into a gene pool. Gene flow can be particularly effective in preventing isolation that can trigger speciation. For example, where *Pristionchus nematode* populations are ecologically interacting, genetic exchange between such populations is capable of causing variability in reproduction strategies and environmental plasticity of shared environments. By analyzing these interactions, researchers are subjected to the reality that gene flow imposes resilience in evolutionary processes, acting on both genetic diversity and evolutionary lines of species. The interaction between these forces, gene flow, genetic drift, and natural selection, then is a system generating complexity and propelling evolution depending on environmental forces and population dynamics (Burger et al, 2024) (Figure 1.8).

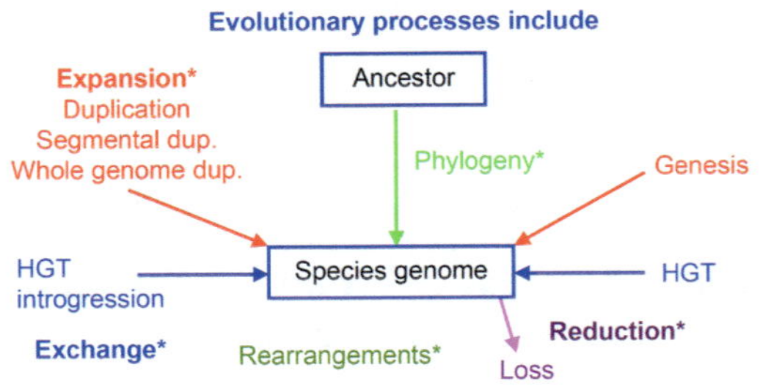

Figure 1.8. Evolutionary processes.

Source: https://www.researchgate.net/profile/Fredj-Tekaia/publication/297292133/figure/fig2/AS:342539539370000@1458679198884/evolutionary-processes-Notes-This-figure-illustrates-some-significant-evolutionary.png.

1.3.1. Speciation and the Formation of New Species

Speciation, or the formation of new species, is one of evolutionary biology's central ideas for explaining the existence of diversity in life on our planet. Speciation frequently results from processes such as allopatric, sympatric, and parapatric speciation. The most well-defined of the three is allopatric speciation, which occurs when populations become geographically separated, ending gene flow and resulting in divergent evolution. For instance, if a river splits one population of a species into two, after numerous generations, they will have evolved into different environments and will be two different species. These processes give an understanding of how evolution is dynamic in nature and how conditions in the environment lead to an effect on genetic differences in populations. Recent advances in genomic sciences have enhanced our understanding of speciation to a significant extent by bringing to light the genetic processes that cause these events. Researchers are now able to analyze vast collections of genomic data, so they are able to trace genetic markers associated with speciation events. With the ferns, genomic studies have emphasized the importance of DNA sequence information in phylogenetic information, which may be utilized for reconstructing the evolution of the nuclear and plastid genomes. This is not only peculiar to fern research but may even be applied in comprehending flowering plants as well as how they evolved. Genomic data allow for the calculation of speciation rates and processes and thus enhanced knowledge of new species adaptation and formation to the environment (Barker et al., 2010).

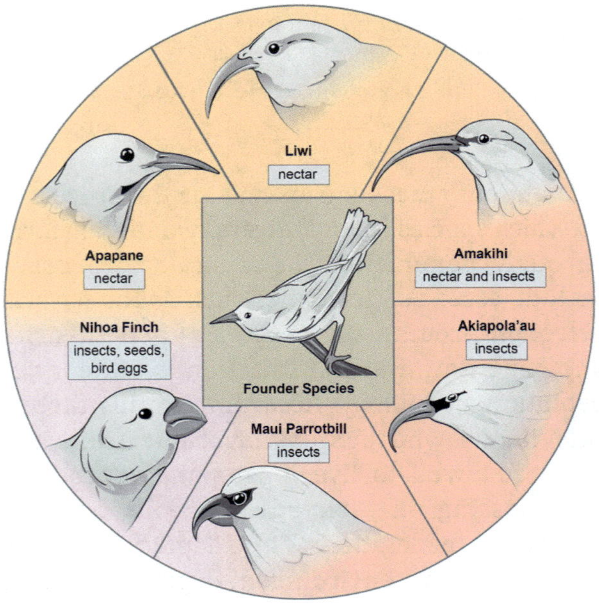

Figure 1.9. Formation of new species.

Source: https://files.mtstatic.com/site_4463/54644/0?Expires=1747973088&Signature=efGl0bvjvkFC-TmucpFkHnbqTTIkhvkTTwHc2r-ON57hgIw7sBfyaCn2aAWgih5goc5jIabuzWSpVBSxViNsMdo3qDAmnPxeqXRAw1rEGGTeSMWIFECJLiCpyUi9yuFG2ZC5i6O0ToY6CPk50ztxMy0TslKr-Or0r5KiI0Iziuk_&Key-Pair-Id=APKAJ5Y6AV4GI7A555NA.

Secondly, the reticulated model underscores the difficulty in interpreting speciation. Unlike linear evolutionary processes, there are species with certain complicated relationships, and hybridization and gene flow between lineages complicate species definition. This reticulated complexity can be envisioned by envisioning Darwin's metaphor of the tangled bank, which describes the non-linear interaction that is accountable for biodiversity. Methods such as the phylogenetic analysis for comparing trees (PACT) enable scientists to examine these complex interactions and unravel the evolutionary histories that are accountable for speciation. By integrating linear and reticulated models of evolution, the field of speciation becomes more whole with a richer appreciation of the processes of creating new species (Brooks & van Veller, 2008) (Figure 1.9).

1.3.2. Adaptation and Survival in Changing Environments

The adaptation process is essential to the survival of species in changing environments. With changing climate and altered landscapes, the ability of organisms to adapt to this change determines whether they survive or perish. Evolutionary theory contends that species gradually adapt in response to the environment to develop traits that enhance their survival. For instance, during the Quaternary period, climate change significantly impacted ecological communities, driving the evolution, adaptation, and extinction of various species. Evolutionary processes were not just a response to environmental changes but also defined the very fabric of ecosystems, showing how deeply embedded the relationship between adaptation, survival, and adversity (Stewart, 2010). It is these dynamics that must be understood to comprehend how present and future environmental changes will influence biodiversity and ecosystem resilience. Genetic differences are important for an organism's capacity to respond to environmental change. For example, polyploidy, a state in which organisms have more than two sets of chromosomes, has been found to be beneficial to plant survival. With the sequenced genomes of numerous plants, over 30 instances of paleopolyploidy have been discovered, most of which occurred at significant events throughout history, such as the K/Pg boundary 66 million years ago. It has been suggested that the coincidence of this event means that species that had genome duplication during this time of crisis would be more likely to survive following one of Earth's most significant mass extinction events. Through the study of such trends, we learn to appreciate the ability genetic variability provides for adaptability and resilience, which brings into perspective the importance of evolutionary processes in life development (Lohaus et al., 2016). A study of such events shows the complexity of life's reaction to environmental pressure in an evolutionary process. Niche construction also illustrates how organisms directly affect their environments, allowing them to survive and continue adapting. This occurs when organisms modify their environments in such a way as to create new opportunities for themselves and others, thus redirecting evolution for various species.

Figure 1.10. Adaptation and survival in changing environments.

Source: https://texasgateway.org/sites/default/files/h5p/content/4996/images/image-576af4c472758.jpg.

For example, beavers construct dams that create aquatic environments that benefit many other species in the process. This interaction emphasizes that adaptation is not merely a passive response but can actually be an active strategy that can drive ecological processes. With species navigating complex and dynamic environments, their ability to reshape their worlds can improve their survival prospects and foster biodiversity. The better understanding of such relations further assists us in realizing the complexity of adaptation and survival in a dynamic world, illustrating the multidimensionality of evolution itself (Figure 1.10).

1.3.3. Coevolution and Interactions Between Species

Species interactions are at the center of coevolution research, which examines how two or more species' evolutionary histories influence each other. The interactions occur in various forms, including predator-prey, mutualism, or parasitism, each having an impact on the interacting species' adaptive traits. Examples of adaptations like prey species' increased speed are often formed under predation pressure, with the predators benefiting from improved hunting strategies. These reciprocal alterations are indicative of the dynamic nature of biodiversity in which evolutionary trajectories are dependent on each other. By investigating these interaction-mediated selection pressures, scientists can gain insight into mechanisms driving ecological balance and evolution. To analyze these processes is not only to gain insight into the adaptations of specific species but also to gain a generalized understanding of ecosystem function and resilience. The coevolution work is very evident when invasive species are in question, which have the tendency to evolve very fast after being introduced into a new range. Invasive species, upon interaction with native fauna and flora, can cause significant ecological changes that lead to both good as well as undesirable effects. In biological invasions, understanding these evolutionary processes matters because invasive populations may undergo divergence from their native relatives and hence modify the local ecosystems (Aragon et al., 2016). Basically, the phenotypic changes observed in invasive species illustrate the intricate interactions between coevolution and ecological effects.

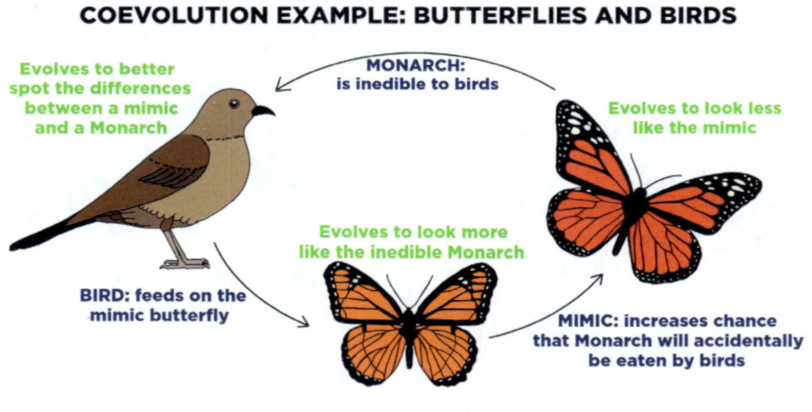

Figure 1.11. Coevolution and interactions between species.

Source: https://d20khd7ddkh5ls.cloudfront.net/coevolution_butterflies.jpg.

The interaction between invasive species and their novel communities demonstrates the need for a combined framework that considers evolutionary history, ecological interactions, and adaptive mechanisms in order to approach the complexities of biological invasions more effectively. The use of coevolution with an eco-evolutionary perspective allows the consideration of species interactions and the subsequent effects on community dynamics with a greater depth of understanding. This strategy highlights that species are not fixed objects but ever-evolving because of their interaction with the environment as well as with other species. For example, research on plankton communities illustrates how evolutionary development at a fast pace can trigger biodiversity and affect ecological function (Domenico, 2017). Through such analysis of species interactions in different habitats, scientists can determine the ecological and evolutionary drivers that control population dynamics and community structures. These questions underscore the need for coevolution in understanding ecological adaptation and resilience, particularly with regard to climate change and habitat alteration. Thus, integrating coevolutionary thinking into ecology is imperative in solving issues in biodiversity conservation in a changing world (Figure 1.11).

Test Your Knowledge

1. What is speciation and the formation of new species?
2. What is adaptation and survival in changing environments?

1.4. The Role of Genetics in Evolution

Learning Outcome

1. Mendelian genetics and inheritance;
2. Mutations and their impact on evolution;
3. Population genetics and allele frequency.

Genetics comes into play in the process of evolution to define the way species change and evolve through time. Within populations, genetic diversity comes from mutations, gene flow, and sexual reproduction to create different traits that enhance survival and reproduction within certain environments. Natural selection acts on these differences, leading to the spreading of those individuals whose traits are positive, hence the enhancement of the positive traits within the population. The complexity of such genetic processes shares much in similarity with developmental systems theory, which views grasping biological change as a collection of interacting processes rather than fixed attributes. Evolutionary change observed through the dynamic life cycles according to Griffiths and Russell D. Gray offers an enormously explanatory force, placing genetic processes as interwoven strands in the sweeping saga of evolution (Griffiths et al., 2018). Besides, genetics forms the foundation for hereditary transmission laws, which define how traits are passed from one generation to the next. Inheritance systems define how genetic information encoded in DNA shapes the phenotypic properties of an organism, determining its potential to thrive and reproduce. This genetic basis offers the underpinnings of evolutionary adaptation, particularly when in conjunction with environmental pressures. Species exposed to changing climates or resource availability are forced to evolve quickly in a bid to survive, emphasizing the relationship between genetic variation and environmental factors. As noted in the descriptions of Quaternary evolutionary processes, climate change significantly influences species adaptation and extinction, emphasizing the role of genetics in such ecosystem changes (Stewart, 2010). As researchers increasingly venture deeper into the discovery of the complexities of genetic influence on evolution, it is increasingly apparent that genetics underlies and propels evolutionary change. With molecular genetics and advances in genomic technologies, researchers are now able to investigate the genetic processes of adaptation and speciation with unprecedented precision. This ongoing discovery not only sheds light on the evolutionary history of existing species but also assists in predicting future evolutionary directions owing to natural and anthropogenic processes. Moreover, the integration of evolutionary biology and genetics necessitates an interdisciplinary exercise that further informs life on our planet. Through blending developmental systems theory insights and ecological dynamics, the discipline is capable of developing a more complete picture of evolution that accommodates the complex interaction between genetics and environmental factors (Griffiths et al., 2018; Stewart, 2010) (Figure 1.12).

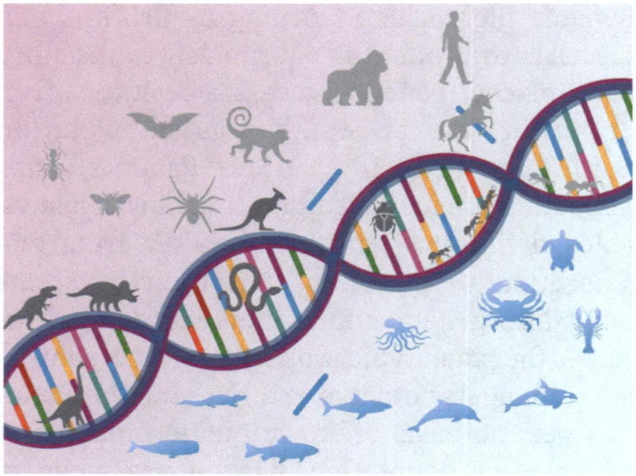

Figure 1.12. The role of genetics in evolution.

Source: https://external-content.duckduckgo.com/iu/?u=https%3A%2F%2Fgeneticeducation.co.in%2Fwp-content%2Fuploads%2F2020%2F08%2FGenetic-polymorphism-and-evolution.001.jpeg&f=1&nofb=1&ipt=485114deb520d758f43cfd26f4c6b436bc7967ee430c4ba3845548e140a9001c.

1.4.1. Mendelian Genetics and Inheritance

It is elementary to the science of biology that a comprehension of Mendelian genetics serves as the basis for modern theories of inheritance and evolution. Gregor Mendel's experiments with pea plants revealed the laws of segregation and independent assortment, which reveal how parents transmit characteristics to offspring. They illustrate how characteristics are transmitted from parents to offspring in discrete units, or alleles, which exert control over differences in populations. From the evolutionary point of view, Mendelian inheritance provides insight into genetic variation, one of the ingredients necessary for natural selection. From observing dominant and recessive traits, Mendelian genetics illustrates how variations may be selected or eliminated over time, determining the trajectory of evolution in species. The Mendelian theory of classical inheritance also serves as a foundation for examining more complex genetic interactions, such as polygenic inheritance, which play important roles in survival and reproductive success traits. Mendelian genetics is used beyond the simple inheritance of characteristics; they form the core of modern evolutionary theory. One of the key dimensions is how gene mutation, both spontaneous and environmentally induced, may result in new phenotypic expression within a population. These mutations are the very raw material of evolution, offering the genetic variation on which natural selection has something to work. In addition, contemporary controversy surrounding the modern evolutionary synthesis and possible elaboration of this theory points towards the interface of classical genetics and contemporary evolutionary thought. How empirical evidence meets theoretical constructs highlights the importance of philosophy to drive evolutionary biology and challenge researchers to think outside classical Mendelian constructions (Finkelman et al., 2014). In this light, as we integrate Mendelian theory with modern genetic theory, the complexity of evolutionary processes becomes increasingly obvious. Moreover, the emerging field of genetics increasingly employs advanced mathematical models for interpreting Mendel's accounts of inheritance mechanisms. The relationship between the mutation pattern and its shape in genetic graphs can simplify the complex nature of inheritance. By applying graph theory, scientists can plot and manipulate genetic variation, enabling a more effective explanation of the control and transmission of traits across generations.

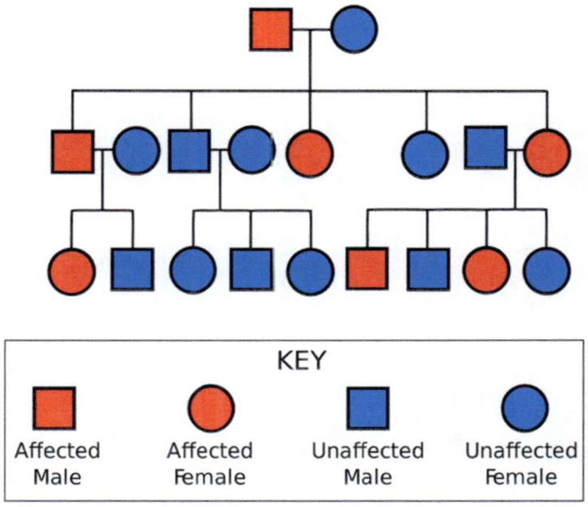

Figure 1.13. Mendelian genetics and inheritance.

Source: https://external-content.duckduckgo.com/iu/?u=https%3A%2F%2Fhumanbiology.pressbooks.tru.ca%2Fwp-content%2Fuploads%2Fsites%2F6%2F2019%2F06%2FPedigree_Chart2.svg_-1536x1326.png&f=1&nofb=1&ipt=84630443883009e8b23eb5f631cf3f8dbde6f4d1996abb7b86aa8e6f215f1ccc.

The study of total-colored graphs provides a new finding that offers a deeper understanding of the impact that mutations have on gene pathways in cellular processes like the mitotic cycle (Falcón et al., 2019). In addition to enhancing Mendelian laws, these mathematical approaches show that there is a dynamic interaction between genetics and evolution. In general, the integration of Mendelian genetics into the broader dialogue of evolution educates us on the science of biological diversity and adaptational strategies and underscores continued inquiry at the intersection of different scientific disciplines (Figure 1.13).

1.4.2. Mutations and Their Impact on Evolution

Mutations are the cause of genetic diversity that drives evolution. They may arise spontaneously during the process of replication of DNA or due to exposure to environmental pressure. Mutations may manifest in numerous different ways such as point mutations, insertions, deletions, or duplications. Mutations bring with them new traits that may help an organism survive and adapt to new environments. Where a mutation is beneficial, it can spread through populations in subsequent generations by the forces of natural selection. Mutations thus contribute importantly to generating diversity and complexity in biological systems today as populations evolve to fill their niches. Ultimately, it is the interplay between mutation and natural selection that causes evolutionary change to occur, allowing species to compete in a constantly changing world where adaptability is a strict necessity. The significance of mutations extends beyond variation; they can have the ability to lead to rapid evolutionary changes under certain environmental pressures. Cryptic or silent mutations are a case in point: they appear to be inconsequential under conditions of stability but prove to be beneficial if conditions alter. Recent studies reveal the ability of mutations to be reutilized in generating composite phenotypic variation. In the immune system, cryptic genetic variation can catalyze infectious adjustments in the T cell response, illustrating how subtle genetic differences can spur essential evolutionary consequences. Rapid evolution expresses the ability of genetics to drive rapid evolutionary changes, and this significance underscores the importance of understanding the role of mutations as dynamic variables in evolutionary science, demonstrating how they have the potential to enable instant changes to occur through environmental changes and challenges. Furthermore, we explore further into the genomics and genetics of many organisms, where mutations become more sophisticated in mechanisms and significance, as is the case with research done on Plasmodium spp., which are malarial causative agents. Genome sequencing has shown how mutations can play a role in determining evolutionary directions under selective pressures like drug therapy. *Plasmodium falciparum*'s *kelch13* gene, for example, has mutations associated with resistance to artemisinin, which identifies the immediate impact of genetic change on survival and reproduction. This feature exemplifies how rapidly mutation and selection can lead to radical changes in populations, creating evolutionary pressures that necessitate ongoing research and new disease control approaches. It is important to understand the relationship between evolution and mutation if one is to appreciate not only the adaptation of species but also how these changes impact public health and conservation (Figure 1.14).

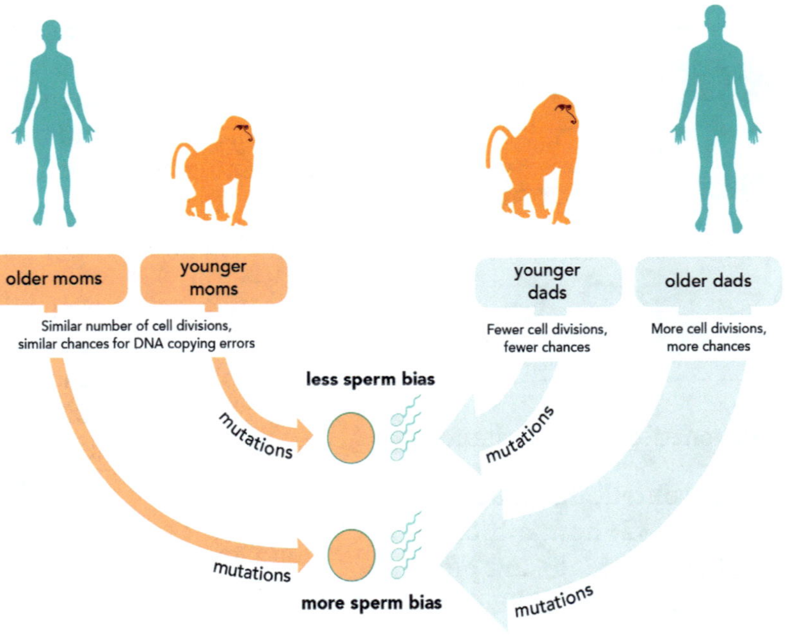

Figure 1.14. The causes of mutations.

Source: https://evolution.berkeley.edu/wp-content/uploads/2021/09/sperm-bias.jpg.

1.4.3. Population Genetics and Allele Frequency

It is essential to learn about population genetics in understanding the mechanisms of evolution, particularly allelic frequency changes in populations. Allele frequency refers to the frequency of a single allele in a gene pool relative to the frequency of other alleles for the same gene. It can be altered by various evolutionary pressures such as natural selection, genetic drift, and gene flow. Natural selection supports those alleles that maximize the fitness of an organism, thereby increasing their prevalence over generations. Genetic drift has the potential to bring about unpredictable allele frequency changes in populations, particularly in small populations, thus the loss of genetic variation. Interactions of such mechanisms outline the genetic structure of populations, and through them, we can see evolutionary processes. One of the most significant theoretical models used in population genetics is the Wright-Fisher model, which covers allele frequency dynamics under varying selection pressures. Under the model, allele frequencies may shift not only by stochastic reproductive events but also by the effect of fitness. Using the Wright-Fisher process with selection, researchers are able to simulate cases illustrating how environmental pressures drive allele distribution over time. This process provides the capacity to conclude the adaptive population behavior and their response to differing conditions. Furthermore, new analytical methods, such as the path integral formulation employed to study this process, continue to enlighten us about the genetic alterations evolving through selection. It provides fine-scale measurements of allele frequencies that aid in understanding the complexity involved in evolutionary transformations (Schraiber et al., 2013). The use of concepts from game theory can yet further improve our understanding of allele frequency changes in a population. Specifically, evolutionary game theory examines how individual strategy influences collective behavior, particularly in social situations. As an example, in multi-player games in which group payoffs are influenced by personal choices, the incidence of cooperative versus self-serving behavior can mirror changes in allele frequency even when not predicted by traditional models. This process highlights the non-linear influences between individuals of a population, hypothesizing that factors such as social interactions and networks must be incorporated while studying allele frequencies. By examining this intricate influence, researchers are able to better tackle the complexity of population genetics and unravel the influence in the context of evolutionary biology (Gokhale & Traulsen, 2014) (Figure 1.15).

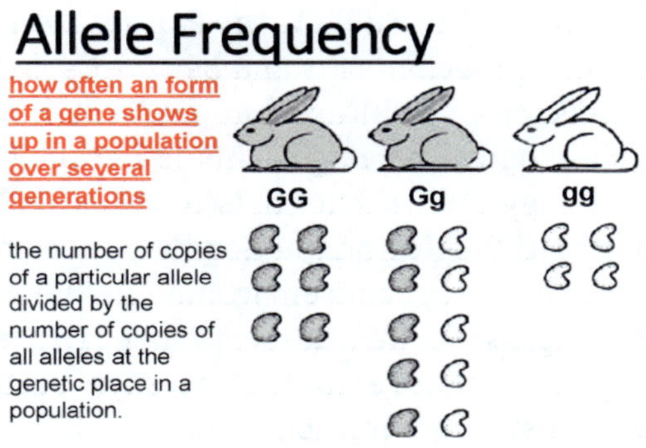

Figure 1.15. Population genetics and allele frequency.

Source: https://external-content.duckduckgo.com/iu/?u=https%3A%2F%2Fbiologydictionary.net%2Fwp-content%2Fuploads%2F2016%2F12%2FAllele-Frequency.jpg&f=1&nofb=1&ipt=945d972fb2d574023a6593f166e2d7aef4e4e5287663a08adb6bece5fcddf5c0.

Test Your Knowledge

1. What are Mendelian genetics and inheritance?
2. What are mutations and their impact on evolution?

1.5. Evolutionary Psychology

Learning Outcome

1. The influence of evolution on human behavior;
2. Natural selection and social behaviors;
3. Critiques and controversies in evolutionary psychology.

Evolutionary psychology is a field that provides an explanation of human behavior within the context of evolution, paying specific attention to how mechanisms and psychological attributes evolved for survival and reproduction. Fundamental to the field is the supposition that our thinking mechanisms and behaviors are not only creations of culture but find their roots in our biological heritage. This is because this methodology allows scientists to research how various psychological adaptations have developed as a response to specific environmental forces. It also highlights the intricate relationship between biology and culture in that it says cultural forces can strongly influence our psychological processes but are also being influenced by our evolution history. This exchange is central to the study of behavior in diverse sociocultural environments, as attested to by cross-cultural psychology's claim that evolutionary principles should be added to the study of human behavior and decision-making (Chasiotis, 2011). The emphasis on natural and sexual selection as a force that shapes human behavior is one of the fundamental axioms of evolutionary psychology. Natural selection operates through the survival of those that are most appropriate to their surroundings, while sexual selection is used to explain the traits that maximize reproductive fitness. These mechanisms are used to explain not only why some traits, such as altruism or aggression, may have evolved but also how they are expressed differently across different societies. The ongoing debate regarding the interaction between biological imperatives and cultural pressures is a sign of the complication of human psychology. Significantly, as scientists explore such issues, they reveal the importance of considering both the biological and the sociocultural aspects to gain a holistic appreciation of behavior (Hanisch & Eirdosh, 2020). This integration ultimately increases our understanding of human psychological evolution in various settings. Additionally, evolutionary psychology brings its implications across disciplines like medicine, anthropology, and even economics, proving to be interdisciplinary. By explaining how evolutionary background shapes behavior, this discipline can help shed light on present-day social matters such as mental illness, family formations, and social relationships. In line with scholars' calls for the infusion of evolutionary thought into psychological studies, it is evident that these methods can enhance our knowledge about human distinctiveness and diversity. For example, studying psychological adaptations may shed light on how contemporary social change interacts with old biological tendencies to produce resilience and conflict in human relations. Therefore, an investigation of these processes within evolutionary psychology not only illuminates theoretical discussion but also has the potential for real-world application, leading ultimately to a more sophisticated understanding of human action (Chasiotis, 2011).

1.5.1. The Influence of Evolution on Human Behavior

The intricate relationship between human behavior and evolution is a captivating field of study, shedding light on the biological determinants of social life. Human behavior has been shaped by evolution and natural selection to ensure survival and reproduction. Cooperative behavior, for instance, has flourished in contexts where group action increased the chances of acquiring resources and offspring survival. This evolutionary development highlights the importance of social behavior in evolution, proving that human behavior cannot be fully understood without consideration of its evolutionary origins. By analyzing these dynamics, scientists illustrate how evolutionary pressures have laid the groundwork for complex social organizations among humans, dictating such things as parenting behaviors and group cohesion. Further, the application of inclusive fitness theory (IFT) has significant discoveries in human cooperative behavior, in that our affiliations appear to transcend kin. W. D. Hamilton's original IFT is responsible for explaining cooperative disposition in societies, in that people will prefer group well-being in addition to personal fitness. The growth of IFT, as posited in recent studies, includes direct and indirect benefits acquired from transactions with related and unrelated entities, providing a more nuanced explanation of human behavior (Jaffe et al., 2015). The expanded framework facilitates comparative testing for cooperation across different fields, showing obviously how evolution not only favored altruistic behaviors but also facilitated complex social structures. The ongoing evolution of cooperation among groups is indicative of the flexibility of human behavior under social and environmental pressures. In addition, understanding the evolutionary roots of human interactions yields significant implications for addressing contemporary social challenges. Evolution has endowed humankind with psychological mechanisms that enhance cooperation, trust, and loyalty to the group, all key conditions for society's survival. As evidenced in discussions of insect population management tactics, evolutionary principles may inform strategies that limit resistance and facilitate cooperation (Bolton et al., 2015). The incorporation of such principles into modern conservation and public health policy demonstrates the potential for evolution-informed tactics to better shape human behavior. By invoking the learned knowledge from evolutionary biology, we can develop new strategies for promoting cooperation and reducing behavior that influences communal success, eventually architecting societies towards more efficiency in addressing complex environmental challenges (Figure 1.16).

Figure 1.16. The influence of evolution on human behavior in biology.

Source: https://holistic.news/en/wp-content/uploads/2023/11/f9632784-81b5-414c-8ba5-8e9d91adfab7.webp.

1.5.2. Natural Selection and Social Behaviors

The relationship between natural selection and social behavior is pivotal in understanding evolutionary dynamics. Social behavior, ranging from altruism to aggression, has a strong influence on individual fitness and population composition. Natural selection acts on these behaviors, favoring those that ensure reproductive success either directly through the action of an individual or indirectly through social interaction. For instance, social behaviors that contribute to the survival of groups can lead to an increased likelihood of gene transmission to future generations. Furthermore, the conditions under which social behavior occurs, e.g., group size and group composition, can also have an influence on evolutionary outcomes by impacting the degree of relatedness within members. Through scrutiny of various social behaviors in the light of natural selection, there are complexities and strategies for survival imposed by adaptation across various species. The IFT is the very foundation for understanding how natural selection acts on social behavior. Formulated by W. D. Hamilton, this theory suggests that one's genetic fitness also depends on assisting relatives and hence fostering cooperation over competition. In recent years, though, the theory has been further advanced through research, leading to the formulation of extended inclusive fitness theory (EIFT). This broader strategy incorporates not just kin selection but interactions with non-kin as well and emphasizes cost-benefit relations for cooperative behavior. Emphasizing sustainable levels of cooperation, EIFT allows for a larger overall treatment of social dynamics, including division of labor and symbiotic relationship development. This model demonstrates how social actions, seemingly altruistic, can be extremely linked with evolutionary advantages in natural selection. Group process study within social behavior further explains the natural selection process. Variation in group sizes greatly impacts social interactions and helping versus harming. Research reveals that while larger groups will provide more opportunities for cooperation, smaller groups can lead to larger competition, affecting the evolution of traits under the selective pressures of natural environments. More specifically, variation in group size can be mediators for harming and helping behaviors, where larger groups will promote helping behavior and smaller ones will favor harming behavior. This social environment manipulation illustrates the extent to which natural selection is not merely a physical attribute-favoring force but a potent force that drives social dynamics, which form part of species evolution and how they interact with one another (Gardner et al., 2013; Jaffe et al., 2015) (Figure 1.17).

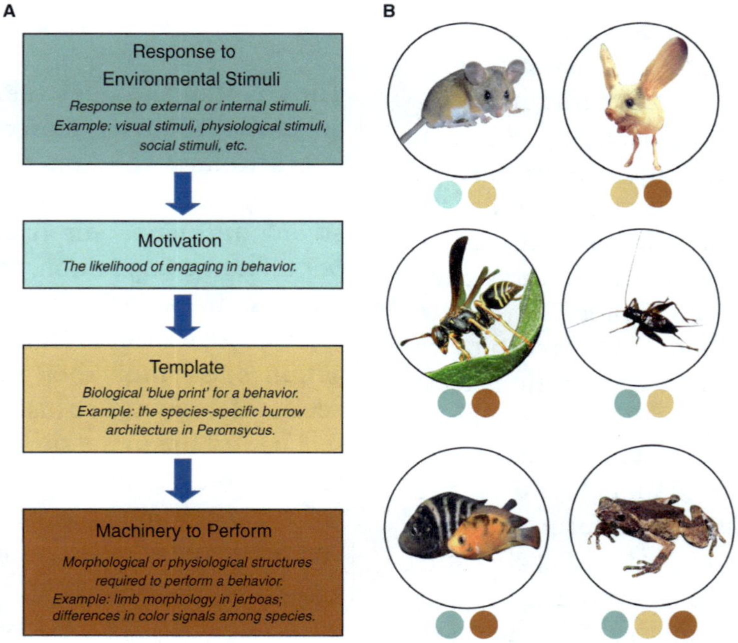

Figure 1.17. Natural selection and social behaviors.

Source: https://ars.els-cdn.com/content/image/1-s2.0-S0960982217314616-gr1.jpg.

1.5.3. Critiques and Controversies in Evolutionary Psychology

Evolutionary psychology has drawn considerable attention, though at some cost of controversy and criticism. At the center of the criticism is the approach taken by evolutionary psychologists, which draws more on theoretical assumptions than empirical investigations. These assumptions are accused of producing sweeping generalizations about human behavior that fail to account for cultural diversity or the nuance of social context. These criticisms observe a need for more balanced accounts of human behavior that consider both biological impulses and sociocultural influences. The reliance on anecdotal evidence for using evolutionary explanations has also raised questions among researchers as to the validity of such an approach, finally requesting a greater scientific methodology to empirically support what is argued within the discipline (Wildman, 1998). Further, the implications of evolutionary psychology extend beyond scientific arguments into wider societal debates, particularly into its understandings of gender roles and behavioral standards. According to critics, by attributing particular behaviors to evolutionary forces, such as aggression or mating behavior, evolutionary psychology can end up reinforcing deadly stereotypes and validating inequality. This becomes particularly contentious within the debate over gender concerns, where evolutionary explanations actually serve to perpetuate patriarchal systems by outlining certain traits as innate. The inclination towards the emphasis on biological determinism risks undermining individual autonomy and augmenting socio-political injustices. It is thus the problems that evolutionary psychology poses that necessitate critical examination to identify its impact on society's values and norms (Cassidy et al., 2016). Finally, popular science media's public broadcasting has assisted in securing the contentious status of evolutionary psychology. In the 1990s, for instance, there was a sensationalized explosion in media coverage of evolutionary psychology, where scholars themselves were actively engaging with the public to communicate their findings. The situation is two-edged: it increases the visibility and popularity of evolutionary explanations but can also blur complicated scientific debates. The mass popularity can be sensational rather than scientific, creating misconceptions about the nature of human behavior and evolution. Therefore, the public opinion-academic credibility interface presents an additional challenge to the discipline in maintaining academic integrity in the context of meeting public interest in the study of human behavior (Wildman, 1998).

Test Your Knowledge

1. List the influence of evolution on human behavior.
2. What are natural selection and social behaviors?

1.6. Applications of Evolutionary Theory

Learning Outcome

1. Evolution in medicine and public health;
2. Conservation biology and biodiversity;
3. Evolutionary algorithms in technology.

The applications of evolutionary theory extend well beyond the limits of strict academic discussion, permeating all fields of medicine, conservation, and genomics. In medicine alone, the understanding of evolutionary forces of pathogen resistance underpins successful vaccine and therapy development. By examination of microbial population evolutionary trajectories, scientists can predict novel outbreaks and develop adaptive responses in the context of antibiotic resistance. Additionally, conservation practices have increasingly utilized evolutionary principles to encourage the conservation of biodiversity. Knowledge of evolutionary relationships and the adaptive potential of species can inform habitat restoration and species reintroduction efforts. This transdisciplinary integration highlights the significance of evolutionary theory as a central component that drives advancements in human health and ecological stability. Among all the uses of evolutionary theory, evolutionary genomics has been one of the most interesting. The birth-and-death process mathematical model demonstrates this application, with applications in population dynamics, speciation, and genome evolution. It describes the way gene family dynamics evolve over time through mechanisms such as HGT and innovation within organisms. Through empirical observations, such as paralogous gene family size distributions, biologists can select models that simulate properly historical evolutionary trends. The mathematical clarity and flexibility of such models make them indispensable to answer complicated biological questions. Thus, evolutionary genomics can unravel the complications of life, permitting a deeper understanding of genetic variation and evolutionary history across all taxa (Karev et al., 2005). The relationship between information theory and evolutionary biology also facilitates a greater insight into biological processes. Information held in an organism's genome is vital to its development, functioning, and adaptation. Where populations are confronted with environmental alterations, information processing and storage are essential for survival. Evolutionary adaptations, such as protein structure differences or neural processing differences, rely on information theory. Research on how simulated agents learn to deal with intricate activities illustrates the general implications of information processing within ecological systems. Through relating information dynamics to evolutionary success, scientists can facilitate more advanced understanding of how organisms still evolve to fit changing environments (Adami et al., 2011). Such an exploration of such topics is thus crucial for theoretical understanding and practical applications in numerous disciplines.

1.6.1. Evolution in Medicine and Public Health

The intersection of evolution and medicine is the secret to unlocking the mechanisms of infectious diseases alongside the strategies developed to combat them. One such case is in the use of the bacterium Wolbachia in controlling dengue fever, whereby evolutionary principles are employed in disrupting the lifecycle of the virus's mosquito vector, *Aedes aegypti*. Three Wolbachia control strategies have been proposed, all relying on the evolutionary advantages conferred to the infected female mosquitoes. Of all the interventions, releasing incompatible males with the local females is one that takes a repetitive process of population control. However, the evolutionary fitness of the virus and mosquito is a challenge because such measures are under threat of development of resistance strains, in which there exists an important evolutionary perspective in the practice of public health (Bull et al., 2013). Evolution also surfaces in the cold, hard realities of diseases such as malaria, particularly as environments become modified. Recent studies in East Africa document the spread of malaria into new territories not previously infected, a trend that has sparked an open discussion about the contribution of climate change. Just as there is ample evidence implicating the contribution of environmental factors to this trend, it is important to recognize that attributing the resurgence of malaria to climate change alone overlooks intervening factors. Evolution is a multifaceted process that includes not only abiotic factors like climate but also vector and host dynamics and resistance trends.

Therefore, understanding malaria's evolution calls for an integrated approach considering these various drivers so that researchers and public health practitioners can implement more effective and responsive interventions (Chaves et al., 2010). These examples underscore the utility of evolutionary thinking in guiding medical and public health practice. The dialogue between pathogen evolution, vectors, and environmental evolution demonstrates that medicine is unable to move without biological evolution. By seeing what happens in the situation of Wolbachia and malaria, predicting evolutionary organismic responses can facilitate more sustainable health interventions. These are important for making overall public health responses since they enable proactiveness in measures that take into consideration potential evolutionary changes in pathogens. Lastly, an acceptance of inclusive knowledge about evolution provides health workers and policymakers with the skills to combat diseases more efficiently, confirming the interconnectedness of biology and public health in countering global health threats (Figure 1.18).

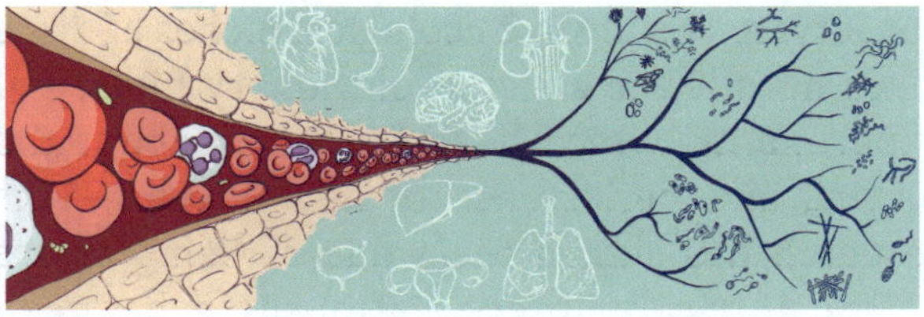

Figure 1.18. Evolution in medicine and public health.

Source: https://iiif.elifesciences.org/journal-cms/collection%2F2021-07%2Felife-evolutionary-medicine-2.jpg/0,1,10738,3576/2000,/0/default.jpg.

1.6.2. Conservation Biology and Biodiversity

Conservation biology is a significant component of understanding biodiversity, particularly as it pertains to the evolutionary processes that shape the ecosystem. The intricate web of existence is subject to constant natural and human-generated pressures; thus, the role of conservation biology becomes critical in mitigating these impacts. Conservation biologists try to explain how different species live or die due to environmental pressures by the study of interactions among species, the preservation of their habitats, and the effects of climate change. As stated, Climatic change affected the composition of ecological communities during the Quaternary, which was probably both the result of, and the result caused by, evolutionary processes such as species evolution, adaptation, and species and population extinction (Stewart, 2010). This interconnectedness is an ideal for the preservation of biodiversity since it makes ecosystems resilient and capable of adapting in the face of adversity. The East African Great Lakes may be regarded as a microcosm for the examination of the meeting place of conservation biology and evolutionary science in biodiversity. Renowned for their rich endemic fish faunas, particularly the specialized cichlids, these lakes show us the dynamic processes by which species diversification is fostered and why they must be conserved. The diverse ecological niches of these lakes have led to a rich evolutionary history of experiments, which have taught us about the origins and persistence of new species. This research has shown the value of being aware of the intricate inter-relationships between species, their environments, and forces acting upon them. As indicated, the East African Great Lakes are nowadays well known for (i) their fisheries, of vital importance to their rapidly expanding riparian human populations; and (ii) as biodiversity hotspots with stunning endemic faunas (Lowe-McConnell et al., 2003).

This backdrop highlights the need for conservation practices that not only preserve biodiversity but also preserve human livelihoods intrinsically enmeshed within such water systems. As conservation biology becomes a maturing discipline, it is increasingly focused on becoming proactive towards stemming biodiversity loss. Coordination between scientists, policymakers, and local populations plays a pivotal role in ensuring sustainable endeavors that foster ecological integrity. Conservation measures against the dangers that species are facing, such as loss of habitat, overfishing, and global change, are essential in order to sustain the delicate equilibrium in ecosystems. The findings of various studies on the East African Great Lakes underscore the necessity of taking conservation action against their unique fish fauna and the overall environmental condition. By uniting scientific research with Indigenous people and global conservation action, we can both enrich our understanding of evolutionary mechanisms and promote conservation of biodiversity. Thus, the two themes of biodiversity and conservation biology are inextricably entwined in the grand narrative of evolution, and it is incumbent on all of us to conserve our planet's natural inheritance (Figure 1.19).

Figure 1.19. Conservation biology and biodiversity.

Source: https://blogger.googleusercontent.com/img/b/R29vZ2xl/AVvXsEh T8cRpP8fY39NpwwvZax856ga1CZEzbPlvFn1UJZT121g-fPOEOrS NJ1epkxXSCf8ZNCSajfnjyQqLatV4N-nCfbsT_06Sb1SmSN3 zVJWsky47H mT8VqkuT3BGdEkyN 4nfef0b6dmLjzA/s1600/conservation.jpg.

1.6.3. Evolutionary Algorithms in Technology

The application of evolutionary algorithms in technology is a culmination of computer innovation and nature. Evolutionary algorithms are founded on the premise of natural selection, wherein the solutions to problems evolve over time through such mechanisms as reproduction, mutation, and selection. Evolutionary algorithms have been applied in fields such as optimization and AI to derive solutions that not only optimize but also adapt based on changing environments. By emulating competitive dynamics in biological systems, these algorithms are able to efficiently search tremendous amounts of space and discover optimal or near-optimal solutions within highly complex problem spaces. Biological analogies and algorithmic construction further encourage our understanding of adaptability in both natural and artificial systems and further enhance the discussion regarding biological evolution and technology applications. A fascinating development here is the concept of

Digital Ecosystems, designed to mimic biological self-organizing properties. By leveraging the laws of ecology, scientists have successfully created a system in which software programs evolve in a decentralized network that promotes robustness and scalability. The Digital Ecosystem operates at two levels: decentralized agents' migration and localized evolutionary computations intended to address particular users' requirements. This two-layered optimization approach is a reflection of the intricate dynamics of biological ecosystems, where interactions among and between populations and their biotic and abiotic environments produce sophisticated adaptations. Experimental measurements, grounded in theoretical ecology, have confirmed the use of this model, asserting its ecological succession behavior, a key aspect in assessing the maturity and responsiveness of the Digital Ecosystem to evolving user requirements (Briscoe et al., 2011). Also, the use of evolutionary algorithms extends beyond mere problem-solving, but is applied to understand the intricacies of biological evolution itself. By simulating such factors as population dynamics and spatial distributions that are responsible for self-organization in nature, researchers are beginning to define how these biological attributes may be employed to develop artificial systems. For instance, the resilience and adaptability of biological ecosystems provide a blueprint for constructing technologies that can manage dynamic and complex environments. The iterative nature of evolutionary algorithms mirrors the gradual changes observed in biological evolution, emphasizing not only the technological advancements achievable through these methods but also their philosophical implications regarding our understanding of evolution. This search for biological properties in technology reveals a most fertile line of research in the future, testifying to the untapped potential of evolutionary theory in designing stable artificial systems (Briscoe et al., 2007) (Figure 1.20).

Figure 1.20. Futuristic depiction of a genomics research lab featuring a central 3D DNA helix surrounded by digital interfaces and data analysis tools, representing advanced bioinformatics and genetic engineering.

Source: https://easy-peasy.ai/cdn-cgi/image/quality=80,format=auto,width=700/https://media.easy-peasy.ai/21050371-16ad-4990-be21-1e086641c9a6/99d63ab3-7431-4878-b40f-a5bf8c43ff95.png.

Test Your Knowledge

1. What is conservation biology?
2. What are evolutionary algorithms in technology?

SUMMARY

The evolutionary research in biology creates a broad context for understanding the complicated processes that have developed life on Earth. The synthesis of genetic, environmental, and ecological processes in the story of evolution highlights the complexity of organisms and their history of adaptations. Recent research on viral evolution, particularly of recent global health crises such as COVID-19, demonstrates the ever-changing interface between evolution and the subject matter at hand. The understanding of viral mutation processes and mechanisms of viral genetics not only advances our knowledge of evolution but also dictates the guidance of public health interventions aimed at avoiding viral disease (Moghaddasi et al., 2020). This live environment underscores the fact that evolution mechanisms extend beyond theoretical frameworks and affect real-world implications in medicine, ecology, and biotechnology. Additionally, the syncretism of evolution with inter-disciplinary approaches, such as project-based learning in teaching, broadens our knowledge and interest in biological concepts. Modern pedagogical approaches, which facilitate active, student-centered learning, result in a deeper understanding of complex topics like evolution, genetics, and ecology. 21st-century critical thinking and problem-solving skills may be taught to students by integrating biological concepts with applicable contexts. As well demonstrated by the recent research, interdisciplinary procedures and methods ensure not only learning but also interest in biological sciences (Bernini et al., 2019). Thus, the implementation of new pedagogical methods is essential during the education process of future biologists who will be capable of comprehending the nuances of the theories and results of evolution. Lastly, ongoing exploration of evolution in biology is imperative for science and humanity. As we struggle with pressing threats such as climate change, loss of habitats, and new infectious diseases, evolutionary biology has something to learn from resilience and adaptation. It reiterates the need to enhance scientific literacy and interdisciplinarity among students, teachers, and researchers. By facilitating the establishment of core areas of scientific investigation directly related to evolutionary research, communities are better able to meet future biological challenges and offer sustainable ecological stewardship. Finally, the evolutionary process is not nearly as much an intellectual exercise; it is a model for moral stewardship of our planet and its diverse citizens, in that it keeps us mindful of our interdependence and responsibility to generations to come.

Evolutionary process research finds a collection of significant concepts that regulate the mechanism of biological diversity. They include one of the causes of evolutionary change through genetic variation. Genetic variation supplies populations with the raw material on which natural selection acts. Over time, survival and reproduction of favorable traits become more common in the population. In addition, the interactive nature of the forces of the environment and genetic drift supports the process through which organisms evolve as a response to the environment. The interactive nature of such relations supports the validity of the study of evolutionary processes, which provides evidence for the complexity of life on Earth. The use of evolutionary game theory to explain social life and interactions of organisms is another major issue discussed. This contemporary approach widened the scope of evolutionary biology by incorporating concepts from other fields like ecology and social science. Multiplayer interactions are studied by researchers to see how cooperation and competition between players influence evolutionary processes. Outcomes of evolutionary game theory serve as a basis to study complex social behavior and organization, thus making the tale of evolutionary success greater than mere survival. This inter-disciplinary book not only enhances theoretical understanding but also encourages greater scrutiny of nonlinear dynamics in nature, as noted in recent books (Gokhale & Traulsen, 2014). Finally, studies of eusocial animals, such as bumblebees, yield fascinating dynamics between social structure and reproductive modes. In particular, workers versus queens are an excellent example of how rival functions in a colony translate into rival pressures of evolution. An examination of the egg-laying switch points of bumblebee queens illustrates the measured extent of queen control over these critical life-history switches, and this suggests that social interaction could not have had significant effects on reproduction timing as has been widely assumed (Andrew et al., 2013). It entails a revision of traditional cooperative behavior by members in colonies and challenges the reconsideration of hierarchy and self-interest impacts on the evolution process. These examples indicate the complex interplay of the evolutionist discourse that includes the core of biological adaptation and survival.

KEY TERMS FROM THIS CHAPTER

Adaptation	Analogous structures	Artificial selection
Biogeography	Common ancestor	Descent with modification
Embryology	Evolution	Fitness
Fossil record	Gene flow	Gene pool
Genetic drift	Genetic variation	Homologous structures
Mutation	Natural selection	Speciation

CRITICAL THINKING QUESTIONS

1. **Who is known as the father of the theory of evolution by natural selection?**
   a. Gregor Mendel
   b. Charles Darwin
   c. Louis Pasteur
   d. James Watson
2. **What is evolution?**
   a. A sudden change in an organism's appearance
   b. A theory that life never changes
   c. A gradual change in a species over time
   d. The development of an individual organism
3. **Which of the following provides the best evidence for evolution?**
   a. Weather patterns
   b. Fossil records
   c. Astrology
   d. Magnetic fields
4. **Natural selection acts on an organism:**
   a. Acquired traits
   b. Learned behavior
   c. Inherited traits
   d. Daily habits
5. **What is a common result of natural selection over many generations?**
   a. Decrease in genetic variation
   b. Increase in harmful mutations
   c. Adaptation to the environment
   d. Elimination of all variation

Short Answer Questions

1. What is evolution in biological terms?
2. Who is Charles Darwin, and what was his major contribution to biology?
3. What is natural selection, and how does it drive evolution?
4. What is a population in terms of evolution, and why is it important to study populations rather than individuals?
5. What are adaptations, and how do they arise in a species over time?

Answers to Critical Thinking Questions

1. (b); 2. (c); 3. (b); 4. (c); 5. (c)

REFERENCES

1. Adami, C., Ash, A., et al. (2011). The use of information theory in evolutionary biology. http://arxiv.org/abs/1112.3867 (accessed on 13th June 2025).
2. Aldrich, H. E., Aunger, R., Bagehot, W., Baum, J. A. C., Boyd, R., et al. (2014). Untangling the conceptual issues raised in Reydon and Scholz's critique of organizational ecology and Darwinian populations. https://core.ac.uk/download/29841735.pdf (accessed on 13th June 2025).
3. Aleksandrova, G., Ding, L., Herman, A., Jin, Y., et al. (2017). Paleogene monsoons across India and South China: Drivers of biotic change. https://core.ac.uk/download/153444247.pdf (accessed on 13th June 2025).
4. Alfonso-Durruty, M. P. (2024). *Becoming Humans: An Introduction to Biological Anthropology*. https://core.ac.uk/download/619720763.pdf (accessed on 13th June 2025).
5. Altermatt, F., Astegiano, J., Baguette, M., Becks, L., Bell, G., et al. (2017). Eco-evolutionary dynamics in fragmented landscapes. https://core.ac.uk/download/131025034.pdf (accessed on 13th June 2025).
6. Aragon, R., Burgess, T. I., Crous, C. J., Dickie, I. A., et al. (2016). Evolutionary dynamics of tree invasions: Complementing the unified framework for biological invasions. https://core.ac.uk/download/365068425.pdf (accessed on 13th June 2025).
7. Bacher, S., Blackburn, T. M., Essl, F., Genovesi, P., et al. (2017). Socio-economic impact classification of alien taxa (SEICAT). https://core.ac.uk/download/286567949.pdf (accessed on 13th June 2025).
8. Barker, M. S., & Wolf, P. G. (2010). Unfurling fern biology in the genomics age. https://core.ac.uk/download/32552645.pdf (accessed on 13th June 2025).
9. Barnett, R., Brace, S., Cucchi, T., Dobney, K., et al. (2013). Divergent evolutionary processes associated with colonization of offshore islands. https://core.ac.uk/download/20664364.pdf (accessed on 13th June 2025).
10. Barrientos, Gustavo; On the problem of identifying homologies in lithic artifacts; Archaeopress; 24; 2016; 49-66(accessed on 13th June 2025).
11. Bernini, M. A. M., Carlos-Bender, J., & Carvalho, A. L. (2019). Project-based learning: Video game development as a tool to integrate ecology, genetics, evolution and informatics. https://core.ac.uk/download/544083957.pdf (accessed on 13th June 2025).
12. Bolton, M., Chapman, T., & Leftwich, P. (2015). Evolutionary biology and genetic techniques for insect control. https://core.ac.uk/download/41991375.pdf (accessed on 13th June 2025).
13. Bourke, A. F. G., Aron, S., Guidat, F. S., Holland, J. G., Maynard Smith, J., & Oster, G. F. (2013). Queen control of a key life-history event in a eusocial insect. https://core.ac.uk/download/19085620.pdf (accessed on 13th June 2025).
14. Briscoe, G., De Wilde, P., & Sadedin, S. (2011). Digital ecosystems: Ecosystem-oriented architectures. http://arxiv.org/abs/1112.0204 (accessed on 13th June 2025).
15. Briscoe, G., Paperin, G., & Sadedin, S. (2007). Biology of applied digital ecosystems. http://arxiv.org/abs/0712.4153 (accessed on 13th June 2025).

16. Brooks, D. R., & van Veller, M. G. P. (2008). Assumption 0 analysis: Comparative phylogenetic studies in the age of complexity. https://core.ac.uk/download/pdf/29259838.pdf (accessed on 13th June 2025).

17. Bull, J. J., & Turelli, M. (2013). Wolbachia versus dengue: Evolutionary forecasts. https://core.ac.uk/download/323070711.pdf (accessed on 13th June 2025).

18. Byron (formerly Baker), J. M. (2006). Whence philosophy of biology? https://core.ac.uk/download/11921247.pdf (accessed on 13th June 2025).

19. Calsin*a, A., S*asaki, A., Koonin, E. V., Karev, G. P., Price, G. R., Gudelj, I., Metz, J. A. J., et al. (2013). A simple mathematical model of gradual Darwinian evolution: Emergence of a Gaussian trait distribution in adaptation along a fitness gradient. https://core.ac.uk/download/12827691.pdf (accessed on 13th June 2025).

20. Cassidy, A. (2016). Evolutionary psychology as public science and boundary work. https://core.ac.uk/download/77032079.pdf (accessed on 13th June 2025).

21. Chasiotis, A. (2011). Evolution and culture. https://core.ac.uk/download/10687091.pdf (accessed on 13th June 2025).

22. Chaves, L. F., & Koenraadt, C. J. M. (2010). Climate change and highland malaria: Fresh air for a hot debate. https://core.ac.uk/download/pdf/29238217.pdf (accessed on 13th June 2025).

23. Chitneni, S. K. (2008). Development and evaluation of diagnostic tools for nucleopolyhedroviruses (NPVs) infecting major lepidopteran pests of legume crops in the semi-arid tropics. https://core.ac.uk/download/211008554.pdf (accessed on 13th June 2025).

24. D'Alelio, D. (2017). Biological complexity behind plankton system functioning: Synthesis and perspectives from marine long term ecological research. https://core.ac.uk/download/572832186.pdf (accessed on 13th June 2025).

25. Finkelman, L., & Pigliucci, M. (2014). The extended (evolutionary) synthesis debate: Where science meets philosophy. https://core.ac.uk/download/186330782.pdf (accessed on 13th June 2025).

26. Gardner, A., & Rodrigues, A. M. M. (2013). Evolution of helping and harming in viscous populations when group size varies. https://core.ac.uk/download/30317564.pdf (accessed on 13th June 2025).

27. Gill, J. (2012). An extra-memetic empirical methodology to accompany theoretical memetics. https://core.ac.uk/download/9051161.pdf (accessed on 13th June 2025).

28. Gokhale, C. S., & Traulsen, A. (2014). Evolutionary multiplayer games. http://arxiv.org/abs/1404.1421 Dyn Games Appl DOI 10.1007/s13235-014-0106-2. (accessed on 13th June 2025).

29. Griffiths, P. E., & Stotz, K. (2017). Genetic, epigenetic and exogeneity information. https://core.ac.uk/download/131214859.pdf (accessed on 13th June 2025).

30. Griffiths, P. E., & Stotz, K. (2018). Developmental systems theory as a process theory. https://core.ac.uk/download/131212934.pdf (accessed on 13th June 2025).

31. Guan, J., Wang, X., & White, S. C. (2020). A new genus and species of sabretooth, Oriensmilus liupanensis (Barbourofelinae, Nimravidae, Carnivora), from the middle Miocene of China suggests barbourofelines are nimravids, not felids. https://core.ac.uk/download/286728871.pdf (accessed on 13th June 2025).

32. Hanisch, S., & Eirdosh, D. (2020). Conceptual clarification of evolution as an interdisciplinary science. EdArXiv. https://doi.org/10.35542/osf.io/vr4t5Falcón Ganfornina, R. M., Falcón Ganfornina, Ó. J., & Núñez Valdés, J. (2019). An application of total-colored graphs to describe mutations in non-Mendelian genetics. https://core.ac.uk/download/286565273.pdf (accessed on 13th June 2025).

33. Harding, A., Lin, J., & Whitacre, J. M. (2*012). T cell adaptive i*mmunity proceeds through environment-induced adaptation from the exposure of cryptic genetic variation. https://core.ac.uk/download/pdf/8680282.pdf (accessed on 13th June 2025).

34. Henchman, A. A. (2018). Charles Darwin's final book on earthworms, 1881. https://core.ac.uk/download/215930410.pdf (accessed on 13th June 2025).

35. Herrmann, M., Mayer, W. E., & Sommer, R. J. (2006). Sex, bugs and Haldane's rule: The nematode genus Pristionchus in the United States. https://core.ac.uk/download/pdf/4058454.pdf (accessed on 13th June 2025).

36. Jaffe, K. (2015). Extended inclusive fitness theory bridges economics and biology through a common understanding of social synergy. http://arxiv.org/abs/150*9.02745 (acc*essed on 13th June 2025).

37. Karev, G. P., Koonin, E. V., & Novozhilov, A. S. (2005). Biological applications of the theory of birth-and-death processes. http://arxiv.org/abs/q-bio/0507026 (accessed on 13th June 2025).

38. Kendig, C. (2016). Homologizing as kinding. https://core.ac.uk/download/333651030.pdf (accessed on 13th June 2025).

39. Kirchner, S., Power, B. J., & Waters, A. P. (2016). Recent advances in malaria genomics and epigenomics. https://core.ac.uk/download/46580201.pdf (accessed on 13th June 2025).

40. Kosolosky, L. (2011). 'Causa efficiens' versus 'Causa finalis': Functional explanations in biology. https://core.ac.uk/download/55732344.pdf (accessed on 13th June 2025).

41. Lohaus, R., & Van de Peer, Y. (2016). Of dups and dinos: Evolution at the K/Pg boundary. https://core.ac.uk/download/55855277.pdf (accessed on 13th June 2025).

42. Lowe-McConnell, R. (2003). Recent research in the African Great Lakes: Fisheries, biodiversity and cichlid evolution. https://core.ac.uk/download/11020199.pdf (accessed on 13th June 2025).

43. Mattick J, Amaral P. RNA, the Epicenter of Genetic Information: A new understanding of molecular biology. Abingdon (UK): CRC Press; 2022 Sep 20. Available from: https://www.ncbi.nlm.nih.gov/books/NBK595939/ doi: 10.1201/9781003109242 (accessed on 13th June 2025).

44. Moghaddasi, H., Rezaei Tavirani, M., & Samadi, L. (2020). The role of systems biology and synthetic biology in appearing and managing COVID-19. https://core.ac.uk/download/483640779.pdf (accessed on 13th June 2025).

45. NC DOCKS at The University of North Carolina at Greensboro, & Remington, D. L. (2003). Candidate genes, quantitative trait loci, and functional trait evolution in plants. https://core.ac.uk/download/345089383.pdf (accessed on 13th June 2025).

46. Oskar Burger; Ronald Lee; Rebecca Sear.2025). Human evolutionary demography. https://doi.org/10.11647/OBP.0251 (accessed on 13th June 2025).

47. Podgórski, J. S. (2010). Humberto Maturana's view on the theory of evolution: From autopoiesis

to natural drift metaphor. https://core.ac.uk/download/214936779.pdf (accessed on 13th June 2025).

48. Schraiber, J. G. (2013). A path integral formulation of the Wright-Fisher process with genic selection. http://arxiv.org/abs/1307.7756 (accessed on 13th June 2025).

49. Stewart, J. R. (2010). Understanding evolutionary processes during past Quaternary climatic cycles: Can it be applied to the future? https://core.ac.uk/download/4898204.pdf (accessed on 13th June 2025).

50. Tabaczek, M. (2014). An Aristotelian account of evolution and the contemporary philosophy of biology. https://core.ac.uk/download/189339565.pdf (accessed on 13th June 2025).

51. UiTM, Faculty of Plantation and Agrotechnology. (2017). *AGC702: Molecular Entomology / Faculty of Plantation and Agrotechnology*. https://core.ac.uk/download/635301090.pdf (accessed on 13th June 2025).

52. Warschefsky, E. (2018). The evolution and domestication genetics of the mango genus, *Mangifera* (Anacardiaceous). https://core.ac.uk/download/212918337.pdf (accessed on 13th June 2025).

53. Wildman, W. J. (1998). A theological challenge: Coordinating biological, social, and religious visions of humanity. https://core.ac.uk/download/142035389.pdf (accessed on 13th June 2025).

MECHANISMS OF EVOLUTION

2

LEARNING OBJECTIVES

After studying this chapter, you will be able to:

1. Discuss natural selection;
2. Define genetic drift;
3. Explain about mutation;
4. Describe gene flow;
5. Explain speciation;
6. Focus on the term coevolution.

Introduction

The explanation of the mechanisms of evolution is a foundation for understanding the diverse processes that shape life on our planet. Evolution, an elementary biological concept, is made up of a wide variety of phenomena, including natural selection, genetic drift, and gene flow. Through the study of the mechanisms, the complexity of how species evolve and adapt in their environment through time is unveiled. It is by this combination of theory and evidence that the evolutionary paradigm is fully developed, with a demonstration that life does not stand still but is continually changing in response to an awe-inspiring series of ecological pressures and gene variation. This context is central to framing for further debate about the meaning of evolution in both past and present contexts, to highlight its importance as part of the study of biology. Besides natural selection, other mechanisms of evolution pertinent to this study provide valuable insights into the mechanisms that facilitate speciation and adaptation. Genetic drift, for instance, accounts for random allele frequency changes in a population, typically causing significant evolutionary changes in the long term, particularly in small populations. Similarly, gene flow introduces to populations new genetic information, yet again modeling genetic diversity as well as evolutionary patterns. An understanding of these processes contributes towards a more advanced awareness of evolution, the emphasis being on the cooperation of necessity and chance to mold biological diversity. The cooperation between these processes demonstrates that evolution is a multifaceted process, reliant on a synthesis of genetic and environmental pressures that, in combination, regulate the evolving development of life upon our planet. Furthermore, ongoing research into the operation of evolution serves to stress the dynamic nature of scientific inquiry. For instance, recent findings in astrophysics indicate that the early cosmic events played a pivotal role in the development of structures like supermassive black holes, as they destabilize traditional frameworks of evolution on both cosmic and biological levels (Ferrara et al., 2016).

Figure 2.1. The mechanisms of evolution.

Source: https://external-content.duckduckgo.com/iu/?u=https%3A%2F%2Fjournalhow.com%2Fwp-content%2Fuploads%2F2021%2F04%2FWhat-are-the-mechanisms-of-evolution-768x463.jpg&f=1&nofb=1&ipt=7bc67fbaea26e128d84ba86018b589e83f13eb9cdb92ec59e7acba7e8f7c125a.

These findings prompt scientists to rethink their concepts about how evolution is perceived in a broader scheme, extending past what happens on the Earth and out into cosmic interactions as well. This interconnectedness demonstrates that evolutionary processes are not separate thoughts; rather, they are an integrated concept of change and adaptation across different systems. The investigation of these intersections, therefore, contributes to a deeper understanding of evolution and its impact on other fields of research, such as ecology, genetics, and evolutionary biology (Patel et al., 2005) (Figure 2.1).

Definition of Evolution

Evolution theory depends on the theory of the gradual evolution of species over time through natural selection, genetic drift, and mutation. The process is not only a biological change mechanism but also an explanatory system for the complexity of life on Earth. Evolution is the change in biological populations' heritable traits over generations, with such changes potentially resulting from various genetic differences. Evolution is therefore an active process that targets organisms adapting to their environments, which in turn affects their survival and reproductive success. Further, the redefinition of organisms in the context of new advances in biology raises questions about the multifaceted relationships between entities and the evolutionary complexities underlying the relationships between them, as seen in the contextual organismality framework posited in recent work (Boddy et al., 2016). This vision highlights the imperative of understanding the evolution not just as a mechanism but also as a process. Understanding evolution involves considering its mechanisms underlying the observable changes in populations. Natural selection, a fundamental concept in evolutionary theory, is the differential survival and reproduction of individuals based on their advantageous traits. The mechanism operates based on the hypothesis that creatures with traits better suited to the environment are likely to survive and transmit such traits to future generations. In addition, genetic drift introduces an element of randomness into the process of evolution, where random changes in allele frequencies can greatly impact populations, especially in small populations. Once more, mutation is the material for evolutionary change, which introduces new variations of genes that can give rise to favorable characters in future generations. Together, these processes illustrate the intricacy of evolution as they show the subtle interaction of the genetic factors and environmental pressures that control biodiversity (Bich et al., 2016). Besides evolutionary processes, defining evolution also entails responding to more general philosophical questions about life and its manifestations. Scholars have held differing views on the nature of life itself, for instance, in a lack of consensus regarding the required and sufficient conditions to determine what constitutes a living organism.

Figure 2.2. Evolution.

Source: https://wol.jw.org/ko/wol/mp/r8/lp-ko/lf/2010/68.

This section-wide debate reflects the importance of operational definitions, which include several theoretical and practical applications in scientific fields like synthetic biology and astrobiology (Bich et al., 2016). Such debates establish that not only does the term evolution need to encompass biological processes, but also the philosophical and epistemological dimensions of the term life. With scientists delving deeper into the phenomenon of evolution and its mechanisms, the conjoining of biological, ecological, and philosophical paradigms will be integral in advancing knowledge on both the topic of evolution as well as the notion of life itself (Figure 2.2).

Importance of Studying Evolutionary Mechanisms

The study of evolutionary processes lies at the heart of comprehending diversity on Earth. By examination of the forces underlying evolution, such as mutation, natural selection, and genetic drift, scientists are able to untangle the complexities of biological diversity. Comprehension of these processes is vital not just for the science of biology but also for medicine, agriculture, and conservation applications. For instance, knowing how some traits evolve, researchers can combat disease and strengthen crop resistance. Without an explicit grasp of evolutionary processes, efforts to solve these worldwide issues would lack the foundational knowledge necessary for effective solutions. The implications of evolution studies thus extend much beyond academic interest, affecting actual practices that have direct consequences for human health. One of the most fascinating aspects of evolutionary processes is the dynamic view of genetic codes and their influence on the physiology and adaptability of an organism. Current research indicates some of the mechanisms whereby genetic code evolution significantly influences the evolution of species, as shown by the genetic changes within the Candida genus. This important result challenges previous presumptions of stability in genetic code, pointing out that the changes can happen through mechanisms of functional significance rather than merely by neutral drift. These results underscore the necessity of rigorous research in this area, suggesting experimental designs incorporating molecular genetics and comparative genomics. By investigating the complexities of genetic code evolution, scientists will know about how organisms evolve and survive in changed environments, and through the process, enhance our understanding of biodiversity. The research on evolutionary mechanisms also converges with the understanding of complex behaviors, such as the evolution of language in humans. Studying the evolution of communication involves developmental and evolutionary factors, as this shows how sophisticated behavior emanates from primal origins. The role of self-organization and exploratory behavior stimulated by curiosity in characterizing speech patterns is a classic case of the relationship between evolution and behavioral development. These studies highlight the need to learn not only biological evolution, but also the cognitive and social evolution of living organisms. Operating in such multi-dimensional aspects of evolution facilitates a better-integrated comprehension of the involved mechanisms for favoring the necessity of a multidisciplinary approach to study evolutionary processes, as explained by Oudeyer et al. (2015).

2.1. Natural Selection

Learning Outcome

1. Concept of survival of the fittest;
2. Examples of natural selection in action;
3. Role of environmental pressures.

Natural selection is an evolutionary concept that explains how living organisms adapt over generations. Natural selection basically states that there are variations in traits among individuals in a population, some of which will be an advantage to them in their environment. These advantageous traits make it more likely for an individual to survive and reproduce, therefore doubling the frequency of these traits in subsequent generations. Natural selection, however, never occurs in isolation; other processes of evolution accompany it, such as genetic drift and mutations. It is only by considering the dynamic interplay between these multiple forces that one can arrive at a complete understanding of how natural selection influences the process of evolution. Most significantly, while natural selection has been widely lauded for its role in bringing about adaptation, it is vital to realize that the parameters of the evolutionary process itself can change, resulting in complex outcomes over time, as noted in recent studies (Clark et al., 2015). Although it is significant, a sole emphasis on natural selection can conceal the broad evolutionary biology context. In class, for instance, students tend to misconceive other forces of evolution, and consequently, they misconceive the whole spectrum of evolutionary processes. Evidence indicates that undergraduates struggle to distinguish between natural selection and other elementary mechanisms, such as genetic drift and evolutionary developmental biology (evo-devo). This ignorance congeals a pedagogic deficit, whereby learning remains short of embracing the multi-faceted character of evolution beyond adaptationist thinking. Identifying a more integrative mode of teaching evolution and embedding natural selection within other mechanisms may improve students' understanding of evolutionary dynamics. By integrating these ideas within the school curriculum, instructors can help learners appreciate the full depth of evolutionary processes, ultimately improving their knowledge of biological diversity (Price & Perez, 2016). Developmental, ecological, and reproductive organizations evolve that alter how natural selection operates. As organisms evolve, the interaction between and within these organisms influences the selective pressures on them. The emerging discipline of evolutionary connectionism posits that natural selection operates not only on discrete characters but also on the interactions between characters, generating complex behaviors at higher levels of biological organization. It is incompatible with the conventional picture of evolution as a linear process held together by survival of the fittest. By integrating insight from cognitive learning theories, scientists are now finding links to the adaptive ability of systems in the long term, hence demystifying the intricacy of natural selection. These findings bring out the need for a broadened theoretical framework of evolution that addresses these systemic interactions, especially during significant evolutionary transitions that redefine how natural selection functions across scales of organization (Clark et al., 2015) (Figure 2.3).

Figure 2.3. Natural selection.

Source: https://images.nationalgeographic.org/image/upload/t_edhub_resource_related_resources/v1638890178/EducationHub/photos/darwins-finches-diagram.jpg.

2.1.1. Concept of Survival of the Fittest

Survival of the fittest, often attributed to Charles Darwin and his natural selection theory, explains how certain traits become increasingly prevalent in a population over time due to pressures from the environment. The concept does not merely suggest that the strongest individuals survive; instead, it proposes some dynamic interaction among several adaptive traits that enhance an organism's reproductive success. Environmental pressures, availability of resources, and competition between species contribute significantly to what is considered beneficial. Under this perspective, survival does not mean only physical strength but includes behavioral traits, mating strategies, and even social organization in species. These nuances need to be grasped in order to comprehend the full extent of how organisms respond to their changing environment in the broader framework of mechanisms of evolution, as explained in present evolutionary theory, e.g., the Darwinian law of survival of the fittest (Sammut-Bonnici et al., 2002). Additionally, the application of Darwinian theory is further extended from biological traits to social structures and processes, thereby presenting a broader framework for the investigation of mechanisms of evolution. Evolutionary psychology, for instance, promotes the idea that cognitive human processes and human behavior are also explainable in terms of evolutionary considerations, explaining how our mind's functionality is a result of adaptive forces over time. This is the type of thinking that makes synergetic study into both genetic evolution and the evolution of social behaviors possible. The applicability of such ideas requires a reinterpretation of survival mechanisms among human societies and social forms, with bearing on the ways that such adaptation sustains the fabric of social stability and innovation. Accordingly, an understanding of survival of the fittest requires a study of such collective processes, demonstrating how social structures evolve under similar pressures to biological ones (Aldrich et al., 2009). Moreover, the historical development of the idea of survival of the fittest calls for critical reflection on its meaning today across biology and the social sciences. The recent popularity of applying these ideas to various contexts, organizational change, say, witnesses an attestation of evolutionary forces much wider than those biological. They have cited in support of a complementary research model that unites evolutionary analysis with organizational theories to describe how organizations adapt to environmental pressures through transformation, thriving, or extinction. This interplay between competition and resilience encapsulates not only basic biological truths but also resonates through social evolution's fabric. Thus, the survival of the fittest concept transcends its biological origins to offer a valuable model through which to understand transformation and change within a broad spectrum of contexts (Sammut-Bonnici et al., 2002) (Figure 2.4).

Figure 2.4. Survival of the fittest.

Source: https://external-content.duckduckgo.com/iu/?u=https%3A%2F%2Fwww.collegenp.com%2Fuploads%2F2024%2F02%2FSurvival-of-the-Fittest.jpg&f=1&nofb=1&ipt=9f3d50f2c8fbb92ba16b90d1174ee63f24bbce9ecf886588d6a3afe5bd0b034a.

2.1.2. Examples of Natural Selection in Action

Natural selection is demonstrated to life in the evolutionary history of molluscan eyes; a particularly strong example of how complex features can appear incrementally. In the past century, these eyes have been cited time and again in accounts of convergent evolution, by which unrelated species have evolved similar characteristics in response to similar selection pressures. For instance, the camera-like eyes of cephalopods and vertebrates show parallel evolution of elaborate eye systems in response to the necessity of interacting successfully with their environment. This feature brings to focus the significance of exploring the variety of mollusk eye types because knowledge of their unique structure and the mechanisms that control them could assist us in comprehending the evolutionary process, particularly the dynamics of natural selection in the fixation of sensory adaptations. Thus, the above examples demonstrate the ability of natural selection to work not only on the current traits of organisms but also on the intricate paths towards those developments (Packard et al., 2008). Equally compelling evidence of natural selection in action would be witnessed through the repeated recurrent evolution of adaptations to demonstrate its ability to overcome ecological issues. Crop mimicry among certain weeds illustrates how natural selection can favor traits that enhance the fitness of an organism in a specific environment. By imitating the appearance of cultivated crops, these weeds undercut agricultural harvests through the ability to avoid herbicide or hand-pulling detection. This adaptive mechanism illustrates how natural selection drives evolutionary change with significant implications for agriculture and food security. Additionally, the origin of CO_2-concentrating C_4 photosynthesis is another innovative solution that has evolved under the pressures of the environment, exhibiting the cleverness of natural selection. Such adaptive chains are evidence that natural selection is no random process but an effective mechanism that optimizes survival strategies in different species by following the constraints of earlier selection pressures (Denison, 2011). The relationship between trade-offs and evolutionary success also further illustrates natural selection's role in species shaping. For example, within agricultural contexts, plant-plant fitness is often set against overall crop performance and reveals a trade-off between human agricultural and natural selection goals. Through an understanding of the effects of this trade-off, scientists are able to identify and reverse some of the effects of past selection that have detrimental consequences for crop performance. This exchange emphasizes the idea that what was previously advantageous traits can become detrimental under other conditions. Similarly, the exchange between legume hosts and their symbionts also follows the same pattern of selection. These examples emphasize how natural selection continually re-evaluates fitness and adaptation as conditions in the environment shift and thus how evolution is an active process shaped by both past and present selective forces (Denison, 2011) (Figure 2.5).

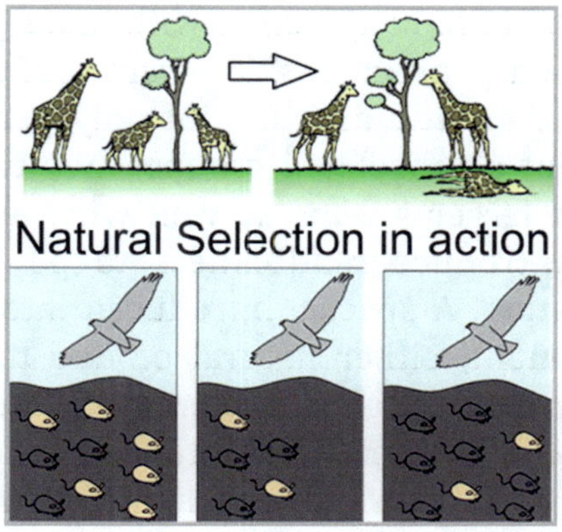

Figure 2.5. Examples of natural selection in action.

Source: https://www.solpass.org/science6-8-new/s7/images-life/Natural-Selection-Giraffes.png.

2.1.3. Role of Environmental Pressures

Environmental pressures play a crucial role in determining the evolutionary course of species, particularly in choosing which characteristics are useful or deleterious to a species within a given environment. Natural selection, driven by pressures such as predation, competition, and availability of resources, is a force that often drives evolution. For instance, organisms that react to their environment with favorable characteristics will be likely to survive and reproduce and thus pass on their characteristics to the next generation. Such behavior is observed within various ecosystems, where adaptation, like camouflage or behavioral adaptation, is selected under specific environmental pressures. Because these pressures vary over time and across habitats, they are the ones making life forms diverse and enhancing biological diversity. This complex interaction among environmental factors and evolutionary forces points to the necessity of examining both sides in order to understand evolution as a whole. Biotic factors aside, abiotic stresses from the environment also exert significant influences on evolutionary outcomes. Climate, geological disturbances, and habitat disturbance can create selective pressures that drive species to evolve or die. The Quaternary period is a great case in point where ecological communities were affected by climate change and leading to major adaptations and the rise or decline of numerous species. Whenever the climate shifted, organisms needed to shift, adapt, or perish, testifying to the powerful force of environmental oscillations on evolutionary processes. These pressures do not just have a history; today's environmental pressures, like global warming, are transforming ecosystems and forcing species to evolve at a rapid rate. As such, understanding the role of abiotic and biotic pressures is important in how they make evolutionary mechanisms work and how they can be implemented when addressing today's global changes, making history paramount in today's evolutionary debate. Furthermore, environmental stress can induce unforeseen effects, like the emergence of antibiotic resistance in causative pathogens. Research has shown that minute concentrations of antibiotics in ecosystems can potentially facilitate the evolutionary development of resistant bacterial genotypes, which present a new problem to public health (Amado et al.,

2022). This evolutionary process is a prime example of how human-induced modification of natural ecosystems can inadvertently initiate evolutionary processes at a very rapid pace. Moreover, documented patterns of ecological disturbance in the past, observed in the Quaternary period, reaffirm that evolving environmental conditions can instigate emergent evolutionary responses, from extinction of species to adaptation (Stewart, 2010). Under the continued action of evolutionary forces driven by environmental pressures, past and present, the realization keeps increasing that information on such forces is a prerequisite for solving modern ecological and health crises. From this viewpoint, the interactive dynamics of environmental pressures and evolutionary forces tell a multidimensional tale essential to comprehending the whole of evolution.

Test Your Knowledge

1. What is the concept of survival of the fittest?
2. What are the examples of natural selection in action?

2.2. Genetic Drift

Learning Outcome

1. Definition and explanation of genetic drift;
2. Impact of population size on genetic drift;
3. Case studies illustrating genetic drift.

Genetic drift is a fundamental evolutionary process that occurs due to random change in allele frequencies within a population, particularly important in small populations. It is a random process that results in the loss of genetic variation over time, which can significantly influence a population's potential to adapt and survive. For instance, in isolated populations, genetic drift may result in certain alleles becoming fixed while others are completely lost, reducing the overall genetic diversity necessary for long-term adaptation to changing environments. Here, the process of genetic drift is evident during episodes of abrupt ecological change, when the resistance of populations is tested by novel selection pressures, as indicated by the belief that climate change influenced species evolution and adaptation across the Quaternary, thereby illustrating how genetic drift can act in conjunction with more general evolutionary processes (Stewart, 2010). The effects of genetic drift do not only pertain to small, but also isolated groups; they can take effect in larger groups as well when external forces introduce randomness to population composition. A species introduced into a new environment, either natural or due to man's activity, can establish a condition known as the genetic paradox. This paradox is that small groups of animals can result in a genetically variable population when multiple genetically disparate lineages interbreed together when they are introduced. A case is the work on the UK Eastern grey squirrel, illustrating how introduced populations have varying genetic diversity from native ones, showing how human-mediated dispersal and genetic drift combine to shape the genetic architecture of invasive species (Carbone et al., 2016). The implications of these processes are broad, extending not only to the evolutionary course of invasive species but also to the ecological equilibrium of introduced habitats. Moreover, effects of genetic drift may extend beyond mere genetic variation; they may play an important role in speciation processes. When genetic drift separates populations to the degree that gene flow is reduced, it can solidify reproductive isolation. Through generations, such processes could result in the evolution of new species, enhancing biodiversity. The process demonstrates the intricate relationship between genetic drift and natural selection, with drift sometimes hindering or promoting adaptive processes under evolving environments. The significance of genetic drift is emphasized in studies examining population structures using the history of past events, such as climate change and demonstrating its power to bring about significant evolutionary adjustments, including adaptation and extinction (Stewart, 2010). Genetic drift is basically a powerful evolutionary force that affects the genetic plan from which biodiversity and relationships among species evolve (Figure 2.6).

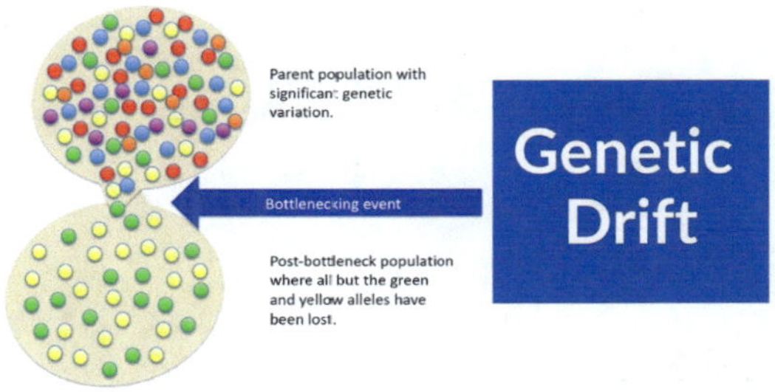

Figure 2.6. Genetic drift.

Source: https://www.biologyonline.com/wp-content/uploads/2019/10/Genetic-drift-definition.jpg.

2.2.1. Definition and Explanation of Genetic Drift

Genetic drift is one of the simple mechanisms of evolution, in which the random change of allele frequencies among populations occurs with the passage of time. The mechanism operates primarily in small populations, where random events have more effect on alleles that are passed from generation to generation. Unlike in natural selection, where reproduction and survival are encouraged by traits favorable under environmental pressures, genetic drift will continue irrespective of the adaptive value of the alleles. It follows that alleles may become fixed or lost within a population purely by chance, decreasing genetic diversity. This is a demonstration of how stochastic variables are capable of having a significant influence on controlling the evolutionary direction of a species, illustrating the complexity of randomness and evolution in producing genetic topography. A proper illustration of genetic drift is the founder effect, where a small group of individuals establishes a new population elsewhere. In these cases, the novel population's genetic composition may vary from that of the source population in terms of allele frequencies. Consequently, alleles can become more frequent or even lost, affecting the resultant group's genetic variation. This may lead to increased susceptibility to environmental factors and disease, particularly in diminished populations where genetic variation is limited. Moreover, the effects of genetic drift can typically be integrated with other mechanisms of evolution, including migration and mutation, into intricate intermeshed dynamics of population genetics that are deserving of careful attention, as documented in the examination of indirect modeling in evolutionary scenarios (Toussaint et al., 2001). The effects of genetic drift extend beyond direct allocation of alleles across populations; they encompass the general evolutionary landscape. For instance, genetic drift can cause speciation by permitting populations to diverge when they have become isolated from one another. As populations accumulate differential genes through drift, the populations can eventually become distinct species or else if coupled with direct selective pressures or environmental changes. This process provokes important questions about the relative roles of random vs. deterministic processes in evolution that compel reevaluation of traditional Darwinian paradigms. An integrated perspective that factors in both natural selection and genetic drift as key forces in evolution can give better insights into processes that build biodiversity, hypothesizing that integration of various theories of evolution can give a broader perspective of life's complexity on Earth (Sydow et al., 2001).

2.2.2. Impact of Population Size on Genetic Drift

The concept of genetic drift is inextricably linked with the population size. In large populations, the effect of randomly occurring allelic changes is minimized because the numerous individuals average out the changing occurrences. In small populations, however, genetic drift is particularly vulnerable to randomly occurring events modifying allele frequencies in future generations. This can lead to loss of genetic variation, where some alleles may become fixed or lost outright simply at random and not through selection. The effects of this alteration can be dramatic, affecting a population's ability to adapt and the general evolutionary direction, creating a feedback mechanism that amplifies preliminary genetic differences among populations. Because genetic drift is more efficient in small populations, it has direct consequences on evolutionary processes such as speciation and extinction. With smaller populations, for example, brought about by environment- or human-imposed pressures, their genetic variability becomes limited. That is, the ability to adapt to new environments decreases. Consequences of such an occurrence can be seen in most species experiencing bottlenecks caused by climate fluctuations. For example, some species have gone through population extinctions in the course of previous climatic events. As stated by Stewart et al. (2010), these changes in the environment not only impacted the growth of the population but became linked with evolutionary processes to form the potential for adaptation or extinction due to the lack of genetic variance. Furthermore, the effects of genetic drift on population size illustrate the interdependence of evolutionary processes. Reduced population size by genetic drift leads to increased inbreeding, with closely related individuals breeding and passing on deleterious traits. This impact also supplements the impacts of genetic drift by introducing additional gene pool homogeneity. Such genetic homogeneity in combination with the randomness of drift contributes to lower adaptability of populations to environmental changes, as seen in most species facing extinction. These encounters capture how population size and genetic drift interact to establish the evolutionary trajectories of organisms and address the complexity of evolutionary mechanisms in general, as it is advanced historically by the intersections of genetic and ecological forces delineated by Altermatt et al. (2017) (Figure 2.7).

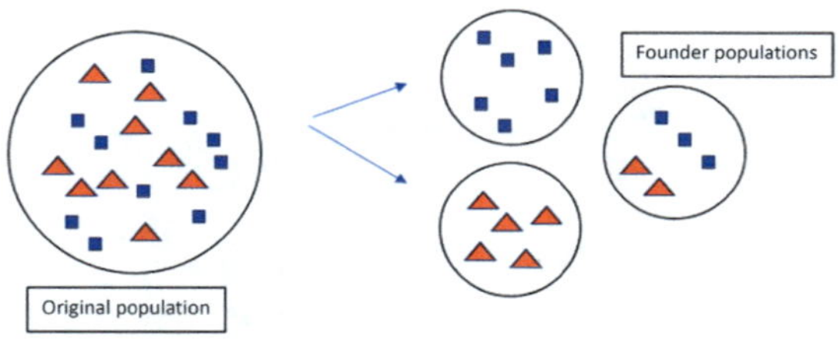

Figure 2.7. Impact of population size on genetic drift.

Source: https://www.vedantu.com/question-sets/4a2c9933-8a90-44f6-842e-a2c7b61f4b8a3737065159762805537.png.

2.2.3. Case Studies Illustrating Genetic Drift

Genetic drift, one of the simplest mechanisms of evolution, can be illustrated by a number of case studies that demonstrate its influence on populations. One good example is the influence of genetic drift on small, isolated populations such as those found in island biotic communities. Random fluctuations of allele frequencies in these environments may lead to loss or fixation of some traits through generations, particularly where the populations are prone to serious founder effects. For instance, the colonization of a new environment with only a handful of individuals will leave a genetic lineage that becomes exceedingly divergent from the initial population due to random sampling of genes, whose impacts become stronger with each passing generation. As Stewart (2010) outlines, the environmental transformations induced by climate change, combined with genetic drift, may influence the evolutionary history of species, leading to new adaptations and potential extinction events as populations respond to environmental stressors. A second famous illustration of genetic drift is the Northern Elephant Seal, which had previously been reduced to just 20 individuals due to hunting in the 19th century. The sudden population bottleneck resulted in a severe loss of genetic diversity, showing that genetic drift has the power to irreversibly alter the adaptive potential of a species. The current population, although bouncing back to over 200,000 animals, retains genetic traits that can constrain its long-term survival and capacity to evolve in a changing environment. This process emphasizes the new interaction between genetic drift and environmental adaptation, and populations thus affected may have alleles less adapted in new environmental situations. The implications of such genetic limitations serve to further highlight the importance of understanding how genetic drift functions within the general framework of evolutionary mechanisms, as is also the case with other studies examining evolutionary stressor response and adaptation potential, such as in the findings included by Giraud et al. (2009). When examining the case examples illustrating genetic drift, it is important to consider the relevance for conservation biology. By recognizing the role of genetic drift, scientists and conservationists can make genetic diversity between threatened species a priority. Low genetic diversity is a feature of species that are most susceptible to extinction, and it further dictates tactics that maximize outbreeding and genetic transfer between isolated populations. The Northern Elephant Seal is both a message of warning and a beacon of hope, demonstrating that despite genetic drift potentially causing huge losses of genetic variation, wise management can lead to ecological restoration. As studies look back to the importance of genetic diversity acquired through organized breeding programs, the recent research into adaptive capacity through evolutionary processes such as genetic drift reaffirms the intricate relationship between evolution and conservation efforts, further highlighting the importance of genetic mechanisms in shaping future biodiversity outcomes.

Test Your Knowledge

1. Write the definition and explanation of genetic drift?
2. What is the impact of population size on genetic drift?

2.3. Mutation

Learning Outcome

1. Types of mutations and their effects;
2. Role of mutations in genetic variation;
3 Examples of beneficial mutations in species.

Mutation is a critical process of the mechanisms of evolution and is responsible for most of the genetic variation that exists within a population. In creating random changes in DNA sequences, mutations can lead to phenotypic variations that can affect the adaptation of an organism towards its environment. This adaptive relationship is best illustrated in RNA viruses where high mutation frequencies combined with short generation times enable rapid rates of evolutionary change. These attributes make possible the quick adaptation of RNA viruses to selective pressures from host immunity or antiviral treatment and hence their augmented survival and spread. The emergence of emerging viral strains through mutation is a great example of how this mechanism not only creates genetic diversity but also drives evolutionary processes, underscoring the sheer contribution of mutation toward shaping biodiversity and species hardiness against environmental harshness (Ahmadi et al., 2019). The complexity of mutation is also evidenced in the modern synthesis of evolutionary biology, where genetics and natural selection are brought together. This synthesis emphasizes the importance of mutation as a random process at work in generating genetic variability, a condition necessary for natural selection to act upon. As synthesis writers explain, the role of chance in evolution can be appreciated through various different ways, including the randomness of mutations which introduce variety into a population. These mutations are not necessarily beneficial or harmful in themselves; their effects are contingent on environmental circumstances. Thus, modern synthesis describes a dual role for mutation, namely both as a stochastic force of alteration and as a key agent of adaptive evolution, whereby specific mutations may convey advantages that promote survival and reproduction (Matthews et al., 2016). Even though the concept of mutation as an evolutionary process is firmly rooted, its application in larger evolutionary and ecological landscapes needs to be understood. Mutations cause phenotypic variations that lead to evolutionary trajectories based on environmental pressures. For instance, if the mutation confers resistance to illness, those with this mutation will be more likely to live and reproduce and consequently transfer those adaptive traits to later generations. Such favorable mutations over time can become prevalent within a population of organisms, demonstrating the close overlap of mutation rates and evolutionary success. Furthermore, the study of mutation not only lends itself to the comprehension of evolutionary processes but also explains disease and the evolution of new traits in other organisms, which testifies to its pivotal role in the evolutionary narrative (Ahmadi et al., 2019) (Figure 2.8).

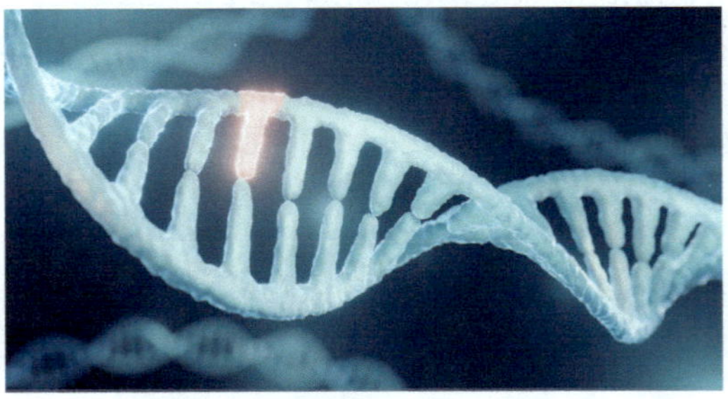

Figure 2.8. Mutation.

Source: https://www.yourgenome.org/wp-content/uploads/2023/10/100-shutterstock_2046452252.jpg.

2.3.1. Types of Mutations and Their Effects

Mutations in evolutionary biology are the starting material for genetic variation, giving natural selection something upon which it can subsequently act. Mutations can be divided into many types, including point mutations, insertions, deletions, and large chromosomal changes in excess of a nucleotide. Point mutations are a modification to one nucleotide and can be silent, missense, or nonsense mutations depending on their action on codons and protein function. These alterations typically bear large effects on organisms, leading not only to phenotypic differences but also influencing fitness in specific environments. The role of mutations in the generation of diversity is of primary significance, as indicated through existing research on microbial systems, which illustrates how mutations facilitate adaptations to environmental stressors, thus driving evolutionary processes forward. Insertions and deletions, combined under the label indels, are particularly problematic since they can cause frameshift mutations, which significantly alter downstream amino acid sequences of proteins. Such alterations render proteins inactive or, in some cases, generate new functions altogether, which would impact the adaptability and survival of an organism.

At a larger scale, chromosomal mutations like duplications and inversions can affect entire gene sequences and regulatory elements, with effects on developmental processes and evolutionary processes. The ability of organisms to bring in novel genetic information through mechanisms such as HGT makes the landscape of mutation even more daunting. The interaction between these types of mutations and their effects constitutes the dynamic integrity of genomes in evolutionary processes, the intricacies of which illustrate the intimate relationship between mutation, adaptation, and natural selection. The CRISPR-Cas system is a fascinating illustration of the intersection between mutations and evolutionary pressure within prokaryotic cells. This adaptive immune response enables bacteria and archaea to defend themselves against viral invasion by introducing bits of viral DNA into their genome, creating a molecular memory of previous invaders (Briner et al., 2017). Such mechanisms include several that enable organisms to co-evolve together with their threats, showing the interplay between mutation and selection. Resolution of singular mutations enables bacteria to form targeted countermeasures, demonstrating an unparalleled ability to adapt. The more we understand such processes, the way specific mutations may confer advantages or disadvantages in microbial populations will illuminate more pervasive principles of evolutionary processes, the intricate dance of survival and adaptation within ecosystems (Bonomo et al., 2018) (Figure 2.9).

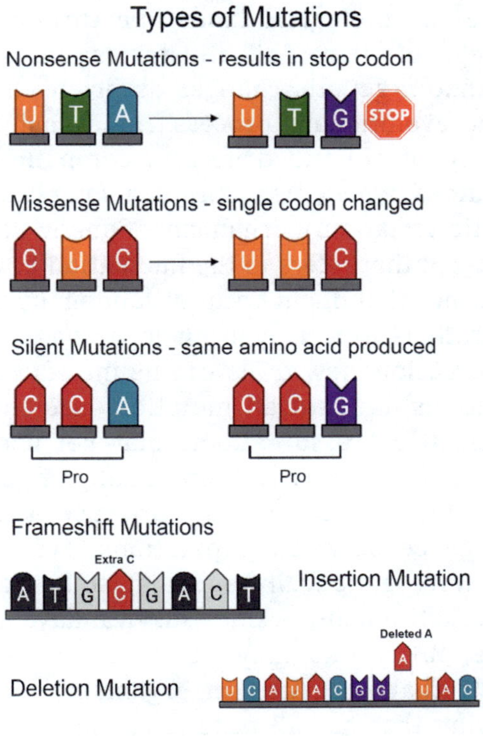

Figure 2.9. Types of mutations and their effects.

Source: https://study.com/cimages/multimages/16/types_of_mutations4925220602252006814.jpg.

2.3.2. Role of Mutations in Genetic Variation

Mutations are at the heart of the mechanism for the delivery of genetic variation among populations, a foundation of the evolutionary process. Mutations are alterations in the DNA sequence that can lead to differences in phenotype, which may enhance the ability of an organism to adapt within its environment. If mutations occur in germ cells, they are passed from one generation to the next, and thus they determine the future population's genetic makeup. The accumulation of these mutations is the basis for natural selection because the characteristics become more beneficial in a specific environmental context. Therefore, the combination of mutation and natural selection not only generates diversity in the population but also drives the evolutionary forces that shape the biological landscape through the passage of time. This illustrates the utility of random genetic changes as prime drivers of the evolutionary process. Additionally, the activity of genetic drift is accompanied by mutations, which bring an extra complexity to genetic variation mechanisms. Genetic drift is a concept that refers to the fluctuation in allele frequency that might occur at random between generations, especially in small populations. On this occasion, new traits can be introduced by mutations, but not all mutations are subject to selective pressure; some may get fixed or lost just by chance. The intermixing of genetic drift and mutation may produce sudden divergence between populations, even ones that were genetically equal at some earlier time. Additionally, besides survival advantages alone, mutations may produce deleterious effects that influence an organism's fitness, showing that not all genetic changes are not adaptive. An understanding of these varied outcomes enriches the argument about how mutations are a fundamental mechanism in the broader framework of evolution. Also, recent controversies surrounding the Extended Evolutionary Synthesis highlight the complexity of the effects of environmental pressures on mutation and genetic variation beyond the older traditional frameworks of neo-Darwinian evolution. Traditional evolutionary theory primarily explains genetic mutations as the primary driving mechanism of phenotypic variation. Recent studies, however, suggest that non-genetic factors can dominate the process. For instance, environmental epigenetics has proved that environmental changes can induce heritable phenotypic variation independent of classical gene mutations. The discovery reveals a more intricate interaction between the environment and genetics that signifies that the mechanisms of evolution are involved with a set of parameters influencing genetic variation, including epigenetic modifications. This subtler comprehension is what makes clear how mutations, important as they are, are part of a larger system of evolutionary influences that drive organisms' adaptation processes in relation to their ever-changing environments (Sánchez et al., 2016; Nilsson et al., 2021) (Figure 2.10).

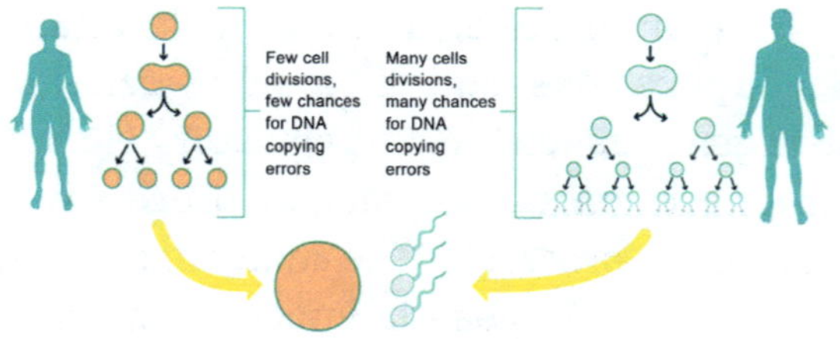

Figure 2.10. Role of mutations in genetic variation.

Source: https://evolution.berkeley.edu/wp-content/uploads/2021/09/cell-divisions.jpg.

2.3.3. Examples of Beneficial Mutations in Species

The study of adaptive mutations is at the core of the understanding of the evolutionary process, as such mutations have a tendency to confer organisms with improved survival or reproductive advantage. One of the best examples of this is the mutation that allows for lactose tolerance in human beings. Lactose was digested after weaning in our ancestral populations; yet a mutation that allowed adults to digest lactose allowed some to utilize milk products as an enduring supply of nourishment. This evolutionary trait provided an advantage for survival in pastoralist populations, and with the passage of time, there was a greater number of such mutations present in populations that practiced cattle rearing. Such examples illustrate how particular genetic variations might lead to adaptive traits that are essential to the survival and well-being of a species over generations, thus rewriting the evolutionary history and contributing to human diversity and robustness. In plant species, adaptive mutations can significantly expand adaptability to changing environmental conditions. Consider the mutations of the wildflower *Mimulus guttatus*, which facilitate flower color changes that are tailored to meet special pollinator requirements. Such changes not only increase reproduction through the proper attraction of pollinators but also illustrate the importance of gene control in evolutionary processes. Experiments have also proven that mutations within the transcription factor binding sites can enhance the expression of genes related to floral color and pattern, thus influencing the beauty of the flowers to attract pollinators. This expression dynamic is also mirrored in natural selection theory, where positive traits double within populations, demonstrating how adaptive mutations generate diversity and adaptability in response to environmental stress, affirming ideas proposed in adaptive evolution models (Barton et al., 2015). Alongside adaptations for food availability and pollinator relationships, positive mutations will similarly transform physiological traits that enhance survival in alternative environments.

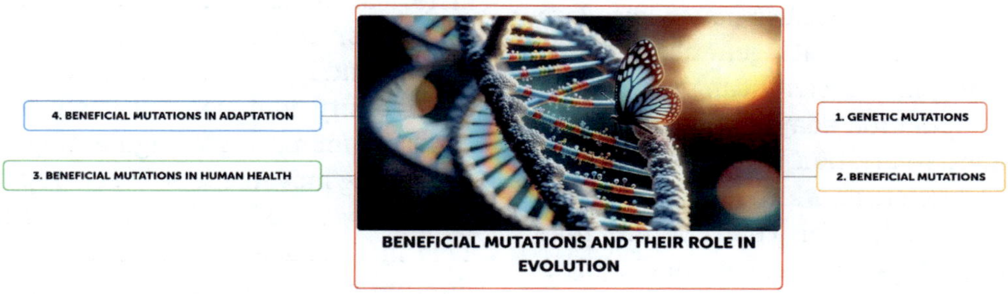

Figure 2.11. Examples of beneficial mutations in species.

Source: https://cards.algoreducation.com/_next/image?url=https%3A%2F%2Ffiles.algoreducation.com%2Fproduction-ts%2F_S3_d39522cc-53d1-4655-b1ab-cc3caeae26d4&w=3840&q=100.

One such instance is the gene mutation responsible for the darker pigmentation in certain species of lizards such as Anolis carolinensis, so that they can better camouflage themselves in the environment and reduce predation risk. Such a mutation resulting from an alteration in the expression of melanin-producing cell genes is a prime example of natural selection in action. Darker-colored lizards are more likely to survive and reproduce in dark environments and thus transmit this advantageous mutation to future generations. The association between environmental pressures and genetic mutation raises questions regarding the intricacy of evolutionary mechanisms, outlining the multi-dimensionality of adaptive evolution (Joly, 2010). All these instances reflect the extent to which beneficial mutations play an important role in shaping the evolutionary trajectory of species (Figure 2.11).

Test Your Knowledge

1. What are the types of mutations and their effects?
2. What is the role of mutations in genetic variation?

2.4. Gene Flow

Learning Outcome

1. Definition and significance of gene flow;
2. Mechanisms of gene flow between populations;
3. Effects of gene flow on genetic diversity.

Gene flow, the transfer of genetic information from one population to another by reproduction or immigration, plays a significant role in the evolutionary background of species. By diffusing genes, gene flow can introduce new alleles into a population, expanding its genetic diversity. This gene flow plays a significant role in the retention of population adaptability as a result of environmental variation since it provides a larger set of characters on which natural selection can act. For example, for yeast populations that have been long-term evolved under high-ethanol conditions, the complex interaction of mutations and genetic changes demonstrates how gene flow can play a significant role in the adaptation of adaptive phenotypes. Populations' ability to accumulate adaptive mutations and eliminate deleterious ones reveals the significant contribution of gene flows in mechanisms of evolution, as populations are improved against environmental stress (Arslan et al., 2015). Moreover, gene flow is a balancing force relative to the impact of inbreeding and genetic drift, particularly in small or isolated populations. Because small populations suffer from loss of gene diversity due to the founder effect or population bottlenecks, gene flow from other populations introduces new variability into them. This genetic variation is necessary to ensure long-term survival and responsiveness of populations and enables them to have a many-sided arsenal of characters for enduring dynamic stresses, such as global climate change. It has been evidenced that previous climatic changes during the Quaternary period had a significant effect on gene flow between species, thus dictating evolutionary trajectories through the evolution of species, adaptation, and extinction (Stewart, 2010). Gene flows therefore not only influence present populations but also reflect historical connectivity and divergence patterns. In addition to its evolutionary relevance, gene flow is intriguing with regard to controlling biodiversity and conservation. With human activities increasingly fragmenting habitats and restricting movement between populations, the possibility of reduced gene flow is heightened. This may lead to lower genetic diversity, and populations may become more vulnerable to environmental stresses and disease. An understanding of the patterns of gene flow can guide conservation efforts, calling for maintaining ecological corridors along which movement and interbreeding are feasible among populations. Gene flow can also shed light on adaptive strategies implemented by various species in response to anthropogenic challenges. Thus, the study of gene flow not only contributes to the study of evolutionary processes but also throws light on the urgent need for concerted conservation efforts to protect genetic diversity in an age of rapid development (Figure 2.12).

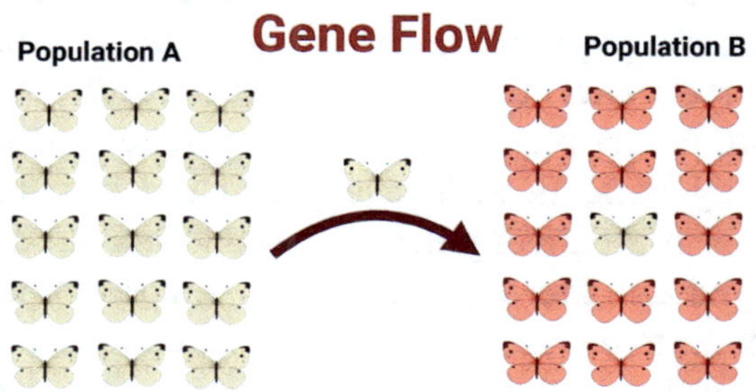

Figure 2.12. Gene flow.

Source: https://microbenotes.com/wp-content/uploads/2022/10/Gene-Flow.jpg.

2.4.1. Definition and Significance of Gene Flow

The process of gene flow is significant in explaining mechanisms of evolution since it establishes the transfer of genetic material from one population to another within the same species. Gene flow is through any of several methods, such as the transport of organisms, gametes, or by human activities, causing the transfer of genes. Gene flow is a significant link between evolution and population genetics as it has the potential to add novel alleles to a gene pool of a population, thereby affecting genetic diversity and adaptability. High levels of gene flow can prevent populations from diverging, leading to greater genetic similarity among populations. Thus, gene flow is something to be understood to understand how species evolve over time since the genetic information passed on through this process directly affects ecological trajectories and even impacts speciation events in complex communities. Gene flow's function in evolutionary processes is not just stopping at gene exchange; it also promotes genetic variation maintenance that is central to ecological resilience. Populations subject to high gene flow are likely to have a higher level of response to environmental change, as new alleles introduced by gene flow could have desirable traits. For instance, in case of being threatened by climatic or habitat change, genetically diverse populations are likely to have individuals with traits beneficial for reproduction and survival. As described in contemporary evolutionary frameworks, for instance, the Extended Evolutionary Synthesis, gene flow is an enabling function to the systems of genetic inheritance, both of which operate synergistically to realize the dynamic interplay between adaptation and evolution (Griffiths et al., 2017; Blancke et al., 2018). Thus, the comprehension of gene flow emphasizes genetic connectivity as a factor for maintaining biodiversity and promoting evolutionary success. Furthermore, gene flow diverges from mechanisms of isolation that can lead to speciation, and hence it has a crucial function in evolutionary processes. When populations are isolated from one another, either geographically or in behavior, they get genetically distinct with time, which can lead to the development of new species. Such conditions are then addressed by gene flow as a balancing force against isolating forces, so that there can be genetic exchange that can revitalize populations and cut down inbreeding and genetic drift risks. Such interaction further forms another set of questions about evolutionary mechanisms within species and how the interaction between gene flow and isolation shapes species-level biodiversity. Therefore, an in-depth analysis of gene flow demonstrates its pivotal place within the evolutionary framework, where its effects reverberate throughout ecological processes, genetic diversity, and adaptive measures to environmental change, augmenting evolutionary process complexity (Blancke et al., 2018).

2.4.2. Mechanisms of Gene Flow Between Populations

An understanding of the process of gene flow between populations is necessary to understand the broader context of evolution. Gene flow, or the transfer of genetic information from one population to another, can be achieved via a range of mechanisms, including hybridization and migration. These processes permit the mixing of genetic traits, creating genetic diversity within and between populations. Interestingly, gene flow plays a significant role in reversing the effect of genetic drift and stimulating adaptability when the environment evolves. For instance, the influence of climatic change on ecological community composition throughout the Quaternary illustrates how shifting environmental conditions stimulate changes in gene flow, thereby influencing evolutionary directions like species adaptation and extinction (Stewart, 2010). This relationship between gene flow and environmental stress merely strengthens its role in shaping evolutionary tendencies in various ecosystems. Gene flow is not only a plain migration phenomenon but also involves complex interactions amongst populations in parallel evolution. With different populations invading corresponding environments, they may experience different evolutionary responses, driven by genetic forces led by gene flow. These processes not only result in the divergence of populations but also emphasize the significance of environmental context to evolutionary outcomes. Experiments reveal that while some populations create very analogous adaptations due to gene flow and environmental stress, other populations develop divergent traits even though they have similar ancestral origins. Parallel evolutionary analysis allows researchers to quantify the extent of genetic adaptation equivalence and ideally justify the conditions under which nonparallel or parallel evolutionary developments take place (Barrett et al., 2018).

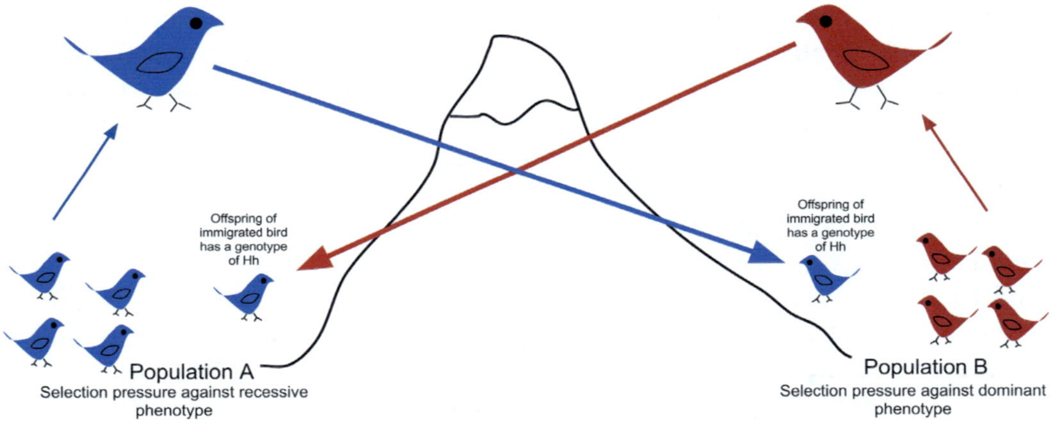

Figure 2.13. Mechanisms of gene flow between populations.

Source: https://upload.wikimedia.org/wikipedia/commons/4/47/Gene_flow_final.png.

Gene flow is therefore a key aspect to study in explaining evolutionary processes within various ecological environments. In summary, gene flow between populations offers an opening through which evolutionary processes can be observed and studied. Complexity in these processes reflects a multifaceted relationship in which climate factors, genetic variance, and migration patterns converge to dictate evolutionary directions. As populations respond to climate change and other stressors within the environment, gene flow facilitates the exchange of positive features, enhancing overall fitness and resilience. Furthermore, knowledge of parallel and nonparallel evolution within the context of gene flow improves our concept of adaptation and also reveals the dynamic relationship between the environment and genetics (Barrett et al., 2018; Stewart, 2010). Lastly, this complete understanding of gene flow mechanisms also teaches us more about the intricate processes powering evolution and species diversity on the planet (Figure 2.13).

2.4.3. Effects of Gene Flow on Genetic Diversity

Gene flow, or the passage of genes from one population to another, is key to the establishment of genetic diversity within and between species. Gene flow, through various mechanisms such as migration, intermating, and hybridization, brings novel alleles to a population, which contributes to its genetic diversity. Such a boost in genetic diversity provides a population with increased ability to evolve under shifting environmental stress and survive ecological hardship. For instance, gene flow would be particularly significant in fragmented habitats where isolated populations would be at risk of inbreeding depression, a decrease in fitness and adaptability. Through genetic material mixing, gene flow offsets some of these negative effects, so that populations remain with a pool of adaptive traits to be used in responses to environmental stress, which, as noted, usually relies on greater climatic change affecting ecological communities (Stewart, 2010). Though gene flow is generally a positive factor for genetic diversity, under certain circumstances, it can be negative. It is particularly so in the context of speciation processes. Gene flow among populations at an incipient stage of divergence results in homogenization of genetic characters, which undermines the development of reproductive isolation. It has the effect of retarding speciation by preventing the populations from evolving to their own particular environment or ecological niches. While in other cases, maladaptive characters introduced through gene flow can lead to lower fitness within surrounding populations and render them less fit. Understanding these interspecific genetic links is critical in comprehending the complex mechanisms of evolution and ongoing processes of speciation, genetic research, particularly with scientists studying the facilitators and brakes of gene flow among incipient species (Serrato-Capuchina et al., 2019). Moreover, gene flow carries consequences that go beyond its direct genetic effects to influence ecological relationships and the patterns of evolution. Gene flow, for example, can facilitate the spread of useful characters among populations, to make a species more capable of adaptation overall. Conversely, it can precipitate dilution of local adaptations to the point of preventing Indigenous populations from being able to maintain their identity and fitness in the presence of an importation of foreign genetic material. Such interactions are especially significant within the framework of environmental rapid change, where genetic diversity must be maintained to provide for resilience. Gene flow gives insights into how environmental stressors (e.g., climate change and habitat destruction) can impact genetic landscapes and community structures, making it of utmost significance to evolutionary theory and conservation biology (Stewart, 2010).

Test Your Knowledge

1. What are the mechanisms of gene flow between populations?
2. What are the effects of gene flow on genetic diversity?

2.5. Speciation

Learning Outcome

1. Definition and types of speciation;
2. Mechanisms leading to speciation;
3. Examples of speciation events in nature.

Speciation, the process by which emerging species arise from the founding populations, is a fundamental evolutionary process that accentuates the dynamic nature of biodiversity. Speciation is a complex process comprising multiple mechanisms, including allopatric, sympatric, and parapatric speciation. Allopatric speciation is usually triggered by geographical barriers that isolate populations and result in divergent adaptation as the isolated populations adapt to different environmental conditions. Such barriers could be caused by tectonic movement or climatic changes, creating basic ingredients in the context of adaptive ecology. Ecological community dynamics and their meaning to species evolution is critically important, especially during periods of immense environmental change like that felt during the Quaternary period, explaining how species can survive or go extinct with changing environments (Baudry et al., 2014). Sympatric speciation, by contrast, occurs without the geographical separation, where reproductive barriers emerge within a shared environment. This often occurs through polyploidy, particularly in plants, or through behavioral changes that reduce interbreeding among subpopulations. Ecological factors are involved as the usage of resources and niche differentiation can induce reproductive isolation. This demonstrates the intricate interplay between speciation and ecological dynamics and reveals that environmental pressures can influence not only individual species but entire communities. Moreover, the function of interactions between species, particularly in competitive and co-operative relationships, confirms the complexity of speciation. It calls for an understanding of evolutionary forces that can lead to diversification, a topic relevant to both biological and software systems, as outlined in efforts like DIVERSIFY, exploring adaptive ability through principles drawn from biological evolution (Baudry et al., 2014). Finally, the speciation process is inherently intertwined with the processes of adaptation and natural selection. When environmental conditions shift, selective pressures on populations will cause evolutionary adaptation and therefore allow speciation. This dialogue between the environment and forces of evolution enables species to draw on genetic variation as a source of survival. Quaternary epoch is a case in point for this intersection, where climate change not only played a role in shaping species structure but also acted to initiate evolutionary forces that drove adaptation, extinction, and speciation (Stewart, 2010). Last conclusion: Speciation as an evolutionary tool deepens our knowledge of biological diversity and resilience and promotes investigation into how species respond, adapt, and evolve in perpetually shifting environmental ecosystems (Figure 2.14).

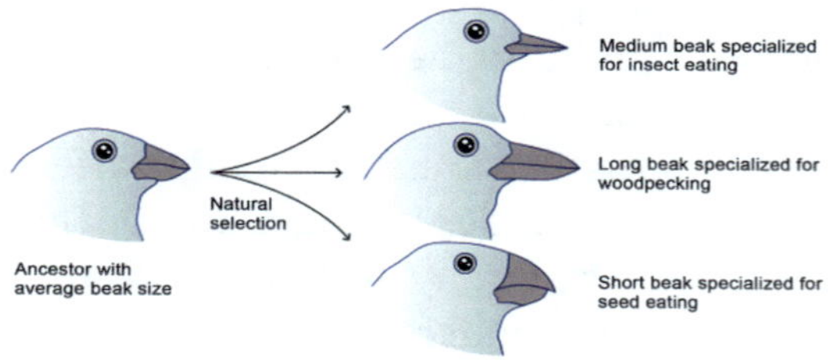

Figure 2.14. Speciation.

Source: https://i0.wp.com/cms.jackwestin.com/wp-content/uploads/2020/03/Adaption.png?resize=624%2C327&ssl=1.

2.5.1. Definition and Types of Speciation

Speciation theory is key to explaining the mechanisms of evolution and describes the mechanism by which a new biological species arises. It usually entails the creation of reproductive barriers to prevent interbreeding among populations that are derived from a common ancestor. Speciation may occur through a variety of mechanisms, largely categorized as allopatric and sympatric speciation. Allopatric speciation occurs when populations become geographically separated and create genetic divergence due to environmental factors and ultimately result in reproductive isolation. This process, as explained by Charles Darwin for the first time, indicates that isolation causes the variation of different characteristics, and this ultimately results in the creation of new species. However, the concept that reproductive barriers are a passive consequence of geographic isolation has been put into question among scientists (Joly, 2010). Conversely, sympatric speciation is the opposing hypothesis that eliminates the traditional account by arguing species can evolve without geographical separation. It would evolve frequently depending on behavior, polyploidy, or ecological niches in which some traits become advantageous to a population. For instance, if a section of the population begins to prefer some new food source or environment, it might end up separating to the level of reproductive isolation even as there is potential for intermating. This process shows the adaptive dynamic capacity and how it is capable of producing speciation without the requirement of physical barriers. The mechanistic complexity of speciation draws attention to the multidimensional nature of evolutionary processes, pointing out that selection and genetic drift may be powerful initiators in the formation of new species (Joly, 2010). Furthermore, analysis of speciation draws attention to the significance of genetic factors in the formation of reproductive barriers. In general, advantageous recessive mutations preserved due to inbreeding can lead to speciation, since these traits become fixed in separated groups of organisms. This perspective is notable in the sense that it emphasizes that speciation is not a simple outcome of drift or isolation but more an outcome of selection on unique phenotypic traits, which emerge in specific environments. This model proposes that reproductive barriers, instead of being accidental, are perhaps subject to group-facilitating selective pressures along with the differential reproduction of individuals within a group (Joly, 2011). Overall, the theory of speciation extends the paradigm of evolutionary biology, representing complex interactions among genetic, environmental, and behavioral influences that shape the diversity of life on our planet (Figure 2.15).

Figure 2.15. Types of speciation.

Source: https://i.ytimg.com/vi/YACR6WkQUGw/maxresdefault.jpg.

2.5.2. Mechanisms Leading to Speciation

Processes such as allopatric, sympatric, and parapatric speciation initiate the process of speciation, therefore, the evolution of new species. Allopatric speciation occurs when populations become physically isolated, thus facing different evolutionary histories since the environments are distinct. Isolated populations tend to accumulate genetic differences over time, ultimately leading to reproductive barriers that prevent from interbreeding. This spatial isolation may be brought about by a variety of factors, e.g., climatic processes or physical terrain, e.g., rivers or mountains. The concept of spatial isolation emphasizes the importance of space dynamics in evolution and illustrates how environmental occurrences may force diversification within a line to favor the establishment of new species. Sympatric speciation, however, happens within one geographical area, and the new species develops from a common ancestor and inhabits the same habitat. Sympatric speciation most often constitutes behavioral, diet, or hostplant change that leads to reproductive isolation, as there is habitat overlap. Niche differentiation, which is how the different resources are occupied by the different species so that competition is minimized and specialization becomes more pronounced, can be initiated by adaptive radiation. Furthermore, mechanisms like polyploidy, particularly in plants, can create immediate reproductive isolation because of the development of organisms of varying chromosome numbers.

The intricate relationship between ecological adaptation and speciation shows how wide-ranging evolutionary processes are influenced by environmental forces and the availability of resources, resulting in organism diversity. Climate change and ecological change play major roles in the speciation processes. In events like the Quaternary period, climatic oscillations drastically altered ecological community structure, creating novel selective pressures that imposed evolutionary change, e.g., adaptation and extinction (Stewart, 2010). Such changes result in variable environmental conditions that have the ability to isolate populations and influence gene flow, setting the stage for speciation. Furthermore, biological systems and ecosystems' adaptation processes enhance our sensitization to the intersection of organismal evolution with environmental dynamics. Not only does identification of such mechanisms enhance our understanding of species development, but also it facilitates easier exportation of ecological concepts to other domains, and it identifies ways through which adaptive ability can be engineered in complex systems like software development (Baudry et al., 2014) (Figure 2.16).

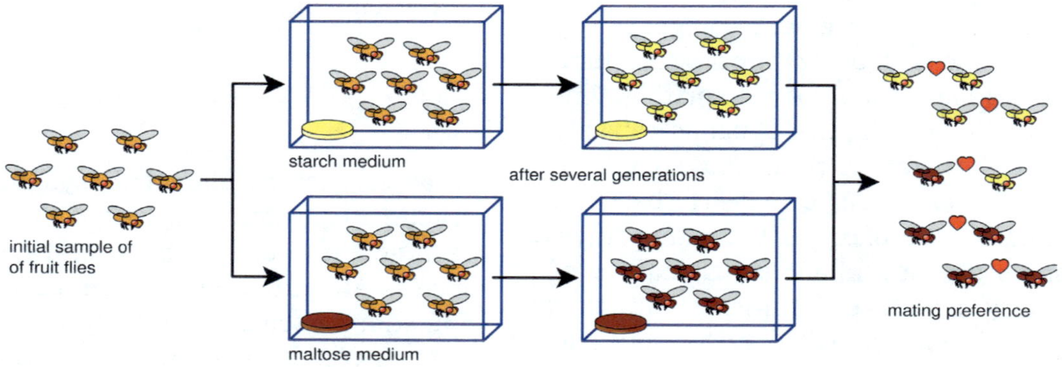

Figure 2.16. Mechanisms leading to speciation.

Source: https://dr282zn36sxxg.cloudfront.net/datastreams/f-d%3A62b4faf6968a12de0b53130f63bae27222af5a76957d05e9a0c1b93a%2BIMAGE_THUMB_POSTCARD_TINY%2BIMAGE_THUMB_POSTCARD_TINY.1.

2.5.3. Examples of Speciation Events in Nature

The speciation process, or the creation of the reproductive barriers between a group of populations that have a common ancestral population, can be illustrated using a number of examples in nature.

These include Darwin's finches on the Galápagos Islands, which, from a single ancestral species, evolved into many species, each being adapted to a specific ecological niche.

Beak shape and size differences in these finches are adaptations to different food sources on the islands. With the passage of time, geographic isolation, along with natural selection, favored reproductive barriers, allowing these populations to diverge considerably. This case shows the adaptive radiation process, whereby one line of descent gives rise to numerous species that are highly specialized into diverse environmental situations, thus offering a clear image of speciation following evolutionary processes that create biodiversity within isolated ecosystems.

Another good example of speciation is found with cichlid fish species in the African Great Lakes. They have an incredible diversity of form and behavior, typically related to their respective habitats and feeding methods. This radiative speciation has occurred within a geologically relatively short period of time as a result of sexual selection and ecological differentiation, leading to the derivation of various species from one stock. The complexity of mechanisms for reproductive isolation, such as behaviorally or color-specifically based mate choice or mate preference, has also contributed to increased speciation in the fish. The example of the cichlid fish illustrates how intense levels of competition and niche specialization can drive speciation processes to develop rich biodiversity patterns that underscore the dynamic interaction between environmental forces and reproductive strategies in nature (Joly, 2011).

In addition, the polyploidy process of plant speciation shows how diverse mechanisms of evolution can produce new species. Polyploidy is a state in an organism that possesses more than two complete sets of chromosomes, typically the product of hybridization between species. The process is most prevalent in flowering plants such as wheat and cotton, leading to spontaneous speciation. The new resulting species are typically endowed with features that confer adaptive advantages, allowing them to occupy novel ecological niches. Empirical evidence suggests that polyploid species are more genetically variable and tolerant of environmental stresses. Such speciation processes, therefore, shed light on the mechanisms of evolutionary change, suggesting that hybridization and chromosomal shifts can play key roles in conjunction with natural selection in shaping the biodiversity we observe today (Joly, 2010).

Test Your Knowledge

1. What are the types of speciation?
2. What are the mechanisms leading to speciation?

2.6. Coevolution

Learning Outcome

1. Definition and examples of coevolution;
2. Interactions between species and evolutionary changes;
3. Importance of coevolution in ecosystems.

Coevolution is a key element within the broader framework of evolutionary dynamics, affecting not only species adaptation but also the intricate dynamics of ecological relationships. As species interconnect, their evolutionary trajectories become intertwined, leading to two-way adjustments that may promote or complicate survival. This co-dependence is found in systems in which predator and prey evolve to accommodate one another, usually in the form of enhanced hunting prowess or avoidance strategies. One of the more fascinating aspects of coevolution is its potential to maintain diversity in ecosystems because the constant arms race between competing species spurs new adaptations. As populations adapt to respond to the shifting nature of their opponents, such a process can produce a rich tapestry of life, wherein the extinction of one species can produce ripple effects throughout the ecosystem and illustrate the profound interconnectedness of life on our planet. Coevolutionary dynamics are also illuminated further by evolutionary game theory, which illustrates the nuance of cooperative behavior emanating from self-interested actors. Such nuances challenge the classical conception of natural selection, which goes for individual fitness at the cost of cooperative benefits. Interestingly, recent studies indicate that the simple evolution of strategies within groups is not sufficient to sustain cooperation in egoistic players. Instead, coevolutionary rules need to be implemented, wherein not only strategies but also the resulting interaction patterns and the reproductive capacity evolve together with them. This dual evolution encourages stability in cooperative behavior, suggesting that cooperation in social dilemmas should often be ensured through a multi-faceted approach in terms of both individual strategies and evolutionary dynamics. To this end, coevolution broadens our understanding of cooperation by focusing on the necessity of dynamic frameworks through which such interactions occur (Abramson et al., 2009). In addition, the forces of coevolution, such as frequency dependence and spatial structure, emerge as strong drivers in the facilitation of cooperation. For instance, coevolving random networks could potentially allow certain cooperative strategies to dominate, with a significant influence on social interaction dynamics. In evolutionary games, being able to adapt not only individual strategy but also the network through which interactions occur can improve resilience against exploitation by defectors. According to recent evidence, such adaptive processes can lead to cooperation stability even in challenging settings, and so effectively counteract the selfish drift normally observed with natural selection (Perc et al., 2009). This tunability represents the coevolution's flexibility, which means that it is a powerful toolkit for organisms to adapt to the complexity of their ecological niches and social structures and hence ensuring the stability of cooperation over generations (Figure 2.17).

Figure 2.17. Coevolution.

Source: https://allyouneedisbiology.wordpress.com/wp-content/uploads/2015/10/pasarea-colibri-hummingbird-wallpaper.jpg?w=1038&h=576&crop=1.

2.6.1. Definition and Examples of Coevolution

Coevolution is a central concept in evolutionary biology that describes the reciprocal adaptation processes among interacting species. It is a phenomenon whereby two or more species exert selective pressures on one another, leading to reciprocal evolutionary modification. A very well-known example of coevolution is between flowering plants and their pollinators, such as bees and hummingbirds. When plants evolve traits such as flower form and color to attract certain pollinators, pollinators also evolve to the plants by forming specialized systems for feeding. This co-influence forms diversity and encourages biodiversity, since adaptations can cumulatively lead to the creation of new species in the long run, emphasizing the dynamic nature of ecological associations. Coevolution, therefore, emphasizes the intricate dependence of species within ecological communities, noting that evolution is often an interactive process and not a single occurrence. Besides mutualistic associations, coevolution also plays a major role in antagonistic interactions, such as predator and prey. An example is the evolutionary arms race between cheetahs and gazelles. Cheetahs have developed enhanced speed and acceleration to chase after their prey, while gazelles have developed excellent agility and endurance to evade their predators. The adaptation cycle creates a dynamic process in which the two species are continually influencing each other's evolutionary tracks. Such interactions clarify the complexity of ecological networks and the processes by which competition shapes trait evolution. Moreover, such interactions illustrate the extensive ramifications of coevolution in not only forming individual species but the ecological community as a whole. For this reason, coevolution becomes a central process in the explanation of evolutionary processes and ecological resilience. The impact of coevolution is not specific to individual species interactions but to greater ecological and environmental systems. As species evolve through coevolution, the changes that result may alter ecological niches and habitat structures and thus contribute to community composition changes. Herbivore evolutionary changes, for instance, can significantly affect plant populations, leading to fertility patterns and growth. This interaction occurs in the changing ecosystems brought about by climate change, where coevolutionary events can trigger species extinction alongside new adaptation in communities, as noted in broader evolutionary processes governed by environmental changes (Stewart, 2010). Further, coevolution also overlaps with socio-technical dynamics, such that in the evolution of sustainable societies, technology and institutions must co-evolve to address environmental challenges (Marletto et al., 2025). Thus, an understanding of coevolution broadens our understanding of the evolutionary mechanism and its consequences on biodiversity and sustainability.

2.6.2. Interactions Between Species and Evolutionary Changes

Species interactions play a significant part in the evolution of evolutionary change, particularly through mechanisms like co-evolution and natural selection. With two species interacting, either predator and prey, or even mutualists, the forces of evolution that act through this interaction can lead to profound evolutionary adaptation in the long term. For instance, during predator-prey interactions, the prey can evolve such defenses as enhanced camouflage, agility, or toxic qualities for avoidance from predation. Meanwhile, the prey population can also change to develop enhanced evasion skills or characteristics to fend off the predator better. These mutual changes are a good representation of co-evolution, where development in one species has a direct influence on the evolutionary course of another and depicts the complexity and interdependence of ecological processes that guide organismal evolution. The evolutionary foundation of these modifications at the genetic level frequently involves changes in regulatory regions governing the expression of genes. The research reveals that TFs and their corresponding binding motifs can become altered with time, influencing a wide range of traits across an organism's genome. Studies indicate that such changes are not random occurrences; they demonstrate orchestrated genome-wide changes that are associated with significant evolutionary events, for instance, the speciation of closely related species, e.g., mice and humans (Thorne et al., 2010).

Such an interaction between evolution and genetic control underscores that many traits are not only the products of natural selection but also involve subtle alterations at the molecular level that could lead to various phenotypic outcomes in response to environmental pressure. Moreover, the science of microbial ecosystems offers a fascinating window from which to view these interactions and their evolutionary effects. Predatory relationships in microbial communities show that evolution can be extremely fast and have significant impacts on community structure over ecological timescales. In these systems, microbe-microbe dynamics, such as bacteria and their predator ciliates, are an exemplar of evolutionary interaction with the predator and prey co-evolving in response to each other's adaptations (Becks et al., 2020).

This co-evolutionary process reflects common ecological principles, demonstrating that ignoring evolutionary change when considering community dynamics can warp our understanding of ecological relationships. Thus, the interaction of evolution and ecology emphasizes the dynamic process of species interaction, initiating a deeper understanding of the adaptive landscape on which these organisms exist (Figure 2.18).

Figure 2.18. Co-evolution interactions between species and evolutionary changes.

Source: https://cdn.britannica.com/53/6553-050-953F7469/yucca-moth-mutualism-plant-seeds-larvae-feed.jpg?w=400&h=225&c=crop.

2.6.3. Importance of Coevolution in Ecosystems

Coevolution is key to ecosystems' structuring in that it determines the interactions and adaptations between different species. As species are forever dancing around each other, adapting and responding, coevolution makes for a dynamic scenario in which species are forced to coevolve at the same time in order to sustain their ecological niches. This complex interaction can lead to specialist relationships like pollinators and flowering plants, where mutualistic benefits fuel evolutionary processes. Moreover, coevolutionary processes can also be extended beyond species, with effects on entire communities and their accompanying biodiversity. As an example, the disappearance of a single species can cascade across the environment, deranging interactions and altering evolutionary trajectories, as in the case of major climatic events like those in the Quaternary period when coevolution was a driving and resultant force of ecological change (Stewart, 2010). The Red Queen hypothesis is the concept that best explains the coevolutionary idea, which posits that species must keep evolving to survive and, additionally, to remain ahead of other developing organisms with regard to relative fitness. This continuous evolution creates a running in place where populations constantly change and evolve as a result of changes in predator-prey relationships or competition. As each species develops, it can strongly influence the direction of evolution of others in its environment, and this illustrates the interconnectedness of life. Genetic changes that involve coevolutionary associations are also strongly linked with interspecific interactions, and they can induce new adaptations and evolutionary innovations to emerge, illustrating the profound impact of these interactions on ecological fitness landscapes (Hastings et al., 2013). Understanding coevolutionary processes intensifies our observation of ecosystem resilience and stability. Appreciating how species coevolve causes us to cherish the subtle web that preserves biodiversity. This dependence becomes even more vital under environmental pressures such as climate change, habitat destruction, and invasion by alien organisms. Apart from endangering specific species, these problems also disrupt the coevolutionary processes involved in upholding ecosystem integrity. By studying and extrapolating these interactions, we know what the consequences of human activities on ecological linkages can be and thus inform conservation actions and policy decisions that seek to preserve these critical evolutionary processes (Stewart, 2010; Hastings et al., 2013).

Test Your Knowledge

1. Write the interactions between species and evolutionary changes.
2. What is the importance of coevolution in ecosystems?

SUMMARY

A journey into the process of evolution, it is necessary to appreciate the interplay between environmental pressures and genetic variation. Evolution, fueled by forces of natural selection, mutation, and genetic drift, functions within a framework that can be described as both a scientific theory and a history of life's complexity. By empirical data and theoretical models, scientists can better understand how such mechanisms contribute to the adaptive character of species. However, evolutionary research is not free from methodological limitations. For instance, simulation models seeking to model evolutionary mechanisms are empirically irrelevant and question their ability to capture real evolutionary mechanisms, as evidenced in reference to Axelrod's pioneering work (Hubert et al., 2014). Furthermore, new developments in neuroscience have presented insightful insights into the cognitive aspect of memory, and a biological approach to understanding evolutionary mechanisms. Research indicates that the architectural structure of the neural network, responsible for the storage of long-term as well as short-term memory, is governed by the survival and adaptation laws of evolution. To be more specific, in the study of short-term memory mechanisms, the excitatory to inhibitory neuron ratio should be preserved for optimal cognitive processes (Hubert et al., 2014). This is a comparison that brings evolutionary theory and neurobiology together, suggesting that mechanisms of memory and cognition are central to the explanation of larger evolutionary adaptations. These intersections demonstrate how complicated evolution is and how much interdisciplinary effort is necessary to properly comprehend how disparate mechanisms interact with each other in creating life. Generally, the machinery of evolution is multifaceted, having genetic, environmental, and even cognitive components. Evolution thrives on the understanding that change is not only a biological process but also cognitive and social. As our academics learn more about these mechanisms, it becomes even clearer that evolution is a process grounded in the material world of species and the intellectual world of thought. After this, future research must look beyond simulation models and towards implicit empiricism, gaining a better knowledge of how the evolutionary dynamics operate in the world (Arnold et al., 2013). Ultimately, recognizing the interconnectedness and complexity of these mechanisms opens a larger picture of evolution, allowing for ongoing exploration both in biology and adjacent fields.

Evolution occurs through a series of initial mechanisms, all based on gradual change of species over generations. One of these underlying processes is natural selection, whereby individuals with advantageous traits have greater chances of survival and reproduction and thereby transmit those traits to their offspring. This process, fundamental to Darwin's theory, illustrates how environmental pressures shape the genetic variation within a population. Mutations that enhance the fitness of an organism in a specific environment become more frequent, leading to evolutionary adaptation. Such mechanisms uncover the reciprocity of adaptation and survival, working in a dynamic symbiosis between organisms and their surroundings. Natural selection brings about the recurring process of evolution, duplicating adaptation strategies evolving in response to environmental pressures. Genetic drift is another significant force of evolution, particularly applicable in small populations. This is how random allele frequency change takes place, which can lead to biased changes in traits over time. Unlike natural selection, where survival-positive traits prevail, genetic drift can favor traits that are not survival-positive. This randomness has the ability to lead to fixation or loss of alleles and genetic diversity, which can culminate in speciation events. The understanding of genetic drift enlightened the intricate dance of uncertainty that is part of evolutionary processes, suggesting randomness could play a critical role in shaping the biological tapestry of life. The interaction between genetic drift and other mechanisms ensures that evolution, as much as it is a process of systematization, is also a process of uncertainty. Lastly, gene flow is an important mechanism in the sense that it brings new genetic information into a population through migration and interbreeding. The transfer may add genetic diversity and flexibility, which allow populations to adapt more effectively to environmental change. Hybridization between species, for instance, can bring helpful traits that will allow survival in a changed environment. Gene flow may also reverse genetic drift and maintain genetic diversity. Seeing the significance of gene flow heightens the interrelatedness of environments and the penetrable character of species boundaries. The convergence of these three mechanisms, natural selection, genetic drift, and gene flow, is the way to bring out the richness of evolutionary dynamics, and it highlights how numerous factors converge to tell the unfolding story of life on Earth.

KEY TERMS FROM THIS CHAPTER

- Adaptation
- Allele frequency
- Balanced polymorphism
- Bottleneck effect
- Coevolution
- Directional selection
- Disruptive selection
- Founder effect
- Hardy-Weinberg equilibrium
- Macroevolution
- Microevolution
- Neutral evolution
- Nonrandom mating
- Selection pressure
- Sexual selection
- Stabilizing selection

CRITICAL THINKING QUESTIONS

1. **Which of the following is a mechanism of evolution?**
   a. Gravity
   b. Natural selection
   c. Photosynthesis
   d. Homeostasis
2. **What is genetic drift?**
   a. The movement of genes between populations
   b. A change in allele frequencies due to chance
   c. Selection for the best traits
   d. Mutation of all genes at once
3. **Gene flow occurs when:**
   a. One species goes extinct
   b. DNA is replicated incorrectly
   c. Individuals migrate between populations and breed
   d. Organisms stop reproducing
4. **A mutation is best described as:**
   a. Always harmful
   b. A random change in DNA
   c. Caused by natural selection
   d. The result of evolution
5. **What type of selection favors individuals with average traits?**
   a. Disruptive selection
   b. Directional selection
   c. Stabilizing selection
   d. Artificial selection

Short Answer Questions

1. What are the four main mechanisms of evolution?
2. How does natural selection lead to evolutionary change?
3. What is genetic drift, and in what types of populations is it most significant?
4. Compare and contrast the founder effect and the bottleneck effect.
5. What is gene flow, and how can it affect genetic variation in a population?

Answers to Critical Thinking Questions

1. (b); 2. (b); 3. (c); 4. (b); 5. (c)

REFERENCES

1. Abramson, A., Alonso-Sanz, R., Altrock, P. M., Assenza, S., Szolnoki, A., Axelrod, R., et al. (2009). Coevolutionary games–A mini review. http://arxiv.org/abs/0910.0826 (accessed on 13th June 2025).
2. Ahmadi, E., Bamdad, T., & Soleimanjahi, H. (2019). Introduction of RNA virus evolution. https://core.ac.uk/download/483639435.pdf (accessed on 13th June 2025).
3. Aldrich, H. E., Alexander, S., Archer, M. S., Aunger, R., Bagehot, W., Baldwin, J. M., et al. (2009). Learning from early attempts to *generalize Darwinian prin*ciples to social evolution. https://core.ac.uk/download/1639801.pdf (accessed on 13th June 2025).
4. Altermatt, F., Astegiano, J., Baguette, M., Becks, L., Bell, G., et al. (2017). Eco-evolutionary dynamics in fragmented landscapes. https://core.ac.uk/download/131025034.pdf (accessed on 13th June 2025).
5. Arnold, E. (2013). Simulation models of the evolution of cooperation as proofs of logical possibilities: How useful are they? https://core.ac.uk/download/158370503.pdf (accessed on 13th June 2025).
6. Arslan, A., Das, A., De Maeyer, D., DeLuna, A., et al. (2015). Adaptation to high ethanol reveals complex evolutionary pathways. https://core.ac.uk/download/55800864.pdf (accessed on 13th June 2025).
7. Barrett, R. D. H., Bolnick, D. I., Oke, K. B., Rennison, D. J., et al. (2018). (Non)parallel evolution. https://core.ac.uk/download/323311175.pdf (accessed on 13th June 2025).
8. Barton, N. H., Paixão, T., Tkačik, G., Tuğrul, M., et al. (2015). Dynamics of transcription factor binding site evolution. https://core.ac.uk/download/268224596.pdf (accessed on 13th June 2025).
9. Baudry, B., Chauvel, F., Clarke, S., Fleurey, F., et al. (2014). DIVERSIFY: Ecology-inspired software evolution for diversity emergence. https://core.ac.uk/download/49704396.pdf (accessed on 13th June 2025).
10. Becks, L., Hiltunen, T., Kaitala, V., Scheuerl, T., et al. (2020). Co-evolution as an important component explaining microbial predator-prey interaction. https://core.ac.uk/download/359922397.pdf (accessed on 13th June 2025).
11. Bich, L., & Green, S. (2016). Is defining life pointless? Operational definitions at the frontiers of biology. https://core.ac.uk/download/131211791.pdf (accessed on 13th June 2025).
12. Blancke, S., Braeckman, J., & De Tiège, A. (2018). The modern versus extended evolutionary synthesis: Sketch of an intra-genomic gene's eye view for the evolutionary-genetic underpinning of epigenetic and developmental evolution. https://core.ac.uk/download/187307814.pdf (accessed on 13th June 2025).
13. Boddy, A. M., Bronstein, J. L., Dantas, G., Diaz-Munoz, S. L., et al. (2016). Contextual organismality: Beyond pattern to process in the emergence of organisms. https://core.ac.uk/download/129069601.pdf (accessed on 13th June 2025).
14. Bonomo, M. E., & Deem, M. W. (2018). The physicist's guide to one of biotechnology's hottest new topics: CRISPR-Cas. http://arxiv.org/abs/1712.09865 (accessed on 13th June 2025).

15. Briner, A. E., Andersson, A. F., Levin, B. R., Pourcel, C., Sun, C. L., Bhaya, D., Pennisi, E., et al. (2017). How the other half lives: CRISPR-Cas's influence on bacteriophages. http://arxiv.org/abs/1711.09113 (accessed on 13th June 2025).
16. Carbone, C., Lurz, P., Reuman, D., Signorile, A. L., et al. (2016). Mixture or mosaic? Genetic patterns in UK grey squirrels support a human-mediated 'long-jump' invasion mechanism. https://core.ac.uk/download/77008143.pdf (accessed on 13th June 2025).
17. Clark, A., Crombach, A., Davies, A., Mesoudi, A., Traulsen, A., Wagner, A., et al. (2015). Evolutionary connectionism: Algorithmic principles underlying the evolution of biological organization in evo-devo, evo-eco and evolutionary transitions. https://core.ac.uk/download/30615145.pdf (accessed on 13th June 2025).
18. Denison R. F. (2011). Past evolutionary tradeoffs represent opportunities for crop genetic improvement and increased human lifespan. Evolutionary applications, 4(2), 216–224. https://doi.org/10.1111/j.1752-4571.2010.00158.x(accessed on 13th June 2025).
19. Eves-van den Akker, S., Goss, E. M., & Sacristán, S. (2021). How do pathogens evolve novel virulence activities? https://core.ac.uk/download/388542546.pdf (accessed on 13th June 2025).
20. Ferrara, A., & Latif, M. A. (2016). Formation of supermassive black hole seeds. http://arxiv.org/abs/1605.07391 (accessed on 13th June 2025).
21. Gelepithis, P. A. M. (2009). Outline of a new approach to the nature of mind. https://core.ac.uk/download/pdf/87224.pdf (accessed on 13th June 2025).
22. Giraud, A., Sali, A., Shaver, A. C., Li, B., Wylie, C. S., Klimov, D. K., Dekel, E., et al. (2009). Interplay between pleiotropy and secondary selection determines rise and fall of mutators in stress response. https://core.ac.uk/download/8388553.pdf (accessed on 13th June 2025).
23. Griffiths, P. E., & Stotz, K. (2017). Genetic, epigenetic and exogenesis information. https://core.ac.uk/download/131214859.pdf (accessed on 13th June 2025).
24. Hastings, A., Hoffman, A., Ilachinsky, A., Kerr, A., Berryman, A. A., Agrawal, A. F., et al. (2013). Red Queen coevolution on fitness landscapes. http://arxiv.org/abs/1303.5633 (accessed on 13th June 2025).
25. Hernando Amado, S., López Causapé, C., Martínez, J. L., Oliver, A., et al. (2022). Low ciprofloxacin concentrations select multidrug-resistant mutants overproducing efflux pumps in clinical isolates of *Pseudomonas aeruginosa*. https://core.ac.uk/download/541456202.pdf (accessed on 13th June 2025).
26. Hubert, J., & Ikegami, T. (2014). Short-term memory through persistent activity: Evolution of self-stopping and self-sustaining activity in spiking neural networks. http://arxiv.org/abs/1411.6912 (accessed on 13th June 2025).
27. Joly, E. (2010). On the close relationship between speciation, inbreeding and recessive mutations. https://core.ac.uk/download/pdf/289652.pdf (accessed on 13th June 2025).
28. Joly, E. (2011). The existence of species rests on a metastable equilibrium between inbreeding and outbreeding. https://core.ac.uk/download/pdf/290207.pdf (accessed on 13th June 2025).
29. Marletto, G. (2025). Which conceptual foundations for environmental policies? An institutional and evolutionary framework of economic change. https://core.ac.uk/download/pdf/6799087.pdf (accessed on 13th June 2025).

30. Matthews, L. J., Molter, D., Plutynski, A., Vernon, J. A., et al. (2016). Chance in the Modern Synthesis. https://core.ac.uk/download/186329988.pdf (accessed on 13th June 2025).

31. Miranda, I., Moura, G., Santos, M. A. S., Silva, R. M., et al. (2004). Yeast as a model organism for studying the evolution of non-standard genetic codes. https://core.ac.uk/download/294779910.pdf (accessed on 13th June 2025).

32. Nilsson, E. E., & Skinner, M. K. (2021). Role of environmentally induced epigenetic transgenerational inheritance in evolutionary biology: Unified evolution theory. https://core.ac.uk/download/603337074.pdf (accessed on 13th June 2025).

33. Oudeyer, P. Y. (2015). Open challenges in understanding development and evolution of speech forms: The roles of embodied self-organization, motivation and active exploration. https://core.ac.uk/download/49439677.pdf (accessed on 13th June 2025).

34. Packard, A., Hurley, A. C., Henchman, A. P., Morton, B., Budelmann, B. U., Lalli, C. M., Darwin, C. R., et al. (2008). Charting evolution's trajectory: Using molluscan eye diversity to understand parallel and convergent evolution. https://core.ac.uk/download/38934996.pdf (accessed on 13th June 2025).

35. Patel, M. (2005). The transition from patented drugs to generics: TRIPS constraints and policy options. https://core.ac.uk/download/76354.pdf (accessed on 13th June 2025).

36. Perc, M., & Szolnoki, A. (2009). Resolving social dilemmas on evolving random networks. http://arxiv.org/abs/0910.1905 (accessed on 13th June 2025).

37. Rebecca M. Price, Kathryn E. Perez; Beyond the Adaptationist Legacy: Updating Our Teaching to Include a Diversity of Evolutionary Mechanisms. The American Biology Teacher 1 February 2016; 78 (2): 101–108. https://doi.org/10.1525/abt.2016.78.2.101 (accessed on 13th June 2025).

38. Sammut-Bonnici, T., & Wensley, R. (2002). Darwinism, probability and complexity: Market-based organizational transformation and change explained through the theories of evolution. https://core.ac.uk/download/17180819.pdf (accessed on 13th June 2025).

39. Serrato-Capuchina, G. A. (2019). Exploring the role of transposable elements in reproductive isolation. https://core.ac.uk/download/286142638.pdf (accessed on 13th June 2025).

40. Stewart, J. R., 2010. Understanding evolutionary processes during past Quaternary climatic cycles: Can it be applied to the future? In: Grant, F., Mergeay, J., Santamaria, L., Young, J. and Watt, A.D., eds. Evolution and Biodiversity: The evolutionary basis of biodiversity *and* its potential for adaptation to global change. Report of an e. Mallorc*a, Spain: Centre for Ecology & Hydrology, INBO, IMEDA, CSIC-UIB.(accessed on 13th June 2025).*

41. *Sydow, M. von. (2001). Sociobiology, uni*versal Darwinism and their transcendence: An investigation of the history, philosophy and critique of Darwinian paradigms, especially gene-Darwinism, process-Darwinism, and their types of reductionism towards a theory of the evolution of evolutionary processes, evolutionary freedom and ecological idealism. https://core.ac.uk/download/6116908.pdf (accessed on 13th June 2025).

42. Sánchez, A. (2016). Why Toll-like Receptors genes are so polymorphic? https://core.ac.uk/download/296280449.pdf (accessed on 13th June 2025).

43. Thorne, J. L., Wray, G. A., & Yokoyama, K. D. (2010). Coordinated genome-wide modifications within proximal promoter cis-regulatory elements during vertebrate evolution. https://core.ac.uk/download/pdf/8514380.pdf (accessed on 13th June 2025).

THE ORIGIN OF LIFE AND EARLY EVOLUTION

3

LEARNING OBJECTIVES

After studying this chapter, you will be able to:

1. Explain the theories of the origin of life;
2. Define the role of chemical evolution;
3. Discuss the first living organisms;
4. Explore the evolution of photosynthesis;
5. Identify the rise of eukaryotic cells;
6. Focus on the major evolutionary milestones.

Introduction

The quest for the origin of life is a rich and complex scientific endeavor, weaving together strands of biology, chemistry, and theory of evolution. It begins with the realization that life, in its first form, had commenced from non-living chemical compounds in an early setting that offered the necessary conditions for such transformations. An ever-growing body of evidence points to several environmental factors, including energy sources in the form of volcanism and solar irradiation, playing important roles in this chemical-to-biological transition. A distinction from cellular life needs to be made, since the latter, while abundant, are often peripheral to the main narratives of evolutionary theory. A recent synthesis challenges conventional paradigms, proposing a TOL which combines both cellular and viral worlds, emphasizing the co-evolution of both these organisms as a key element of life's evolutionary story (Claudiu, 2009). Another contribution to life's pluralistic richness is the emergence of homochirality, an inherent quality of biochemical molecules. Modern models, such as the activation-polymerizationepimerization-depolymerization (APED) model, provide insight into how homochiral architectures may have arisen on prebiotic Earth through specific chemical dynamics.

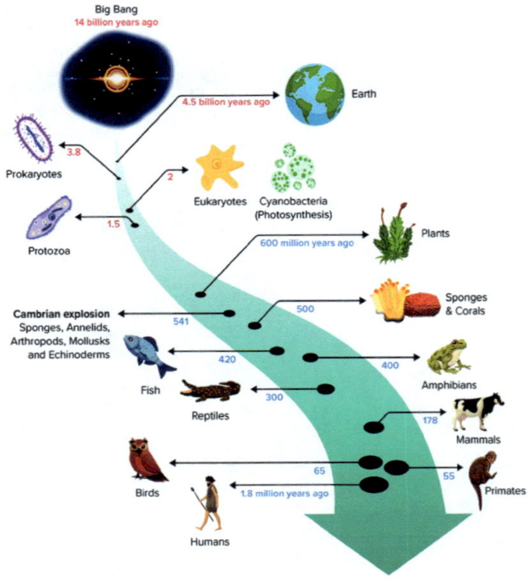

Figure 3.1. The origin of life and early evolution.

Source: https://dr282zn36sxxg.cloudfront.net/datastreams/f-d%3A9f5a2959c1d36d8e9b5b056c30506950cc6147d31135ed8cee06f0ee%2BIMAGE_TINY%2BIMAGE_TINY.1.

The APED model predicts that interactions of reaction parameters in optimally conducive environments can result in the generation of isolated homochiral domains, which are critical to the emergence of life's molecular chirality. The influence of external forces—such as tidal flows or evaporation from water bodies—may enhance these proto-domains' stability, drawing them towards self-organization. These assemblies are now believed to be an important phase of the formation of the earliest protocells, laying the groundwork for future cellular life (Brack et al., 2009). In studying the origin of life, it is essential to investigate multilevel hypotheses with an understanding of the part that environmental interactions play in shaping evolutionary processes. The interaction among biotic and abiotic components has consequences that suggest life is not an independent phenomenon but rather a continuous process influenced by a myriad of forms of matter and energy. As scientists learn more about these processes, they try to learn more about the origin of life and how it evolved. Looking at biological and viral perspectives as well as exploring chemical processes like homochirality, we can begin constructing a more comprehensive narrative that encompasses the complexity of the origin of life and early evolution. This desire highlights the importance of interdisciplinary investigation in a work examining life's biological and historical forms (Figure 3.1).

Definition of Life and Its Significance

Life is more than mere biological criteria of growth, reproduction, and metabolism; it is a complex interaction of biochemical processes that facilitate adaptation and evolution. Life can be defined in terms of its ability to respond to environmental stimuli and reproduce, a sign of a highly developed genetic infrastructure that enables species to pass on traits from generation to generation. This definition emphasizes the significance of life as a dynamic, adaptive creature responding to its environment and so affecting ecosystems and biodiversity. Moreover, the definition of life is significant in evolutionary biology, particularly as it informs our conceptualization of phenotypic plasticity. This is a term referring to how life adapts to deal with varied environments, a requirement in terms of endurance and adaptability to environmental alterations, thereby affirming the intricate relationship of life to its surroundings and evolutionary trajectories (Martins et al., 2018). The significance of life also lies in its intrinsic worth to ecological systems. Life organisms engage in complex ways with one another and the environment, resulting in ecosystem stability and function. Organisms are involved in energy transfer through food webs, enabling nutrient cycling and the development of complex habitats. Ecological resilience depends substantially on elevated biological diversity; hence, the existence and interdependence of varied life organisms enhance the general stability of ecosystems." Evolutionarily speaking, the relevance of such interactions is the foundation of conservation practice, emphasizing that understanding the evolutionary past and ecological roles of species can determine management. The principles are underscored by the recognition of how global changes in the environment can impact not only individual species but the entire web of life they are dependent on, illustrating the interdependence of ecological and evolutionary processes (Martins et al., 2018). In establishing terms for the debate on life's definition and meaning, it is important to note the opposing positions that contribute towards public acceptance of evolution, particularly the tension between scientific theory and creationist beliefs. While scientific literacy heightens a sense of evolutionary biology, the resilience of creationist perceptions reveals that education processes must tread these delicately. Engaging students with scientific facts can make them better understand the natural world and the processes of life. As a way of demystifying evolution, teachers play a crucial role in building critical thinking, bridging the gap between recognized scientific theory and public perception. This discussion emphasizes the importance of scientific inquiry when it comes to differing perceptions and makes sure that the essence of life and its line of evolution are realized on a large scale throughout.

Overview of the Scientific Inquiry into Life's Origins

Scientific exploration of the origin of life has long been an issue of concern to scientists, giving rise to interdisciplinary research that draws on chemistry, biology, and earth sciences. Foremost in this debate is the abiogenesis hypothesis, which maintains that life emerged from non-living material through natural processes. The first experiments, e.g., the Urey-Miller experiment of 1953, simulated conditions on early Earth, demonstrating that simple organic molecules could be produced from inorganic precursors. These are the foundational breakthroughs on which subsequent work was based, extending our knowledge of metabolic pathways. Although in this quest to define the exact steps of abiogenesis much remains to be done, such breakthroughs on this front have gone a long way in determining potential environments, e.g., hydrothermal vents or primordial ooze, under which life's building blocks may have assembled. Scientists also turned to extremophiles, organisms that thrive in extreme environments, to discover more about life's ability to adapt and survive. Not only do these organisms challenge the traditional definition of habitability, but they also offer analogs for potential extraterrestrial life, making the search for scientific exploration off our planet even more valid. The search for the origins of life is not merely a review of Earth's past conditions but also an assessment of life's versatility in other worlds. Furthermore, theoretical models such as the RNA world hypothesis suggest the idea that replicating RNA molecules may have been precursors to the first cellular forms of life. An understanding of these models contributes to a better knowledge of the underlying simple biological processes involved in all life known, facilitating the incorporation of evolutionary theory into molecular biology. As this scientific investigation goes on, the convergence of frontier technologies such as synthetic biology and AI will enhance our understanding of life's origins. These technologies allow researchers to manipulate biological systems and engineer life-like structures, providing a rich platform to experiment on theories of the origin of life. Furthermore, the importance of the findings extends beyond the understanding of Earth's biological past; they have potential implications for other areas of study, including biotechnology and astrobiology. However, the controversy has not been unscathed by any means, particularly on the philosophical implications and ethics of synthetic life. As this rich fabric of inquiry progresses, it is increasingly clear that the quest to understand where life came from lies at the heart of comprehending our place in the universe and the nature of existence itself, as highlighted within present discourse on scientific education and standards (Lynch et al., 2012) (Figure 3.2).

Figure 3.2. The scientific inquiry into life's origins.

Source: https://scitechdaily.com/images/Origin-Emergence-of-Life-Concept-Art.jpg.

Importance of Understanding Early Evolution

Knowledge of early evolution is crucial to understanding the complexity of biological systems that have evolved over a billion years. The simple information is what lets us understand how life evolved to adapt to many environmental stresses, thereby laying the ground for the amazing biodiversity that is prevalent today. It is impossible to comprehend fully the mechanisms of natural selection and genetic variation without the knowledge of early evolutionary processes. As evidenced in arguments against Darwin's writings, the natural selection metaphor is an important component of the delivery of such evolutionary concepts; it can be misunderstood, though, unless delivered appropriately. Thus, a complex understanding of early evolution gives scholars the capacity to communicate such difficult concepts well, and the capacity to see that students grasp both science and history, an important issue in teaching evolution today (Blancke et al., 2013). Furthermore, exploring early evolutionary processes is paramount to guiding current scientific inquiry and uses. New theoretical insights in evolutionary biology are increasingly necessary to leverage the vast amounts of genomic and phenotypic data scientists are gathering, particularly as these data illuminate the interactions among species. Appreciating the origins of life and the earliest evolutionary adaptations not only explains our history but also makes us more capable of comprehending contemporary organisms, including ourselves. Such information can be utilized to explore evolutionary relationships that eventually inform medicine, conservation, and biotechnology. In this way, using the knowledge of early evolution can further our ability to solve today's biological issues with innovative solutions based on genetic evolution and adaptation that have taken centuries (Arnold et al., 2013). Finally, understanding early evolution encourages a more general understanding of life as interconnected, challenging students to consider biology in an integrated context. These findings can potentially create interdisciplinary connections with other sciences already established that resonate beyond the science of evolutionary biology, pointing to the relevance of evolutionary ideas in ecology, genetics, and anthropology. It places individuals for shared action toward the resolution of global issues such as climate change and biodiversity loss, which call for an exact understanding of how organisms evolve and adapt to their environment over time. Hence, emphasizing the importance of early evolution in the classroom setting not only makes students science-conscious but also nurtures the role of responsibility in managing life on this planet so that generations to come are ready to manage the vital challenges they will be facing.

3.1. Theories of the Origin of Life

Learning Outcome

1. Primordial soup hypothesis;
2. Hydrothermal vent hypothesis;
3. Panspermia theory.

As researchers delve into the mystery of life's origin, a variety of theories emerge, each seeking to disentangle the complexity of life's birth on Earth. The primordial soup theory is one general hypothesis that attempts to say organic molecules evolved in ancient oceans from chemical reactions initiated by energy sources such as lightning and UV radiation. This hypothesis came into prominence after the Miller-Urey experiment in 1953, in which the production of amino acids was demonstrated under early Earth conditions. Despite this, the criticisms of the hypothesis touch on the unlikelihood of producing complex compounds based merely on random interactions in an environment lacking nutrients. Hence, this theory's explanation demands questioning about the synergy of the various environmental conditions within the emergence of life and further examination of how these molecular building blocks evolved into the first organisms. The deep-sea vent hypothesis represents a leading hypothesis suggesting the possibility that life began within the extreme conditions of hydrothermal vents. These vents provide a new environment that is filled with mineral catalysts and energy sources, so there is potential to form the first organic compounds. The theory is that the ancient metabolic processes can be maintained by the high temperatures and pressures of these environments, with the eventual evolution of life forms. Of particular interest is the recent finding of extremophiles—organisms that survive in such extreme conditions.

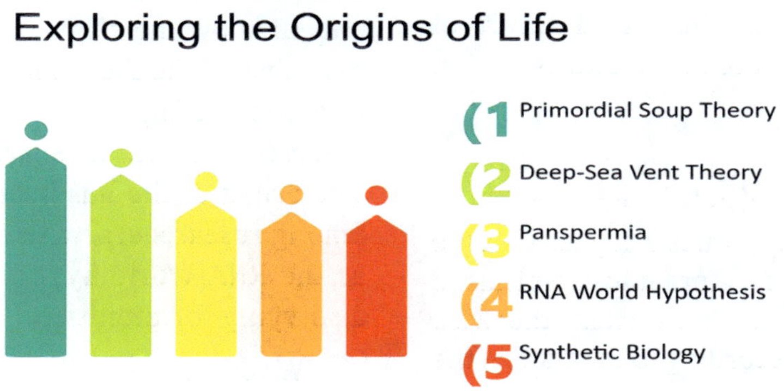

Figure 3.3. Theories of the origin of life.

Source: https://fastercapital.co/i/Primordial-Soup--Cooking-Up-Life--Big-Bang-and--Exploring-the-Origins-of-Life.webp.

This supports this model, as it suggests that not only can life survive but also thrive in conditions previously considered hostile. While scientists explore geochemical trends within these vent ecosystems, they learn about the basic processes that could have driven life's development and hence continue to advance the conversation regarding the flexibility and resilience of early biological systems. The development of the RNA world hypothesis introduces yet another central perspective on life's origin, with the concept of RNA as the first molecule to replicate itself, prior to DNA. The hypothesis is supported by RNA's ability to code genetic information as well as catalyze biochemical reactions, suggesting it could have facilitated the formation of early organisms. The model's detractors, though, believe that the transition from simple RNA molecules to the cellular life forms of today remains inadequately explained. In response to these criticisms, researchers explore the potential functions of simple peptides and other organic molecules in the early life of primordial organisms. Furthermore, the suggestiveness of David Hume's paradox of induction in understanding the contingency of existence provides a philosophical flavor to the science of origin, and that necessitates relentless investigation into the processes which have led to the origin of life (Aranda-Anzaldo et al., 1998). This blend of empirical research and philosophy indicates the varied nature of theory in the context of the origin and evolution of life (Figure 3.3).

3.1.1. Primordial Soup Hypothesis

The Primordial Soup Hypothesis argues that life originated from a variety of organic chemicals in the early conditions of Earth, which is otherwise referred to as a soup. According to the theory, the organic molecules, under the right conditions, such as in primordial oceans, might perform complex chemical reactions that ultimately lead to simple life. One distinctive aspect of this hypothesis is its suggestion that energy sources such as ultraviolet radiation, lightning, and volcanism were the vital forces responsible for these chemical transformations. Scientific laboratory experiments, such as the famous Miller-Urey experiment, have demonstrated that amino acids and other organic molecules can be synthesized artificially from inorganic precursors under early Earth-like conditions, thus confirming the theory that life can evolve naturally from the environment around it. Recent studies have also advanced this theory further by suggesting novel mechanisms for the replication of molecular systems in the primordial soup.

A single large study suggests that RNA replication might have been supported by self-catalyzed thermochemical oscillations. This model suggests that oscillating reactions, possibly in the presence of hydrogen peroxide-rich environments, would facilitate replicating complementary strands of RNA, thereby catalyzing life's molecular evolution. The simulation of these systems is promising, showing how, under optimally chosen parameters, RNA replicants would be capable of massively outcompeting their ancestors, eventually leading to increased complexity in early Earth's molecular environment (Ball et al., 2014). This work points toward the potential for energetic interaction in the primordial soup not only to allow the origin of life but also enable it to evolve. While the Primordial Soup Hypothesis is a compelling explanation of the origins of life, it comes with its share of rival hypotheses. The origin of life has been a contentious topic among researchers, with rival hypotheses such as the RNA World hypothesis and panspermia also vying for attention.

Whereas panspermia suggests that life-carrying organic molecules could have been transported to Earth via comets or meteorites, it fails to discuss the manner in which the molecules could have gained life from non-life on Earth (Gibbons et al., 2012). Similarly, whereas the RNA World hypothesis places great emphasis on using RNA as a precursor to life, it fails to fully address the environment under which these original molecules existed.

As more studies of the interaction between the theories are being conducted, the Primordial Soup Hypothesis remains a cornerstone of the search for understanding of the workings of early evolution (Figure 3.4).

Figure 3.4. Darwin's primordial soup theory explains the origins of life.

Source: https://hips.hearstapps.com/hmg-prod/images/stromatolites-on-early-earth-royalty-free-image-1670963451.jpg.

3.1.2. Hydrothermal Vent Hypothesis

The hydrothermal vent hypothesis is a plausible theory for the origins of life predicated upon the potential that life might have begun in the harsh environments of hydrothermal vents on the seafloor. Hydrothermal vents emit warm, mineral-rich fluid from beneath the Earth's crust, a high-chemical-potential environment. This offers the energy and resources necessary for early life forms that must synthesize complex organic molecules. As suggested, the unique function of energy production from chemical gradients is at the core of metabolic processes that might have led to the origin of the earliest living things. The unique chemical and geothermal conditions in the vents might have provided the environment for the development of essential biomolecules, initiating the evolution from chemically pure systems to biological replicators with metabolisms (Lanier et al., 2017). Besides, the stability of hydrothermal environments offers the potential that these environments could have favored the formation of organic compounds for extended periods. The heat and pressure conditions and the variety of available minerals can have the potential to supply the chemical reactions required for life. In this respect, the hydrothermal vent hypothesis is in line with other hypotheses, such as the metabolism-first model, which takes precedence in metabolic processes over replication mechanism formation. The hypothesis also aligns with the overall debate in the origin of life, which requires that there be a dynamic interaction between early biological systems and abiotic chemistry (Gibbons et al., 2012).

With empirical evidence showing an association between vent chemistry and biomolecule synthesis, the hypothesis itself provides a worthwhile avenue of study for scientists interested in the history of life. Nevertheless, although the hydrothermal vent hypothesis is significant, it is not problem-free. Critics cite the challenge of bridging simple chemical reactions to the intricacies of modern biochemistry. The dependency on special functions, as explained in various models of life's emergence, obliges scientists to account for the large steps taken in this process. Each model's reliance on separate chemical precursors or systemic overhauls introduces levels of complexity that can fog our understanding. Moreover, the interaction among panspermia and abiogenesis origin hypotheses complicates it further, requiring no monolithic hypothesis that can completely capture the complexity of life's emergence (Gibbons et al., 2012; Lanier et al., 2017). Therefore, although hydrothermal vents offer a convincing clue regarding potential directions for life's origins, ongoing study and compilation of plural hypotheses remain required to uncover the true narrative of life's early evolution (Figure 3.5).

Figure 3.5. Life originated in hydrothermal vents.

Source: https://cdn.sci.news/images/enlarge6/image_7772e-Hydrothermal-Vent.jpg.

3.1.3. Panspermia Theory

The Panspermia theory is a very fascinating paradigm on the origin of life, which states that microbial life and organic compounds might not have evolved on the Earth, but instead they might have been delivered here by other celestial bodies. According to this theory, life may be planted throughout the universe by means of comets, meteoroids, or spaceships. This hypothesis is strengthened when taking into account that some extremophiles on Earth show a high tolerance to harsh conditions, indicating that should life were able to survive the extreme conditions of space, then there is the possibility that it could travel between stars. Furthermore, recent evidence of complex organic molecules on comets and in interstellar clouds adds credence to the idea that the building blocks of life are widespread throughout the galaxy, potentially serving as prebiotic material in various environments (Gibson et al., 2010). Moreover, the active movement of the bodies in the sky, e.g., the transfer of meteorites between Mars and Earth, might have also contributed to the transfer of microbial life, which also speaks in favor of the principles of panspermia. The hypothesis is also consistent with hydro-gravitational-dynamics cosmology that anticipates a universe fraught with planets that contain critical chemical elements that support life. According to this model, elements like carbon, nitrogen, oxygen, and phosphorus, which are fundamental to biological processes, may have been produced and widely disseminated through supernova events (Gibson et al., 2010). Consequently, panspermia as a concept not only assumes that life is spread throughout the universe but also sheds light on a process through which it may get favorable conditions to flourish in various habitats. The panspermia theory has more implications than just being a speculation, as it seeks to diminish other conventional stories about the beginning of life on Earth. In adopting this view, scientists can narrow the astrobiological search parameters. For instance, using probabilistic cellular automata (PCA), scientists can model the distribution and proliferation of life across the Milky Way, shedding light on its potential prevalence throughout the Galactic Habitable Zone. With further improvements in observational technology and the exploration missions, the empirical study of panspermia may reveal vital information regarding our neighbors in the universe. These studies could tell us if life, as we know it, is a common thing or a rare one, which will radically change our perspective on the course of life in the universe (Figure 3.6).

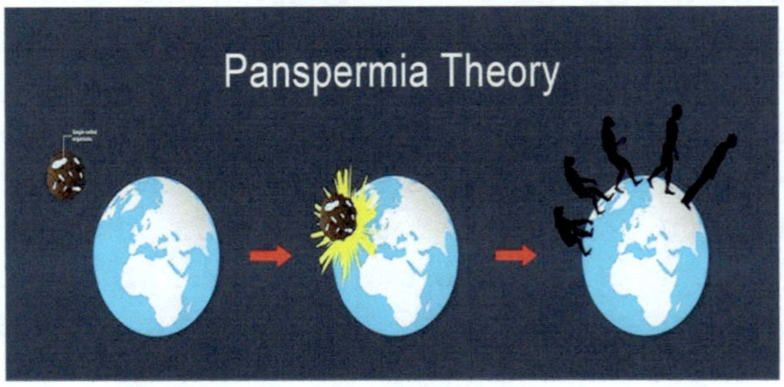

Figure 3.6. Panspermia theory.

Source: https://i.ytimg.com/vi/RHw7kMRXpG8/hq720.jpg?sqp=-oaymwEhCK4FEIIDSFryq4qpAxMIARUAAAAAGAElAADIQj0AgKJD&rs=AOn4CLAUqLDuHzWnKl3lQdj2mX4_4WSa8Q.

Test Your Knowledge

1. What is the primordial soup hypothesis?
2. What is the panspermia theory?

3.2. The Role of Chemical Evolution

Learning Outcome

1. Formation of simple organic molecules;
2. Development of complex macromolecules;
3 The significance of RNA in early life.

Chemical evolution is the most important explanation for life's origin because it entails the process from simple organic molecules to complex biological molecules. It is characterized by the emergence of information-containing systems that replicate themselves and can evolve to fit the external environment. The ability of such systems to process and manage information has significant implications, suggesting that information flow and dissemination were critical in the course of early life. This is further supported by emerging research that offers the evolution of life as a reconfiguration of causal structure and information flow, with a focus on the crucial changes that occurred in chemical evolution.

This paradigm change creates our comprehension of the way straightforward chemical reactions evolved to complex biological systems and thereby the basis for life's diversity on Earth (N/A, 2014). Secondly, research on extraterrestrial ancient materials, such as carbonaceous chondrites, provides crucial data on chemical precursors to life. These meteorites, packed with organic compounds, serve as a tangible reminder of the organic chemistry that seeded the early solar system. The circumstance that these intricate molecules are embedded in the inorganic matrix makes it possible for scientists to trace the paths that might have led to the construction of life's building blocks. Notably, the interaction between meteoritic organic material and clay minerals suggests that these clays were potentially the drivers for catalyzing chemical reactions, thereby making the complicated process toward the origin of life possible. This understanding of this interdependence between organic and inorganic compounds also accounts for chemical evolution to thrive on early Earth (Alpern et al., 2002). In summary, the role of chemical evolution in the origin of life is an ongoing dance between ancient organic material and the physical forces that drove their complexity. The transition from abiotic chemistry to biotic systems is not a random event but rather a guided evolution facilitated by the presence of resources as well as the early Earth's physical environment. With research into organic molecules and mineral substrates' interaction continuing to unravel, the potential for alternative routes to life's origins becomes increasingly promising. Such research fosters an understanding of the complexity involved in chemical evolution while emphasizing the need for an interdisciplinary approach in deciphering the processes of life's origin. Ultimately, the understanding of such complex processes will significantly contribute to our knowledge of the early stages of biological evolution and the development of ecosystems we observe today (Figure 3.7).

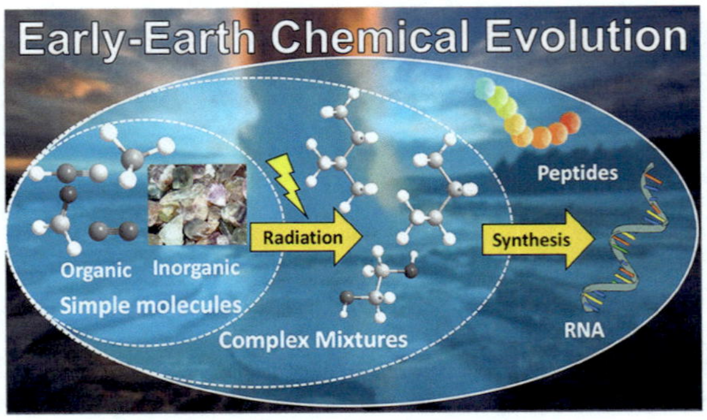

Figure 3.7. The role of chemical evolution.

Source: https://www.researchgate.net/profile/Ruiqin-Yi/publication/348561390/figure/fig1/AS:1003100055818243@1616169098874/Chemical-evolution-for-generating-substances-that-are-essential-for-life-such-as-RNA-and.ppm.

3.2.1. Formation of Simple Organic Molecules

Formation of uncomplicated organic molecules is a pivotal step towards understanding the origin of life and the emergence of its initial forms. Research has shown that the molecules, such as amino acids and nucleotides, may be produced through various abiotic processes on prebiotic Earth. One of the most widely held hypotheses postulates that the conditions in early Earth environments, such as volcanic activity and lightning strikes, provided the energy necessary to catalyze chemical reactions among simpler compounds. Such reactions would be most likely to yield more complex organic molecules as a basis for life to have emerged.

The significance of these early organic molecules cannot be overstressed because they served as the precursors to more complicated biological macromolecules, which later were the cause of the formation of rudimentary cellular structures and metabolic processes basic to life's evolution. During the examination of the production of simple organic molecules, it is essential to value the role of extraterrestrial events and processes that controlled their production. One of the hypotheses suggests that dendritic crystallization might be one physico-chemical process whereby molecules might have been organized into complex structures. This hypothesis involves branching dendritic structures playing a role in the self-organization of simple organic compounds and providing a basis for the evolution of living systems. This process demonstrates how abiotic processes might have resulted in self-replicating biological systems with adaptive and evolutionary capabilities. Furthermore, the suggestion of organic chemical evolution as a natural continuation of matter's evolution in the universe underscores the interdependence of chemical processes and biological processes in suggesting that life emergence may extend beyond Earth and could be feasible elsewhere in the universe, as presented by Umantsev (2011). The study of various hypotheses about the origin of life also supports the complexity of the evolution of simple organic molecules. For instance, models such as abiogenesis, the RNA World, and panspermia attempt to explain the mechanisms of life's emergence on the planet. While panspermia, in which life-bearing molecules were delivered on comets to Earth, has been supported, recent findings show that such delivery was insufficient to initiate the origin of life independently. Instead, Earth's processes played a greater role in initiating biological evolution through synthesizing organic molecules. This perspective encourages a holistic understanding of the emergence of life, as it emphasizes the necessity of consolidating knowledge from various scientific disciplines, including biochemistry and astrobiology, towards uncovering the mysteries of the origin of life on Earth, as proposed by Gibbons et al. (2012) (Figure 3.8).

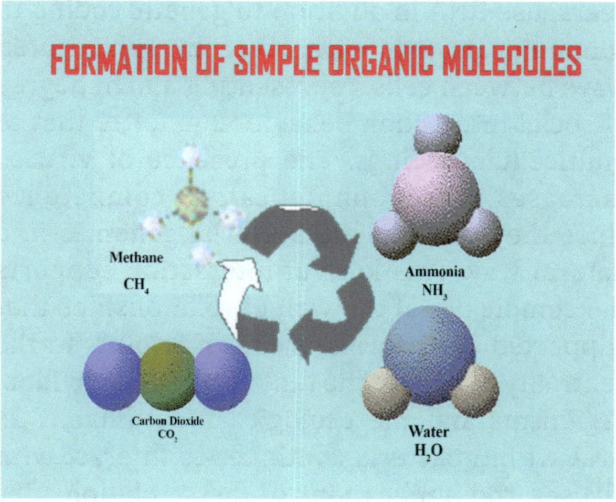

Figure 3.8. Formation of simple organic molecules.

Source: https://slideplayer.com/slide/2864446/10/images/8/FORMATION+OF+SIMPLE+ORGANIC+MOLECULES.jpg.

3.2.2. Development of Complex Macromolecules

The creation of intricate macromolecules is a cornerstone in the explanation of life's origin and further development on Earth. Initially, low-molecular-weight organic molecules are thought to have existed in prebiotic environments, which gradually evolved towards developing more intricate structures. Of these structures, nucleic acids such as RNA and DNA are of utmost significance since they contain and transfer genetic information. The general theory is that these macromolecules were formed from rudimentary units, possibly as simple as being readily synthesized. Experiments like the Miller-Urey experiment illustrate the possibility of how Earth's conditions support the formation of RNA's required precursors, providing an accessible avenue for life to have developed from inorganic material. One of the main arguments used in this debate is the role of nitrogen heterocycles, which are suggested to be the most primitive building blocks of the first genetic polymers. Research indicates that these compounds could have been prevalent on ancient Earth, opening a simple pathway towards RNA and its precursors' evolution. In consonance with reactivity results on N-heterocycles, the majority of these molecules have been found to form adducts that mirror the building blocks to more elaborate structures, such as peptide nucleic acids (PNAs) (Callahan et al., 2019). These results essentially change our understanding of the conditions of macromolecule development, suggesting that under some atmospheric conditions, the paths to genetic material were feasible, thereby aligning with abiogenesis theory. In addition, the results of this study transcend Earth's origin of life and extend into a broader context of extraterrestrial life. Studies focused on the identification of compounds of biological importance in different surroundings may exhibit analogous processes at work. The quest for life beyond our world increasingly depends upon our understanding of how complex macromolecules can be produced under convergent circumstances. By expanding the knowledge of nitrogen heterocycles and their formation in other atmospheres, scientists contribute to the current discourse of the possibility of the universality of life in the universe, as cited in the interests of extraterrestrial mission explorations (Hoffman & Kontaratos, 2025). Thus, the formation of complex macromolecules is not only vital in deciphering the past but in envisioning the future in our concepts of life on a cosmic scale.

3.2.3. The Significance of RNA in Early Life

The importance of RNA with regard to early life is difficult to overestimate, as it is commonly viewed as one of the key actors in the rise of cellular complexity. The combination of the capabilities of RNAs to play the role of a genetic material carrier and a biochemical catalyst provides a special niche for it in the scheme of early evolutionary events. This catalytic and sequence specificity is an important part of the RNA world hypothesis, the idea that early life was based on RNA molecules that were able to reproduce themselves and perform the needed biochemical tasks. RNA in this model would have played the role of allowing a transition between simple molecular structures to complex forms of life. The evolutionary implications of RNA's capabilities underscore its role in the origin of life, potentially laying the groundwork for modern biological systems that rely on DNA and proteins (Claudiu, 2009). The interaction between RNA and viruses has proved to be informative in the study of the evolutionary trend of life. Studies suggest that the co-evolution of viruses and cellular organisms has resulted in major metabolic innovations and cell complexities. Of particular interest are RNA viruses, which demonstrate the flexibility of interrelation between genetic material and cellular hosts. Such encounters must have compelled primitive cellular life forms to evolve complex defense systems and resulted in the diversification of the genetic scope of these primitive life forms. These evolutionary forces may have been responsible for the development of more organized cellular structures, eventually developing into the archaebacterial forms of bacteria and eukaryotes. The viral influence on genomic evolution emphasizes that RNA's role in early life was not merely foundational but rather an ongoing evolutionary catalyst that shaped the diversity of living organisms (Witzany, 2008). Moreover, another aspect of the significance of RNAs is reflected in the communication functions of the bacterial RNA systems. Bacteria, dating back almost 4 billion years, use RNA in addition to genetic coding to communicate with each other in sophisticated ways between cells, representing a high degree of social interaction behavior similar to that of multicellular beings. The presence of viruses increases this communicative competence since they can edit the bacterial genomes in a natural way. This kind of interaction supports the complexity of the web of relationships that supported early forms of life and points to the centrality of RNA to the functioning of individual organisms and the ecologies they inhabit. As RNA within bacteria continues to interact with viruses, the implications of these relationships reveal critical insights into the evolutionary pathways that led to the current biodiversity of life on Earth (Witzany, 2008) (Figure 3.9).

Figure 3.9. Artistic 3D rendering of RNA strands, representing their structure and role in molecular biology and gene expression.

Source: https://www.science.org/do/10.1126/science.adc9663/abs/_20220511_on_rna.jpg.

Test Your Knowledge

1. Write the steps of the formation of simple organic molecules.
2. What is the significance of RNA in early life?

3.3. The First Living Organisms

Learning Outcome

1. Characteristics of prokaryotic life;
2. The emergence of anaerobic organisms;
3. Evidence of the earliest life forms in the fossil record.

The emergence of the first living organisms marks the culminating point in the evolution of life on Earth, encapsulating the evolution from abiotic chemical reactions to the complexity of biological organisms. This is buttressed by chemical evolution theory, which postulates that life emerged when biogenic components came into contact with one another under natural mechanisms to produce more complex molecules. Of special interest, the environments on early Earth and on Mars are similar in ways conducive to such chemical evolution, leading to speculation that similar processes occurred on each planet. Martian environments, particularly its potential existence of liquid water and a more Earth-like atmosphere in the past, offer intriguing similarities with Earth. This credits the theory that building blocks of life could have been common throughout the universe, widening the scope for where life could have begun other than our Earth and supporting the fact that complex, replicable organic molecules are necessary to life's origins (Klein et al., 2025). The idea of earliest life is more than simple chemical operations; it is the formation of cellular structures that mirror the complexity of life. Cell theory, one of biological science's pillars, demonstrates that the earliest forms of life likely arose as simple prokaryotes, existing in a world teeming with potential for stress of evolution. This perspective turns the traditional notion on its head by suggesting that even the first living organisms had already built sophisticated interactions with their surroundings. Non-cellular, non-living containers are imagined by the consolidation scenario for the origin of life to have been incubators of crucial biological processes toward the emergence of ancient genetic systems. Coupled replication, transcription, and translation processes are supposed to have been evolving inside these containers and ultimately led to more sophisticated forms of life. This dynamic interplay between evolving cellular functions and the environment underscores the significance of structure and systems in the earliest living forms (Claudiu, 2009).

Figure 3.10. The first living organisms.

Source: https://www.popsci.com/wp-content/uploads/2019/03/18/2FWBZWLDLUJQJPAWP7YAH3UVRQ.png?quality=85.

The evolutionary process from these most ancient living forms set the stage for the diversification of living forms that followed. As cellular domains began to co-evolve, the foundation was formed for the origin of three primary lineages: Bacteria, Archaea, and Eukarya. Such diversification was not achieved through a linear route, but rather consisted of intricate interactions with viruses, which were crucial determinants of genetic evolution and the structural outline of life in its current form. The evolution of cellular life forms created genetic exchange and mutation mechanisms that allowed them to be compatible and survivable in changing environments. Thus, the study of the early life forms and their evolution provides in-depth information on the origin of complex life. Understanding how the early things evolved through processes into different biological entities assists in comprehending the big picture of life development on the planet Earth, providing insights that continue to inform the biological sciences (Claudiu, 2009) (Figure 3.10).

3.3.1. Characteristics of Prokaryotic Life

Prokaryotic organisms, such as bacteria and archaea, are the focus of research on the origin and early evolution of life. A cellular structure that lacks membrane-bound organelles or a fixed nucleus is one of the key characteristic features of prokaryotes. The minimal nature of their structure allows prokaryotes to proliferate rapidly and adapt rapidly to changes in their environment. Their small size, typically 0.2 to 10 micrometers, makes them efficient at taking in nutrients and eliminating wastes and also helps them survive in diverse environments. As the first forms of life on Earth, prokaryotes also assisted in setting biochemical pathways and systems, which subsequently were the foundation for more complex organisms. The simplicity of the cell could also be a sign of an early period in evolutionary times, providing us with information on how life started and evolved from individual cells to the complex organisms that inhabit the world today. Another fascinating aspect of prokaryotic organisms is their great amount of genetic flexibility, which occurs primarily through HGT mechanisms. In contrast to eukaryotes, prokaryotes can transfer genetic information directly by means of transformation, transduction, and conjugation mechanisms. This capability not only promotes genetic diversity but also allows for quick adaptation to environmental stress, e.g., antibiotic resistance. This kind of adaptability has important interpretations when considering evolutionary dynamics. The exponential growth of biological complexity seen in prokaryotes predicts that early life forms could have had a similar potential for genetic exchange and cooperation. The pace of complexity increase during early evolution should have been at least as rapid as in prokaryotes and eukaryotes and therefore suggests a complicated intertwine of interactions between the earliest life forms that conditioned the path of evolution over tens of billions of years (Sharov et al., 2006). Besides, prokaryotic life is characterized by a very high level of communicative competence, allowing for sophisticated social interaction in microbial communities. Such complexity can be seen in processes like biofilm formation, where bacteria coordinate their behavior to build multicellular groups that enhance their survival and resistance. Bacteria employ chemical signals to communicate and respond to their environment, engaging in what is called bio-communication. These are required for cooperative action, exchange of nutrients, and defense against predators. Current knowledge of the role of viruses in bacterial evolution suggests that viral interactions are capable of altering bacterial communication and genetic diversity and thus add yet another aspect of complexity to prokaryotic life. The realization of these characteristics not only emphasizes the evolutionary significance of prokaryotes but also highlights their underlying role in the process of ecosystems and biogeochemical cycles, maintaining the fabric of life on the planet Earth (Witzany, 2008).

3.3.2. The Emergence of Anaerobic Organisms

The appearance of anaerobic organisms marks a milestone in the evolution of life on our planet. The very first organisms to live without oxygen were an integral part of the biosphere long before the planet's atmospheres began to shift. The very first anaerobes developed chemical reactions that did not require oxygen, such as fermentation and anaerobic respiration, which enabled them to tap energy sources within their ancient environments. Since oxygen was scarce at the time, these organisms had evolved to survive in an atmosphere that was reducing, which shaped their metabolic systems. This development was essential for the subsequent development of more complex life forms as anaerobes laid the groundwork for an array of ecological systems, allowing for microenvironments where varying living organisms could act and interact, hence promoting biological complexity. Anaerobic life evolved directly due to many geological and chemical processes on early Earth. Free oxygen was missing in the atmosphere, so the free-oxygen-poor biosphere housed mostly anaerobic life such as methanogens, sulfate-reducing bacteria, and fermentative bacteria. These microbes metabolized major elements in the form of carbon, sulfur, and nitrogen, which affected Earth's geochemical landscape significantly. One of the most significant characteristics of this stage was the development towards oxygenic photosynthesis, which some evidence shows occurred around 2.7 billion years ago. But before this, anaerobic microorganisms could reproduce and not be competed against by oxygen-based life and continue to thrive in a variety of niches. Such adaptability would finally be challenged by the discovery of oxygen-evolving cyanobacteria, which transformed the atmosphere and consequently rendered most anaerobic routes infeasible, illustrating the dynamic rivalry among different forms of life throughout the Earth's history. Recent research has proposed intriguing hypotheses regarding the evolutionary development of phototrophy that likely include initial anaerobic organisms utilizing primitive light-harvesting strategies. Specifically, hypotheses suggest that retinal-based phototrophy could have appeared prior to the sophisticated systems of photosynthesis that currently dominate the scene. Research indicates that these primitive light-harvesting arrangements, such as those in halophilic Archaea, could have been at their pinnacle in early evolutionary stages before the evolutionary quantum leap of cyanobacteria and their oxygenic photosynthesis. The intricacy of these processes illustrates how anaerobic life contributed to the metabolic tapestry of early existence and how their existence predisposed the universe to the eventual development of more elaborate oxygen-based life. Such evolutionary developments are a mirror of the revolutionary developments in the energy sources of life, which themselves led the path of biological evolution on Earth back to the very pioneering role of anaerobic organisms in the early life (Kirschvink et al., 2008; DasSarma et al., 2018) (Figure 3.11).

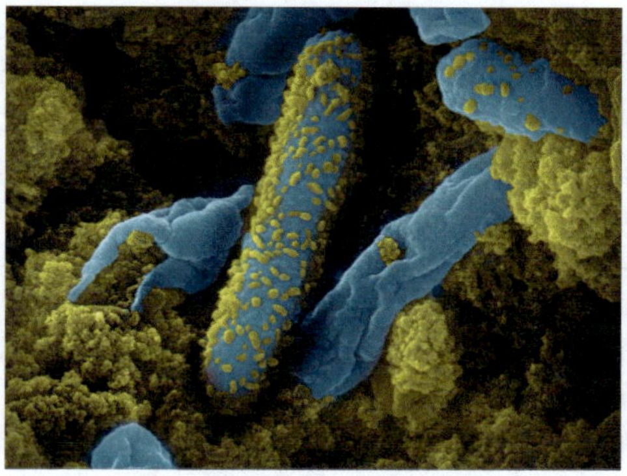

Figure 3.11. The emergence of anaerobic organisms.

Source: https://yerun.eu/wp-content/uploads/2021/06/SO-10ml-5-color-yellow-v3-1.jpg.

3.3.3. Evidence of the Earliest Life Forms in the Fossil Record

The fossil record has been a very important source of data about the first living organisms on earth, and it sheds light on the biological and ecological complexity of those living organisms. The earliest known microfossils are stromatolites and other prokaryotic structures dating back perhaps to 3.5 billion years ago. Such algae-like structures are characteristic of ancient microbial organisms that lived in the ocean. The existence of stromatolites indicates that photosynthetic cyanobacteria were also among the first organisms to change the atmosphere of the earth to produce oxygen as a byproduct of their metabolic activities. This ancient life form not only had a significant influence on the biosphere of the planet but also helped to create the Great Oxidation Event, which had a major effect on the climate and evolutionary trend of the earth, an example of how biological evolution and environmental change dynamically interact. Additional analysis of the fossil record throughout the Precambrian time shows a more complicated view of early life, especially in the Neoproterozoic era. The fossils like Ediacaran biota contain organisms of diverse shapes, sizes, and inferred lifestyles, which show the evolutionary novelty that preceded the Cambrian explosion. It is a transitional period because these multicellular organisms are a clue to the earliest experiments with biological complexity and multicellularity. The variety depicted by the Ediacaran fossils has cast doubt on the evolutionary relationship to the modern animals and has complicated the previous assumptions of the simplicity of the early life forms. The importance of this discovery in regard to the subsequent development of the bilaterian organisms is critical to the understanding of the pathways that resulted in the major morphological innovations of the later organisms, implying that the history of life is much more complex than the previous models assumed. The convolution of meanings regarding these ancient fossils presents the extant controversies in paleobiology and evolutionary biology. With new fossil evidence being discovered by the researchers, the challenge lies in the need to harmonize the anatomical similarities and differences of the early organisms. This can be equated to the debate on genetic homologies that are observed in contemporary species as pointed out in different literature. Such as the evolutionary tree that assumes a last common ancestor of protostome-deuterostome which implies that even simple organisms were not so simple. These discoveries stretch our knowledge regarding how and why life evolved in the first place and enable scientists to improve their methods of investigating the process of biological evolution within the history of Earth. The fossil record thus remains a vital source for reconstructing the profound narrative of life's origins and its early evolutionary steps, as scientists strive to piece together the intricate puzzles of our biological past (Burgess et al., 2015; Erwin & Davidson, 2002) (Figure 3.12).

Figure 3.12. Evidence of the earliest life forms in the fossil record.

Source: https://petrostrat.com/wp-content/uploads/2022/04/Living-colony-of-stromatolites-at-Lake-Thetis-near-Cervantes-Western-Australia-Photo-By-Ruth-Ellison-CC-BY-2.0.jpg.

Test Your Knowledge

1. What are the characteristics of prokaryotic life?
2. Focus on the emergence of anaerobic organisms.

3.4. The Evolution of Photosynthesis

Learning Outcome

1. The significance of cyanobacteria;
2. Impact on Earth's atmosphere;
3. The role of photosynthesis in the evolution of life.

The evolution of photosynthesis is the turning page of the book of life on Earth, which has its effects not only on biological systems but also on the geochemistry of the Earth. Photosynthesis tends to be divided into anoxygenic and oxygenic types. Anoxygenic photosynthesis, which is widespread in certain bacteria, makes use of alternative electron donors and releases no oxygen, while oxygenic photosynthesis uses water as an electron donor, with the result being oxygen liberation. The discovery of oxygenic photosynthesis is particularly significant, as not only did it remake the atmospheric composition, but it also enabled oxygen-dependent organisms to become prolific, thereby creating the complexity of life that exists today. Scholars such as Sam Granick have underscored that the development of these processes is complicated and contentious, leading to a host of speculative ideas that attempt to explain their origins and evolutionary paths (Londono et al., 2019). This, despite the existence of photosynthesis as the backbone of the biosphere, leaves questions about its origins. The timing and mechanistic processes of the evolution of oxygenic photosynthesis, traditionally attributed to Cyanobacteria, remain controversial among scientists. Recent findings imply that uncovering the history of photosynthesis may hinge upon the discovery of significant genetic breakthroughs, such as the development of elaborate reaction centers with the capacity to oxidize water. The unique structure of Photosystem I, comprising a heterodimeric core with PsaA and PsaB subunits, is a significant evolutionary innovation. This development is forecasted to allow for greater capacities for resistance to ROS that could have facilitated more effective photosynthetic functions (Londono et al., 2018).

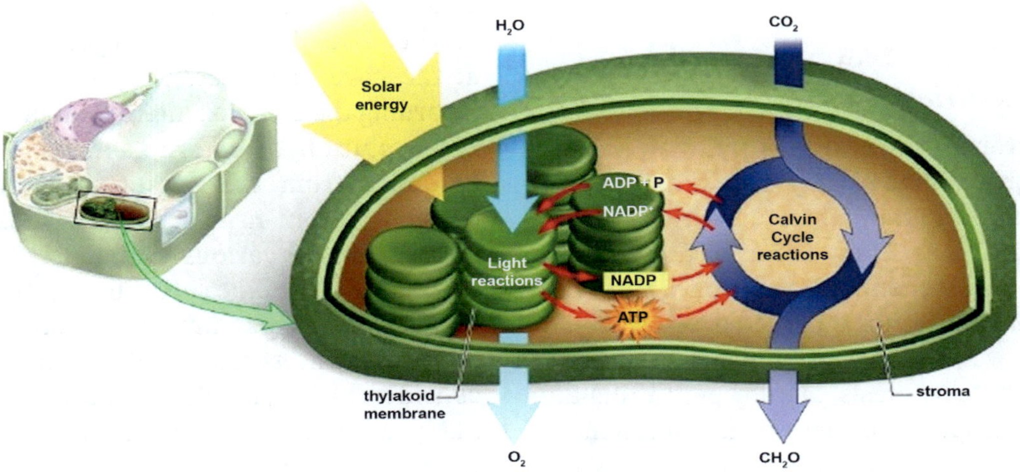

Figure 3.13. Evolution of photosynthesis.

Source: https://fos.cmb.ac.lk/blog/wp-content/uploads/2023/01/511700FG0010.jpg.

This consequently implies that the evolutionary history of photosynthesis is inextricably entwined with the larger narrative of life's adaptation to fluctuating environmental forces. Ongoing research into the evolution of photosynthesis requires there to be a critical assessment of existing theories to separate speculative concepts from fact-based propositions. Granick's analysis of assumptions regarding the phylogenetic history of oxygenic and anoxygenic photosynthesis illustrates the necessity of rigorous empirical testing. The usual assumption, for instance, that anoxygenic photosynthesis preceded oxygenic processes is suspect, according to new evidence potentially reinterpreting these timeframes. With ongoing research through the application of molecular clocks and genetic sequencing, they may uncover the previously evolved stages of photosynthesis and the role it played in the development of complex life. This ongoing research not only adds to our knowledge of photosynthetic evolution but also reflects the intricate web of interactions that typify the development of life itself, shedding light on the path to a deeper grasp of our planet's past and future (Londono et al., 2019) (Figure 3.13).

3.4.1. The Significance of Cyanobacteria

Cyanobacteria, or more popularly known as blue-green algae, play the fundamental role of shaping the history of life's origin and ancient evolution because they singly imparted biogeochemical cycles on Earth. The microorganisms, which emerged over 2.5 billion years ago, were the first organisms to perform oxygenic photosynthesis, thereby permitting the planet's atmosphere to change. The evolutionary significance of cyanobacteria extends beyond oxygen production; they were the function of primitive organisms to form nutrient-rich habitats that are conducive to more complex life forms. Cyanobacteria, through the process of metabolism of sunlight, water, and CO_2, formed organic material and released oxygen as waste, leading to the Great Oxidation Event. This groundbreaking transformation not only altered atmospheric conditions but also set the stage for the evolution of aerobic organisms and complex multicellular life, thus marking the biological landmark in Earth's history. Cyanobacteria's metabolic capabilities also illustrate the close association between life and non-equilibrium thermodynamic processes. The evolution of organic pigments in these organisms may be understood in terms of thermodynamic dissipation, where the need to optimize the dissipation of solar photons resulted in the building of more effective photosynthetic machinery. The original pigments absorbed light at specific wavelengths to optimize the collection of energy, a fundamental adaptation to the environmental conditions of the Archean Earth. The relevance of these metabolic activities is underscored by their ongoing function in modern ecosystems, where cyanobacteria are still key actors in the cycling of nutrients and productivity across habitats ranging from the ocean to freshwater. Furthermore, the phylogeny of cyanobacteria is a window into the early evolution of photosynthesis, confirming their status as a keystone clade in the TOL. In addition, cyanobacteria are at the core of evolutionary biology research regarding the complex interaction underpinning the evolution of chloroplasts in plant cells. Evidence from the evolutionary history of the D1 protein in the PSII, as substantiated in research on various cyanobacterial strains, emphasizes their role in photosynthesis innovation. As put into perspective by some of the recent research, D1 forms exhibited considerable variability, some of which most likely antedated the emergence of efficient water oxidation processes. This evolutionary history illustrates how cyanobacteria not only adapted to their environments but also brought about the fundamental processes that later influenced other evolutionary pathways. These findings place into context necessary background for appreciating the significant biochemical advances that have been behind diverse ecosystems of planet Earth and further affirm cyanobacteria role as a central figure in the story of the history and evolution of life (Figure 3.14).

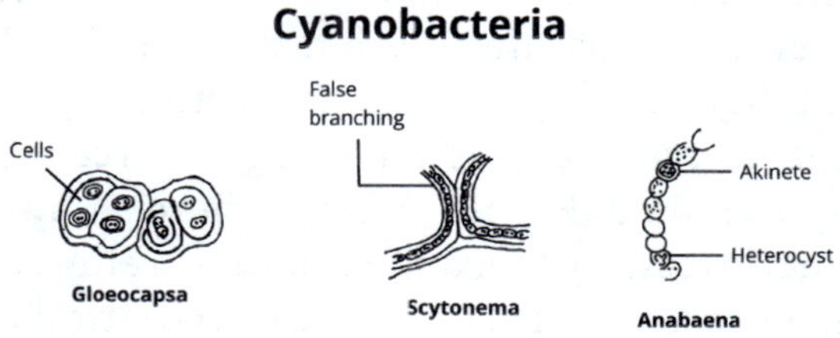

Figure 3.14. The significance of cyanobacteria.

Source: https://biologylearner.com/wp-content/uploads/2023/12/Cyanobacteria-diagram-1024x536.jpg.

3.4.2. Impact on Earth's Atmosphere

The early Earth's atmosphere also underwent dramatic changes that were crucial in shaping conditions that were conducive to life development. Following cosmic impacts that had pummeled the world, the atmosphere was rich in gases like CO_2, hydrogen (H_2), and methane (CH_4). Volatile compounds initiated a greenhouse effect that was probably responsible for maintaining surface temperatures above 500 K, despite intense cooling episodes following giant collisions. As the planet's surface cooled, it rained minerals and gases that would serve to initiate the geochemical processes that would fuel the development of organic compounds (Zahnle, 2025). The oscillation of oxidizing and reducing conditions of the atmosphere allowed the process of a broad array of chemical reactions that could have been responsible for the origin of prebiotic chemistry and, by process of extension, the origin of life. Surprisingly, atmospheric chemistry in these initial years was neither uniform nor always consistent. Several factors, such as the incident stellar radiation, atmospheric processes of escape, and geochemical evolution of the planet, determined the elemental makeup of the atmosphere. With the evolution of atmospheric composition over time, especially after impacts reduced in rate, the atmosphere evolved towards less reducing conditions. Last but not least, CO_2 was sequestered in the mantle, resulting in atmospheric stabilization, favorable to the evolution of biotic processes (Forget & Leconte, 2013).

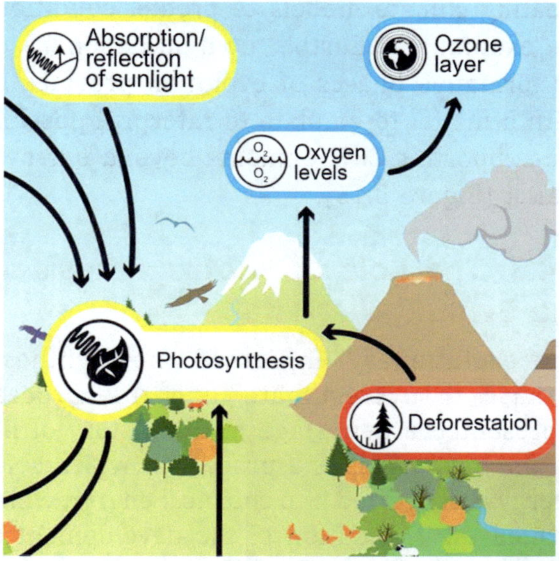

Figure 3.15. The evolution of photosynthesis impact on Earth's atmosphere.

Source: https://ugc.berkeley.edu/wp-content/uploads/2016/03/photosynthesis-sysmodel-cr.jpg.

The varying concentration of greenhouse gases and pressure and temperature fluctuations caused by mega-impacts and geological processes resulted in an active environment that influenced climate, as well as the possibility of life to evolve and survive. Lastly, the intricate interaction between Earth's changing atmosphere and the possibility of life on the planet cannot be stressed too much. The atmosphere initially provided a cradle for organic chemistry that would progressively lead to biogenesis with an almost total reliance on the volatile balance and external cosmic activity (Zahnle, 2025). In addition, as the evolving atmosphere unfolded, it prepared the stage for the more advanced biological systems ahead. Knowledge of these early atmospheric processes is important, not just to comprehend the origin of life on our Earth but also to ongoing research in looking for extraterrestrial homes for life off Earth (Forget & Leconte, 2013). As we go about creating climate models to predict conditions on exoplanets, Earth's own atmosphere in its formative phases of evolution provides us with a model to emulate to interpret possible life-supporting environments beyond our own planet (Figure 3.15).

3.4.3. The Role of Photosynthesis in the Evolution of Life

The evolutionary impact of photosynthesis is hard to overestimate, and this process represented a new stage in the history of life on the Earth. This is a process in which solar energy is converted into chemical energy, which formed the foundation of the development of complex life forms. The process of oxygenic photosynthesis, which was mainly caused by cyanobacteria 2.3 billion years ago, transformed the atmosphere of the planet significantly by raising the amount of oxygen. This transition, sometimes called the Great Oxidation Event, produced an environment favorable to aerobic life and higher multicellularity. The photosynthetic metabolic pathways not only made autotrophic organisms flourish but also created complex food webs that underlie the great diversity of life observed today, making the interdependence of these biological processes in the evolution story of life on Earth. When considering the ancestors to the photosynthetic systems, one can see that the simpler light-harvesting systems were necessary steps in the evolution of the more complex photosynthetic systems. Remarkably, the retinal-based systems, as found in archaea such as bacteriorhodopsin, are possible candidates for ancient phototrophy. The use of these systems probably formed the basis of the evolutionary step leading to the use of chlorophyll-based photosynthesis, as a half-way house to the complexity of oxygenic photosynthesis. This shift does not only concern the evolutionary lineage but also marks the flexibility of early life forms in utilizing the available resources as an energy source. The proposal of a Purple Earth, where simpler retinal-based systems dominated, reiterates the necessity of continuous evolutionary experimentation that eventually led to diverse photosynthetic pathways that we observe today (DasSarma et al., 2018).

This evolution experiment indicates the complexity of the relationship between organismal evolution and environmental change. Moreover, the diversification of photochemical reaction centers also gives vital information about the complicated evolutionary routes that occurred during the evolution of photosynthesis. Extensive phylogenetic investigations indicate that Type I and Type II reaction centers had appeared before the divergence of main bacterial phyla. It was these reaction centers that allowed the development of diverse photosynthetic strategies, increasing the ecological niches that could be exploited by organisms. The details of how chlorophyll synthesis first evolved are somewhat unclear, but the interactions of the various phototrophic bacteria, including the Heliobacteriaceae and Cyanobacteria, illustrate a complex network of evolutionary relationships that formed the basis of photosynthetic life today. The early diversification of these reaction centers underscores how the mechanisms of photosynthesis adapted to changing environmental pressures, thus playing a pivotal role in the evolutionary narrative of life on Earth (Boussac et al., 2014).

Test Your Knowledge

1. Write the significance of cyanobacteria.
2. What is the role of photosynthesis in the evolution of life?

3.5. The Rise of Eukaryotic Cells

Learning Outcome

1. Endosymbiotic theory;
2. Differences between prokaryotic and eukaryotic cells;
3. Evolutionary advantages of eukaryotic cells.

The origin of eukaryotic cells is an important page in the history of life on Earth with respect to which dramatic genomic and physiological alterations prevailed. Eukaryotic cells, with their complex structures, are thought to have resulted from a single prokaryotic endosymbiotic event that occurred about 1.6 to 2.1 billion years ago. This single process entailed the endosymbiotic uptake of a prokaryotic ancestral cell, later developed into mitochondria, into a host cell, which created a new biological paradigm. This endosymbiosis permitted a dramatic increase in cellular complexity because the host cell was able to take advantage of the enhanced metabolic efficiencies provided by the endosymbiont and thereby overcome the energy constraints imposed by prokaryotes. Such evolutionary cleverness not only enabled eukaryotes to thrive but also set the stage for the eventual development of multicellular life forms, which had a profound influence on the direction of life on Earth. The genomic implications of this endosymbiotic event were enormous. The merging of prokaryotic and eukaryotic genomes created a complete restructuring of genome structure and size. Compared to the prokaryotes, with their tight genomic structure subject to harsh selection for gene loss, the early eukaryotes possessed a huge genomic context with the reduced constraints of their endosymbionts. This diversity enabled more diversity of genetic attributes to be accumulated, supporting the evolution of complex cellular functions such as sexual reproduction and complex metabolic pathways. As the first eukaryotic cells began to exchange and combine genes through processes that exhibit the features of gene mixing within a prehistoric gene pool, they became increasingly differentiated from their prokaryotic ancestors. Lane et al. (2011) contend that this genomic differentiation, coupled with the selective advantages of endosymbiotic relationships propelled the evolution of eukaryotic life to diversify it as a crucial evolutionary innovation. Additionally, eukaryote evolution initiated advanced ecological processes that promoted the evolution of complex multicellular organisms and ecosystems. This ecological shift was strained by the unification of different forms of life, like symbiosis among various eukaryote and prokaryote systems. The postulation that eukaryogenesis evolved from a complex social structure of prokaryotic ancestors assumes an exceedingly connected evolutionary landscape. Through horizontal gene exchange and mutualistic activity, eukaryotic cells established the basis for complex life, quite distinct from the more sedentary prokaryotic evolutionary model. Lastly, these eukaryote-emergent turns of evolution not only disclose the chronology of life's origin but also depict the intricate tapestry of interactions that continue to propel biodiversity on Earth, as hypothesized in the emerging concept of an ancient virus world networked with cellular evolution (Dolja et al., 2006) (Figure 3.16).

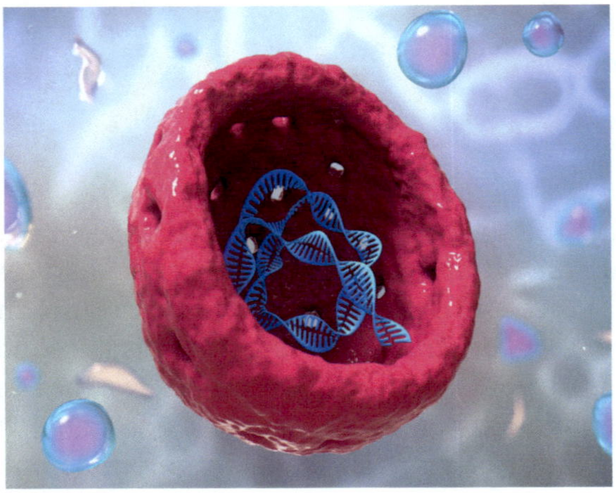

Figure 3.16. The rise of eukaryotic cells.

Source: https://cff2.earth.com/uploads/2019/08/13125027/Scientists-may-have-finally-uncovered-the-origin-of-eukaryotic-cells.jpg.

3.5.1. Endosymbiotic Theory

The theory of endosymbiosis provides a plausible account of how eukaryotic cells evolved, maintaining that the sophisticated organisms arose through symbiotic relationships between early prokaryotes. The theory states that early eukaryotic cells evolved when large host cells engulfed smaller prokaryotic cells, which, instead of being digested, established a symbiotic relationship with each other. These engulfed cells eventually formed into the required cellular components that are chloroplasts and mitochondria, which are responsible for energy production and photosynthesis, respectively. This situation not only describes the evolutionary transition from unicellular prokaryotes to multicellular eukaryotes but also raises the issue of cooperation and competition dynamics that facilitate evolutionary processes. Therefore, endosymbiotic theory has significantly changed the scientific comprehension of early organisms' complexity and integration of various species through evolutionary history. Besides, advancements in genomic and molecular biology have resulted in sufficient evidence to support the endosymbiotic theory. Comparative analyses of DNA sequences show impressive similarities within the mitochondria, chloroplasts, and a few prokaryotic organisms' genomes, which reveal their common heritage. For instance, mitochondria possess their own circular DNA and double up independently, just like bacterial cells. These kinds of findings are fundamental to the recasting of our understanding on life's origins, highlighting the importance of HGT and symbiosis in evolutionary science. In addition, the roles of bacteria in early evolutionary processes, as recorded in evidence of bacterial bio-communication, suggest that complex interdependencies were central to the evolution of multicellular life (Witzany, 2008). This evidence testifies to the reality that evolutionary novelty is not solely the result of random mutations but a sophisticated web of cooperative interactions. The implications of endosymbiotic theory extend beyond the mere portrayal of cellular evolution; they involve outcomes of entrenched notions of evolution defined by neo-Darwinian dogma, which are primarily concerned with genetic variation in populations. While the neo-Darwinian synthesis has been highly successful in describing most mechanisms of evolution, it typically does not address symbiosis and hybridization as significant factors (Beckley et al., 2017). The novel perspective espouses a broader view to comprehend evolutionary dynamics by embracing such ideas as polyploidy, chimerism, and lateral gene transfer. These are crucial to comprehend the complex relationships that enable evolutionary change. As researchers continue to explore the evolutionary significance of endosymbiosis and the intricate network of early life, it becomes increasingly clear that the emergence of life was accompanied by a great deal more cooperation and adaptability than has been heretofore realized (Figure 3.17).

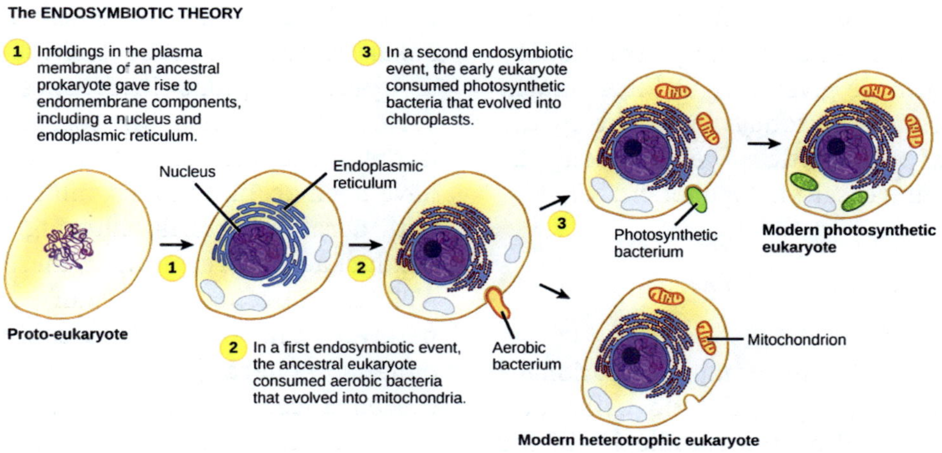

Figure 3.17. The endosymbiotic theory.

Source: https://bpb-us-e1.wpmucdn.com/sites.gatech.edu/dist/6/1810/files/2018/12/Endosymbiotic_theory-1024x510.jpg.

3.5.2. Differences Between Prokaryotic and Eukaryotic Cells

The distinctions that exist between the eukaryotic and prokaryotic cells lie at the heart of the process of the evolution of life on our planet. Prokaryotic cells, with bacteria and archaea as examples, are characterized by their simplicity and lack of a membrane-enclosed nucleus. Instead, their genetic material is located in a region of the cell referred to as the nucleoid, which lacks a membrane. This structural difference produces a more streamlined but less complex process of replication to allow prokaryotes to inhabit a vast variety of environments for nearly 4 billion years. Their remarkable adaptability and resilience are otherwise testified by their potential for HGT, which enables them to evolve and innovate quickly in other ecological niches, as seen by their involvement in such major evolutionary processes as the formation of DNA and the evolution of cellular organisms (Witzany, 2008). Eukaryotic cells, however, are more intricately organized with membrane-bound organelles such as the nucleus, mitochondria, and ER. This compartmentalization allows for specialized functions, thereby optimizing the efficiency of cellular processes in general. The presence of a cytoskeleton complements such complexity, allowing for movement and support. In addition, eukaryotes have a linear organization of chromosomes, a significant departure from prokaryotic circular DNA. This telomere evolution, the ability to perceive natural chromosome ends and contribute to genomic stability, is an adaptation whose genesis lies in ancient retroviral functions (Witzany, 2007). Such structural advancements are not only a direction towards greater biological complexity but also the emergence of multicellular life forms. The difference between prokaryotic and eukaryotic cells illuminates the origin of complex life forms and the evolutionary pathways leading us to today's diversity. Prokaryotic cells are basically unicellular, yet they have remarkable organizational capabilities and at times coordinate behaviors like multicellular animals via complex bio-communication systems. In contrast, the eukaryotic clade has given rise to multicellularity and the development of complex life forms like plants, animals, and fungi. The significance of these differences in evolution cannot be overemphasized; they mirror the differing survival and adaptational strategies of organisms. An understanding of such cellular differences gives clues to the early evolutionary processes that have given rise to life on Earth as we understand it today, highlighting the significance of prokaryotic as well as eukaryotic forms in the universal phylogenetic tree.

3.5.3. Evolutionary Advantages of Eukaryotic Cells

The eukaryotic cells are a major evolutionary development of the prokaryotic cells, and this is mainly based on the complexity of the structural organization. Among the key benefits, one can distinguish the compartmentalization of cellular processes by organelles bounded by membranes. This specialization enables the eukaryotic cells to devote specific cell functions to specific regions increasing efficiency of various biochemical processes. An example is the availability of mitochondria thought to have been acquired in a symbiotic relationship with aerobic bacteria, allowing these cells to conduct aerobic respiration, thereby producing ATP more efficiently than the anaerobic pathways that prokaryotes could utilize. This enhances energy production that facilitates the enhanced metabolism and overall fitness of eukaryotic organisms to enable them to live in a variety of environments over time, which is the evolutionary success that results from the complexity of the cell.

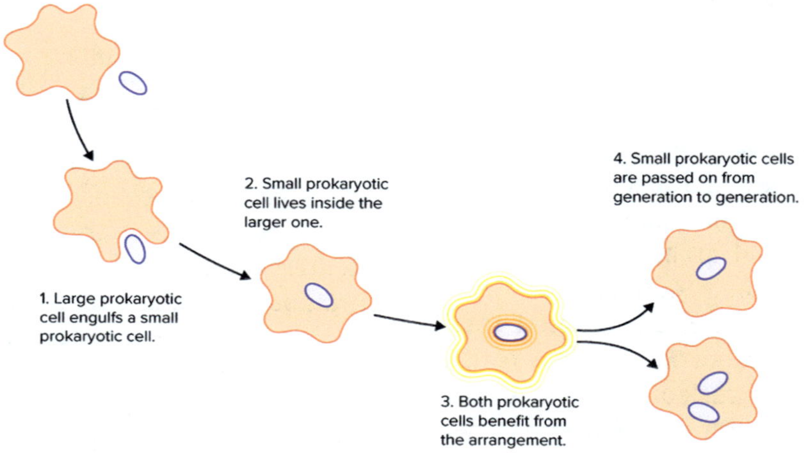

Figure 3.18. Evolutionary advantages of eukaryotic cells.

Source: https://dr282zn36sxxg.cloudfront.net/datastreams/f-d%3A830e9efdcc70f2dc17ef55fba269ad0b2f5d12dd3482bed9d7eb0bbe%2BIMAGE_TINY%2BIMAGE_TINY.1.

The appearance of the eukaryotic cells also simplified the evolution of multicellularity which offered evolutionary benefits regarding survival and adaptation. The size and complexity of multicellular organisms can be larger, and the result is usually an increase in the capacity to utilize different ecological niches. The transition facilitates cellular cooperation and allows cells to act as a unit and to adopt more complex reproductive tactics. Moreover, eukaryotic cells can form mutualistic relationships, which increase the resistance to environmental fluctuations. As illustrated by the mitochondrion and the eukaryotic cell, mutualisms are vital for increasing fitness and fostering co-evolution between various cellular components, leading to the development of innovative biological strategies and responses to ecological pressures (Lanier et al., 2017). These dynamics are typical of how the development of eukaryotic cells opens the door to the progressive forms of life. Besides these benefits, eukaryotic cells have more genetic regulation and transcriptional control, which is essential in their evolutionary success. The availability of a nucleus enables protection and organization of genetic material, resulting in better processes in gene expression and regulation over prokaryotic systems. This complexity gives the possibility of increased phenotypic diversity by alternative splicing and post-translational modifications, enabling eukaryotes to respond more easily to a changing environment. In addition, the ability to undergo HGT, albeit at a lesser degree than prokaryotes, allows eukaryotes to adapt to other organisms' positive features, increasing their adaptability even more. Collectively, these factors illustrate how the evolutionary advantages conferred by eukaryotic cells have been instrumental in shaping the trajectory of life on Earth and underscore their critical role in the origin of more complex life forms (Witzany, 2008) (Figure 3.18).

Test Your Knowledge

1. What is endosymbiotic theory?
2. What are the evolutionary advantages of eukaryotic cells?

3.6. Major Evolutionary Milestones

Learning Outcome

1. The Cambrian explosion;
2. The transition from water to land;
3. The development of multicellular organisms.

The evolution of life on Earth was marked by a series of critical evolutionary events that radically altered the course of biological history. One of these events was the transition from simple molecules to systems of self-replication, which laid the foundation for cellular life. This evolutionary process was facilitated by specific environmental conditions, including the presence of liquid water and multiple energy sources, such as UV radiation and hydrothermal vents. In a surprising twist, such energy flows are not only important for life, but also for the cosmos as a whole, since interdependence between energy and complexity has been observed in numerous systems across the universe. The ability of primitive organisms to exploit available energy significantly enhanced their survival prospects and capacity for adaptation, ultimately leading to more complex multicellular organisms that would dominate the Earth's ecosystems (Chaisson et al., 2015). Another key evolutionary milestone was the evolution of photosynthesis, which revolutionized the planet's atmosphere and made possible the diversification of life. Early cyanobacteria used sunlight to synthesize glucose from CO_2 and water, producing oxygen as a waste product. This 'Great Oxygenation Event' around 2.4 billion years ago had a profound impact on atmospheric oxygen levels, making it possible for diverse aerobic organisms to diversify. The oxygen-rich atmosphere thus created permitted increased complexity of life, with conditions favorable to cellular respiration and hence more efficient energy generation than by anaerobic fermentation. The interplay of the complexity of energy conversion and evolutionary success demonstrates just how central energy management has been to the life of the biosphere of Earth during this pivotal evolutionary time (Abbott et al., 2016). One more landmark of early evolution was the advent of multicellularity, facilitating organisms to evolve specialization and complexity above single-celled states. This process allowed for cells to interact and integrate into each other, and from it emerged tissues and organs, thereby enhancing overall function and adaptability to different environments. Evolution established numerous pathways for multicellularity, most notably in such groups as plants, fungi, and animals, each evolving independently to take advantage of ecological niches. The ability to form multicellular structures results in increased utilization of resources and energy efficiency, required to thrive under competitive conditions. Multicellularity's emergence is a significant step forward in the evolution of life, demonstrating how evolutionary pressures have time and again evolved organisms to maximize the usage of energy to ensure their survival and evolutionary success in simple yet complex ecological situations (Chaisson et al., 2015).

3.6.1. The Cambrian Explosion

The Cambrian explosion is an evolutionary milestone in the history of life on Earth, a burst of diversity and complexity of organisms 542 million years ago. The Cambrian explosion is characterized by the sudden advent of numerous different groups of multicelled animals in the fossil record, including primitive vertebrates and other types of invertebrates. These rapid changes imply a revolutionary evolutionary transition that included not just morphological innovation but also profound genetic advances. While there are a number of theories to explain this event, it is an evocative subject in evolutionary biology, as scientists remain contentious regarding the timing and nature of the origin of metazoans. The interaction between genetic, environmental, and ecological factors has emerged as the core field of study in present times, pointing towards a multi-aspected view to comprehend the complexity of the Cambrian explosion (Adami et al., 2008). Analyzing the genomic developments during the Cambrian explosion throws light on the intricate relationship between species development and the genetic factors. New insights from molecular biology reveal that profound changes in genome size and organizational framework may have been forces behind advanced life form evolution at this period. For instance, the formation of gene regulatory networks and improvement of signaling pathways would likely have permitted a new scale of complexity, allowing diversification of various body plans and functions. The emergence of development-controlling primordial mechanisms in early metazoans may have foreshadowed the evolution of more advanced features in subsequent vertebrates. This genomic process has significant implications for understanding life's evolutionary trajectory, placing focus on the Cambrian explosion as an age of diversification but also as a critical juncture in the evolution of genes (Adami et al., 2008). The appearance of vertebrates in the Cambrian explosion underscores the importance of evolutionary innovations, particularly concerning steroid hormone control by steroid receptors. These receptors were crucial to the evolutionary achievement of vertebrates, which first appeared in the fossil record toward the start of this sudden radiation of life. The presence of adrenal and sex steroid receptors in ancient vertebrate precursors represents an evolutionary advance that made possible high-level regulatory command over physiological functions, remains important to growth, reproductive functions, and responses to stress. Such forms of regulatory systems, though absent in more primitive animals like echinoderms and hemichordates, highlight the complex coupling between gene evolution and adaptation to the environment. Investigating their beginnings and how they developed brings illumination to an expanding tale of adaptive success within vertebrates and forever ties the Cambrian explosion to the fundamental mechanisms that laid out subsequent evolutionary trajectories (Figure 3.19).

Figure 3.19. The Cambrian explosion.

Source: https://static.scientificamerican.com/sciam/cache/file/0FD8D291-6299-419E-A362FD23A21ED41A_source.jpg?crop=1%3A1%2Csmart&w=1000.

3.6.2. The Transition from Water to Land

Terrestrialization was a milestone process in the evolutionary history of life, a document of both adaptive radiations and significant biological innovations. Initially, living organisms flourished in aquatic environments, where physical conditions offered certain advantages, such as buoyancy and uniform temperatures. Afterwards, once some organisms transitioned onto the land, they were faced with novel constraints such as desiccation and the need for structural support. The gradual transition to terrestrial life necessitated enormous physiological and morphological changes. Upon the ancestors of today's vegetation and animals reaching land, shifts in reproduction strategies and developmental processes became crucial, ultimately leading to the proliferation of various ecosystems. Here, the process of homeostasis emerges; it was not merely a synchronic response to environmental modification but a dynamic, multigenerational process that altered organismal form and function in response to living in new environments (Torday, 2015). One of the key evolutionary novelties that accompanied the transition to land was the discovery of sexual reproduction. A shift towards multicellular organisms led to complex interactions among organisms and their environment, transforming life strategies in radical ways. Founded on asexual reproduction initially, species need to develop more sophisticated reproductive processes in adapting to terrestrial environments. The development of meiosis announced an age with greater genetic diversity, able to burst, consolidating populations' resistance to fluctuating environments. As organisms transitioned to dry land, the evolution of sexual reproduction allowed for genome renewal and the establishment of new cycles of reproduction, intricate fertilization systems, and greater organismal interaction. This combination of sexual reproduction and the capacity to respond to the environment opened the door to the major evolutionary diversifications that were to ensue (Willemse, 2009). Lastly, the transition from water to land shaped the early evolutionary history by establishing important biological and ecological associations. Evolution of plant vascular tissues and gas exchange and nutrient uptake adaptations, for instance, exemplify how terrestrial plant life started to utilize its environment more efficiently. Additionally, the shift brought about the creation of new complex interdependencies between taxa, with species starting to interact in novel ways that affected food webs and ecosystem structure. The biotic dispersal machinery in place during this period enabled life to spread among various terrestrial environments. As plants and animals developed specialized adaptations to live on land, this line of development underscored the intricate interdependence of life with its fluctuating environment, emphasizing communication and interaction as the cornerstones of terrestrial development (Willemse, 2009) (Figure 3.20).

Figure 3.20. The transition from water to land.

Source: https://miro.medium.com/v2/resize:fit:6400/1*6otm2cqfJ3flqikJ7AQUYg.jpeg.

3.6.3. The Development of Multicellular Organisms

The development of multicelled organisms can be considered one of the major leaps in the history of life, which had simple unicellular predecessors. This transition is closely connected with the emergence of specialized cellular functions, allowing the complex formation of tissues and organs. The early multicellular organisms had diverse morphogenetic processes, which were vital in the coordination of cell differentiation and cell communication. As an example, cell signaling pathways started to evolve, which enabled cells to act in cooperation, not in isolation, which is a key to the evolution of multicellularity. Research has shown that the formation of these morphogenetic mechanisms played a crucial role in the emergence of multicellular organisms, thereby enhancing survival and adaptability in diverse environments (Dovgal et al., 2010). These breakthroughs highlight how extraordinary the evolution tactics of life forms have been, leading to a more complicated biological structure and interplay. In addition to morphogenetic innovation, HGT among unicellular organisms significantly influenced the evolutionary trajectory of multicellular life. HGT offered the possibility of genetic exchange across conventional reproduction, allowing rapid response to the changing environment. The interplay of HGT and common descent presents a complex network of evolutionary patterns, highlighting that lineage diversification is not exclusively tree-like. Instead, the complexities of life's evolution, as seen in organisms reliant on horizontally transferred genes, reveal a dynamic interplay of genetic information shaping early multicellular development (Espinosa et al., 2010). The study of these processes deepens our knowledge of the way multicellular organisms have emerged and remained in the evolutionary history, and how cooperation and adaptation contributed to the diversification of life. Multicellular development has implications that are not restricted to evolutionary biology, and it has effects on ecological interactions and the progression of life forms on Earth. The innovation of multicellularity enabled the formation of more complicated ecosystems, in which organisms could have various trophic levels, like predators, producers, and decomposers. This shift led to the complex interrelationships between species, which encouraged dynamic equilibrium in the ecosystems. With the diversification of multicellular organisms, they have played a significant role in the formation of environments due to the different processes, such as nutrient cycling and energy flow. The combination of multicellularity and ecological complexity shows the drastic effect of evolutionary innovations, which supports the idea that the evolution of multicellular organisms was not just a biological improvement but a revolution in the history of life on Earth.

Test Your Knowledge

1. What is the Cambrian explosion?
2. What is the development of multicellular organisms?

SUMMARY

The search for the origin of life and early evolution has revealed sophisticated dynamics behind the emergence of biological complexity. The exponential increase of functional non-redundant genome size over time is a testament to the profound interaction between evolutionary time and biological complexity. As predicted by the exponential hypothesis, this increase exhibits correlation among major organismal groups and indicates an overall tendency in the evolutionary process. Exactly, the regression of genome size over time reveals a staggering 7.8-fold increase per billion years in agreement with the proposition that early phases of evolution, particularly pre-prokaryotic life, have not outpaced subsequent advances in prokaryotic and eukaryotic lineages. Consequently, this key observation stimulates a reassessment of the timescales over which life might have originated, even to the point of considering panspermia along with traditional de novo hypotheses for the origin of life on Earth (Sharov et al., 2006). In addition, advances in molecular evolutionary understanding refine the debate around the origin of life by introducing concepts such as chiral selection. The identification of chiral molecules and the resulting enantioselection of preferred enantiomers highlight an intrinsic property of molecular assembly that may have been critical in the origin of early life processes. It is proposed by the graded autocatalysis replication domain (GARD) model that the dynamic nature of chiral symmetry breaking implies that discrimination of one enantiomer over the other may stem from the intrinsic properties of catalytic molecular networks rather than being based on an initially preordained condition. This new insight is perfectly in agreement with the perception that complex biochemical systems can self-organize under certain circumstances, and that the processes which gave rise to life's origin are more subtle and complex than previously thought (Kafri et al., 2010). It follows that integration of information from both the evolutionary history of genome complexity and molecular mechanisms of ancient biochemical interactions produces a strong explanation of life's origin. The exponential hypothesis not only describes the incremental growth of biological complexity but also offers an interdisciplinary exploration of the biophysical and chemical processes that may have been responsible for the emergence of life. Such thoughts suggest that the emergence of life is not an event of punctuated character, but a continuum characterized by intricate molecular interactions and evolutionary pressures. Grasping the two de novo and panspermia hypotheses brings the debate further, facilitating a greater understanding of life's origin and instigating logical scrutiny of alternative models that could more aptly provide solutions for the mysterious road from simple molecular assemblies to complex living organisms.

The realization of the history of life and early evolution raises numerous interrelated themes that shape our knowledge of biological diversity. Central to this discussion is metabolism, which is not just the basis for life but also a hierarchical and modular process that can allow the emergence of complex biochemical processes. Metabolic networks contain a remarkable degree of regularity, in that they have structures that are conserved between species and environments. Structural conservation permits evolutionary adaptation, a sign that metabolic architecture is the basis upon which higher orders of biological organization are constructed. How these metabolic modules operate sheds light on the ancient environment of early forms of life and can be used to account for evolution toward more complex organisms as outlined by Alberts et al. (2012). Archaeological and genetic data are central to the comprehension of early life on Earth, especially regarding the shift from simple to complex forms. An important change in archaeological views reflects a slow transition from hunter-gatherers to agriculturalists, stressing a more nuanced development of domestication instead of a sudden transition. This viewpoint has opened up new prospects for the explanation of genetic data since past explanations sometimes relied on old rapid transition models. The reinterpretation of genetic data as prolonged ecological and social processes requires an advanced understanding of diversity among early domesticated organisms. These interpretations contrast with previous findings and highlight the complexity of integrating multiple lines of evidence, as the results discussed by Allaby et al. (2010) illustrate. Additional study of these areas shows that evolution and the origins of life are also coupled to broader ecological and genetic processes. Evolution of metabolic networks and genetic diversity encompasses the multiplicity of biological change, biochemistry modularity, and environmental constraint. In acknowledging these interdependencies, scientists can weave a more integrated account of life's origins that is consonant with the multiplicity of processes supporting evolutionary innovation. The outcomes of the analysis of metabolic pathways coupled with archaeology provide a more detailed context for understanding the process that regulated early life and diversification. The integration of various data sets calls for a well-matured methodological approach, for instance, Bayesian estimations, indicating the complexity of biological evolution in the discussion board from both metabolic and genetic perspectives, exemplified by Alberts et al. (2012) and Allaby et al. (2010).

KEY TERMS FROM THIS CHAPTER

Abiogenesis	Anaerobic respiration	Cyanobacteria
Endosymbiotic theory	Eukaryotes	Extinction events
Fossil record	Hydrothermal vents	Miller-Urey experiment
Multicellularity	Organic molecules	Oxygen revolution
Panspermia hypothesis	Photosynthesis	Primordial soup
Prokaryotes	Protocells	RNA world hypothesis

CRITICAL THINKING QUESTIONS

1. **What was the likely composition of Earth's early atmosphere?**
   a. Oxygen and nitrogen
   b. Carbon dioxide, methane, ammonia, and water vapor
   c. Pure hydrogen and helium
   d. Nitrogen and ozone
2. **What experiment supported the idea that organic molecules could form under early Earth conditions?**
   a. Hershey-Chase experiment
   b. Griffith's experiment
   c. Miller-Urey experiment
   d. Meselson-Stahl experiment
3. **Which type of organism is thought to be the first form of life on Earth?**
   a. Multicellular eukaryotes
   b. Anaerobic prokaryotes
   c. Photosynthetic algae
   d. Fungi
4. **What does the endosymbiotic theory explain?**
   a. The origin of the DNA double helix
   b. How enzymes evolved
   c. How mitochondria and chloroplasts originated in eukaryotic cells
   d. The extinction of dinosaurs
5. **Which of the following was a major result of the evolution of photosynthesis?**
   a. Increase in volcanic activity
   b. Formation of the ozone layer
   c. Decrease in atmospheric nitrogen
   d. Cooling of Earth's core

Short Answer Questions

1. What conditions on early Earth made the origin of life possible?
2. What is the significance of the Miller-Urey experiment?
3. What are the hypotheses for how the first organic molecules formed?
4. What are protocells, and why are they important to our understanding of the origin of life?
5. What is the RNA world hypothesis?

Answers to Critical Thinking Questions

1. (b); 2. (c); 3. (b); 4. (c); 5. (b)

REFERENCES

1. Abbott, A., Allen, J., Bachman, M., Berghe, E. V., Berthold, A., Braun, M., Brockelman, W. Y., et al. (2016). Differences in early developmental rate and yolk conversion efficiency in offspring of trout with alternative life histories. https://core.ac.uk/download/42369200.pdf (accessed on 13th June 2025).
2. Adami, C., Bousso, R., Bromham, L., Carter, B., Conway Morris, S., Davidson, E. H., et al. (2008). The C-value enigma and timing of the Cambrian explosion. http://arxiv.org/abs/0806.0108 (accessed on 13th June 2025).
3. Alberts, B., Andreini, C., Balche, W. E., Barker, H. A., Benner, S. A., Berge, C., Braakman, R., et al. (2012). The compositional and evolutionary logic of metabolism. http://arxiv.org/abs/1207.5532 (accessed on 13th June 2025).
4. Allaby, R., Badr, R., Fuller, D. Q., Heun, M., Marjoram, P., Matsuoka, Y., Molina-Cano, J. L., et al. (2010). Integrating the processes in the evolutionary system of domestication. https://core.ac.uk/download/46660.pdf (accessed on 13th June 2025).
5. Alpern, B., Aplin, K. P., Becker, L., Bernal, B., Bishop, J. L., Brown, P. A., Bunch, T. E., et al. (2002). Clay mineral–organic matter relationships in the early solar system. https://core.ac.uk/download/12873.pdf (accessed on 13th June 2025).
6. Aranda-Anzaldo, A. (1998). On natural selection and Hume's second problem. https://core.ac.uk/download/131210719.pdf (accessed on 13th June 2025).
7. Arnold, S. J., Bejerano, G., Brodie, E. D., Hibbett, D., et al. (2013). Evolutionary biology for the 21st century. https://core.ac.uk/download/12174890.pdf (accessed on 13th June 2025).
8. Baker, M. E. (2019). Steroid receptors and vertebrate evolution. http://arxiv.org/abs/1901.09944 (accessed on 13th June 2025).
9. Ball, R., & Brindley, J. (2014). Hydrogen peroxide thermochemical oscillator as driver for primordial RNA replication. http://arxiv.org/abs/1402.3875 (accessed on 13th June 2025).
10. Bandea, C. I. (2009). A unifying scenario on the origin and evolution of cellular and viral domains. https://core.ac.uk/download/pdf/288874.pdf (accessed on 13th June 2025).
11. Beckley, C., & Bonillas, U. (2017). The evolution of diversity. https://core.ac.uk/download/131214501.pdf (accessed on 13th June 2025).
12. Blancke, S., Braeckman, J., Schellens, T., Soetaert, R., et al. (2013). From ends to causes (and back again) by metaphor: The paradox of natural selection. https://core.ac.uk/download/55800171.pdf (accessed on 13th June 2025).
13. Boussac, A., Cantrell, A., Guskov, A., Magnuson, A., Quaiser, A., Allwood, A. C., Clarke, A. K., et al. (2014). A fresh look at the evolution and diversification of photochemical reaction centers. https://core.ac.uk/download/76995652.pdf (accessed on 13th June 2025).
14. Brack, A., Brandenburg, A., Commeyras, A., Lazcano, A., Rode, B. M., Shen, C., et al. (2009). Toward homochiral protocells in noncatalytic peptide systems. http://arxiv.org/abs/0810.5398 (accessed on 13th June 2025).
15. Burgess, N. D., Church, A. H., Edwards, D., Fanning, U., Kenrick, P., et al. (2015). The early evolution of land plants, from fossils to genomics: A commentary on Lang (1937) 'On the plant-

remains from the Downtonian of England and Wales.' https://core.ac.uk/download/42046095.pdf (accessed on 13th June 2025).

16. Callahan, M. P., House, C. H., Roberts, M. R., Rodriguez, D., et al. (2019). Nitrogen heterocycles form peptide nucleic acid precursors in complex prebiotic mixtures. https://core.ac.uk/download/237180278.pdf (accessed on 13th June 2025).
17. Cardona Londono, T. (2018). Early Archean origin of heterodimeric Photosystem I. https://core.ac.uk/download/151191757.pdf (accessed on 13th June 2025).
18. Cardona Londono, T. (2019). Thinking twice about the evolution of photosynthesis. https://core.ac.uk/download/195780965.pdf (accessed on 13th June 2025).
19. Chaisson, E. J. (2015). Energy flows in low-entropy complex systems. http://arxiv.org/abs/1512.04981 (accessed on 13th June 2025).
20. DasSarma, S., & Schwieterman, E. W. (2018). Early evolution of purple retinal pigments on Earth and implications for exoplanet biosignatures. http://arxiv.org/abs/1810.05150 (accessed on 13th June 2025).
21. Dolja, V. V., Koonin, E. V., & Senkevich, T. G. (2006). The ancient virus wor*ld and evolution of cells. https://co*re.ac.uk/download/pdf/4072546.pdf (accessed on 13th June 2025).
22. Dovgal, I. V. (2010). The evolutionary reorganization of ontogeny and origin of multicellularity. https://core.ac.uk/download/12086867.pdf (accessed on 13th June 2025).
23. Erwin, D. H., & Davidson, E. H. (2002). The last common bilaterian ancestor. Development (Cambridge, England), 129(13), 3021–3032. https://doi.org/10.1242/dev.129.13.3021(accessed on 13th June 2025).
24. Espinosa, A., & Paz-y-Mino C., G. (2010). Integrating horizontal gene transfer and common descent to depict evolution and contrast it with "common design." https://core.ac.uk/download/56704917.pdf (accessed on 13th June 2025).
25. Forget, F., & Leconte, J. (2013). Possible climates on terrestrial exoplanets. http://arxiv.org/abs/1311.3101 (accessed on 13th June 2025).
26. Gibbons, A., Lal, A. K., Bohler, C., Chyba, C., Huber, C., Manning, C. E., Salter, C. J., et al. (2012). Origin of life. http://arxiv.org/abs/0907.3552 (accessed on 13th June 2025).
27. Gibson, C. H., Schild, R. E., & Wickramasinghe, N. C. (2010). The origin of life from primordial planets. http://arxiv.org/abs/1004.0504 (accessed on 13th June 2025).
28. Hoffman, D. B., & Kontaratos, A. N. (2025). Biological generalizations and the search for extraterrestrial life. https://core.ac.uk/download/pdf/80660612.pdf (accessed on 13th June 2025).
29. Ilachinski, A., Pavlov, A. A., Knoll, A. H., Wood, A. J., Wallace, A. R., Carter, B., et al. (2012). Astrobiological complexity with probabilistic cellular automata. http://arxiv.org/abs/1206.3467 (accessed on 13th June 2025).
30. Kafri, R., Lancet, D., & Markovitch, O. (2010). Spontaneous chiral symmetry breaking in early molecular networks. https://core.ac.uk/download/pdf/8424893.pdf (accessed on 13th June 2025).
31. Kirschvink, J. L., & Kopp, R. E. (2008). Paleoproterozoic icehouses and the evolution of oxygen mediating enzymes: The case for a late origin of Photosystem II. https://core.ac.uk/download/4878076.pdf (accessed on 13th June 2025).

32. Klein, H. P. (2025). Continuing the biological exploration of Mars. https://core.ac.uk/download/pdf/42829304.pdf (accessed on 13th June 2025).

33. Kusch, M. (2019). *Georg Simmel and Pragmatism*. https://core.ac.uk/download/223236234.pdf (accessed on 13th June 2025).

34. Lane, N. (2011). Energetics and genetics across the prokaryote–eukaryote divide. https://core.ac.uk/download/pdf/8604677.pdf (accessed on 13th June 2025).

35. Lanier, K. A., & Williams, L. D. (2017). The origin of life: Models and data. https://core.ac.uk/download/323062617.pdf (accessed on 13th June 2025).

36. Lanier, K. A., Petrov, A. S., & Williams, L. D. (2017). The central symbiosis of molecular biology: Molecules in mutualism. https://core.ac.uk/download/323063216.pdf (accessed on 13th June 2025).

37. Lynch, J., Lerner, L. S., Schwartz, M., Gross, P. R., Schwartz, R., & Goodenough, U. (2012). The state of state science standards 2012. https://core.ac.uk/download/71359935.pdf (accessed on 13th June 2025).

38. Martins, R. P. (2018). The conceptual structure of evolutionary biology: A framework from phenotypic plasticity. https://core.ac.uk/download/286139307.pdf (accessed on 13th June 2025).

39. Michaelian, K., & Simeonov, A. (2014). Fundamental molecules of life are pigments which arose and evolved to dissipate the solar spectrum. http://arxiv.org/abs/1405.4059 (accessed on 13th June 2025).

40. Reiss, M. (2009). The relationship between evolutionary biology and religion. https://core.ac.uk/download/pdf/83602.pdf (accessed on 13th June 2025).

41. Rutherford, A. W., Anisimova, M., Arato, A., Arnold, E., Bader, R., Beanland, T. J., Blankenship, R. E., et al. (2015). Origin and evolution of water oxidation before the last common ancestor of the Cyanobacteria. https://core.ac.uk/download/76996571.pdf (accessed on 13th June 2025).

42. Sharov, A. A. (2006). Genome increase as a clock for the origin and evolution of life. https://core.ac.uk/download/pdf/4011443.pdf (accessed on 13th June 2025).

43. Sydow, M. von. (2001). *Sociobiology, Universal Darwinism and Their Transcendence: An Investigation of The History, Philosophy and Critique of Darwinian Paradigms, Especially Gene-Darwinism, Process-Darwinism, And Their Types of Reductionism Towards a Theory of The Evolution of Evolutionary Processes, Evolutionary Freedom and Ecological Idealism*. https://core.ac.uk/download/6116908.pdf (accessed on 13th June 2025).

44. Torday, J. S. (2015). Homeostasis as the mechanism of evolution. https://core.ac.uk/download/323075985.pdf (accessed on 13th June 2025).

45. Umantsev, A. (2011). On the origin of biological functions. https://core.ac.uk/download/56318720.pdf (accessed on 13th June 2025).

46. Walker, S. (2014). Top-down causation and the rise of information in the emergence of life. https://core.ac.uk/download/97831534.pdf (accessed on 13th June 2025).

47. Willemse, M. T. M. (2009). Evolution of plant reproduction: From fusion and dispersal to interaction and communication. https://core.ac.uk/download/pdf/29248993.pdf (accessed on 13th June 2025).

48. Witzany, G. (2007). Telomeres in evolution and development from biosemiotics perspective. https://core.ac.uk/download/pdf/287482.pdf (accessed on 13th June 2025).
49. Witzany, G. (2008). Bio-communication of bacteria and its evolutionary interrelations to natural genome editing competences of viruses. https://core.ac.uk/download/pdf/287751.pdf (accessed on 13th June 2025).
50. Zahnle, K. (2025). Earth's early reduced atmospheres. https://core.ac.uk/download/pdf/288485289.pdf (accessed on 13th June 2025).

SURVEY OF PROKARYOTES AND PROTISTS

4

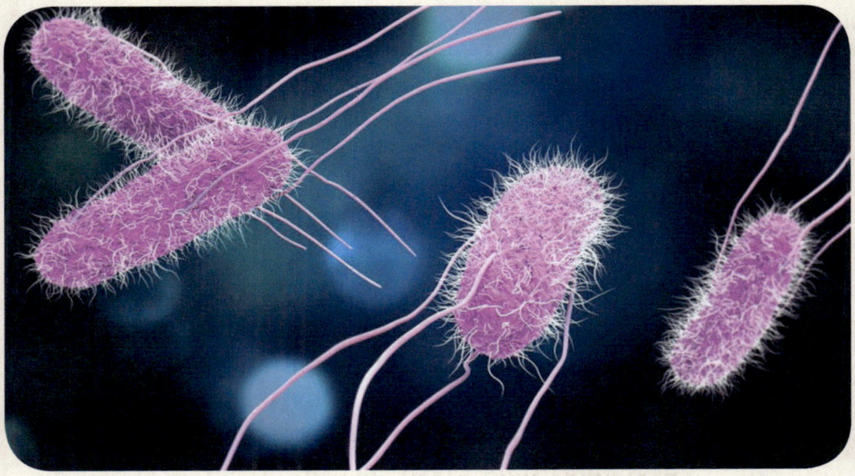

LEARNING OBJECTIVES

After studying this chapter, you will be able to:

1. List the characteristics of prokaryotes;
2. Explore the types of prokaryotes;
3. Define the ecological roles of prokaryotes;
4. Identify the characteristics of protists;
5. Explain the types of protists;
6. Focus on the ecological and economic importance of protists.

Introduction

Understanding prokaryotes and protists is central to understanding the evolutionary and functional diversity of life on our planet. Prokaryotes, including bacteria and archaea, are the most ancient and widespread organisms and play crucial roles in biogeochemical processes, human medicine, and ecosystems. Their simple structure hides their functional sophistication, adapting to fill an immense range of environments from hot to cold, from high salinity to low oxygen. Protists, a miscellaneous group of mostly unicellular eukaryotes, are revealed to hold the secrets of origins and the evolution of multicellular life. By examining these two populations, we can better see how life evolved and diversified over billions of years. Current research has revealed unique and resilient populations of protists and prokaryotes that exist within environments of extremes, providing us with a clear view as to why they are so resilient. For instance, research on the deep-sea hypersaline anoxic basins (DHABs) shows evidence of elaborate microbial communities that challenge previous assumptions about the viability of life within such extreme conditions. In these locations, nematodes and other metazoans populated upper layers of the halocline, indicating an unexpected level of biodiversity within otherwise inhospitable areas (Beaudoin et al., 2015). Moreover, studies of hypersaline environments such as Lake Tyrrell have documented the existence of specific eukaryotic groups that can thrive in a situation of elevated salinity, demonstrating yet again the protists' incredible adaptability to environmental extremes (Andrade et al., 2013).

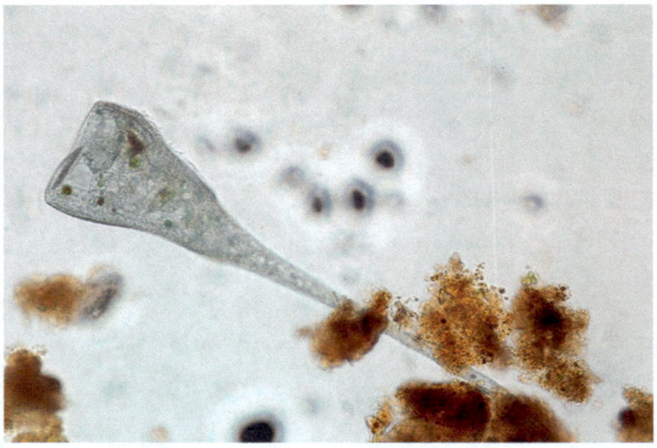

Figure 4.1. Survey of protists.

Source: https://cdn11.bigcommerce.com/s-v5lcc6/product_images/uploaded_images/bio-p-cel-y11-12-13-img1h.jpeg.

All these findings serve to emphasize the need to study prokaryotes and protists as a way to be able to assess the full range of life. Understanding the roles of prokaryotes and protists in their respective environments is a foundation for ecological and environmental science. Interactions among these microorganisms are crucial for nutrient cycling, food web function, and climate control. As an example, bioprocessing by bacteria and archaea in poorly nourished environments makes ecosystem function and resilience possible. Moreover, research on protistan diversity, especially in extreme habitats, adds to our understanding of ecological interactions and evolutionary mechanisms. This study not only reiterates the significance of such organisms in sustaining life on Earth but also compels researchers to consider how global change influences their survival and performance. As we proceed to investigate these fascinating groups, we unwind the intricate tapestry of life that sustains the biospheres of our Earth (Figure 4.1).

Definition of Prokaryotes

In order to define prokaryotes, one must first learn about their basic characteristics that set them apart from eukaryotic organisms. Prokaryotes are unicellular organisms that do not have a true nucleus and membrane-bound organelles. They contain genetic material, usually in the form of a single circular DNA strand, in an area designated as the nucleoid, which is not compartmented. This simplicity of cellular structure results in an extremely efficient method of reproduction and metabolic activity. Prokaryotes are significant members of diverse ecosystems and play significant roles, like nutrient cycling and organic matter decomposition. Given this, while single-celled, prokaryotes have a significant overall impact on biological processes, ranging from local to planetary scales. Prokaryotic life forms are generally classified into two domains: Bacteria and Archaea. While both share the prokaryotic structure, both are characterized by significant biochemical and genetic differences. Bacteria possess a wider variety of shapes and metabolic activities, allowing them to thrive in a range of environments. While most bacteria are beneficial to human beings and the environment, some may cause disease or antagonism through pathogenic interactions. Archaea are extremophiles, surviving in extreme conditions such as temperatures or salinity, and they possess unique biochemical pathways that deviate from bacteria. The identification of these differences increases our appreciation of prokaryotic diversity and their functions in biochemical cycles and ecological interactions (Bar-On et al., 2018). Moreover, prokaryotes make a significant contribution to global biogeochemical cycles, particularly in the ocean. Their biomass is a vital component of the oceanic ecosystem that influences everything from community structure to energy transfer within food webs. There is evidence that prokaryotes, especially bacteria and archaea, have considerable biomass in aquatic ecosystems and terrestrial habitats. Bacteria alone, for instance, contribute an estimated 70 gigatons of carbon to the biosphere, demonstrating their contribution to nutrient cycling (Castillo et al., 2019). Moreover, prokaryote interactions, such as virus-host interactions, are of interest in understanding microbial ecology. These processes create not only the community arrangements but also determine overall productivity and energy flow within an ecosystem, highlighting the necessity of integrating prokaryotic studies into other ecological observations (Figure 4.2).

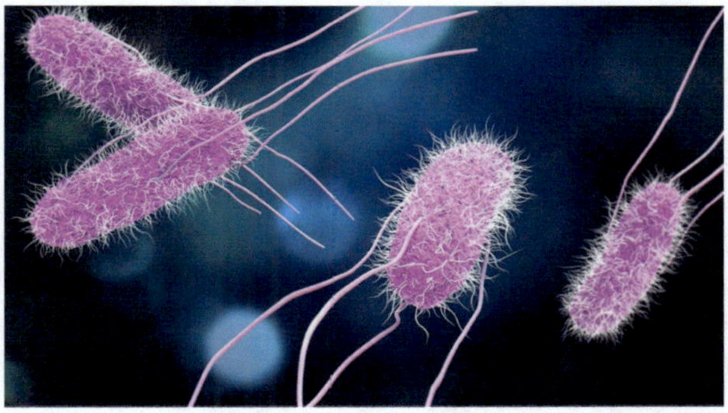

Figure 4.2. Prokaryotes.

Source: https://assets.technologynetworks.com/production/dynamic/images/content/395566/prokaryotic-cells-definition-structure-and-characteristics-395566-960x540.jpg?cb=13274790.

Definition of Protists

In biological taxonomy, protists form a diverse group of eukaryotic organisms that cannot fit into the categories of animals, plants, or fungi. Typically, microscopic in shape and size, these organisms consist of diverse forms and metabolic types. Protists are usually divided into three categories: protozoa, algae, and slime molds. Protozoa are predominantly heterotrophic and very mobile, possessing varying mechanisms of motility, like flagella and cilia. Alternatively, algae are autotrophs and perform photosynthesis, and they contribute significantly to global primary production. Slime molds, with both fungal-like and protozoan-like properties, are of utmost importance in nutrient cycling in ecosystems. Such built-in protist diversity makes them a significant area of discussion, as they are responsible for playing important roles in ecological interaction as well as evolutionary processes, and therefore their influence on the dynamics of the biosphere (Bar-On et al., 2018). The protists have functions in the environment beyond living as individual organisms; they occur within food webs and nutrient cycling within different ecosystems. Within aquatic ecosystems, the phytoplankton, which consist of a number of algal groups, are the base of the marine food web that sustains a variety of marine organisms. The protozoan protists, on the contrary, serve a function as primary consumers, consuming bacteria and other small organisms. This symbiotic relationship highlights the dependence of life forms and the intricacy of biotic interactions. In addition, protists are also important in soil fertility and structure as they decompose organic matter and release nutrients. Understanding the ecological roles protists exhibit enhances our knowledge of large-scale environmental change and its impact on biodiversity. As we go on to examine such organisms, we discover the intricacies of ecosystems and evolutionary processes that form them (Goto et al., 2007). Protist taxonomy and information remain evolving, based on ongoing study and molecular biology advancements. Genetic analysis improvement and phylogenetic studies have revealed intricate interrelations among protists and their evolutionary past. For instance, new findings reveal a rapid evolution of animal-specific domains to emphasize the functional diversification of various eukaryotic clades, for example, some groups of protists (Goto et al., 2007). This continued advancement in classification not only rearranges our understanding of protists but also emphasizes their evolutionary importance. Although scientists strive to develop a comprehensive taxonomy of protists, implications are extended across ecology, evolution, and applied sciences. Closing protist knowledge gaps is critical to meet environmental conservation challenges of climate change and the destruction of habitats. Hence, ongoing protist research is critical to unravel the complexity of life on Earth within the framework of a protracted survey of prokaryotes and protists (Figure 4.3).

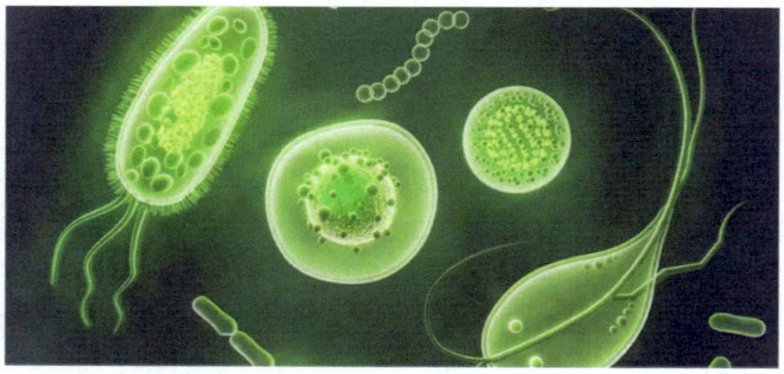

Figure 4.3. Protists.

Source: https://cache.careers360.mobi/media/article_images/2024/8/9/Protista.jpg.

Importance of Studying These Organisms

The importance of researching prokaryotes and protists is crucial to different fields of science since these organisms can provide essential information about the evolutionary mechanisms that formed life on Earth. Prokaryotes, among the oldest of the living organisms, display a phenomenal diversity in metabolic abilities, which has helped them adapt to a wide variety of environments. Evolutionary transitions from Prokaryotic to unicellular eukaryotic organisms demonstrate the important biological innovations, which, besides being essential to understand the complexity of life origins, also guide modern-day studies on metabolism and energetics. By investigating these transitions, researchers can identify the adaptations that allowed organisms to overcome physical constraints, thereby gaining a deeper understanding of ecological interactions and evolutionary mechanics, as described by Okie et al. (2011). This information is critical to areas like evolutionary biology, microbiology, and ecology. The protists or eukaryotic microbes also help in improving our understanding of the complexity of life, especially in their mitotic apparatus diversities and their evolution. The wide variety of cell shapes and modes of action of the protists, in contrast with the relatively homogeneous nature of prokaryotic organisms, was already pointed out in the early classifications by Édouard Chatton. This difference not only highlights the evolutionary novelties of the eukaryotes but also clarifies the ecological functions that these organisms perform in ecosystems, including nutrient cycling and primary production in water ecosystems. For instance, the intricate biology of protists like dinoflagellates raises essential questions about their evolutionary history and ecological significance, demonstrating that their unique chromosome organization and reproductive strategies can inform broader evolutionary theories, as noted by Soyer-Gobillard & Dolan (2015). Therefore, the study of protists is important in enhancing our understanding of cell complexity and ecology. Lastly, the significance of learning about prokaryotes and protists goes as far as their application in real-life practices in medicine, biotechnology, and environmental science. Prokaryotes play an essential role in fermentation, bioremediation, and bioengineering, demonstrating their usefulness in solving human problems, such as waste disposal and renewable energy. At the same time, protists have complicated life cycles that require additional research to create effective therapies and learn more about the dynamics of diseases, since some of them are pathogenic. Through an understanding of the molecular and ecological functioning of these organisms, researchers can develop novel solutions, taking advantage of their unusual characteristics. Such a combination of knowledge not only stimulates a better understanding of the biological diversity but also enables scientists to utilize the potential of prokaryotes and protists in a quickly changing world, which designates their paramount significance in both basic and applied sciences.

4.1. Characteristics of Prokaryotes

Learning Outcome

1. Cell Structure and Organization
2. Reproduction Methods
3. Metabolic Diversity

Prokaryotes, the least evolved and most rudimentary forms of life, are characterized primarily by their lack of membrane-bound organelles, a definite nucleus among them. Their structural naivety does not translate into naivety of function. Prokaryotic cells, that is, bacteria and archaea, possess a hard cell wall that confers shape and protection on them and allows them to survive in different environments. Incredibly, prokaryotes are highly metabolically versatile and can fill an enormous range of ecological niches, from deep-sea vents to the human gastrointestinal tract. Advances in microbial ecology in recent years have shown that these microbes are crucial participants in global biogeochemical processes and thus are ecologically relevant. The estimated biomass of prokaryotes is a reflection of their density and impact, with approximately 70 gigatons of carbon associated with bacteria and another 7 gigatons associated with archaea, according to current research on biomass distribution on Earth (Bar-On et al., 2018). The prokaryotic evolutionary history reflects enormous resilience, as also shown by the observation that cytoskeletal elements are incorporated into their structures. Unlike the initial assumption that prokaryotes were internally disorganized, they have been found to possess proteins that resemble eukaryotic cytoskeletal structures. Such structures are responsible for cell division, maintenance of cell shape, and separation of DNA during replication and are used to demonstrate that prokaryotic cells are far more organized than initially assumed. Prokaryotes are also effective in energy storage as well as storage of nutrients in the form of intracellular storage polymers. This capacity, observed in other species of bacteria, is an adaptive survival mechanism that enables them to adjust according to shifting environmental conditions. This new insight allows scientists to appreciate the intricate dynamics of prokaryotic life, which deviates from the earlier reductionist beliefs subscribed to by biologists such as Stanier and van Niel (Berlanga et al., 2010). Apart from their structural and functional characteristics, prokaryotes are most diverse regarding their mode of reproduction as well as gene exchange. Prokaryotes, under normal circumstances, are asexually reproducing organisms that can reproduce through binary fission, an energetically cheap process that is also extremely rapid. Prokaryotes can, however, undergo lateral gene transfer through transformation, transduction, and conjugation, which brings about genetic diversity among populations. This gene transfer is at the heart of their adaptability, allowing rapid evolutionary adaptation to environmental pressure, e.g., antibiotic resistance. The impact of these adaptations reaches far beyond ecological balance to fundamental public health concerns, as evidenced in the case of antibiotic-resistant bacteria, where the relentless struggle between medical ingenuity and microbial evolution continues unabated. As more research piles up on the functions and implications of prokaryotic characteristics, it is evident that such microorganisms play a focal role in the operation of the biosphere and deserve special scholarly focus (Bar-On et al., 2018) (Figure 4.4).

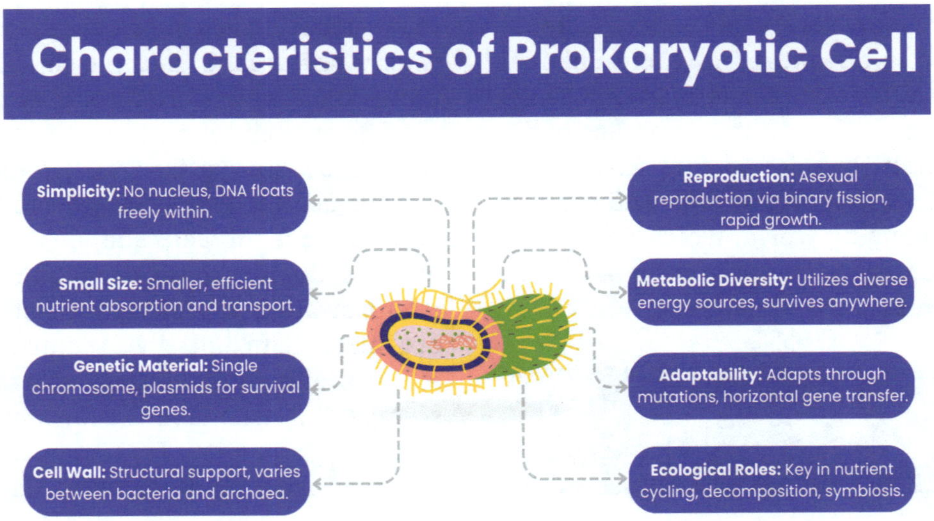

Figure 4.4. Characteristics of prokaryotes.

Source: https://images.examples.com/wp-content/uploads/2024/03/Characteristics-of-Prokaryotic-Cell.png.

4.1.1. Cell Structure and Organization

The variation in prokaryotic and eukaryotic cellular structure is the explanation for the structure and complexity of life. Prokaryotes, which are primarily bacteria and archaea, possess a relatively straightforward cellular structure with no true nucleus or membrane-bound organelles. Their genetic material, organized into a compact form, is located in a region known as the nucleoid. The lack of structure allows for rapid reproduction and adaptability to all kinds of environments. In contrast, eukaryotic cells, of which protist cells are a part, possess a well-defined nucleus enclosed in a nuclear membrane and a range of organelles such as mitochondria and the ER. These structural differences can be observed to reflect the evolutionary divergences that have been experienced by these organisms, and provide testimony to the significance cell structure carries within the broader ambit of biological function and diversity, as has been described in the trail-blazing research of Edouard Chatton (Soyer-Gobillard & Dolan, 2015). Eukaryotic protists display magnificent diversity in cellular composition, particularly in mitosis and chromatin structure. Compared to the relatively uniform prokaryotic cell structure, protists have a variety of cell forms, 36 phyla of which show variations in nuclear arrangement and chromosome morphology. Protist groups differ in the very specific chromatin organization shown. Even more noticeable are dinoflagellates, possessing a peculiar chromosomal organization system not characteristic of eukaryotic models. Their chromatin is quasi-permanently condensed and can be histone-free at times, and it raises intriguing questions about their evolutionary history and cell function. These complexities not only reflect the variability of protists but also their role in evolution as a transition from prokaryotic simplicity and complexity in multicellular organisms (Soyer-Gobillard & Dolan, 2015). The history of cellular organization has been governed by metabolic and energetic imperatives, most notably in the transitions from prokaryotic to eukaryotic forms of life. Recent research suggests that these transitions involved major innovations in metabolic processes that affected cell architecture in return. For example, changes in the scaling of metabolic rate with cell size are a case in point in the way structural changes affect physiological function. Prokaryotes have superlinear metabolic rate scaling with body size, whereas protists have linear scaling. Such evolutionary patterns uncover how surface area scaling innovations, like the formation of specialized organelles, contribute significantly toward increasing metabolic efficiency and functional improvement in organisms. These findings not only add depth to our understanding of cell complexity but also highlight the interdependence of metabolic evolution and cellular structure within the greater context of ecological interaction and evolutionary theory (Figure 4.5).

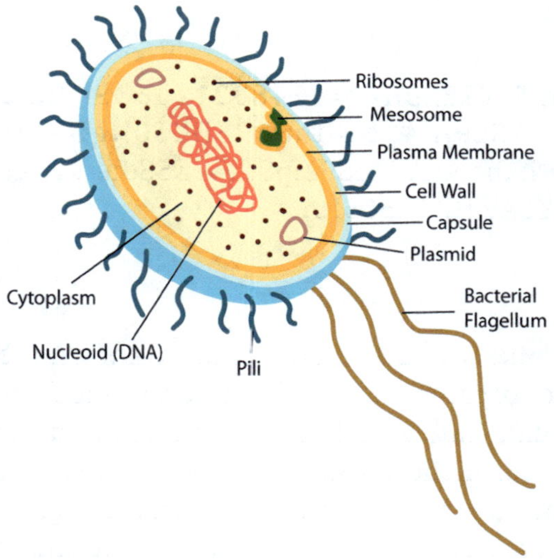

Figure 4.5. Prokaryotic cell structure.

Source: https://lab-training.com/wp-content/uploads/2015/02/Bacterial-structure-1024x995.jpg.

4.1.2. Reproduction Methods

The mode of reproduction in prokaryotes and protists is diverse and adapted to their respective niches. Prokaryotes generally reproduce asexually through binary fission, an operation in which a single cell divides to form two genetically equivalent daughter cells. The process allows for rapid population growth, particularly in niches with ample resources. Protists, in turn, exhibit both sexual and asexual reproduction, which allows them to be flexible to varying environments. Asexual reproduction in protists can also be achieved by budding or fragmentation, mechanisms that can enable immediate colonization of new habitats. The combination of sexual and asexual reproduction helps the protists continue being genetically varied while enjoying stable environments in times of need. These modes of reproduction are more significant than mere survival; they affect the evolution of both groups as well. For instance, the ability to pass genetic information by HGT, in prokaryotes, allows adaptation to environmental change in a rapid manner, creating a vast pool of genetic traits.

The resulting genetic transfers can contribute to antibiotic resistance, thereby impacting ecological processes and human health. Protists, by contrast, employ sexual reproduction to combine genetic material, generating diversity that could enhance resistance to infection and the environment. Recognition of these reproductive strategies is establishing their places in prokaryote and protist life cycles as markers of their ecological importance. In addition, research into rare biosphere prokaryotes is enlightening other reproductive complexities. New research demonstrates a mounting necessity to value low-abundance members of microbial populations and the ways in which their reproduction is contributing to ecosystem function (Costa et al., 2020). These novel prokaryotes may utilize new reproductive strategies that allow them to thrive in novel niches, perhaps being critical for nutrient cycling or bioremediation. The importance of this research is the promise that the identification of their reproductive characteristics may be utilized for future biotechnological uses, improving our capability to cultivate their uniqueness for environmental and industrial utilization. With the progress of research, these findings will illuminate the intricate relationships between modes of reproduction, genetic diversity, and ecological stability of protists and prokaryotes (Figure 4.6).

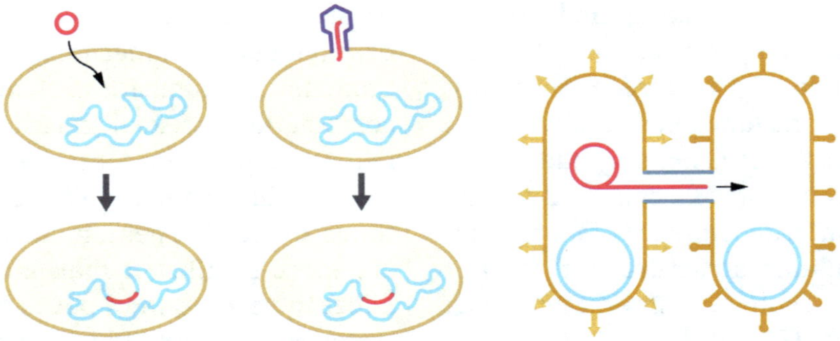

(a) Transformation (b) Transduction (c) Conjugation

Figure 4.6. Prokaryotic reproduction.

Source: https://files.mtstatic.com/site_4463/78615/0?Expires=1748321266&Signature=mhARGgrB7M35SoeE91mYa9YOp03pfvKe-CdSMe255Gz9Jil47cbGtq8pR5BG60NqB~lujmy-nnnkLA1s~FMQcwD5Ruz7a7MIDsdnmEw~1QeWazBDNrGMFcZV38I59efpSTFWK5SduxVWJgdg8r6v8VwsWYcDxrF~k8aJTU76Yuo_&Key-Pair-Id=APKAJ5Y6AV4GI7A555NA.

4.1.3. Metabolic Diversity

Their study of the metabolic diversity of prokaryotes and protists is also vital in knowing their functions in different ecosystems, especially in cycling of organic matter. The different biochemical pathways by which these organisms can gain energy and nutrients is called metabolic diversity and enables them to inhabit a great variety of ecological niches. This diversity is most evident among the prokaryotes, especially in environments that might be regarded as extreme, e.g., deep-sea hydrothermal vents and hypersaline habitats. As an example, certain bacterial strains have evolved distinctive metabolic pathways to allow them to use sulfide and methane as an energy source and demonstrate an impressive adaptation to extreme conditions. Such metabolic flexibility not only aids survival but also leads to the overall operation of biogeochemical cycles within their respective ecosystems, which point to the fact that more research needs to be done on the genes and pathways involved in supporting such metabolic operations. Organic osmolytes are central to the metabolic context of marine microbes, which use them to maintain cellular osmolarity, which may represent major substrates in nutrient turnover. The study conducted on marine bacteria demonstrates that specialized metabolites such as dimethyl sulfoniopropionate (DMSP) can be metabolized by a large proportion of the marine bacterial community, fulfilling considerable nutritional needs (Alexander et al., 2021). This metabolite exchange between the prokaryotes and protists strengthens their interaction in the food webs, allowing the cycling of nutrients and the flow of energy. The genetic capability of osmolyte metabolism suggested by gene surveys in several marine microbial genomes represents a complicated network of interactions that support the productivity of marine environments. The knowledge of such relationships gives an insight into the key roles that these metabolic processes play in the greater scope of marine biology. Moreover, metabolic diversity has also been projected to the deep-sea environments that are scarcely studied but have distinct ecological processes. Recent findings indicate that even in extreme conditions such as DHABs, diverse metazoan and protistan communities have adapted to thrive. For example, research has identified various taxa, with nematodes being among the most abundant, exhibiting distinct metabolic traits that enable them to survive in such hostile habitats (Beaudoin et al., 2015). The metabolic strategies of these organisms are occasionally similar to those of normoxic habitats, indicating that metabolic diversity is a key parameter for their ecological success. With further investigation of these extreme locations, a more complete image will develop regarding how metabolic diversity is preserved and harnessed, adding to the image of the resilience of life in a variety of challenging and varied conditions.

Test Your Knowledge

1. What structural features distinguish prokaryotic cells from eukaryotic cells?
2. How do prokaryotes reproduce, and what makes their reproduction rapid and efficient?

4.2. Types of Prokaryotes

Learning Outcome

1. Bacteria
2. Archaea
3. Cyanobacteria

Prokaryotes are classified primarily into two large domains: Bacteria and Archaea. This is a significant classification since the two domains contain largely different evolutionary lineages. Bacteria, encompassing a multitude of species with a broad diversity of metabolic mechanisms, are characterized by their cell wall structure and mode of reproduction through binary fission. Conversely, Archaea, prevalent in environments that are extreme, have distinct biochemical features, including their different lipid composition within their membranes as well as varying RNA polymerases. These differences help explain the search into prokaryotic function within ecosystems, especially how they provide input to nutrient cycling and interact with eukaryotic organisms across ecosystems from soil to their environment in extreme habitats. The disparity between these two kingdoms also indicates their adaptive changes in evolution, having conditioned them to occupy specialized niches within greater ecological plans, exercising immense impacts on world biodiversity and ecological balance (Bar-On et al., 2018). The amount of biomass produced by prokaryotes is astounding, with bacteria contributing approximately 70 gigatons of carbon to world ecosystems and Archaea adding approximately 7 gigatons (Bar-On et al., 2018). These microbes are responsible for biogeochemical cycling, including nitrogen fixation and organic matter decomposition, thereby playing a significant role in sustaining life on Earth. Their abundance and versatility enable them to thrive in different environments, from deep-sea hydrothermal vents to the termite gut, indicating their environmental adaptability. Particularly, complex symbiosis between prokaryotes and eukaryotes has been reported in studies where prokaryotic symbionts complement host functions. For example, plant-matter-degrading bacteria in the gut of a termite are not only useful in breaking down plant material but also enhance the overall health of the host and nutrient absorption. These interactions show that prokaryotes emerge as influential players in ecosystems. Functional and evolutionary prokaryotic strategies are also incredibly diverse. While the majority of bacteria are advantageous to human well-being, some of them turn pathogenic, thus becoming associated with certain diseases. Prokaryote flexibility is accentuated by their capacity to acquire antibiotic resistance genes, a condition that complicates the treatment of bacterial infections. Likewise, archaea have been recognized to play a pivotal role in the production of methane in anaerobic environments, an activity with unbridled implications on global climatic change and energy production. The discovery of the functions of prokaryotes in diverse manners, both negatively and positively, is at the core of microbiology and environmental science research. Continued investigation of prokaryotic diversity will expand our insight into the complexity of life and provide leads on potential biotechnological advancements upon their unique metabolic capabilities (Bar-On et al., 2018) (Figure 4.7).

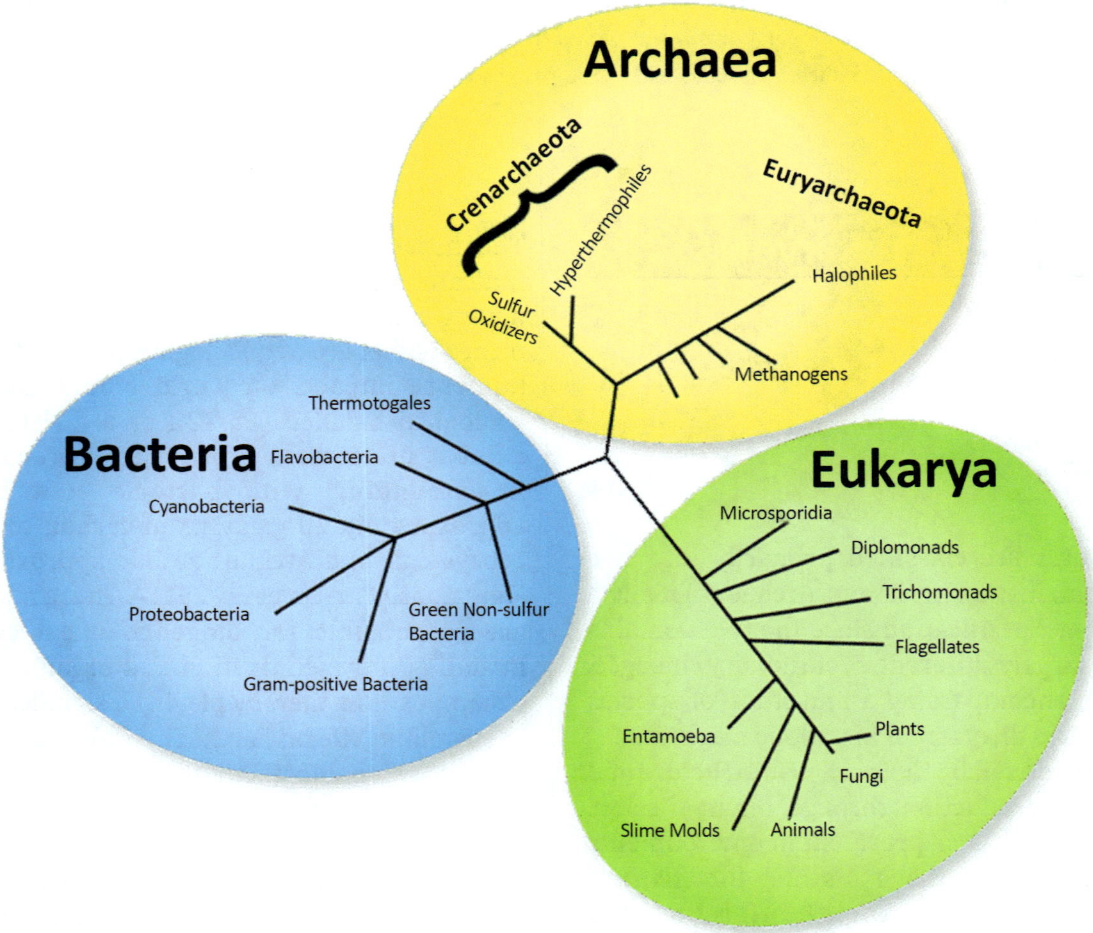

Figure 4.7. Prokaryote classification.

Source: https://dr282zn36sxxg.cloudfront.net/datastreams/f-d%3Aa115f37bf5f8be2569d252c1ab088dc1d29ce7da8dca229f9efef8c2%2BIMAGE_THUMB_POSTCARD_TINY%2BIMAGE_THUMB_POSTCARD_TINY.1.

4.2.1. Bacteria

Bacteria's contribution to the biosphere is extensive and diversified, impacting both ecological processes and human life. The dominant members of the prokaryotic kingdom, bacteria, contribute approximately 70 gigatons of carbon to Earth's biomass, testifying to their numbers and significance in most ecosystems (Bar-On et al., 2018). These microorganisms inhabit an immense diversity of ecosystems, from the deep seas to the human gut, and have shown an impressive ability to adapt to environments ranging from extremely hot to high salinity. Not only is this adaptability a gauge of their success in evolving, but it also indicates the centrality they occupy as decomposers, nutrient cyclers, and primary producers in a broad array of environmental niches. Hence, recognizing bacteria's ecological role is important in recognizing the complex relationships that sustain life on our planet, especially in sustaining biogeochemical cycles. The complexity of bacterial existence should not be dismissed since advances in microbiology have proven that prokaryotic cells possess intricate structures that were previously thought to occur only among eukaryotes. Initially believed to be simple organisms, bacteria have been found to possess all the cytoskeletal components necessary for keeping the cell in shape, DNA segregation, and division (Berlanga et al., 2010). Such a discovery indicates not just that bacteria are evolutionary ancestors but also possess common functionalities that allow them to survive and adapt to environmental changes. Furthermore, intracellular storage polymers within intracellular structures signify the evolutionary advantages that have been acquired by such microorganisms to endure environmental fluctuations. Such characteristics indicate a sophisticated organization and functional diversity within prokaryotic cells, refuting the general perception of bacteria as primitive organisms. In addition, bacteria also play crucial roles in various biotechnological and biomedical processes, which showcase their significance outside ecological considerations. Their metabolic versatility renders them valuable to be use in the processes of waste treatment, bioremediation of contaminated environments, and the production of biofuels. In medicine, certain bacteria are exploited in the production of probiotics and antibiotics that enhance the health of humans and combat infectious diseases. The pioneering work that has been conducted in the area of bacterial genetics has also given birth to genetic engineering, further attesting to their significance to biotechnology. Thus, learning about bacteria in relation to the greater category of prokaryotes and protists not only enhances our understanding of their ecological roles but also reveals to us how societies can leverage these organisms to their benefit in various fields (Figure 4.8).

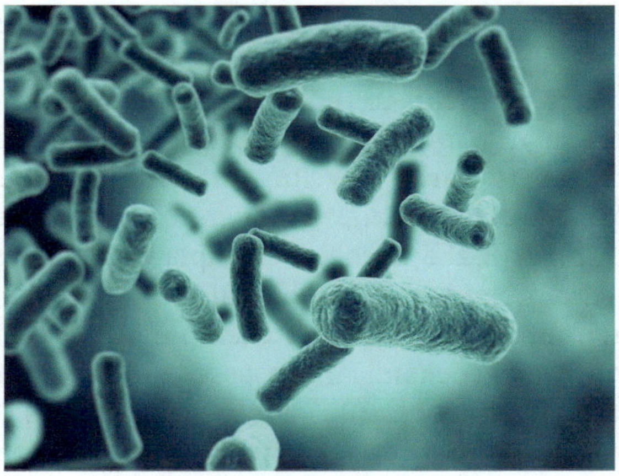

Figure 4.8. Bacteria.

Source: https://www.globalgarden.co/wp-content/uploads/2022/06/bacteria-extract-nutrients-from-many-sources-1568x1192.jpg.

4.2.2. Archaea

The Archaea research reveals a fascinating facet of prokaryotic life that strongly disrupts traditional notions of cellular organization and evolution. Initially believed to be mere vestigial organisms due to a lack of eukaryotic complexity, Archaea prove to contain special adaptations and sophisticated cellular processes. New studies have indicated that these microbes, such as bacteria, possess cytoskeletal structures that are required to preserve cell division and shape, and this has implications that the evolutionary history of the cytoskeleton stretches even deeper than had previously been thought. These features point to the importance of identifying Archaea not just as an ancient lineage of life but as full members in the overall account of evolutionary biology, which renders our understanding of cellular structure and functional capacities in the world of prokaryotes more complicated (Berlanga et al., 2010). Archaea possess a unique ability to thrive in harsh conditions, such as hot geysers and hypersaline lakes, making them ecologically broad and of evolutionary significance. The extremophiles possess specialized proteins and membranes that enable them to thrive in environments that would be lethal to most organisms. The metabolic flexibility in Archaea is also critical to biogeochemical cycling, particularly in methanogenesis reactions that are key to carbon cycling in terrestrial and aquatic ecosystems. This competence and flexibility make Archaea core players not just in their environments but also in planetary environmental health, as exemplified by their role in biogeochemical processes for life on the planet (Goto et al., 2007). Moreover, studies on Archaea have broader implications than the understanding of prokaryotic life. By shedding light on the evolutionary mechanisms of these microbes, researchers can make inferences between their cellular processes and those of more complex life, i.e., eukaryotes. The shared features demonstrated among Archaea, such as the presence of certain metabolic processes and cytoskeletal elements, imply HGT among lineages and enhance our knowledge of evolutionary processes. To this end, research into Archaea offers tremendous insights into the evolutionary origins of complex cell structure and function. As we continue to unravel the intricacies of these extraordinary organisms, prokaryotic and protist diversity will increasingly illuminate our comprehension, posing new issues about the durability and complexity of life (Berlanga et al., 2010) (Figure 4.9).

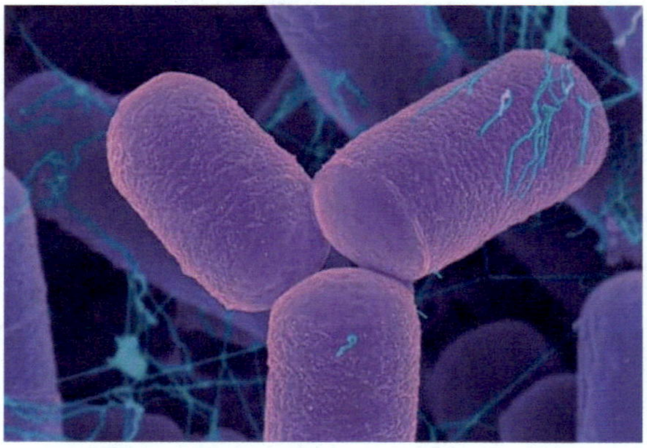

Figure 4.9. Archaea.

Source: https://genetics.org.uk/wp-content/uploads/2017/09/F1.large_-e1563793036383.jpg.

4.2.3. Cyanobacteria

Cyanobacteria are one of the key prokaryotic groups that have created a major impact on the ecological and biochemical profile of the planet. These microorganisms are commonly termed as blue-green algae and are well known to have the unique capacity to carry out photosynthesis, hence to convert light energy into chemical energy quite efficiently. This phototrophic property enables cyanobacteria to live in a wide variety of environments, including freshwater lakes and even in extreme dry deserts. They are important in primary production, helping in the production of oxygen and acting as the basis of aquatic food webs, providing nutrients to support a variety of protist and metazoan communities. By facilitating photosynthesis, cyanobacteria not only fuel local ecosystems but also help regulate atmospheric gases, making them essential to the global biogeochemical cycles (Anderson et al., 2025). In addition to their ecological significance, cyanobacteria are also highly metabolically diverse, which means that they can grow in different conditions. These organisms can alternate autotrophic and heterotrophic lifestyles, depending on the available resources and being supported by the rest of the environmental conditions. Furthermore, a few species contain specialized cells, heterocysts, that allow them to fix atmospheric nitrogen in anaerobic conditions, an important nutrient cycling process in most ecosystems. This nitrogen fixation not only helps them to grow in low nutrient environments but also adds to soil fertility which affects the terrestrial plant community. The ability of cyanobacteria to modify their metabolic pathways demonstrates their evolutionary success, which has implications for their contributions to microbiome dynamics and interactions with other microorganisms, including protists (Deutschmann, 2021). The cyanobacteria-protist interactions comprise a complicated set of ecological interactions that are worth investigating. Recent research emphasizes the importance of protists in controlling the population of cyanobacteria via grazing, thus affecting the community structure and primary production in aquatic environments. Protist bacterivory can significantly control cyanobacterial biomass, demonstrating ecological relevance, particularly in areas where nutrient fluctuations occur (Anderson et al., 2025). Moreover, with climate change and nutrient pollution exerting pressure on marine ecosystems, the interactions between the two become especially important to predict changes in community structures. By employing network analysis methods, researchers aim to elucidate these intricate relationships and enhance our understanding of microbial ecology, revealing the broader implications of cyanobacterial and protist interactions for ecosystem health and function (Deutschmann, 2021) (Figure 4.10).

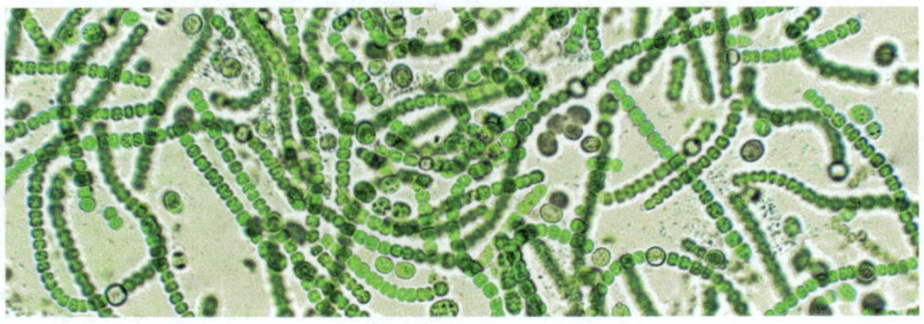

Figure 4.10. Cyanobacteria.

Source: https://www.deq.ok.gov/wp-content/uploads/state-environmental-laboratory-services/Microscope_BGA_2.jpg.

Test Your Knowledge

1. What are the main differences between Bacteria and Archaea?
2. How do autotrophic and heterotrophic prokaryotes differ in their modes of nutrition?

4.3. Ecological Roles of Prokaryotes

Learning Outcome

1. Decomposers in Ecosystems
2. Nitrogen Fixation
3. Symbiotic Relationships

Prokaryotes are important members of many ecological processes, particularly as primary producers and decomposers in terrestrial and aquatic ecosystems. Prokaryotes, such as bacteria and archaea, are responsible for significant nutrient cycling, especially the decomposition of organic materials. By metabolizing dead organic matter, prokaryotes facilitate the recycling of nutrients such as carbon, nitrogen, and phosphorus, which are critical to plant growth as well as other organisms higher on the food web. By decomposing, they not only render soil fertile but also enhance the nutrient availability necessary to promote ecosystem health. Moreover, prokaryotic primary production is at the center of aquatic food webs, particularly oceanic food webs, where they form the core of the food web and illustrate their inherent function in ecological processes and overall ecosystem processes. Prokaryotes account for 70% of biomass in marine ecosystems, a reflection of their involvement in biogeochemical processes. Especially, ocean microbes account for approximately 50% of the Earth's primary production, and it has a major impact on global carbon cycles and climate. The intricate interactions among core marine prokaryotes and protists show how these ecosystems are very complex because some of these microbes rely on others for survival and function. Core marine taxon preserves fundamental ecosystem processes through their interaction, and it reflects both species occurrence and abundance. This networking means that the ecological roles of prokaryotes are not merely of presence, but also actively contribute to shaping their environment through interactive responses that alter community structure and ecosystem resilience, particularly under changing climatic regimes (Ferrera & Krabberød, 2022).

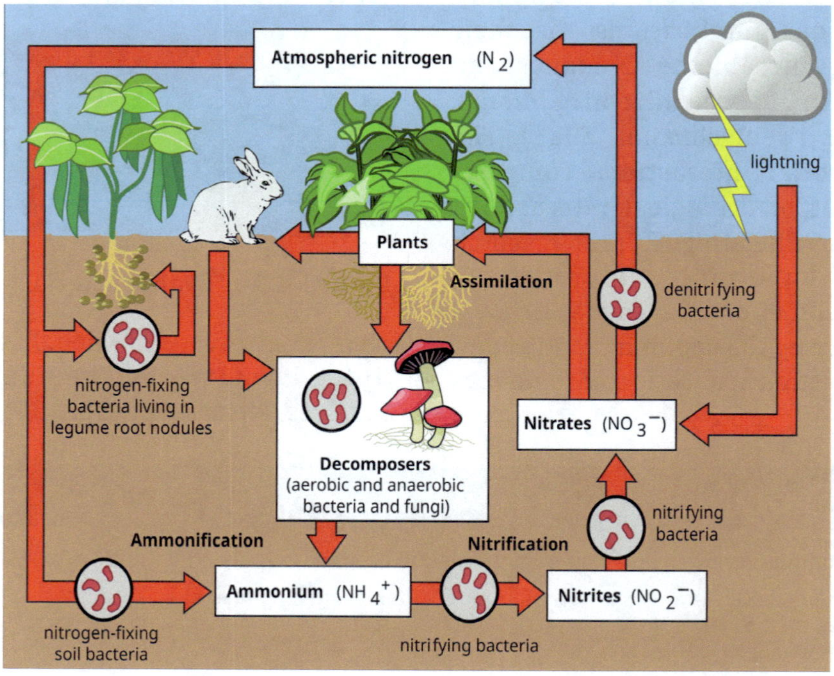

Figure 4.11. Ecological roles of prokaryotes.

Source: https://s3-us-west-2.amazonaws.com/courses-images/wp-content/uploads/sites/1223/2017/01/18213344/Figure_22_03_02.png.

Core microbial community identification demonstrates the potential of specific groups of prokaryotes to dominate during specific seasons, which depicts their adaptive strategies to varying environmental regimes. Research has demonstrated that the core bacteria are typically positively correlated with other bacteria and protists typically interact with those microbial counterparts in a symbiotic relationship (Balagué et al., 2022). The majority of these correlative interactions fluctuate with the season, typically showing an increase in colder, mixed water compared to warmer, stratified water. The seasonal variation of core operational taxonomic units (OTUs) not only indicates the ecological plasticity of prokaryotes but also suggests their intrinsic role in maintaining ecosystem stability. From these correlations, scientists can establish prescriptions to conserve and manage microbial diversity, ultimately maintaining the ecological integrity of aquatic ecosystems and associated biogeochemical processes (Figure 4.11).

4.3.1. Decomposers in Ecosystems

Decomposers are the critical organisms that help maintain ecosystem equilibrium by breaking down organic debris. Decomposers are largely fungi and bacteria, and they have specific enzymes that enable them to degrade complex carbon polymers in dead plants and animals. Decomposition not only returns nutrients to the soil but also helps primary producers grow. As a vital part of food webs, decomposers ensure that energy gets circulated through ecosystems effectively, ultimately sustaining various aspects of life. Through their work, they accumulate the fertility of soil, hence making them a vital part of agriculture and vegetation. Without this continuous process of decomposition, ecosystems would be bogged down in organic waste, reducing overall productivity and diversity, highlighting the essential contribution of decomposers to ecological functioning (Selby & Smith, 2015). Decomposers also have a significant contribution to make in water-based ecosystems, if perhaps one that is less obvious. In oceanic environments at deeper levels, for example, marine snow, or organic particles, is a critical source of nutrition for many microbial communities. Studies have also shown that while prokaryotes overall are more numerous than eukaryotes in marine snow particles, the highly effective organic degraders among them are fungi and labyrinthulomycetes, which prevail in terms of biomass. These cells also show remarkable adaptations to conditions of intense cold and pressure, underlining their function in cycling nutrients from the ocean surface to the ocean floor (Bochdansky et al., 2016). The process contributes its share to the carbon cycle, powering the energy transfer in the marine food chain, and demonstrating that the decomposers are not just fundamental on land but also fundamental when it comes to managing the ecosystem processes in aquatic systems.

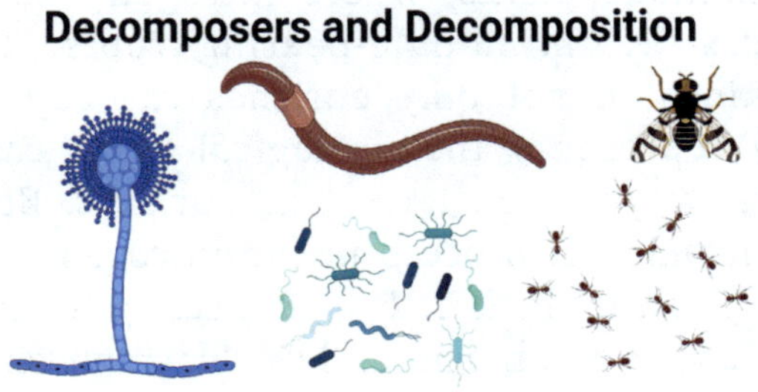

Figure 4.12. Decomposers in ecosystems.

Source: https://microbenotes.com/wp-content/uploads/2022/08/Decomposers.jpg.

The reliance of the decomposers on the remaining trophic levels demonstrates the dynamics of ecosystems. Because decomposers break down organic matter, they return critical nutrients such as nitrogen and phosphorus into the soil or water, which are consumed by primary producers. Through nutrient cycling, the growth of the plants is promoted, which later gets eaten by herbivores, receiving oxygen and food. Decomposers are therefore an important component of the ecosystem, contributing to sustainability and resistance to external shocks. Therefore, aware of the diverse roles of decomposers, their interaction with prokaryotes and protists becomes a necessity to adequately survey the well-being of ecosystems since their occurrence and effectiveness often indicate overall balance and functioning of ecological webs (Selby & Smith, 2015) (Figure 4.12).

4.3.2. Nitrogen Fixation

Nitrogen fixation is a key characteristic in prokaryote and protist surveys because it enables nitrogen (N_2) from the atmosphere to be reduced to ammonia (NH_3), a form that organisms can use. This biological process is predominantly performed by specific prokaryotes, like specific bacteria and archaea, which possess the enzyme nitrogenase for this reaction. Among these, the genera Rhizobium and Frankia are especially famous for their symbiotic association with leguminous plants, where nitrogen fixation occurs within nodule-bearing roots. This relationship not only enriches the soil but also represents the ecological significance of prokaryotes on terrestrial surfaces. Being a fundamental process in the nitrogen cycle, nitrogen fixation directly influences agricultural production and the natural world, emphasizing prokaryotic organisms' role in providing life and maintaining ecological balance, for example, their interactions with various protists and other eukaryotes. Protists, in contrast, also play an important role in contributing to nitrogen cycling, particularly in aquatic environments. Certain protists, such as dinoflagellates and ciliates, engage in bacterivory by feeding on bacteria involved in nitrogen cycling processes. In so doing, protists exert indirect influence on the availability of nitrogenous compounds in their habitats. From experiments on Baltic Sea pelagic environments, protist groups exhibit characteristic structuring on their part tied to their influence on prokaryotic mortality, thereby regulating bacterial populations significant in nitrogen fixation and other biogeochemical processes. This interaction indicates the complex interactions between prokaryotes and protists, suggesting a mutualistic function in nitrogen supply and overall ecosystem wellness. Exploring such associations enhances our understanding of ecological webs and the interdependence of microorganisms among environments. Moreover, the mutualism between protists and prokaryotes is more than a simple interaction; it suggests a complex coevolutionary past crucial to ecosystem function. For instance, various Bacteroidales symbionts exhibit varied affiliations with cellulolytic protists inhabiting the termite gut, stimulating lignocellulose breakdown, and improving nitrogen supply through their metabolic activities. Such symbiosis emphasizes the contribution of prokaryotes to improving nitrogen fixation phenomena and ecological adaptability, which influences not only their host organisms but even their wider ecological associations. The interactions of such associations, ranging from the potential for symbiotic specificity to evolutionary convergence, show the complexity inherent in microbial communities. A knowledge of nitrogen fixation in this regard permits a more informed appreciation of the role prokaryotes and protists play in larger ecological systems, focusing on their important contribution to cycling nutrients and biological productivity that are vital to the viability of many ecosystems, based on the findings of these biological analyses (Hongoh et al., 2009; Anderson et al., 2025) (Figure 4.13).

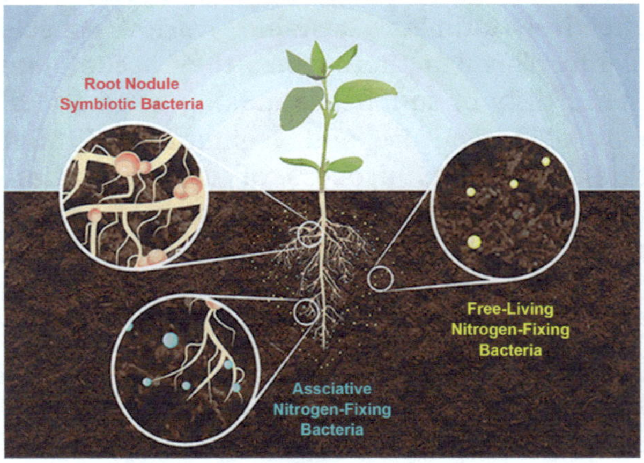

Figure 4.13. Nitrogen fixation.

Source: https://www.pioneer.com/content/dam/dpagco/pioneer/na/us/en/images/agronomy/fertilizers/IMG-diazotroph-nitrogen-Pioneer-NA_US-V1.gif.

4.3.3. Symbiotic Relationships

Two or more biological organisms interact mutually in a relationship that is beneficial to both or all, as a basic element of ecological and evolutionary dynamics. There are several types of such relationships, which can be mutualism, commensalism, and parasitism, and each of them affects the survival and reproduction of the respective organisms in a different manner. Symbiotic relationships in the prokaryotes and protists world are common and result in major adaptations that promote fitness. As an example, the complexity of the relationships established between protists and their prokaryotic counterparts leads to stability and resilience of ecosystems, thus showing the key roles that such relationships perform in environmental sustainability. These interactions are complex and therefore difficult to fit into the conventional biological paradigms, and therefore they show the importance of symbiosis in determining the evolutionary patterns of various organisms, thereby creating a tapestry of life that emphasizes interdependence within ecological communities. Symbiotic relationships, in most instances, provide an understanding of the evolutionary forces that structure biodiversity. These interactions are integrated, reflected in the concept of the holobiont, a term used to describe the overall organism that includes a host and its connected microbiota. According to studies, the holobiome is the genomic landscape that captures the symbiotic interactions between eukaryotic hosts and their microbial communities and, therefore, serves as a unit of selection in evolutionary biology. More controversially, eukaryotic organisms, including most protists, can be viewed as complex, co-evolved systems arising as symbiotic associations between prokaryotes. This integrated viewpoint emphasizes how symbiogenesis has contributed to the diversification of life, positioning symbiotic relationships as not only beneficial but also pivotal in understanding evolutionary patterns, especially in the context of ancient lineages (Margulis et al., 2013). To further emphasize the significance of such symbiotic relationships is the example of the Australian termite, *Mastotermes darwiniensis* which represents just how prokaryotes and protists can work together in mutually beneficial relationships. This wood-feeding termite has a symbiotic hindgut flora comprising several microorganisms, including bacteria, archaea, and protozoa that are vital in the digestion of cellulose, a key component of the diet of termites (lignocellulosic material). The occurrence of certain flagellates in the hindgut not only increases the digestive power of the termite but also represents a closely knit web of evolutionary interactions in which both parties benefit. Consequently, this case study reinforces the notion that symbiotic relationships are crucial for understanding the functional ecology of prokaryotes and protists and their roles in maintaining ecological balance and facilitating energy flow within their habitats (Breunig et al., 2008).

Test Your Knowledge

1. In what ways do prokaryotes contribute to nutrient cycling in ecosystems?
2. How do nitrogen-fixing bacteria benefit plant growth?

4.4. Characteristics of Protists

Learning Outcome

1. Eukaryotic Cell Structure
2. Modes of Nutrition
3. Reproductive Strategies

The kingdom Protista is the classic illustration of stunning diversity, containing a broad range of plant- and animal-like organisms. Protists are a eukaryotic phylum with both unicellular and multicellular forms, ranging from familiar members like the amoebas and paramecia to the highly complex algae. The most typical feature of a protist is a eukaryotic cell structure containing membrane-bound organelles and a nucleus. Unlike prokaryotes, which lack this compartmentalization, protists use their internal organization to carry out their complicated metabolic processes better. In addition, protists can reproduce sexually as well as asexually, adding to their versatility and evolutionary importance. This reproductive versatility, added to their varied modes of nutrition, e.g., photosynthesis in algae and phagocytosis in protozoa—makes protists a basic group to study in ecological processes and evolutionary biology. One of the striking features of protists is how they can exist in different habitats, ranging from freshwater to soil to even extreme environments. Most protists are water-dwelling and form the foundation of aquatic food chains as primary producers. Algae, by photosynthesis, transform sunlight into energy and release oxygen as a waste product while sustaining a variety of life forms. On the other hand, heterotrophic protists such as slime molds and protozoans recycle nutrients by decomposing organic material. Underlying their ecological significance is the intricate relationship that they have with other living organisms, such as symbiotic relationships that facilitate nutrient exchange and pathogenic relationships that affect human health and biodiversity. These processes underscore the importance of conducting more studies on protists and their role in even larger biological systems. The protist grouping is very difficult due to their evolutionary diversity and the continuous discovery of new species. In contrast to well-defined kingdoms such as Animalia or Plantae, the protists are a polyphyletic group since they lack a common origin, hence not easy to group. As such, modern classification tends to utilize genetic information to determine relationships between protist species. Recent studies have presented the rapid evolution of new domains in certain protist lineages, demonstrating the functional diversification of patterns observed in animal evolution (Goto et al., 2007). These findings affirm the adaptive evolution of the protists, suggesting the organisms have evolved to fill various ecological niches intricately. This ongoing research not only increases our understanding of evolutionary processes but also illuminates the intricate tapestry of life where protists form a crucial bridge from prokaryotic simplicity to eukaryotic complexity, further embellishing the broader story of biological diversity and resilience (Berlanga et al., 2010) (Figure 4.14).

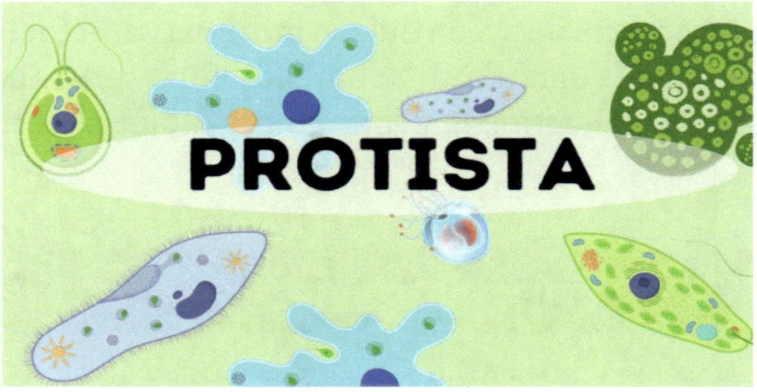

Figure 4.14. Characteristics of protists.

Source: https://i.ytimg.com/vi/DaehcP7GkHI/maxresdefault.jpg.

4.4.1. Eukaryotic Cell Structure

The study of eukaryotic cell structure is undertaken to explain the most important factors regarding the complexity that differentiates such organisms from their prokaryotic counterparts. Compartmentalized eukaryotic cells possess a structural organization that supports specialized functions in sub-cellular compartments. Of special interest is the occurrence of organelles such as the nucleus, mitochondria, and ER, which enhance metabolic efficiency and regulatory control in cells. This compartmentalization aids greater complexity and flexibility in eukaryotic life forms, including protists, fungi, plants, and animals. The emergence of the eukaryotic cell is considered to have marked a milestone in the history of life, providing a foundation for multicellular organisms and complex biological interactions, as discussed in the elucidation of evolutionary connections in a TOL framework considering the eukaryotic domain groups (Choi et al., 2020). Among the eukaryotic phyla, protists comprise a diverse collection of unicellular and simple multicellular organisms that possess the fundamental features of eukaryotic cell structure. Protists encompass a wide range of morphologies and modes of metabolism, which are expressed via their varied types of organelles. For instance, protists such as algae contain chloroplasts that allow them to perform photosynthesis, while others, such as protozoa, employ specialized structures to move and absorb nutrients. The similarity that is apparent in protist communities highlights the versatility of eukaryotic cells to a variety of environmental factors. Moreover, studies conducted on microbial eukaryotes such as those found in Lake Tyrrell reveal intricate community processes, highlighting the role of environmental factors in determining the structure of communities and eukaryotic cell morphology with respect to seasons (Andrade et al., 2013). Understanding eukaryotic cell structure is not just significant to distinguish eukaryotes from prokaryotes but also to examine their role in ecology and evolution. The intricate eukaryotic cell organization facilitates greater complexity of cell processes and interactions that are appropriate for organisms to exploit various niches and resources. This adaptability is observed in the aforementioned protist research, where molecular analysis uncovers unexplored diversity and highlights ecological impact in shaping community structure. These intricate connections among eukaryote species remind us of the holistic nature of life and the evolutionary change narrative that continues to unfold, bringing a richer idea of biological diversity in the eukaryote realm (Choi et al., 2020) (Figure 4.15).

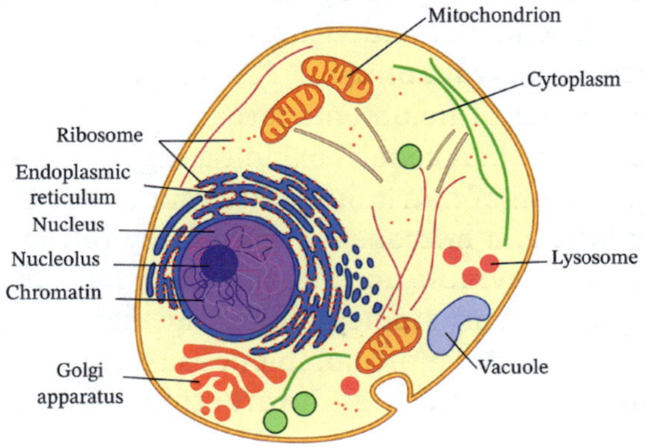

Figure 4.15. Eukaryotic cell structure.

Source: https://images.nagwa.com/figures/explainers/297140534241/2.svg.

4.4.2. Modes of Nutrition

Among prokaryotes and protists, nutritional modes hold the secret to their survival and ecological niches. Members of these groups of organisms can primarily be classified as autotrophs and heterotrophs. Autotrophs, such as certain cyanobacteria, synthesize organic compounds using inorganic substances based on processes like photosynthesis or chemosynthesis. Through this, they can inhabit a variety of environments, contributing significantly to global primary productivity. Heterotrophs, from all types of bacteria and many protists, obtain organic matter by consuming other organisms or organic compounds. This group within nutritional strategies not only influences the organisms' means of deriving energy but also their means of interacting in ecosystems, setting food webs and nutrient cycling key to ecological balance. The prokaryotic and protistan means of nutrition reveal highly efficient evolutionary adaptations. For instance, many protists express mixotrophy, where there is a combination of autotrophic and heterotrophic forms. This amenability makes them optimize energy acquisition in line with their habitat. Other prokaryotic groups, particularly those participating in symbiogenesis, express complex interdependent relationships that enhance their nutritional capabilities. For instance, some microbiota of holobionts, organisms including their microbial cohort, exhibit a highly evolved integration of nutrition modes that favors both the host and microbes. Evolutionary relevance of such symbiotic interactions has been emphasized in studies indicating that natural selection acts on the holobiont as an integrated whole, illustrating interaction between nutrition modes and biodiversity (Margulis et al., 2013). Familiarity with prokaryotic and protistan nutrition modes also illustrates their involvement in global biogeochemical cycles. For example, nitrogen-fixing bacteria are pivotal in the process of nitrogen in the air being converted to forms usable by plants, sustaining entire terrestrial food webs. Photosynthetic protists also contribute to carbon cycles in marine ecosystems, with significant influences on global carbon fluxes. The intricate interactions that take place between such complex organisms not only stabilize ecosystems but also drive nutrient dynamics within varied habitats. In addition, the constant transfer of genetic material between microbes shapes their response to environmental change, stimulating constant evolution. This nutrient-mediated interaction between protists and prokaryotes is a fundamental concept in ecological resilience and stability, with consequences for overall environmental health (Alldredge et al., 2014) (Figure 4.16).

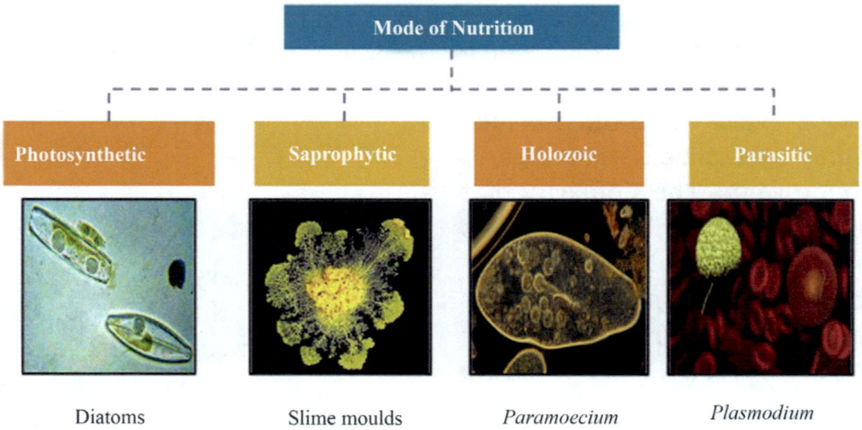

Figure 4.16. Modes of nutrition of protists.

Source: https://dcx0p3on5z8dw.cloudfront.net/Aakash/s3fs-public/inline-images/Modes%20of%20nutrition%20in%20protists.PNG?KBeaXRqC31khTM3ObHD.9aRa1Mx5XQkY.

4.4.3. Reproductive Strategies

The variety of reproduction strategies in prokaryotes and protists has a great impact on their evolutionary success and ecological functioning. Prokaryotes mainly reproduce asexually by means of binary fission, which enables them to grow rapidly in an ideal environment. This process is beneficial because it allows for the fast adjustment to different environmental factors, making sure that the positive genetic features are spread quickly. Also, genetic variation can occur by means of HGT, including transformation, transduction, and conjugation, further facilitating adaptability in prokaryotic populations. Such practices underpin the resilience seen in microbial communities, contributing to their evolutionary dynamics as they exploit ecological niches and respond to environmental pressures, thus allowing a wide array of metabolic and physiological adaptations (Okie et al., 2011). Conversely, the protists display a wider range of reproduction modes with both asexual and sexual reproduction. Asexual reproduction with mitosis is used by many protists, including some algae and protozoa, to rapidly expand their populations, particularly in suitable habitats. Sexual reproduction is however, important in an environment where environmental stresses are dominant, promoting genetic recombination and diversity. This diversity can promote survival since it could provide greater strength in terms of diseases and fluctuating conditions. For instance, certain protists undergo complex life cycles involving alternation of generations, which not only facilitates adaptation but also enables the achievement of complex forms of multicellularity, significantly impacting their ecological interactions and evolutionary trajectories (Okie et al., 2011). The ecological interconnectedness of prokaryotes and protists is further demonstrated by their reproductive strategies, and this can be of great consequence to the global biogeochemical cycles. Among the prokaryotes, it is worth mentioning methanogenic archaea that are involved in the formation of methane using the specific metabolic pathways. This activity not only plays an important role in producing energy in their ecological niches but also leads to the atmosphere with regard to methane, which has the potential consequences of causing climate change. The prevalence of methane emission by these microbial communities along with other anthropogenic sources signifies the significance of microbial reproductive strategies in the backdrop of global environmental issues. Thus, examining these reproductive strategies reveals not only evolutionary advantages but also broader ecological impacts, emphasizing their role as contributors to Earth's metabolic and climatic systems.

Test Your Knowledge

1. Why are protists considered a diverse and polyphyletic group?
2. How do protists differ in terms of locomotion mechanisms (e.g., flagella, cilia, pseudopodia)?

4.5. Types of Protists

Learning Outcome

1. Protozoa
2. Algae
3. Slime Molds

The protist domain is made up of a diverse group of eukaryotic microorganisms with distinct characteristics, hence making up a crucial component of ecosystems around the world. Broadly categorized into protozoa, algae, and fungi-like protists, the group offers a diversity of metabolic strategies as well as environments. Protozoa, typically unicellular and heterotrophic, are exemplified by groups such as amoebas and flagellates that are known for their motility and carnivorous nature. These organisms play important roles in nutrient cycling and serve as a source of food for larger aquatic animals, thus contributing to food webs. Algae, further, sustain global photosynthesis and synthesize organic compounds on which the majority of aquatic food chains are based. Their structural complexity and pigmentation also contribute to ecosystem productivity, once more illustrating the ecological importance of protists despite their classification. Of the interesting protist forms, the most interesting are probably algae since they can photosynthesize, oxidizing and utilizing CO_2. These are made up of a number of subtypes that possess special adaptations to their habitats, such as diatoms, dinoflagellates, and green algae. Diatoms, for instance, possess silica cell walls, making them ubiquitous in the marine habitat, while dinoflagellates cause poisonous algal blooms that impact aquatic life and human health. Green algae, which are closely related to land plants, are major components of freshwater communities and basal producers. The ecological relationships and reactions of such algae indicate the important function of protists to generate ecological stability as well as subsidize superior trophic levels, illustrating their vital inputs into the biosphere as featured in studies of global biomass partitioning between organisms (Bar-On et al., 2018). In contrast to photosynthetic protists, fungi-like protists such as slime molds and water molds are characterized by heterotrophic modes of life with extraordinary life cycles and adaptations. Slime molds, which frequently occur in decaying organic material, form multicellular structures for increased nutrient acquisition and reproduction with extraordinary behavioral adaptations that challenge traditional conceptions of individuality in organisms. Water molds, on the other hand, have profound ecological impacts, particularly in aquatic environments where they can affect fish communities via infectious disease. The study of fungi-like protists contributes to furthering our understanding of decomposition and nutrient cycling processes because their interaction with organisms within ecosystems depicts very complex interdependencies. In general, the diversity of protists indicates their essential position in ecosystem functioning, shedding light on both biodiversity and the evolutionary history of terrestrial life forms (Beaudoin et al., 2015) (Figure 4.17).

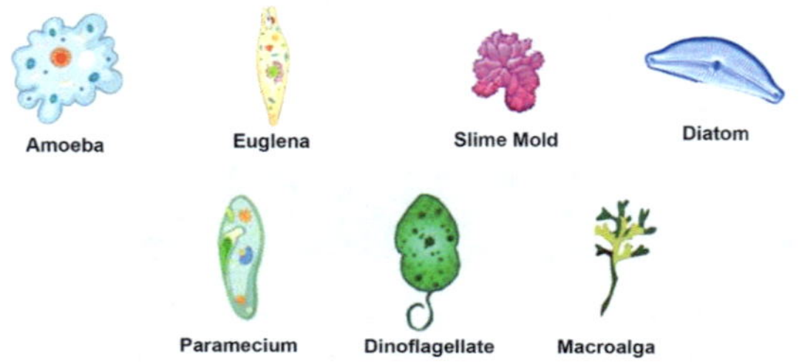

Figure 4.17. Types of protists.

Source: https://media.geeksforgeeks.org/wp-content/uploads/20230420095946/Protista.jpg.

4.5.1. Protozoa

Protozoa, a very diverse class of unicellular eukaryotic organisms, are of critical relevance in aquatic ecosystems, namely within their interactions with prokaryotic life and nutrient cycles of these systems. The classification of protozoa includes several lineages such as flagellates, ciliates, and amoebas, each having particular adaptations and ecological niches. Science unveils that such microorganisms are not merely passive occupants but dynamic participants in the food web, functioning as predators and prey. Recent molecular analysis has highlighted such complex relationships in low-oxygen marine environments where protozoan communities change radically under different conditions of nutrients and oxygen levels, illustrating the complex interdependencies sustaining marine diversity (DeLong et al., 2014). An understanding of the ecological roles of protozoa, especially in response to abiotic factors, is crucial to enhance knowledge of their functions in ecosystem processes. Protozoal-prokaryote interactions are especially significant in most biogeochemical environments, and such interactions underscore their implications for microbial ecology. Protists, for instance, have been shown to exercise strong top-down influence on bacterial communities through predation, governing prokaryotic abundance and diversity in nutrient-rich environments. Evidence shows that in regions of redox gradients, such as the Baltic Sea, protist bacterivory is a controlling force for prokaryotic mortality and hence a determining factor in microbial community structure. Specifically, the decline of protists with increasing levels of hydrogen sulfide shows their role as key players in these chemical transitions and emphasizes the interdependent relationship between eukaryotes and prokaryotes (Anderson et al., 2025). Therefore, protozoa are emphasized by studies as playing an important role in microbial population equilibrium as well as in driving nutrient cycling in diverse environments. In addition to predation, protozoa also exhibit a wide variety of symbiotic relationships with prokaryotes, ranging from mutualism to parasitism. This range ensures their occurrence in diverse environments where they influence microbial dynamics and indirectly affect ecosystem function. Certain species of protozoa have developed the ability to interact with prokaryotic symbionts in order to enhance nutrient uptake and provision of essential substrates to their bacterial symbionts. The interactions enhance the functional capacity of microbial assemblages, especially in oligotrophic environments where resources are limited. Furthermore, the carbon-starvation survival of some protozoa illustrates their capacity to resist and their niche specialization and species-specific weaknesses that play an important role in evaluating community dynamics under stress (Anderson et al., 2025). The protozoa are thus both markers and managers of ecosystem health, and hence their study must be central to comprehending the dynamics of microbial interactions and overall ecological balance (Figure 4.18).

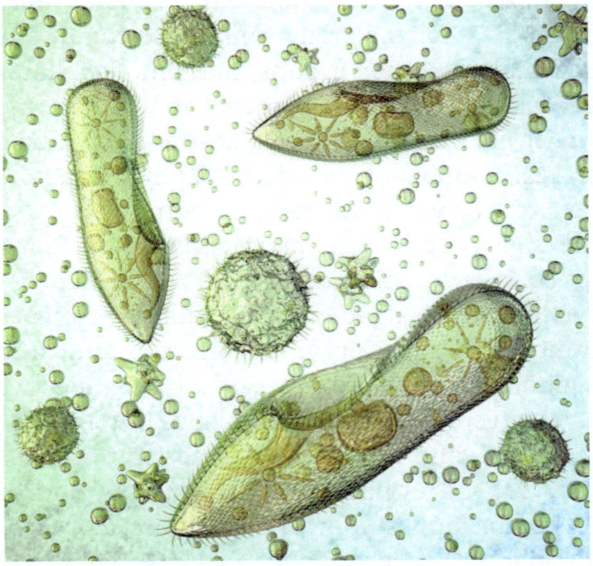

Figure 4.18. Protozoa.

Source: https://media.istockphoto.com/id/610843006/photo/protozoa-under-a-microscope.jpg?s=612x612&w=0&k=20&c=Yjp-NnwyB4SUAoNuLD8fcE82fnAklRury4-U5Z2ylmw=

4.5.2. Algae

Algae, a wide group of photosynthetic organisms generally classified within the protist group, are a vital component in freshwater environments. The organisms form a wide range of forms, from unicellular phytoplankton to multicellular seaweeds, and are of crucial significance in primary production. By using the energy of sunlight and converting CO_2 to organic matter, algae create most of the oxygen in the Earth, and they are also critical in marine and freshwater environments. Algae are a source of base food at numerous trophic levels, from zooplankton to larger aquatic organisms, aside from oxygen production. Improved appreciation of the diversity of algae helps understand aquatic biodiversity and the intricate interaction between such ecosystems, which form a paradigm of their importance in the wider context of both prokaryotes and protists within seawater. The functions of diverse groups of algae are varied and vary considerably depending on their environmental conditions and biology. Other groups, like dinoflagellates and diatoms, have been known to demonstrate strong growth at ideal nutrient conditions, being critical components in blooms of phytoplankton. Conversely, when stressed or nutrient-limited, the algae population can decline, significantly influencing aquatic food webs as well as nutrient cycling. Most algae, besides, also synthesize osmolytes—short organic molecules critical in maintaining cellular osmoregulation. Some of the osmolytes, such as DMSP, also have implications in the marine microbial interactions because they can be utilized by bacteria as a source of carbon and sulfur. This metabolic process highlights the dependence of prokaryotic and algal populations on one another, highlighting not only the algae's role as primary producers but also as active participants in complex biogeochemical activities in marine ecosystems (Alexander et al., 2021). Research has illuminated the evolutionary and ecological sophistication of algal assemblages, particularly their adaptations to various environments, such as oxygen-lacking environments. Current molecular inventories have highlighted the varying eukaryotic algal communities in such challenging environments, where their suspected activities in autotrophy, as well as symbiotic relations with prokaryotes, have been brought to light. Evidence also suggests that specific groups of algae, such as some dinoflagellates and *Syndiniales*, thrive in low-oxygen environments, taking advantage of unique metabolic pathways that enhance their survival. The presence of such organisms may imply tolerance to fluctuating oxygen levels and reactions to nutrient supply, further enhancing our knowledge of the capacity of algal communities to adapt to environmental stressors. Consequently, these ecological role and survival mechanisms persist in the grand narrative of prokaryote and protist interactions, proving their significant contributions to the health and functionality of ecosystems (DeLong et al., 2014) (Figure 4.19).

Figure 4.19. Algae.

Source: https://cdn.mos.cms.futurecdn.net/6ZoRYiwhZBbeTxUJcuvcdY-1200-80.jpg.

4.5.3. Slime Molds

Slime molds are members of the varied kingdom of protists that signify an interesting collection of organisms that defy distinctions between fungi and amoebae. These organisms are generally considered to be moist habitat organisms, and they flourish in leaf litter and other decaying organic materials, where they are known to be very important in nutrient cycling. Slime molds have unusual life cycles, typically alternating between a single-celled and multi-cellular phase. An example of these is the plasmodial slime molds, which may come together to form a large multinucleate cell that acts as a single organism, and which may exhibit amazing behavioral traits, including movement in response to food sources. Their navigational and adaptation skills depict the complexity of microbial life and broaden our knowledge of ecological interaction in the terrestrial ecosystem. Investigations into slime molds are gaining momentum as researchers recognize their significance in soil health and biodiversity, particularly in microhabitats like tropical rainforests (Coelho et al., 2019). The ecological significance of slime molds is hard to overestimate; they are decomposers and help to create soil structure and nutrient cycling. Their occurrence in both land and water habitats indicates that the slime molds could have an impact on the community dynamics through preying on bacteria and other small organisms. A century of research has helped clarify their position in the soil food web, and there is a strong necessity to include slime molds in more ecological research on interactions between protists and prokaryotes. New technologies in microscopy and molecular methods have recently created new opportunities to study slime mold diversity and distribution, which have given hints about the ecological functions they play, and which should be explored in more detail. These efforts are especially crucial in biodiverse regions such as Brazil, where knowledge about microbial species is still developing and comprehensive inventories are badly needed (Creevy et al., 2015). Moreover, the biogeographical distribution of the slime molds may help to reveal their adaptive measures and relations with the environment. The dynamics of slime molds and their environment tell us something vital about the workings of the ecosystem, e.g., their foraging behavior may indicate the state of resources and the habitat. In spite of their significance, a large part of their taxonomy and ecological information, particularly regarding tropical biomes, is understudied with respect to slime molds. As noted in recent research, more work to catalogue and discover new species of slime mold in the diverse Brazilian ecosystems may fill gaps in our knowledge of microbial diversity and its consequences to conservation practices. This comprehensive approach to studying slime molds not only emphasizes the significance of these organisms in ecology but also calls for a reevaluation of existing classifications within the broader context of soil protistology (Coelho et al., 2019) (Figure 4.20).

Figure 4.20. Slime molds.

Source: https://www.nps.gov/articles/000/images/Plasmodium_with-networking-behavior_web.jpeg?maxwidth=650&autorotate=false.

Test Your Knowledge

1. What distinguishes plant-like, animal-like, and fungus-like protists from one another?
2. How do examples like Amoeba, Paramecium, and Euglena represent different types of protists?

4.6. Ecological and Economic Importance of Protists

Learning Outcome

1. Role in Aquatic Ecosystems
2. Impact on Human Health
3. Use in Biotechnology

The ecological significance of protists is brought out by their role as primary producers in various ecosystems, particularly aquatic ecosystems. Certain of these protists, the microalgae such as diatoms and dinoflagellates, form the foundation of the food web by trapping sunlight in chemical form through the process of photosynthesis. Primary productivity by them is not only vital to sustain lush marine life but also to regulate global carbon cycles. For locations such as the Gulf of Maine, rule by microbial populations, i.e., excessive protist populations, is a classic illustration of the intricate web supporting marine environments and contributing to global diversity. The report indicates that cell-based microbes in such environments have been estimated to be as high as 1.7×10^25 organisms, suggesting enormous and largely untapped diversity of life that is worthy of exploration in terms of ecological and environmental wellness (Monier et al., 2011). Other than ecological relevance, protists are also of great economic value across industries. They belong to the aquaculture sector, acting as a provider of essential nutrients and food sources for economically valuable species such as fish and shellfish. The growth of chosen protists, such as phytoplankton and seaweeds, is considered significant in sustainable aquaculture management and food availability assurance in coastal livelihoods. Protists are also very important in biotechnology, where they serve as biofuels, medicines, and in bioremediation processes. By converting organic wastes into renewable energy sources or degrading contaminants, protists provide an opportunity for tackling environmental pressures in conjunction with promoting economic growth. The occurrence and adaptability of these microbes under changing environments expand their possibility of helping sustainable development initiatives (Lear et al., 2018). Recent interest in protist diversity highlights its critical role in ecological dynamics and solving environmental issues. Advances in molecular techniques, such as DNA sequencing, facilitate greater investigation of protist communities and the discovery of intricate relations between ecosystems and how they respond to man-made stress. Such findings are significant in predicting the impact of climate change, loss of habitats, and pollution on marine systems. Protocols for quantifying protist diversity can also stimulate enhanced management and biosecurity procedures for the preservation of marine resources. With their economic and ecological importance, further investment in protist research can result in greater insight and conservation of these critical organisms, supporting ecosystem resilience and, at the same time, human livelihoods (Lear et al., 2018).

4.6.1. Role in Aquatic Ecosystems

It is significant to understand how prokaryotes and protists work in aquatic groups in relation to the cycling of nutrients and flow of energy. Prokaryotes, or more accurately, bacteria, are responsible for decomposing organic matter and recycling nutrients for primary production. They help mineralize organic substances, transforming them into bioavailable forms for phytoplankton, the primary producers for such groups. Additionally, protists, being diverse eukaryotic microorganisms, play a diverse role in the food web as consumers of bacteria and are also used as food by other organisms. That kind of web of relationships intertwined shows the keystone species nature of prokaryotes and protists, maintaining balance in the ecosystem and facilitating the flow of energy across trophic levels. A detailed examination of specific environments, e.g., deep aquatic ecosystems, reveals specific interactions with viral communities that further emphasize the role of prokaryotes in the ecosystem.

In Lake Pavin, for instance, lytic activity of the virioplankton was found to be in correlation with prokaryotic productivity, i.e., there is a dynamic relationship where viruses can influence bacterial populations and, indirectly, nutrient cycling. Seasonal trends in this lytic process were found to be coincident with changes in prokaryotic biomass and production, highlighting the significance of viral predation in controlling bacterial populations in complex aquatic ecosystems. These interactions are central to appreciating the complicated network of life within such ecosystems, as they highlight mechanisms behind prokaryotic and protist function in organic matter cycling and ecosystem processes (Colombet et al., 2012). In addition, biomass measurement across taxa gives important information regarding aquatic ecosystems. Recent estimates indicate that oceanic biomass is largely made up of archaea and bacteria, while they may be less conspicuous than macroorganisms, they are significantly active in energy dynamics and nutrient cycling. The contributions to world biomass by prokaryotes, estimated at approximately 70 Gt C, reveal their relevance compared to the more conspicuous organisms. This distribution points to the idea that marine ecosystems are diverse in life forms and have a remarkable consumer-to-producer ratio that contradicts traditional ecological theory. The explanation of these biomass processes points not only to the role of prokaryotes and protists in aquatic food webs but also to the bringing to the forefront human-driven activities and their impact on these pivotal processes (Bar-On et al., 2018) (Figure 4.21).

Figure 4.21. Protists: role in aquatic ecosystems.

Source: https://www.smorescience.com/wp-content/uploads/2022/11/Protista_collage.jpg.

4.6.2. Impact on Human Health

The impact of protists and prokaryotes on human health is multifaceted, shaping both the disease profile and broader ecological processes. Prokaryotes, or bacteria, can be either positive or negative. Positive bacteria, such as those in the human gut microbiome, help with digestion and are part of immune system processes. Conversely, the pathogenic bacteria that cause tuberculosis and strep throat are harmful to human health. There must be a balance of both the positive and adverse impacts; once disrupted, it has severe consequences on health, such as autoimmune diseases and obesity. It is essential to understand how prokaryotes impact human health to produce effective treatments and prevention measures for bacterial infections, hence, the study of microbiology is vast in scope. Protists are also important to human health, particularly through their ecological roles. The majority of protists are primary producers in aquatic food webs, sustaining fish and other marine animals that are crucial for human nutrition. On this account, they contribute indirectly to nutrition and human food security for billions of the world's population. Furthermore, protist pathogens such as those causing malaria and giardiasis are also extremely direct threats to human health. The interaction of these diverse protists with their environments has important implications. Therapeutics against protist-borne diseases require sophisticated approaches, including the development of vaccines and drugs informed by protist biology and evolution. Today's environmental degradation, as warned, can threaten protist populations, and potentially exert multiplier effects on human well-being and nutrition that must be addressed urgently (Lipps et al., 2006). The increasing issue of antibiotic resistance similarly emphasizes the critical juncture between prokaryotes and human health. Overuse of antibiotics has given rise to strains of resistant bacteria, complicating the treatment of common infections and inducing colossal morbidity and mortality worldwide. This scenario calls for novel approaches, including research into bacteriophages and other new therapies based on the natural virus-carried predation of pathogens. Besides, ongoing studies of both the ecology of protists and prokaryotes are necessary for monitoring and regulation of infectious diseases. A broad overview of these microorganisms' roles in their habitats is crucial to keeping the world healthy, as the implications are not just related to particular pathogens but encompass general ecological wellness (Bar-On et al., 2018). Thus, the incorporation of knowledge of the prokaryotes and protists is essential to address issues of public health within a more interconnected world.

4.6.3. Use in Biotechnology

Biotechnology is beginning to recognize the enormous potential of prokaryotes and protists and is beginning to use their unique characteristics to power innovation in many areas. Prokaryotes especially bacteria, are also essential in various biotechnological processes , such as bioremediation, whereby they are used to decontaminate the environment of heavy metals and hydrocarbons. These microorganisms have great metabolic diversity and therefore can degrade pollutants effectively, thereby offering an environmentally friendly alternative to the conventional chemical techniques. Moreover, new genetic engineering methods enable us to modify those organisms and expand their functions, which will result in more successful bioremediation solutions. All in all, the manipulation of prokaryotic networks changes the way we approach ecological issues, at the same time creating sustainability and broadening our knowledge of microbial ecosystems. Protists are being discovered as important elements in biotechnological studies despite the fact that they are frequently ignored, this is because of their complexity and the variety of biochemical pathways they display. These eukaryotic microorganisms are valuable models to investigate biological processes and have become the centers of attention in the discovery of new drugs and bioactive compounds. As an example, some protists synthesize pharmaceutically useful secondary metabolites and can be used in drug discovery in an attempt to curb disease. The possibility of discovering novel enzymes and metabolic pathways as the researchers explore the genome of various protists increases our biotechnological possibilities. This brings about discoveries in fields like biocatalysis, whereby protist enzymes can be used to perform environment friendly chemical reactions, thereby extending the scope of conventional synthetic routes. More recent developments in sequencing technologies have further established prokaryotes and protists as key players in biotechnology, particularly in biodiversity surveys and environmental surveillance. As noted in recent literature, the isolation and description of the DNA of these microorganisms open up the prospects of better methodology in environmental research. Through standardized procedures of DNA extraction and amplification, researchers will be able to get a better image of microbial communities and their contribution to ecosystem functional processes. These protocols will lead to reproducibility and will increase our knowledge of microbial dynamics in different habitats, which will lead to biodiscovery and biotechnological applications. Such comprehensive genomic approaches not only illuminate the rich biodiversity present in these groups but also enhance our ability to address pressing environmental and public health challenges through innovative biotechnological solutions (Angly et al., 2010; Lear et al., 2018).

Test Your Knowledge

1. How do phytoplankton contribute to marine ecosystems and global oxygen production?
2. What are some economic uses and impacts of protists, such as algae in food or protists in disease?

SUMMARY

The prokaryotes and protists are found to have a wealth of diverse life forms that are essential to understanding the composition and dynamics of the biosphere. The various kingdoms of life, including the bacteria, archaea, and eukaryotic protists, host an interspersed web of interactions and ecological processes. Surprisingly, the biomass count of these organisms reveals that the plant category dominates the trend with a contribution of approximately 450 gigatons of carbon that far exceeds other organisms ($\approx$2 Gt C for animals and $\approx$70 Gt C for bacteria) (Bar-On et al., 2018). This trend hints at the essential functions of these organisms to sustain biological processes, influence nutrient cycling, and form food webs. Thus, the total knowledge of prokaryotes and protists not only reveals their intrinsic biological diversity but also emphasizes their cumulative impact on Earth's biosystems. Moreover, the intricate groupings formed by microbial eukaryotes demonstrate the dynamic nature of protist diversity and its dominant position in various habitats. For example, research in hypersaline ecosystems like Lake Tyrrell has confirmed that microbial eukaryotic communities are subject to specific ecological parameters , such as salinity gradients and season (Andrade et al., 2013). Dominance by specific Alveolate species combined with varying distributions of Chlorophyte genera indicates the complexity and responsiveness of the community structure. In addition, understanding such microbial dynamics helps explain how eukaryotic life responds to environmental stress and competition within their habitats. Beyond improving our knowledge of microbial life, this research suggests the importance of such organisms in attaining ecological balance in response to changing environmental conditions. Overall, a general outline of prokaryotes and protists uncovers their pivotal roles in shaping Earth's biosphere. Since most of the biomass is found in plants, it is clear that the work on lesser visible organisms, , such as prokaryotes and protists, is just as important in terms of understanding larger ecological trends. Paying attention to the complex interactions and processes of these organisms can contribute greatly to the field of ecology, microbiology, and environmental studies. Furthermore, as various anthropogenic processes threaten the stability of these ecosystems, future monitoring of microbial communities and their interactions will be essential in forecasting and mitigating such influences on global biodiversity and ecological health. Through innovative research and enhanced awareness of the roles of all types of life, we can progress toward a harmonious and sustainable coexistence with our world's diverse inhabitants.

Prokaryote and protist studies reveal complex organism interactions that organize the ecosystems of the world. Prokaryotes like bacteria and archaea are vital to nutrient cycling, carbon fixation, and decomposition, demonstrating their global ecological importance. Moreover, their capacity to tolerate extreme conditions is a sign of evolutionary resilience and diverse metabolic capacities. Protists, being a wide variety of unicellular and multicellular forms, have incredible anatomical and way-of-life variability. To understand their functions in providing stability to the environment and in their interactions with other living forms, one needs to investigate their organismic functions, thus laying the foundations for ecological studies in general." Recent research has helped to clarify the intricate dynamics between plant roots and root-associated microbiomes, with a focus on the key contributions of specific root morphologies to prokaryote and protist community assembly. Particularly, root cap activity has emerged as a primary regulator of structuring these microbial communities, with influences not only manifesting in the localized root environment but also having secondary influences reaching deeper root zones and higher trophic levels, as the case of maize root research. Lack of root caps exposed a significant change in the microbiome composition from root hair mutants, thereby demonstrating the significance of various root structures to ecological equilibrium and microbial variety (Bonkowski et al., 2023). This intricate relationship demonstrates how microbial communities in plants are directly linked with plant health and growth. The concept of superorganisms in protists once again deepens the understanding of biological organization and complexity. By focusing on multicellular protists, researchers have established certain parameters that separate simple associations from full integrations as organized superorganisms. This classification includes specificity in attachment structures, co-evolution and obligatory associations between protists and their microbiomes (Gardner et al., 2020). Highly emergent protist lineages like *Symbiontida*, *Oxymonadida*, and *Parabasalia* are excellent examples of these structured superorganisms where the unions are so close that the aggregation mimics tissue or organ structures in the framework of an organism. Such discoveries augment our understanding of ecological interactions and microbial unions' evolutionary significance, shaping future research directions in microbiology and ecology.

KEY TERMS FROM THIS CHAPTER

Algae	Archaea	Bacillus
Bacteria	Binary fission	Cell wall
Cilia	Coccus	Conjugation
Endospore	Flagella	Gram-negative
Gram-positive	Mixotroph	Plasmodium
Prokaryotes	Protists	Protozoa

Critical Thinking Questions

1. **Which of the following is a characteristic of all prokaryotes?**
   a. Membrane-bound nucleus
   b. Multicellular structure
   c. Lack of membrane-bound organelles
   d. Presence of mitochondria
2. **Bacteria and Archaea are two domains of:**
   a. Protists
   b. Prokaryotes
   c. Eukaryotes
   d. Viruses
3. **Which of the following is NOT true about protists?**
   a. They can be autotrophic or heterotrophic
   b. They are all prokaryotic
   c. Most are unicellular
   d. Some are similar to plants, animals, or fungi
4. **Which protist moves using cilia?**
   a. Amoeba
   b. Euglena
   c. Paramecium
   d. Plasmodium
5. **What is a key difference between prokaryotes and protists?**
   a. Protists lack DNA
   b. Prokaryotes are multicellular
   c. Protists have membrane-bound organelles; prokaryotes do not
   d. Prokaryotes have a nucleus; protists do not

Short Answer Questions

1. What are the main differences between prokaryotic and eukaryotic cells?
2. What are the two main domains of prokaryotes, and how are they different?
3. What are the three common shapes of prokaryotic cells?
4. How do prokaryotes reproduce? What is binary fission?
5. What roles do bacteria play in the environment?

Answers to Critical Thinking Questions

1. (c); 2. (b); 3. (b); 4. (c); 5. (c)

REFERENCES

1. Alexander, H., Johnson, W. M., & McParland, E. L. (2021). The osmolyte ties that bind: Genomic insights into synthesis and breakdown of organic osmolytes in marine microbes. https://core.ac.uk/download/483550777.pdf (accessed on 13th June 2025).
2. Alldredge, A. L., Alonso-González, I. J., Bell, T., Boyd, P. W., Calow, P., Clark, M. A., Cossins, A. R., et al. (2014). Microbial gardening in the ocean's twilight zone: Detritivores metazoans benefit from fragmenting, rather than ingesting, sinking detritus. https://core.ac.uk/download/25497686.pdf (accessed on 13th June 2025).
3. Anderson, R. (2025). From single species to the community level: Exploring under-researched avenues in protist ecology and protist–prokaryote interactions. https://core.ac.uk/download/198098419.pdf (accessed on 13th June 2025).
4. Andrade, K., Eisenkolb, N., Emerson, J. B., Heidelberg, K. B., et al. (2013). Characterization of eukaryotic microbial diversity in hypersaline Lake Tyrrell, Australia. https://core.ac.uk/download/323071641.pdf (accessed on 13th June 2025).
5. Angly, F. E., Banfield, J. F., Bergh, Ø., Blazewicz, S. J., Bonetta, L., Brady, S. F., Brehm-Stecher, B. F., et al. (2010). Marine genomics: At the interface of marine microbial ecology and biodiscovery. https://core.ac.uk/download/pdf/8458417.pdf (accessed on 13th June 2025).
6. Balagué, V., Bjorbækmo, M. F. M., Deutschmann, I. M., Ferrera, I., et al. (2022). Long-term patterns of an interconnected core marine microbiota. https://core.ac.uk/download/528380583.pdf (accessed on 13th June 2025).
7. Beaudoin, D. J., Bernhard, J. M., Edgcomb, V. P., Kormas, K. A., et al. (2015). Metazoans of redoxcline sediments in Mediterranean deep-sea hypersaline anoxic basins. https://core.ac.uk/download/74708464.pdf (accessed on 13th June 2025).
8. Berlanga, M., & Guerrero, R. (2010). The Hidden Side of The Prokaryotic Cell: Rediscovering the Microbial World. https://core.ac.uk/download/159084328.pdf (accessed on 13th June 2025).
9. Bochdansky, A. B., Clouse, M. A., & Herndl, G. J. (2016). Eukaryotic microbes, principally fungi and Labyrinthulomycetes, dominate biomass on bathypelagic marine snow. https://core.ac.uk/download/217291150.pdf (accessed on 13th June 2025).
10. Bonkowski, M., Freudenthal, J., Ganther, M., Heintz-Buschart, A., et al. (2023). Root cap is an important determinant of rhiz*osphere microbiome assembly. https://core.ac.uk/download/604742688.pdf (ac*cessed on 13th June 2025).
11. Breunig, A., Brugerolle, G., Dröge, S., Fröhlich, T., et al. (2008). Identifier der endo- und ectosymbiontischen Bakterien und cellulolytic Activiti*es der symbiotic Fl*agellate der Australis Chen Termite Masto Termes Darwinians. https://core.ac.uk/download/14509642.pdf (accessed on 13th June 2025).
12. Castillo de la Peña, Y. M. (2019). Interactions between marine picoeukaryotes and their viruses one cell at a time = Interacciones entre picoeucariotas marinos y sus virus célula a célula. https://core.ac.uk/download/286395302.pdf (accessed on 13th June 2025).
13. Choi, J., & Kim, S. H. (2020). Whole-proteome tree of life suggests a deep burst of organism diversity. https://core.ac.uk/download/323080530.pdf (accessed o*n 13th June 2025).*

14. Colombet, J., & Sime-Ngando, T. (2012). Seasonal depth-related gradients in Viri plankton: Lytic activity and comparison with protistan grazing potential in Lake Pavin (France). https://core.ac.uk/download/49292464.pdf (accessed on 13th June 2025).

15. Costa, R., Magalhães, C., & Pascoal, F. (2020). The link between the ecology of the prokaryotic rare biosphere and its biotechnological potential. https://core.ac.uk/download/288958622.pdf (accessed on 13th June 2025).

16. Creevy, A. L., Valentine, J., & Wilkinson, D. M. (2015). The past, present and future of soil protist ecology. https://core.ac.uk/download/229256046.pdf (accessed on 13th June 2025).

17. DeLong, E. F., Edgcomb, V. P., Ganesh, S., Parris, D. J., et al. (2014). Microbial eukaryote diversity in the marine oxygen minimum zone off northern Chile. https://core.ac.uk/download/222884235.pdf (accessed on 13th June 2025).

18. Deutschmann, I. M. (2021). Disentangling ecological networks in marine microbes. https://core.ac.uk/download/534167647.pdf (accessed on 13th June 2025).

19. Ferrera, I., & Krabberød, A. K. (2022). Long-term patterns of an interconnected core marine microbiota. https://core.ac.uk/download/533906125.pdf (accessed on 13th June 2025).

20. Gardner, A., Stencel, A., Van Lommel, A. T., Leander, B., Dyer, B. D., Leander, B. S., Lynn, D. H., et al. (2020). Superorganisms of the protist kingdom: A new level of biological organization. https://core.ac.uk/download/344433870.pdf (accessed on 13th June 2025).

21. Gomati, V., Kumar, K., Mula, R. V. R., Ramasamy, S., et al. (2011). Greenhouse gas emissions from termite ecosystem. https://core.ac.uk/download/478422278.pdf (accessed on 13th June 2025).

22. Goto, S., Itoh, M., Kanehisa, M., Kuma, K., et al. (2007). Evolutionary history and functional implications of protein domains and their combinations in eukaryotes. https://core.ac.uk/download/pdf/8062304.pdf (accessed on 13th June 2025).

23. Hongoh, Y., Noda, S., Ohkuma, M., Moriya, S., Sato, T., et al. (2009). Complex coevolutionary history of symbiotic Bacteroidales bacteria of various protists in the gut of termites. https://core.ac.uk/download/pdf/8311676.pdf (accessed on 13th June 2025).

24. Lear, G., Dickie, I., Banks, J., Boyer, S., Buckley, H. L., Buckley, T. R., ... Holdaway, R. (2018). Methods for the extraction, storage, amplification and sequencing of DNA from environmental samples. New Zealand Journal of Ecology, 42(1), 10–50A. https://doi.org/10.20417/nzjecol.42.9(accessed on 13th June 2025).

25. Lima Coelho, I. (2019). Species richness and ecological diversity of Myxomycetes and Myxomycete-like organisms in the tropical forests of Brazil. https://core.ac.uk/download/286641135.pdf (accessed on 13th June 2025).

26. Lipps, J. H. (2006). Major features of protistan evolution: Controversies, problems and a few answers. https://core.ac.uk/download/477895230.pdf (accessed on 13th June 2025).

27. Margulis, L., Berlanga, M., & Guerrero, R. (2013). Symbiogenesis: The holobiont as a unit of evolution. https://core.ac.uk/download/159084487.pdf (accessed on 13th June 2025).

28. Monier, A., Monier, A., Longhurst, A. R., Pilskaln, C. H., Johnson, C. L., Pilskaln, C. H., Fautin, D., et al. (2011). Planktonic microbes in the Gulf of Maine Area. https://core.ac.uk/download/pdf/8580329.pdf (accessed on 13th June 2025).

29. Okie, J. G. (2011). Allometric Scaling and Metabolic Ecology of Microorganisms and Major Evolutionary Transitions. https://core.ac.uk/download/151574277.pdf (accessed on 13th June 2025).

30. Selby, S., & Smith, M. (2015). In*structor's Guide to Concepts of Biology, Chapters 12–21. https:// core.ac.uk/download/217306124*.pdf (accessed on 13th June 2025).

31. Soyer-Gobillard, M. O., & Dolan, M. F. (2015). Chromosomes of Protist*s: The crucible of evolution. International microbiology :* the official journal of the Spanish Society for Microbiology, 18(4), 209–216. https://doi.org/10.2436/20.1501.01.252 (accessed on 13th June 2025).

32. Y.M. Bar-On,R. Phillips,*& R. Milo, The biomass distribution on Earth, Proc. Natl. Acad. Sci. U.S.A. 115 (25) 6*506-6511, https://doi.org/10.1073/pnas.1711842115 (2018). (accessed on 13th June 2025).

FUNGI AND PLANT DIVERSITY

5

LEARNING OBJECTIVES

After studying this chapter, you will be able to:

1. Define the role of fungi in ecosystems;
2. Explain the types of fungi;
3. Define plant diversity: an overview;
4. Focus on the symbiotic relationships between fungi and plants;
5. Explore fungi as pathogens;
6. Describe the conservation of fungi and plants.

Introduction

The intricate relationship of fungi and plant diversity is a major factor in the health of ecosystems since the partnership contributes to several ecological processes. Fungi form symbiotic connections with the root systems of plants, enhancing nutrient uptake, particularly phosphorus and nitrogen. Mycorrhizal association allows plants to endure various environments, contributing to resistance to climatic fluctuations. The complexity of these processes underscores the significance of soil biodiversity, which includes the thousands of fungi that reside within it. In a report to the European Commission, soil biodiversity has an essential role in the diversity of soil life, ranging from genes to communities, underscoring its crucial role in the provision of ecosystem services of benefit to human societies and the environment (Van der Putten et al., 2010). It is hence necessary to understand the plant-fungal interaction dynamics to aid biodiversity conservation and sustainable agriculture. Apart from their very important function in nutrient cycling, fungi also contribute to decomposing organic matter, the by-product of which is also improved soil quality and enhanced plant vigor. Through decomposition with the assistance of fungal organisms, essential nutrients are freed that are vital for plant growth. This symbiotic relationship places emphasis on the interrelatedness of organisms in ecosystems, as fungi benefit plants but also possess complex relationships with other soil organisms, including bacteria and invertebrates. Terrestrial ecosystem management practices, especially in agriculture, have immediate implications for these relationships.

A clear review of the soil biological processes points to the importance of maintaining fungal diversity since it is crucial to maintaining and providing support for soil function and ecosystem services, particularly in sustainable agricultural systems (Phillips et al., 2006). Such evidence points to the importance of sustainable agriculture that maintains such significant relationships. The urgent necessity to regain lost biodiversity and protect microbial diversity, i.e., that of fungi, has gained increased awareness in recent years. Urbanization and agriculture, with their continuous loss of native environments, necessitate the preservation of fungal biodiversity not only for ecological balance but for human existence as well. Fungal biodiversity serves as an indicator of ecosystem health and resilience, with significant services , such as carbon sequestration, water regulation, and soil stabilization. A perception of the worth of such biodiversity is essential for informing policy and management responses that generate sustainable ecosystems. Confronted with the issues of land management practices and the need to conserve fungi as well as vegetation will be critical to long-term sustainability, as argued through current discussion regarding the relevance of holistic conservation and ecosystem management strategies (Van der PuttenPhillips et al., 2006; Van der Putten et al., 2010) (Figure 5.1).

Figure 5.1. Fungi and plant diversity.

Source: https://i.ytimg.com/vi/HR_n3ZR6G3s/maxresdefault.jpg.

Definition of Fungi and Plants

The separation of fungi and plants is the basis for understanding the greater diversity that they both share. Fungi form a group of organisms distinct from others, placed under the kingdom Fungi, and characterized based on the type of nutrition followed by them, which includes the engulfing of organic matter through exterior digestion. Fungi do not contain chlorophyll and hence cannot perform photosynthesis. They play critical roles in ecosystems as decomposers, returning dead organic material and nutrients to the ground. Fungi are also characterized by their complex multicellular body, which forms networks referred to as mycelium, which can cover vast expanses of the ground, thereby allowing exchange of nutrients between plant species. This ecological role puts fungi at the basic players in plant diversity and soil health, towards an increasingly wide appreciation for biotic associations within terrestrial communities. Compared to fungi, plants are generally defined through their ability to undergo photosynthesis, where they utilize light energy and CO_2 to synthesize organic compounds. They belong to the kingdom Plantae, which harbors an enormous diversity of form, function, and environmental niche from small herbaceous plants to giant trees. Chloroplasts which contain chlorophyll and enable the trapping of sunlight as chemical energy, occur in plants.

This feature induces a direct reliance of much of the terrestrial ecosystems on plants for energy. In addition, plants frequently have intricate interactions with fungi, especially in mycorrhizal relationships, which increase nutrient uptake. Ectomycorrhizal (ECM) fungi, for example, can gain considerable carbon from hosts while also enhancing the host's ability to acquire nutrients, illustrating a striking interrelationship that emphasizes the interconnectedness of an ecological community (Fransson et al., 2001). Fungi and plants are only comprehensible if the genetic and ecological sophistication generating biodiversity is recognized. The individual genetic heterogeneity (IGH) theory addresses the ways in which genetic diversity within one organism can impact ecosystem and evolutionary processes. This genetic diversity can guide the way fungi and plants respond to environmental change and shape community organizations and stress resistance.

Figure 5.2. Fungi and plants.

Source: https://www.globalgarden.co/wp-content/uploads/2022/06/fungus_extracting_nutrients_from_pine.jpg.

By comparing the interactions and genetic structures of the two groups, scientists can learn about biodiversity quantifications that extend beyond the counting of species. These analyses provide insights on the fundamental roles these organisms play in their ecosystems, highlighting the reality that diversity is not only a matter of quantity but also of the quality and sophistication of biological interactions, as the case of the sophisticated relations developed between plant species and fungal networks illustrates (Cárdenas-Flores et al., 2019) (Figure 5.2).

Importance of Studying Fungi and Plant Diversity

Unraveling the intricate relationship between fungi and plants is important for deciphering the mechanisms that ensure ecosystem stability. Fungi, particularly those forming arbuscular mycorrhizas like the *Glomeromycota*, have key functions in the nutrient uptake of most land plants, facilitating crucial processes like phosphate uptake (Barrett et al., 2005). These symbiotic associations enhance plant growth and resistance, and thus, studying fungi and plant diversity is essential. A more nuanced understanding of these interdependencies can give insight into ecosystem function, particularly how different species contribute to and regulate processes , such as mineral cycling. Further, consideration of these interdependencies can assist pinpoint how biodiversity influences ecosystem stability, particularly when faced with environmental stressors. This data is increasingly vital, particularly with global losses of biodiversity and changing climates. The impacts of plant diversity and fungi extend well into the realm of ecological responses to climate change. The Quaternary period has demonstrated how fluctuating climatic conditions have repeatedly impacted ecological communities, fostering evolutionary processes that lead to species adaptation and extinction (Stewart, 2010). Studying the diversification and resilience of plants and fungi under these changing conditions allows scientists to assess the adaptive strategies employed by different species. It is crucial knowledge that not only highlights the changing conditions in ecosystems but also guides conservation efforts towards maintaining biodiversity. Closing the knowledge gaps as to how specific plants and fungi support climatic adaptation will enable ecologists and conservationists to prepare for setting strategies to mitigate the adverse effects of climate change on the environments. Other than their ecological significance, plants and fungi yield various pragmatic uses, demonstrating their importance in biodiversity research. For instance, a majority of fungi contain bioactive compounds utilized in medicine and farming, whereas diverse plant groups are involved in food security and ecosystem services. It is essential to be aware of the genetic and functional diversity in these groups for bio-prospecting new species and actualization of their potential value to human health. The potential is broadened to agricultural resilience through improved, more vigorous crop breeds and sustainable agriculture. Therefore, the research of plant diversity and fungi contributes not only to our knowledge of ecosystems but also in itself is worthwhile in application, ultimately advocating for further discovery and preservation of these essential organisms (Barrett et al., 2005).

Overview of the Relationship Between Fungi and Plants

The complex fungi-plant interaction is an ecological interaction that forms the basis of several terrestrial ecosystems. This mutualism, especially in mycorrhizal relationships, helps in nutrient exchange that is essential in the health and growth of plants. Mycorrhizal fungi grow their hyphae in the soil and enlarge the surface area of nutrient uptake, especially phosphorus which is commonly a limiting nutrient to plant performance. These fungi, in turn, provide the host plants with carbohydrates and other organic compounds, solidifying a mutualistic relationship that increases the chances of survival of both partners. The flexibility of such interactions as terrestrial ecosystems develops leads to plant diversity, enabling them to grow in diverse conditions, but at the same time also influences the make-up of fungal communities that grow with them. The study of these dynamics helps one to understand the complexity and resilience of ecosystems in which the two organisms coexist. Continuing on the description of this relationship, it has been shown that during the process of ecological succession, fungi-plant interactions may change dramatically. As an example, in an experiment on successional stages in alpine ecosystems, the structure of microbial communities (fungi included) was found to change with changing environmental conditions. Even though the diversity of different taxa was generally higher in the succession, the fungal interactions might get simplified with time. Namely, with the succession, the number of interactions of mycorrhizal fungi with plants decreased in comparison with the interactions with the soil microfauna. These findings highlight that while diversity in plant and fungal taxa can grow, the intricacies of their interactions do not necessarily follow the same trajectory, suggesting a nuanced understanding of ecological networks is essential for interpreting plant-fungal relationships (Bueno et al., 2019). The specificity of mycorrhizal associations is even more evident in the case of orchids, where unusual dynamics of associations can be highlighted, which emphasizes the role of fungi in the development of plants. Genus like *Tulasnella* and *Ceratobasidium* are especially essential in the germination and development of orchid species which in many cases depend on fungal associates to establish initially. More recent phylogenetic analyses have begun to illuminate the evolutionary relationships among these fungi and highlighted the significance of appropriate taxonomic classification to enhance our comprehension of their roles. The research indicates potential novel species and highlights how ongoing taxonomic revisions could affect conservation efforts for endangered orchid species that depend on these mycorrhizal connections (Barry et al., 2019). This shows an important feature of plant diversity, that the existence of particular plant taxa is frequently closely linked to their associated fungal symbionts, revealing the complex networks of life that support ecological integrity (Figure 5.3).

Figure 5.3. Relationship between fungi and plants.

Source: https://plantamerica.com/wp-content/uploads/2023/11/Difference-Between-Plants-and-Fungi-Plant-America.jpg.

5.1. The Role of Fungi in Ecosystems

Learning Outcome

1. Decomposition and nutrient cycling;
2. Mycorrhizal associations with plants;
3. Impact on soil health and structure.

The intricate interactions between fungi and ecosystems are far from simple decomposition; they play central roles in cycling nutrients, most notably with regard to soil health and plant diversity. Various fungal species, e.g., ECM fungi, play a central role in the sustenance of forest ecosystems through the establishment of symbiotic associations with tree roots. Symbiotic association enhances nutrient uptake, especially nitrogen and phosphorus, with direct consequences for plant growth and community composition. Recent studies have shown that variations in environmental conditions, , such as increased snow depth, can alter ECM fungal richness and community composition. For instance, under extreme snow conditions, ECM fungal richness can decline, negatively affecting the nutrient mobilization mechanisms critical to plant vigor. This illustrates how shifts in fungal communities as a result of climate change can cascade throughout ecosystems, influencing not only fungal populations but also plant diversity and land plant health (Bahram et al., 2016). Arbuscular mycorrhizal fungi (AMF) also illustrate the significance of fungi to plant diversity and ecosystem resilience, especially in agricultural environments. These fungi form symbiotic relations with the majority of crop plants, increasing nutrient acquisition and plant health. Plant relationships with AMF are increasingly important as the soil environment is changing with global climate change. Fluctuations in atmospheric CO_2 concentrations have been found to impact the effectiveness of AMF in plant-soil microbial nutrient transfer. For instance, while AMF are still enhancing the uptake of nitrogen and phosphorus even when CO_2 is high, cultivar variations mean that some plants are assisted more than others. Such dynamics of AMF in the context of a changing climate are therefore crucial for the design of sustainable agriculture methods that can leverage such symbiotic relationships (Field et al., 2020). The implications of fungi's role in an ecosystem are widespread, particularly in a heating planet. Fungal diversity is critical to make ecosystems stable, given that shifts in the make-up of fungal communities have direct impacts on nutrient cycling and decomposition of organic matter, key processes to maintaining ecological balance. Because climate change is altering how precipitation falls and temperature varies, the resistance of fungal species is compromised. It has been shown through research that shifts in fungi involved with organic matter and plant roots have the capability to destabilize existing nutrient cycling networks, which in their own right effect diversity in plants. Understanding these cascading effects is sure to help conservation policy and ecosystem management strategies in preserving the intricate interactions of fungi with the larger ecological environment. In summary, the roles played by fungi are priceless in ensuring plant diversity and rendering ecosystems sustainable amidst global environmental changes (Bahram et al., 2016) (Figure 5.4).

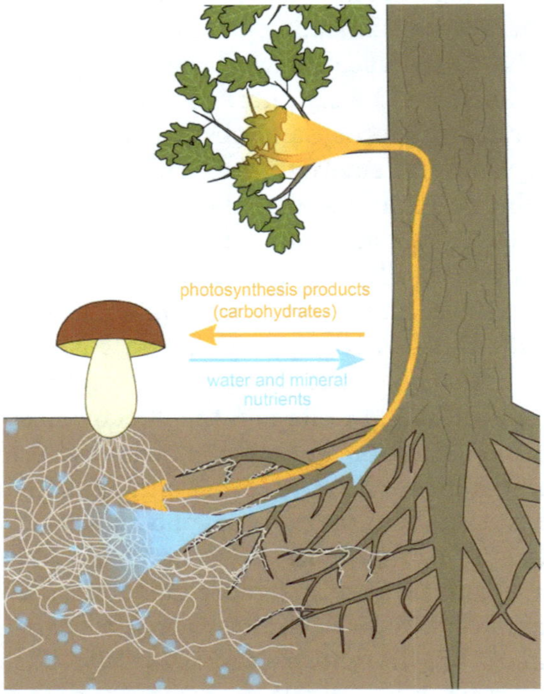

Figure 5.4. The role of fungi in ecosystems.

Source: https://library.fiveable.me/_next/image?url=https%3A%2F%2Fstorage.googleapis.com%2Fstatic.prod.fiveable.me%2Fsearch-images%252F%2522Ecological_roles_of_fungi%253A_decomposers%252C_mycorrhizae%252C_endophytes%252C_pathogens%252C_and_food_sources_in_ecosystems.%2522-607px-Mutualistic_mycorrhiza_en.svg.png&w=640&q=75.

5.1.1. Decomposition and Nutrient Cycling

The role of decomposition in an ecosystem's health is a cornerstone of nutrient cycling, greatly helped by fungi. Decomposition is the process of the degradation of organic matter by a variety of organisms in which fungi play pivotal roles in its success. These organisms, particularly saprotrophic fungi, possess specialized enzymes capable of degrading complex organic substances , such as cellulose and lignin in plant tissues efficiently. As they decompose organic material, fungi recycle important nutrients back to the soil, making them available to plants. Nutrient cycling is pivotal in maintaining soil fertility and promoting plant community diversity. Without decomposing fungi, nutrient cycling would be significantly impaired, and nutrient loss would be experienced in soils of lower plant diversity, indicating intricate interactions in plant and fungal communities (Gopalakrishnan et al., 2024). Fungal diversity is also essential to ecosystem stability, especially in terms of nutrient cycling. Different fungal species make unique contributions to the degradation of organic matter and the mobilization of nutrients. For instance, mycorrhizal fungi develop symbiotic relationships with plant roots, enhancing nutrient uptake and providing plants with crucial soil nutrients , such as phosphorus and nitrogen. This mutualistic interaction not only benefits plant wellbeing but also enhances the overall diversity of plant species within a population. With various fungi adapted to different environments, their capacity to degrade different organic matter is priceless when it comes to ensuring ecosystem resilience in terms of climate change. Fungal diversity maintenance thus seems to be a fundamental element of maintaining productive and diverse plant communities, as they are closely related to decomposition processes (Ambrogi et al., 2024). With the situation of ongoing climate change, the contribution of fungi in decomposition and nutrient cycling is more crucial than ever. Environmental alterations , such as increasing temperatures and altered precipitation patterns, can impact decomposition and nutrient release rates in soils. These can disrupt prevailing fungal communities, leading to potential diversity and functionality losses. As indicated, it is critical to preserve such populations of fungi not only for nutrient cycling but also for the services they provide in maintaining ecosystems stable and resilient. Practices , such as conservation agriculture and integrated nutrient management may enhance soil health by stimulating healthy fungal action. Through their implementation, we can safeguard the essential role that fungi play in decomposing organic matter and cycling nutrients, thus enhancing plant diversity and general ecosystem health in an ever-changing world (Ambrogi et al., 2024; Gopalakrishnan et al., 2024) (Figure 5.5).

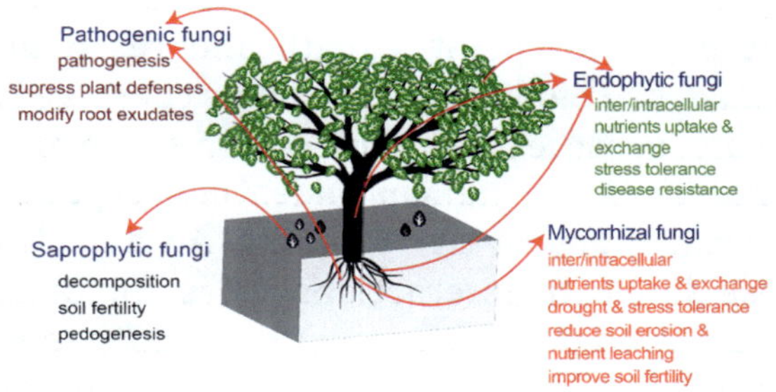

Figure 5.5. Decomposition and nutrient cycling of fungi.

Source: https://www.researchgate.net/profile/Muhammad-Adnan-61/publication/360127134/figure/fig2/AS:1169235233513480@1655778811567/Fungal-diversity-involved-in-nutrient-cycling-decomposition-and-pathogenicity-Soil.png.

5.1.2. Mycorrhizal Associations with Plants

Mycorrhizal associations are a critical element of plant ecology influencing nutrient uptake and interactions among various plant species. Symbiotic association of mycorrhizal fungi with the roots of plants enhances the uptake of vital nutrients, phosphorus, and nitrogen, especially, which are often deficient in most soils. Such a mutually beneficial arrangement not only enhances the health and growth of plants but also renders them more tolerant to environmental stresses. Further, the mycorrhizal network, likened to a wood wide web, allows for plant communication, allowing them to trade nutrients and respond as an association to threats, thus enhancing community stability. , such associations underscore the important role of the mycorrhizal fungi in plant diversity and ecosystem functioning, highlighting their importance in natural and agroecosystems. The implications of mycorrhizal association extend far into invasive species dynamics. Studies have suggested that AMF can be utilized as an important factor determining non-indigenous invasive plant success. The benefits derived from enhanced nutrient and moisture uptake enable these invaders to dominate native-dominated ecosystems. Thus, understanding the relationship between AMF and invasive plant species is essential in developing effective management plans. Furthermore, the research indicates that such fungi can disrupt pathogenic microbes, giving invaders an extra advantage to establish populations. Therefore, understanding how AMF associations influence plant invasion processes has the potential to unlock the greater consequences for ecosystem integrity and diversity (Bongard et al., 2012). New molecular research has established the diversity of AMF in varied environments, further assessing their ecological significance. For instance, research conducted in tropical forests has established extensive AMF diversity, whereby 30 various fungus types have been shown to be linked to tree seedlings. An astonishing 17 of them were not previously documented, which is indicative of the great and yet vast wealth of uninvestigated mycorrhizal fungal biodiversity in tropical populations. Moreover, significant differences in fungal communities were found over time and between hosts of various plants, which indicates that AMF populations are not only diverse but ecologically differentiated as well. These findings illustrate the role of mycorrhizal fungi in plant community recruitment and dynamics, further emphasizing the intricate interactions between fungi and plant diversity in terrestrial ecosystems (Bever et al., 2002) (Figure 5.6).

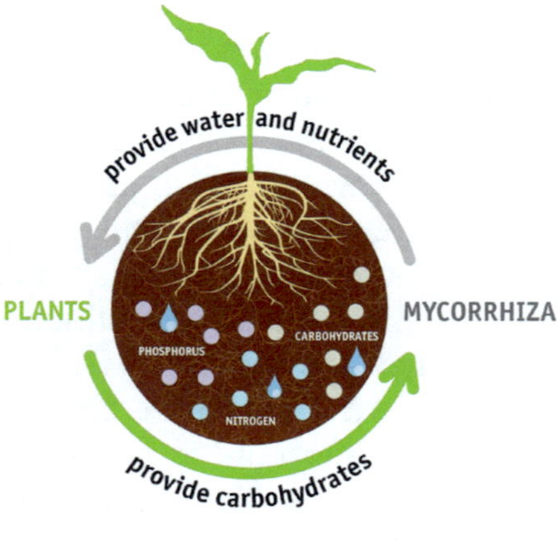

Figure 5.6. Mycorrhizal associations with plants.

Source: https://www.hechenbichler.com/img/containers/assets/news/Mykorrhiza-website-en-RGB.jpg/5d9854dbb9e84917dd1822beeaf92243.jpg.

5.1.3. Impact on Soil Health and Structure

The fungi and soil health associations are very essential with regard to the sustainability of ecosystems. The fungi, especially mycorrhizal species, are very important in improving the soil structure and fertility. They create a complicated network of hyphae which stabilizes soil aggregates, resulting in enhanced aeration, moisture retention, and nutrient availability. This mutually beneficial relationship promotes the wellbeing of plants and leads to the overall strength of soil against erosion and degradation. Urban environments, often characterized by disturbed soils and reduced biodiversity, can still support diverse ECM communities. Research illustrates that ECM fungal richness remains high in urban parks, although community composition varies between conifer and broadleaf trees, hinting at the adaptability and ecological significance of these fungi in modified landscapes (Francini et al., 2017). Therefore, a diverse population of fungi is important in the integrity of soil both in natural and urban ecosystems. Soil biodiversity is not limited to fungi but includes a whole soil biome of organisms that all play a part in soil function and health. Soil biodiversity, defined as the variation in soil life, significantly influences ecosystem services , such as nutrient cycling and water filtration (Van der Putten et al., 2010). The complex interrelationships of microorganisms , such as fungi, bacteria and protozoa, form the basis of biochemical processes that are required to make the soil alive. This microbial diversity is important in healthy ecosystems to decompose organic matter and release nutrients at a rate that promotes plant growth and resistance. Besides, soil biodiversity should be conserved at all costs because it has a direct correlation to the sustainability of human endeavors, especially in farming and in city planning. Awareness of these intricate interactions gives important information to environmental management efforts that target the creation of healthy soils. Moreover, it is hard to overestimate the effects of human activities on soil health. The main threat to soil biodiversity and structure is anthropogenic activity represented by urbanization, agricultural activity, and climate change. Soil compaction, declining organic matter and microbial richness that frequently accompany land-use changes, impede the diverse services that healthy soils can offer. As evidenced by the shifts in ECM fungal communities in urban parks, the resilience of these ecosystems is not solely determined by soil properties but is significantly influenced by the presence of flora (Francini et al., 2017). Thus, a variety of plant life is a key to preserving the quality of soil and creating an environment in which the beneficial fungi may flourish. The solution to these problems should be based on the combined attack; in particular, the focus should be on the conservation measures that would enhance the biodiversity of plants and soil as the united front in the maintenance of the ecosystem health.

Test Your Knowledge

1. What is decomposition?
2. Discuss mycorrhizal associations with plants.

5.2. Types of Fungi

Learning Outcome

1. Ascomycetes and their significance;
2. Basidiomycetes and their ecological roles;
3. Zygomycetes and their contributions to biodiversity.

Fungi offer an unparalleled array of form and function, and becoming familiar with the various groups is the secret to understanding how they function in ecosystems and interact with plants. The four broad categories of fungi are chytridiomycetes, zygomycetes, ascomycetes, and basidiomycetes, each having different reproductive traits and life cycles. Chytridiomycetes, for example, are aquatic and are often important in nutrient turnover in aquatic environments. On the other hand, zygomycetes are soil fungi living in decaying organic matter, with functions in decomposition processes. These forms indicate the functions that fungi have for maintaining environmental well-being in terms of soil fertility and being significant facilitators of plant diversity. Ascomycetes, or sac fungi, are the most diverse group in the fungus kingdom and comprise yeasts to truffles. Their reproductive organ, the ascus, produces spores for reproduction. This division is especially significant in farming conditions due to the fact that some of the ascomycetes form symbiotic associations with plants, , such as mycorrhizal fungi, which enhance nutrient uptake in the host plant. This mycorrhizal interaction is a clear illustration of how fungi can enhance plant diversity by enabling plants to thrive in otherwise poor soils. Furthermore, their roles involve ecosystem services , such as the regulation of soil health necessary for sustaining natural and human-modified landscapes, as emphasized by the importance of soil biodiversity in delivering ecosystem functionality (Van der Putten et al., 2010). Another notable category of fungus is Basidiomycetes, distinguished by their fruiting bodies known as mushrooms. These fungi are instrumental in decomposing complex organic materials, including wood and plant debris, thereby facilitating nutrient cycling within ecosystems.

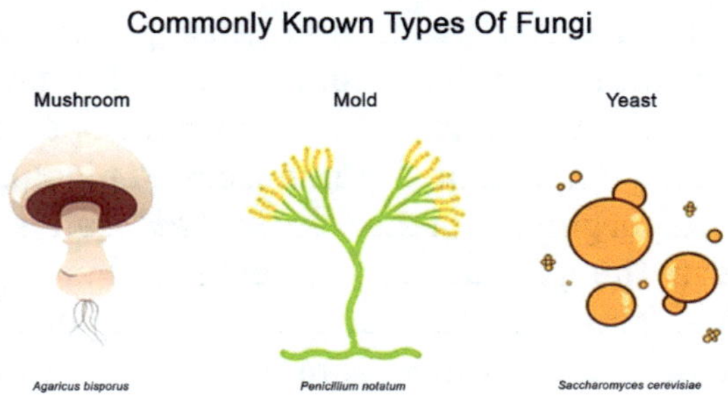

Figure 5.7. Types of fungi.

Source: https://www.shutterstock.com/image-vector/commonly-known-types-fungi-science-260nw-2316433563.jpg.

They can form symbiotic relationships with plants through mycorrhizal associations, significantly impacting plant health and diversity by improving soil structure and nutrient availability. Besides, basidiomycetes support specific ecosystems, , such as grasslands, by ensuring soil fertility and health through their interactions with the soil microfauna (Murray et al., 2007). By these intricate processes, fungi enhance plant diversity and ecosystem stability, consolidating their priceless contributions towards environmental stability and human health (Figure 5.7).

5.2.1. Ascomycetes and Their Significance

The functional and ecological richness of Ascomycetes, which make up one of the greatest groups of fungi, attests to their ecologically significant roles. Ascomycetes are renowned for their ability to form intricate associations with other organisms and plants and contribute to plant nutrition and nutrient cycling through diverse symbiotic relationships. Mycorrhizal association, particularly ectomycorrhizae, assists in nutrient and water uptake by the plant and provides carbohydrates in return. This symbiotic relationship sustains forest ecosystems via soil structure improvement and biodiversity promotion. Also playing an important role in the decomposition of organic matter, the vast array of Ascomycetes species, including important decomposers Penicillium and Aspergillus, further enrich the soil and govern the productivity of ecosystems as a whole. Thus, Ascomycetes are an important part of the intricate system of associations sustaining plant diversity and ecosystem integrity. Ascomycetes contribute to the soil-plant community interaction vital to sustaining biodiversity, particularly for various forest ecosystems. For instance, their associations with tree roots are critical in nutrient delivery, which in turn links fungal diversity with host plant development and health. Research indicates that boreal forests are highly diverse in microbes, including Ascomycetes, involved in decomposing and cycling nutrients (Rajala et al., 2008). Further, variation in Ascomycete diversity and composition may have effects on the overall health of forest societies as they may exert pressure on plant communities through pathogenic relationships or facilitate plant growth by symbiotic associations. Consequently, this oscillating relationship highlights the relevance of fungal ecology in the context of forest conservation and management and indicates the long-standing role of Ascomycetes in maintaining ecological balance.

The significance of Ascomycetes extends beyond their direct ecological roles, as it refers to their possible applications in agriculture, biotechnology, and environmental conservation. For example, the use of Ascomycetes in bioremediation has gained a lot of interest, where Ascomycetes can degrade pollutants and organic waste products, thus contributing to the restoration of ecosystems. In another arena, Ascomycetes are greatly known for their participation in antibiotics production, enzymes, and fermentation products, which refers to their economic importance. With advancing climate change and ecological degradation, Ascomycetes are sentinel species of ecosystem health and resilience. As science continues to advance, an understanding of their complex dynamics with plant diversity will be crucial in the development of strategies that optimize ecosystem sustainability in the face of an evolving environment, a reminder of the critical role played by fungi in the biosphere of our planet (Figure 5.8).

Figure 5.8. Ascomycetes.

Source: https://cache.careers360.mobi/media/article_images/2024/8/9/Ascomycetes.jpg.

5.2.2. Basidiomycetes and Their Ecological Roles

Basidiomycetes are a fascinating group of fungi with varied ecological roles in various environments, greatly influencing plant diversity and ecosystem health. As one of the major decomposers of woodland ecosystems, the fungi play an important role in nutrient cycling, as they are involved in the breakdown of lignocellulosic material in dead and dying vegetation. By decomposing complex organic substances into simpler ones, Basidiomycetes enrich the soil, making it more conducive to plant community development. Their extent of substrate colonization and degradation enables them to interact with an array of environmental factors, which contributes to the initiation and diversity of plant life. The host plant symbiotic association, for example, ECM associations, also demonstrates their ecological importance by enhancing the rate of nutrient acquisition while providing plants with essential micronutrients, thereby promoting a diverse plant community in the majority of ecosystems (Buscot et al., 2000). Their plasticity is evident under environmental change conditions. Fungi respond to alterations in the environment, for example, soil chemistry or disturbance changes, with high resilience. For instance, studies have shown that Basidiomycetes demonstrate differential community variation according to site-specific characteristics , such as substrate and light (Cornejo et al., 2018). Such flexibility not only aids their application as bioindicators of ecosystem health but also demonstrates their resilience in the field of restoration ecology. Through the knowledge of the specific responses of these fungi to environmental changes, ecologists would be able to utilize their capabilities toward restoration planning and implementation of effective restoration activities. Such facts are highly applicable in disturbance-prone ecosystems, illustrating Basidiomycetes' interactions with plant diversity in continuously changing environmental conditions. Basidiomycetes' ecological roles also go beyond nutrient cycling and interactions with plants, providing implications for general ecosystem processes. Their abundance increase could be an indicator of ecosystem resilience, particularly after disturbance, as they are key to soil rehabilitation and vegetation recovery. Through mycorrhizal interaction, Basidiomycetes facilitate the movement of nutrients and water, which improves plant drought resistance and fitness. The interaction provides a matrix for plant diversity, as healthy plants nurture biodiversity-supportive wildlife and other microbial communities. With the pressing imperatives of habitat loss and climate change, the ecological roles of Basidiomycetes must be understood in a bid to preserve ecosystem stability and biodiversity. By fostering such relationships, we can realize the potential of Basidiomycetes in the growth of resilient ecosystems that can sustain both fauna and flora in a world of climatic changes (Figure 5.9).

Figure 5.9. Basidiomycetes.

Source: https://cdn0.thedailyeco.com/en/posts/8/6/4/what_are_basidiomycetes_fungi_468_orig.jpg.

5.2.3. Zygomycetes and Their Contributions to Biodiversity

The Zygomycetes group, with their distinct reproductive structures, has a diversified contribution towards biodiversity. This group of fungi with a main saprophytic mode of life occupies many ecological niches and is important in nutrient cycling as a result of degrading organic matter. Their decaying capabilities greatly improve the soil quality and aid the growth of plants and therefore, they are very essential to the land-based ecosystems. Moreover, Zygomycetes , such as Rhizopus and Mucor have applications in the agricultural setting, where they may affect the health and productivity of crops. These fungi promote an active interrelationship of various organisms towards a balanced ecosystem, which is essential in the maintenance of biodiversity in both natural and managed ecosystems. Besides, Zygomycetes play an important role in the establishment of mycorrhizal associations, which are symbiotic relationships with plant roots. , such associations not only benefit the plants by increasing the water and nutrient uptake capacity but also increase the soil structure and enhance the resilience to environmental stresses. These interactions are an example of the intricate interdependencies that occur in ecosystems and show how Zygomycetes diversify the plant diversity by allowing the prosperity of communities of plants and animals. Research outcomes from soil biodiversity programs highlight the intricate connections between soil fungi and plant health, indicating that Zygomycetes contribute significantly to biological diversity by underpinning the productivity of terrestrial ecosystems (Finlay et al., 2005). Therefore, they do more than just decompose, as they have other ecological implications that drive biodiversity. The Zygomycetes also have a high level of adaptability to different environmental conditions, further increasing their levels of contribution to biodiversity. Their capacity to live in a wide range of substrates, including soil and decomposing organic materials, allows them to live in diverse ecosystems and promote an extensive species interaction. For instance, studies suggest that certain Zygomycetes can impact the presence and abundance of other microbial communities in their vicinity, thereby influencing overall ecosystem dynamics (Broussard et al., 2014). This flexibility also translates to their application in biotechnology where Zygomycetes are being tapped to discover their potential in bioremediation and waste management. These initiatives do not only demonstrate their ecological value but also reflect their significance to the human activities and environmental sustainability. Therefore, Zygomycetes importance in biodiversity is complex and entangles ecological well-being with human concern in the stability of the environment, which is desirable to be improved (Figure 5.10).

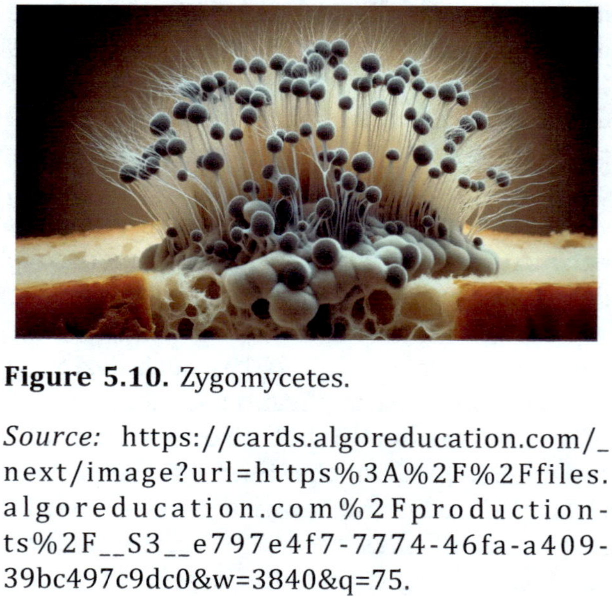

Figure 5.10. Zygomycetes.

Source: https://cards.algoreducation.com/_next/image?url=https%3A%2F%2Ffiles.algoreducation.com%2Fproduction-ts%2F__S3__e797e4f7-7774-46fa-a409-39bc497c9dc0&w=3840&q=75.

Test Your Knowledge

1. What are the ascomycetes and their significance?
2. What are the zygomycetes and their contributions to biodiversity?

5.3. Plant Diversity: An Overview

Learning Outcome

1. Classification of plants;
2. Importance of plant diversity for ecosystems;
3. Threats to plant diversity.

The intricate fabric of plant diversity is central to the ecological resilience of our planet, with innumerable species contributing to many different ecosystem structure and function. In addition to being the major producers, transducing sunlight into energy through photosynthesis, plants also provide vital habitat and resources for other organisms, including fungi. Fungi-plant symbiotic relationships are vital because fungi enhance nutrient uptake, particularly phosphorus, through their extensive mycelial systems. , such associations underscore the role of plant and fungal diversity in maintaining ecosystem health. In addition, the extremely high metabolic diversity of plants makes ecosystems robust and capable of surviving under different environmental conditions. This dynamic relationship means that the protection of plant diversity is necessary not only for the plants themselves but also for the numerous organisms that are dependent on these plants to live, underscored by the imperative for protection. In addition, plant diversity is also now recognized for its function in solving ecological problems , such as global warming and habitat degradation. Plant communities with species diversity have a higher carbon sequestration potential and improve soil health which supports increased flood and drought resilience. Soil erosion is reduced by the mix of species and water holding is maximized, in return, resulting in enhanced larger ecological processes. In urban areas, the application of ecotechnologies, including green roofs, illustrates how multiple plant species increase city biodiversity. Whereas green roofs will tend to provide fewer species than natural ground-level habitats, they can enhance structural diversity, while the latter can facilitate greater species richness. Moreover, wetlands within urban areas with plant-diversity-based designs reflect growing awareness of the way functional diversity is realized in overall biodiversity, highlighting the importance of environmentally friendly approaches that respect plant diversity as an essential component of healthy ecosystems (Knapp et al., 2019).

Figure 5.11. Plant diversity.

Source: https://slideplayer.com/7249171/24/images/slide_1.jpg.

Fungal and plant diversity interaction is not simply symbiotic but also encompasses the ecological processes and interactions defining whole biomes. Fungi contribute to decomposition and nutrient cycling of organic matter, processes critical for healthy plant populations. In most cases, fungal communities exchange nutrients with individual plant species through mycorrhizal associations to benefit both the associates. Microbial module study also reveals how the interactions within plant and fungal networks regulate ecosystem processes. Cooperative dynamics of these associations, deconstructing the complexity of plant diversity as well as demonstrating the possibilities of biotechnological intervention, , such as bioremediation, are aided by understanding them. With ecosystems coming under greater threat from pollution and climate change, an integrated knowledge of how plant and fungal diversity are intertwined will be critical to informing the design of sustainable measures to build ecological resilience and remediating degraded habitats (Folwell et al., 2012) (Figure 5.11).

5.3.1. Classification of Plants

Classification of plants is an elementary knowledge that one needs in order to research the intricate connections between plant life forms and their associated fungi. Plants are most commonly categorized into two groups: vascular and non-vascular plants. Non-vascular plants, , such as mosses, lack specialized tissues for water and nutrient conduction, whereas vascular plants, , such as ferns, conifers, and flowering plants, possess these special systems, thus enabling them to adapt to different environments. The consequences of plant classification are not merely taxonomic; they have ramifications in ecosystem function and the role different species play in their ecosystems. Recognition of such classifications is imperative as it delimits the interaction of plants with fungi, leading to symbiotic relations that can create necessary symbiotic connections that enhance nutrient capture and ecosystem stability, highlighting the dependency of plants and fungi to support Earth's biodiversity (Van der Putten et al., 2010). Within vascular plants, there is further division into gymnosperms and angiosperms, each with a different reproductive strategy. Gymnosperms, , such as pines and cycads, produce seeds not in an ovary, whereas angiosperms, which include the majority of flowering crops, produce seeds in a fruit. This classification is significant because it not only determines the trajectory of evolution of these plant lineages but also how they associate with fungi. For instance, AMF are usually most frequently associated with angiosperms, which assist them in accessing such important nutrients like phosphorus that contribute to better plant growth and survival in nutrient-poor soils. Plant classification hence, directly influences the types of fungi with which they develop symbiotic relationships, thus presenting the complex network of interaction that sustains environmental balance and increases plant diversity in varied ecosystems (Amselem et al., 2020).

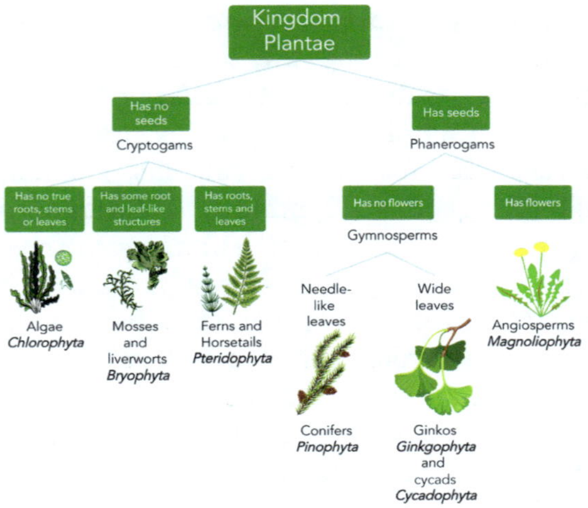

Figure 5.12. Classification of plants.

Source: https://letstalkscience.ca/sites/default/files/2023-01/Plant-classification.png.

An appreciation of the classification of plants also includes the impact of environmental change on diversity. When habitats are destroyed or altered by human processes, plant classifications are important to the prediction of the strength and capacity of plant groups to adjust. The diversity derived in plant taxonomies implies their ability to fill different ecological niches, whose effects affect soil quality and ecosystem functions , such as food generation and carbon storage. Conservation of plant biodiversity is crucial to the maintenance of healthy ecosystems due to its strong association with the functionality of soil organisms, , such as important fungi (Van der Putten et al., 2010). Accurate classification of plants will allow ecologists and conservationists to improve on ways to conserve these species and their associated fungi, ultimately helping to mitigate the impacts of habitat destruction and climate change on world biodiversity (Figure 5.12).

5.3.2. Importance of Plant Diversity for Ecosystems

Plant diversity is important for ensuring healthy ecosystems, having impacts on other ecological processes and services that are essential for life. An ecologically diverse plant community enhances productivity, stability, and resilience, positioning ecosystems to respond to change, including climate variability and human-induced stress. Diverse plant species occupy different ecological niches, thereby optimizing the use of resources , such as sunlight, water, and nutrients. This plant diversity also adds to the soil health via processes like root architecture variation, which stabilizes the soil and enables nutrient cycling. In addition, such diversity maintains a rich food web, where diverse habitats and food are formed for a wide range of organisms, including fungi. Consequently, the high plant diversity creates a stronger ecosystem with the ability to support the varied life forms that depend on it, indicating how life is interconnected within these ecological webs. The relationship between mycorrhizal diversity and plants shows the symbiotic relationships that maintain healthy ecosystems. Mycorrhizal fungi form symbiotic relationships with the roots of plants to facilitate enhanced uptake of water and nutrients, a basic necessity in poor soil conditions. Mycorrhizal colonization is species-specific and varies across plant species, as evidenced by field violet, soybean, and maize. These studies reveal that while higher soil phosphorus levels could inhibit the diversity of arbuscular mycorrhizal (AM) fungi, plant host species are more crucial to the determination of the community structure of these fungi. This procedure supports the need to maintain plant diversity, which in turn supports a diverse mycorrhizal community, leading to healthier soils and efficient ecosystems with beneficial cascading impacts throughout the ecosystem (Abdi et al., 2013). In addition to direct ecological benefits, plant biodiversity is also important to human societies. Plant diversity contributes to ecosystem services, i.e., food production, fiber generation, and climate regulation, which are critical for sustainable human existence. Rich plant diversity in healthy soil is necessary for the consistent supply of these resources. Plant diversity has a direct effect on the capacity of the soil to provide essential services, i.e., clean water and air, and hence underscores interdependence between soil health and biodiversity. Secondly, as highlighted by scientists, soil diversity is the dominant driver of governing these services, thus the need to ensure plant and soil diversity for long-term sustainability (Van der Putten et al., 2010). Thus, the conservation of plant diversity not only preserves ecological integrity but ensures the survival of important ecosystem functions upon which human societies have come to rely (Figure 5.13).

Figure 5.13. Aerial view of a circular forest formation, illustrating natural symmetry and biodiversity in forest ecosystems.

Source: https://earthbuddies.net/wp-content/uploads/2019/11/trees-o.jpg.

5.3.3. Threats to Plant Diversity

The modern world of environmental protection is witnessing a growing interest in the threats to plant diversity, as this issue has an extensive impact on the stability of the ecosystem and human lives. One of the most significant challenges is climate change, where the increase in temperature and changes in precipitation patterns cause a cascade of imbalance in the plant communities. Those species that cannot adapt or move are at a high risk of extinction, causing a reduction in biodiversity. This loss is not just worrying to the plants alone, but also to the myriads of organisms, which include fungi that establish important symbiotic relationships with plants. This plant diversity loss in turn threatens ecosystem services, including nutrient cycling and soil health, upon which human survival and environmental sustainability depend. In this regard, the realization of the dependence of both the plants and fungal diversity would be of utmost importance in designing effective conservation measures. Invasive species also increase the risks to plant diversity, causing shifts in ecosystem-level interactions and possibly causing the extirpation of native plants. Exotic species usually win the native plants in the competition for resources and break the original ecological relationships. They are also able to alter the soil microbial communities that may affect the functioning of the whole ecosystem. As an example, when invasives are stopped, they may provide a legacy effect, which makes it hard to recover the native plants even after the management operation. Other recent work has demonstrated that some management approaches, including herbicide use, can fail to adequately mitigate these legacy effects, resulting in long-term shifts in soil microbiota that prevent the establishment of native plants. This highlights the need for comprehensive approaches that also consider the underlying soil ecosystem when attempting to restore plant diversity in invaded landscapes (Barberán et al., 2020). Biodiversity in the soil, which is part of the maintenance of plant communities, is also under threat, which may disrupt the recovery of diverse plants. The relationship between soil health and plant diversity is immense; healthy soils support diverse plant growth, whereas degraded soils may result in low biodiversity. The diversity of soil biota, as recent reports have stressed, is the difference between thriving ecosystems and those that are vulnerable to collapse. Fungi are beneficial in nutrient cycling and plant communication, but these functions are impaired as the soil biodiversity reduces. With the ever-increasing anthropogenic activities that are lowering the quality of soil, the consequences run across the entire plant kingdom. Consequently, embracing a holistic approach to preserve both fungi and plant diversity, alongside soil health, is essential for maintaining the ecological balance and the services that sustain human societies (Van der Putten et al., 2010).

Test Your Knowledge

1. What is the classification of plants?
2. What is the importance of plant diversity for ecosystems?

5.4. Symbiotic Relationships Between Fungi and Plants

Learning Outcome

1. Mycorrhizae and plant nutrient uptake;
2. Lichens as a model of mutualism;
3. Fungal endophytes and plant health.

Fungal-plant symbiosis is a critical component of ecological processes, particularly as mutualistic relationships. , such associations, often referred to as mycorrhizae, enable intricate networks of fungal mycelia and plant roots for nutritional exchange. Fungi are capable of receiving important minerals from the ground, , such as phosphorus and nitrogen, unavailable to plants in isolation. As a trade, the plants provide fungi with carbohydrates synthesized through photosynthesis, and this encourages symbiotic relationships that both organisms utilize for better survival and growth. Fungi contribute significantly to soil health and plant productivity by facilitating the development of stability and resilience in ecosystems. Such interactions shift as they point towards adopting an integrated land management approach oriented towards the benefits of developing such symbiotic relations in agricultural and natural ecosystems. Moreover, the mutualistic concept comprises a wider spectrum of interactions among species where there is a gain to both the species. Mutualistic relationships range from obligatory to facultative, showing varying degrees of dependence between plants and fungi. For instance, some plants, mycoheterotrophs, obtain their total nutrients from mycorrhizal fungi, illustrating a more extreme, obligatory symbiotic mode. Certain plants may have facultative association, whereby the development is supplemented by mycorrhizal fungi but not necessarily essential for survival. The range of these associations shows the versatility of fungi and plants to diverse ecological niches and environmental stresses. An awareness of mutualism diversity not only indicates the richness of ecological interactions but also the importance of preserving and protecting such interactions against environmental change (Boucher et al., 1982). Further, more recent studies emphasize how mycorrhizal fungi are involved in the delivery of important ecological services like carbon sequestration and soil stabilization.

Figure 5.14. Fungal symbiosis.

Source: https://dr282zn36sxxg.cloudfront.net/datastreams/f-d%3A2dcb8d36823ea53ae64df0e0b86e64dcb1916599f8ae79a624f817e6%2BIMAGE_TINY%2BIMAGE_TINY.1.

As highlighted in biomass forestry debates, these fungi develop enormous mycelium networks that link various plants, enhancing the exchange of nutrients and the health of soils (Hodge et al., 2009). In mitigating climate change, these have also drawn interest due to their capacity for making carbon-neutral measures a possibility. By integrating mycorrhizal fungi into land management practices, we can further the ecological resilience of our landscapes, providing not only food security in the form of enhanced crop production but also environmental sustainability. These approaches emphasize the critical role of fungi in plant diversity, solidifying the symbiosis between biotic communities underpinning our planet's ecosystems (Figure 5.14).

5.4.1. Mycorrhizae and Plant Nutrient Uptake

The association of mycorrhizal fungi with plants is symbiotic and plays a significant role in nutrient uptake, a key factor in plant health and productivity. Mycorrhizae increase the root system of plants through an extraradical mycelium network, facilitating an increased volume exploration in the soil, estimated by Sieverding (1991). This prolonged period of influence not only grants the plants access to essential nutrients, most significantly, phosphorus, a sequestered soil nutrient, but also facilitates the absorption of other minerals like nitrogen and potassium. Even more significantly, the potential for nutrient translocation from the fungal network to the host plant is one of the most pronounced benefits of this symbiotic relationship. The higher the dependence of plants on mycorrhizal association for nutrient uptake, especially in soils of low nutrient status, the higher the necessity to understand mechanisms underpinning the same, especially for improving agricultural productivity and ecosystem sustainability, as identified in recent studies (Gollotte et al., 2008). In addition, mycorrhizae possess benefits beyond the mere use of nutrients. Aside from improving assimilation of nutrients, mycorrhizal fungi are also involved in the development of plant resistance to water stress.

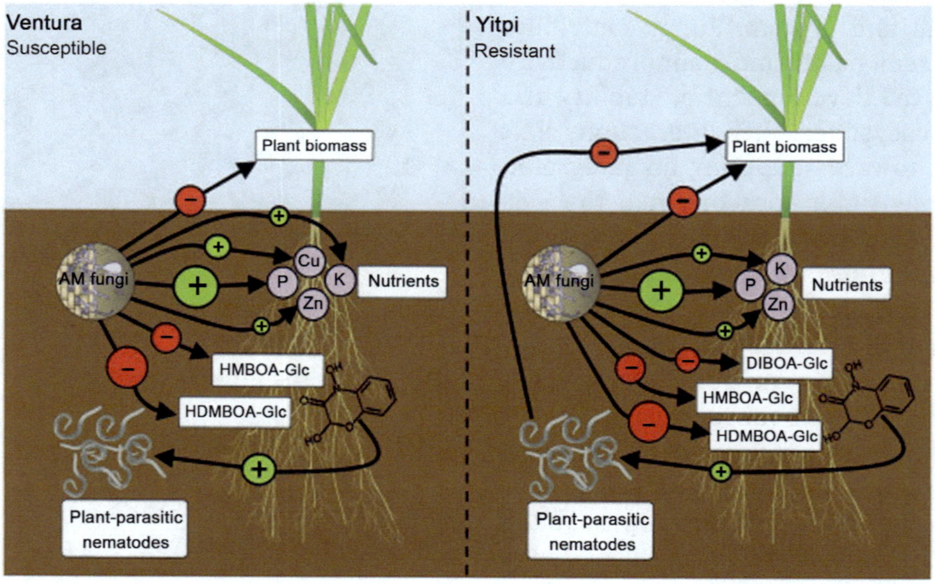

Figure 5.15. Mycorrhizal fungi enhance nutrient uptake.

Source: https://ars.els-cdn.com/content/image/1-s2.0-S003807171830275X-gr6.jpg.

For instance, the mycelium networks can be employed to absorb water during dry spells, effectively eradicating stress to the plant. This is confirmed by the fungus's function to protect root systems from disease-causing organisms, thereby reducing root diseases. The symbiosis facilitates a defense system in which plants can thrive in suboptimal conditions and ultimately advance terrestrial ecosystem health.

The mycorrhizal associations' stress resistance comes in particularly handy in agricultural settings where environmental stresses are the norm, even more so in stressing the significance of these fungi in promoting plant diversity and sustainability across ecosystems (Ianson et al., 2010). Nevertheless, despite the clear advantages of bringing mycorrhizal fungi into agriculture, there are issues in harnessing their maximum capacity. Mycorrhizal inoculation techniques for the enhancement of mycorrhizae in crops are typically marred by economic and practical limitations, , such as high production costs and limited expertise for their proper application. Quality control of mycorrhizal products supplied to the market is also needed to ensure that they are free from disease and well compatible with specific soil and plant conditions. Thus, while inoculation of native mycorrhizal communities by good agricultural practice would be of significant benefit, localized, situational assessments need to be made before large-scale inoculation. Such prudence will make more effective use of mycorrhizae, hence promoting more sustainable agriculture and higher overall diversity of plants within ecosystems (Gollotte et al., 2008) (Figure 5.15).

5.4.2. Lichens as a Model of Mutualism

Lichens present a highly intriguing example of mutualism with the intricate mutualism of fungi and photosynthetic partners that are typically algae or cyanobacteria. Such a symbiotic association showcases how two disparate organisms can coexist in a relationship beneficial to both. The fungus partner provides a protective structure and access to water and available nutrients from the environment, which is major in usually harsh habitats in which lichens can be commonly found. In return, the photosynthetic partner delivers photosynthetic organic carbon compounds and thereby supports the fungal component. Two benefits emphasize the significant ecological role of lichens within ecosystems in which they can inhabit stressful substrates and play roles in soil formation, nutrient cycling, and stabilization of extreme environments. Their ability to survive under abiotic conditions further proves mutualism's adaptability as a survival mechanism within diverse ecological niches (Boucher et al., 1982). Mutualism classification has evolved, and lichens are examples of fluid relationships that defy classical categorizations. While mutualism is typically defined as an association in which both involved species benefit, lichen symbiosis records a spectrum of dependency. For instance, some associations of lichens are obligate, i.e., one partner cannot survive in the absence of the other, while others exhibit facultative relationships where partners can survive independently. This range in symbiotic associations adds sophistication to our understanding of the symbiotic association in lichens, and it suggests that ecological environments and evolutionary pressures shape these associations. Consequently, lichens are an important model for examining symbiotic partnerships, adding to our general knowledge of symbiosis, diversity, and ecological resilience. The appreciation of diversity in mutualistic relationships, especially among lichens, is key to understanding functional biodiversity in ecosystems (Lumbsch et al., 2017). Additionally, research on lichens sheds light on fungal and plant diversity, noted especially in relation to their significant role in ecosystem health. Lichens are not only primary producers but also bioindicators of environmental quality, which is sensitive to changes in air quality and climate. Their sensitivity makes them an invaluable tool for ecologists monitoring environmental change.

The symbiotic association of lichens is also a model for the potential of cooperative strategies in optimizing survival under stress. This is then leveraging conservation's potential to inform conservation strategies in the aftermath of climate change and deforestation. By factoring in lichens' functions into broader discussions on fungi and plant diversity, scientists stand to better appreciate the interconnectedness of life forms and how they underpin ecological resilience. Lichens are thereby the very peak of mutualism that offers necessary insights to enhance biodiversity in our changing world (Boucher et al., 1982) (Figure 5.16).

Figure 5.16. Lichens as a model of mutualism.

Source: https://upload.wikimedia.org/wikipedia/commons/d/d0/Green_Lichen.jpg.

5.4.3. Fungal Endophytes and Plant Health

Fungal endophytes are crucial in boosting the health and resilience of plants in the complex network of ecosystems. These microorganisms colonize plant tissues without causing any detrimental effects and establish mutualistic relationships with their plant hosts, which bring many benefits to them. Among the most interesting benefits of fungal endophytes is the possibility to enhance plant tolerance to abiotic stresses, including drought, salinity, and high or low temperatures. With climate change posing unprecedented challenges, it is important to gain insights into the role of these endophytes in alleviating the stress responses. Recent studies indicate that endophytes can modulate various physiological processes in plants, thereby enhancing growth and survival under stressful conditions (Ruiz-González et al., 2022). They are an important resource to the wellbeing of plants by helping them to acquire nutrients and elicit hormonal reactions, which is particularly important in the fast-evolving environment. Besides, the benefits of fungal endophyte-host plant interaction are not limited to abiotic stress tolerance but also involve increased resistance to pathogenic challenges. The complex symbiosis enables the plants with endophytes to stage superior immune responses. This is most supreme in agricultural environments where a crop disease may reduce yields significantly. Endophytes have the potential to create secondary metabolites that suppress the growth of pathogens, thus serving as natural biocontrol. Research demonstrates that plants harboring specific endophyte communities exhibit reduced incidences of disease and enhanced overall fitness (Ruiz-González et al., 2022). These plants, which are enriched with microbiomes, not only sustain their health but also lead to the stability of the broader ecosystem through the promotion of biodiversity. Decoding these intricate interactions paves the way to sustainable agricultural systems to exploit the relationship between microbes to increase food security. With the disruption of natural habitats due to urbanization and anthropogenic activities, the significance of conservation of fungal endophyte diversity is growing more and more important. It has been found that urban conditions can have an effect on the endophyte community structure, which has a prospective impact on the plant health and diversity. As an example, in a study on seed fungal endophytes linked to *Banksia ericifolia,* it was found that overall community composition was similar, but urban locations had higher species richness and diversity than natural environments. This observation highlights the capability of some endophyte communities to live in disturbed landscapes, which is a factor in plant flexibility and ecosystem stability. Nevertheless, significant knowledge gaps remain with regard to the processes that underlie these dynamics. The overall picture of fungal endophytes in natural and urban environments is needed because this knowledge will be used in the development of conservation measures to preserve biodiversity and promote the health of plants in the face of the rapid environmental change (Figure 5.17).

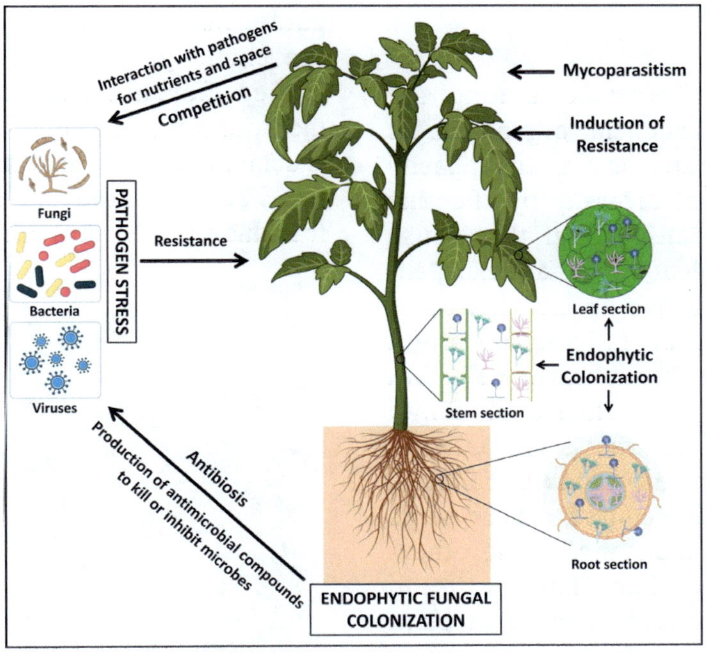

Figure 5.17. Fungal endophytes and plant health.

Source: Gowtham HG, Hema P, Murali M, Shilpa N, Nataraj K, Basavaraj GL, Singh SB, Aiyaz M, Udayashankar AC, Amruthesh KN. Fungal Endophytes as Mitigators against Biotic and Abiotic Stresses in Crop Plants. Journal of Fungi. 2024; 10(2):116. https://doi.org/10.3390/jof10020116

Test Your Knowledge

1. What are mycorrhizae and plant nutrient uptake?
2. What are lichens as a model of mutualism?

5.5. Fungi as Pathogens

Learning Outcome

1. Impact of fungal diseases on plant populations;
2. Case studies of significant plant pathogens;
3. Strategies for managing fungal diseases in agriculture.

In the intricate web of ecosystems, fungi are both pathogenic and beneficial organisms. As pathogenic organisms, fungi possess the capacity to drastically alter plant diversity by affecting plant survival and health. Fungi that cause wheat rusts as well as leaf blights are excellent examples of the destructive impact that these pathogens are capable of inflicting on crops. These diseases not only reduce farm yields but also can hinder the long-term sustainability of cropping systems. The association between fungi and their host plants tends to follow a complex co-evolutionary course, where some hosts have developed resistance while others continue to be susceptible. This variation in susceptibility can create changes in plant community structures, eventually changing overall ecosystem stability and functioning. It is essential to comprehend such interactions to maintain plant diversity as well as guarantee strong agricultural practices that are resistant against such attacks by fungi (Van der Putten et al., 2010). The pathogenic role of fungi is even more significant in the agricultural sector in the context of climate change. As global temperatures rise and precipitation patterns change, the interaction between fungi and their hosts is most likely to be changed fundamentally. For instance, fresh plant pathogens like Sudden Oak Death could become more prevalent under altered climatic conditions, threatening whole forest ecosystems. These pathogens depict how global warming makes some host plants more susceptible to infection by fungi and, consequently, makes them more difficult to control. The dynamic interplay of heterogeneous climatic factors and their capacity to influence the phenology of vegetation and the life cycle of pathogens demands innovative approaches that integrate ecological awareness and stakeholder participation. In the future, scientists suggest that several management strategies will be necessary that consider the intricate influence of climate change on plant disease and fungal pathogens (Dobson et al., 2012). Overcoming the challenge of fungal pathogens entails a unified effort in scientific research, agricultural methods, and preservation of ecosystems. One crucial field of research in the past few decades has been the study of soil biodiversity, the foundation of healthy plant communities.

Figure 5.18. Fungi as pathogens.

Source: https://cordis.europa.eu/docs/article/images/2024-10/454269.jpg.

Healthy soils are crucial not only for sustaining plant growth but also for providing a healthy soil organism community, including beneficial fungi that have a competitive advantage or suppress pathogenic fungi. As has been discussed in recent appraisals, sustaining soil biodiversity is a valuable tactic to increase ecosystem resistance to invasion by fungi. Furthermore, integrating enhanced soil management practices into farming systems can potentially help maintain crop health and minimize reliance on chemical fungicides. By encouraging an interdisciplinary approach to the role of fungi as a pathogen within the larger ecosystem of soil biodiversity and health, we can achieve well-informed policies that encourage plant diversity and ensure food security (Van der Putten et al., 2010) (Figure 5.18).

5.5.1. Impact of Fungal Diseases on Plant Populations

The impact of fungal diseases on plant populations is widely researched and referred to as an important factor influencing agricultural production as well as natural communities. Fungal pathogens are capable of decreasing crop production and quality immensely and thus causing growers economic losses as well as impaired food security globally. For instance, *Bipolaris sorokiniana*-fungal disease has been reported to devastate barley crops and result in widespread foliar necroses that undermine the well-being and viability of the plant. This interaction highlights the complex interactions between hosts and fungi, where the disease not only undermines the immediate growth of the plant but can also create long-term problems in soil health and biodiversity. Appreciation of these dynamics is necessary in an effort to create effective management approaches that can limit the impact of such pathogens and initiate sustainable agricultural practices in agricultural systems. The ecological relevance of fungal diseases is also high, as they can transform plant community structures and diversity. Fungal pathogens preferentially infect certain species, leading to imbalances in plant populations. This can reduce competition so that there are more tolerant or resistant species to establish and thereby disrupt existing ecological relationships. For instance, AM fungi, which are beneficial to the majority of plants, may be impacted by the presence of disease-causing fungi , such as *Mycosphaerella graminicola*. This interaction indicates that while some fungi have the capability to enhance plant health, their relationship with plant pathogens will shift, which will result in an uncertain future for affected plant populations in the environment (Cabello et al., 2014). Such changes in diversity and composition will propagate effects, with further effects on insect populations, soil microbiota, and ecosystem resilience. Effective management of diseases in fungi is essential for plant diversity and sustainability. Sustainable strategies, as has been demonstrated by recent studies, hold the capacity to potentially play a major role in the suppression of disease. It has been observed, for example, that AM fungi have been found to suppress the growth of diseases like *B. sorokiniana*, ultimately making them more resistant (Sjöberg et al., 2005). It is this symbiotic relationship that demonstrates the necessity of adding biocontrol measures in the management of fungal disease, so there could be a dual benefit: reducing the reliance on chemical fungicides and promoting diversified vegetative communities. Through the encouragement of these wholesome relationships, it becomes feasible to create a more stable community that not only deals with the immediate threat of pathogens but also increases the long-term wellness of plant populations and their associated communities (Figure 5.19).

Figure 5.19. Fungal and fungal-like diseases of plants.

Source: https://ohioline.osu.edu/sites/ohioline/files/01-31-hyg-fact-PP401_07.pdf.pdf-2_2.jpg.

5.5.2. Case Studies of Significant Plant Pathogens

The link between plant diversity and fungi is very well illustrated in several case studies of some of the key plant pathogens that have posed threats to ecosystems around the world. For instance, the epidemic of *Phytophthora infestans*, which leads to late blight in potatoes, reconfigured not only agricultural practices but also socioeconomic trends in the nineteenth century. This pathogen enabled the Irish Potato Famine, exemplifying the ability of fungal pathogens to catalyze widespread starvation and migration. Beyond direct devastation, the influence of *P. infestans* also reaches into the future, as its genetic diversity has made it necessary to continually research resistant crop stock. These mechanisms also mirror broader ecological shifts, demonstrating how plant pathogens can impact agricultural diversity and even the socio-political landscape, suggesting that efforts to combat such pathogens must be taken ecologically as well as genetically resistant (Nuala et al., 2015). Another fascinating case is the white rot fungus, *Heterobasidion annosum*, which is a significant conifer forest pathogen. This fungus infects the vascular tissue of trees and, through this, causes weakening and, most often, death of the plants. *H. annosum* has also drawn attention since it potentially threatens timber economies as well as forest management. More than any other aspect, perhaps, its lifecycle and pathogenesis mechanisms give one a profound insight into host-fungus dynamics. The incidence of *H. annosum* appeals to scholars to revisit biodiversity conservation methods since forest structures of complexity can dampen the prevalence of pathogens and facilitate an ecosystem's resilience. This emphasizes the complex pathogen-plant diversity relationship, necessitating comprehensive forest management strategies in combating the danger of such pathogens while increasing ecological stability (Jonàs et al., 2014). In addition to these historical plant pathogens, emergent threats , such as rust fungi of the Puccinia genus embody the ongoing plant pathology challenges and the intricacies of fungal interactions among plant communities. Pathogens such as *Puccinia graminis*, the agent of wheat stem rust, demonstrate how the emergence of pathogens poses global food security threats. The rust fungi life cycle is characterized by complex sporulation, which allows them to easily adapt to changing environmental conditions and resistant host plants. As climate change reorganizes ecosystems, the extent to which these pathogens propagate and impact ecosystems becomes more relevant. These emerging threats put emphasis on the role of extensive surveillance and research activity in plant pathology, diverting attention back to the relation between fungal pathogens and plant diversity in the build-up of sustainable agriculture and ecosystem resilience (Nuala et al., 2015).

5.5.3. Strategies for Managing Fungal Diseases in Agriculture

Control of fungal diseases in agriculture is also becoming very important because of the increased cases of fungal diseases that are associated with climate change. With the increase in the global temperature and changes in the weather patterns, the relationships between plant pathogens and their hosts are getting more complicated. The best management practices should include predictive models that take into account the implications of climate on the prevalence of the disease and the susceptibility of crops. With such knowledge on the climatic conditions that affect the life cycle of pathogens, the agricultural stakeholders can come up with integrated management plans. Such plans can incorporate cultural methods of crop rotation, intercropping and resistant plant variety selection. Emphasizing a holistic approach allows for the adaptation of practices that not only mitigate fungal threats but also promote overall plant health and diversity, thereby ensuring sustainable agricultural productivity amidst changing environmental conditions (Dobson et al., 2012). Biological control is an emerging potential in the control of fungal diseases in the quest to attain sustainable agricultural systems. This is an approach that involves the use of natural enemies of the pathogens, like the beneficial fungi and bacteria and even predatory organisms, to bring down the population of pathogens. Incorporation of the biocontrol agents into the current farming systems can help to reduce the use of synthetic fungicides, which could lead to the selection of resistant pathogens. In addition, research has been improved, and certain strains of the microorganisms have been identified that are effective against certain fungal diseases. The adoption of such biologically-based practices will not only result in environmental sustainability but will also increase the sustainability of agricultural systems by increasing plant diversity. As the agricultural community continues to seek alternatives to chemical interventions, exploring the potential of biocontrol agents will be vital for ensuring long-term success in managing fungal diseases while sustaining plant diversity (Rahmann et al., 2006). Nutritional management of crops is also another approach that can be used to manage fungal diseases, and this can be a great support to the plant defense. Supplying the necessary nutrients may boost the natural resistance of plants to pathogens, resulting in a reduced rate of occurrence of fungal infections.

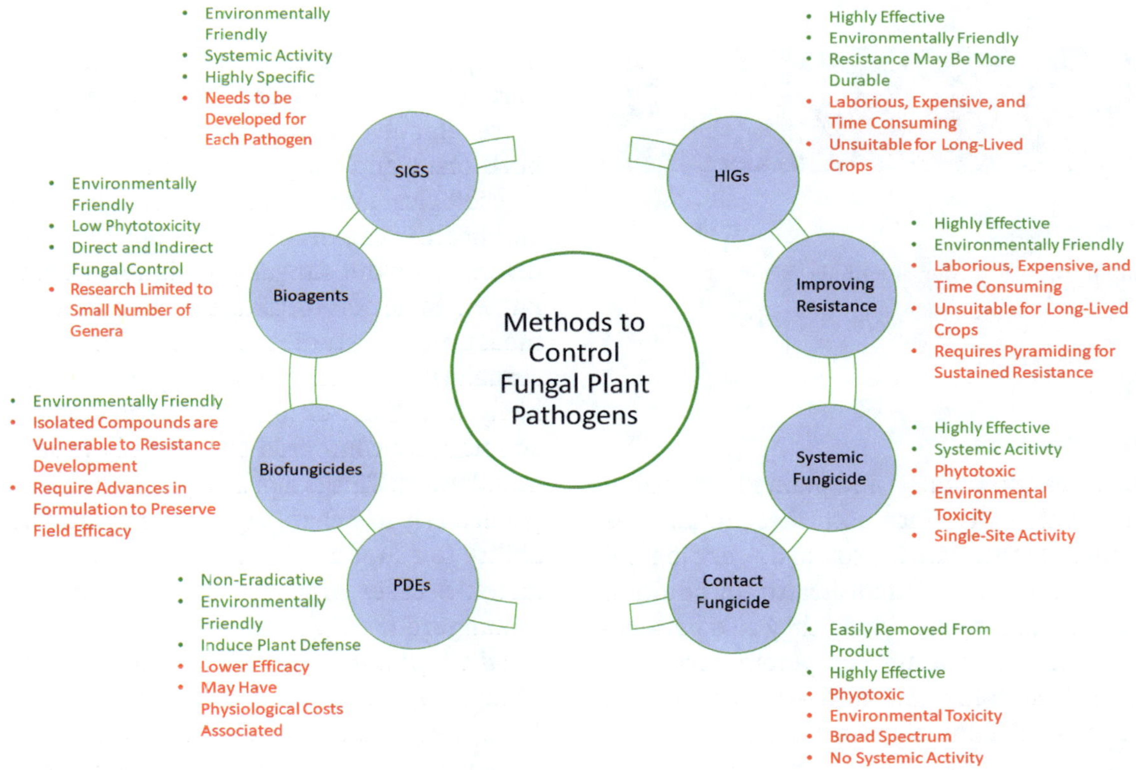

Figure 5.20. Managing fungal diseases in agriculture.

Source: McLaughlin MS, Roy M, Abbasi PA, Carisse O, Yurgel SN, Ali S. Why Do We Need Alternative Methods for Fungal Disease Management in Plants? Plants. 2023; 12(22):3822. https://doi.org/10.3390/plants12223822

Fertilizers that are used to maximize the health of plants can also enhance growth, as well as boost the synthesis of natural antimicrobial compounds. On the same note, the use of some practices like foliar feeding and soil amendment can enhance the general health and vitality of crops, which would make them less vulnerable to the disease pressure. Studies have also suggested that healthy plants are more resistant to environmental stresses, thereby confirming that they are also resistant to fungal infections that are a product of climate change. Sustainable fertilization techniques that synergistically improve plant health while minimizing ecological footprints will be essential in the ongoing effort to manage fungal diseases effectively and protect the biodiversity of agricultural landscapes (Dobson et al., 2012) (Figure 5.20).

Test Your Knowledge

1. What is the impact of fungal diseases on plant populations?
2. What are the strategies for managing fungal diseases in agriculture?

5.6. Conservation of Fungi and Plants

Learning Outcome

1. Importance of preserving fungal diversity;
2. Conservation strategies for endangered plant species;
3. Role of biodiversity in ecosystem resilience.

The conservation of fungi and plants is crucial to provide biodiversity since both these organisms contribute significantly towards ecological processes as well as human health. In addition to cultural and aesthetic values, they also perform key roles in maintaining ecosystem services , such as soil stabilization, nutrient cycling, and gas exchange. For instance, fungi associate with plant roots and enhance the uptake of nutrients that are critical to plant well-being. Soil diversity, which consists of many species like fungi, is the backbone of ecosystem functioning, offering supporting services like clean water and climate regulation. The need to raise awareness for soil diversity is underscored by its central position in ensuring sustainability for human activities, particularly as the world becomes plagued with more global environmental issues. Therefore, both plant and fungal conservation is not only an ecological imperative, but also human survival and health (Van der Putten et al., 2010). Genetic diversity of the fungal population needs to be known in order to take effective conservation measures. In studies, it has been found that sexually reproducing fungal populations are more likely to consist of higher genetic diversity and adaptability compared to asexually reproducing populations. This is a primary genetic distinction, particularly in a changing environment where the ability to adapt can be the survival or extinction factor. A case from research conducted on the endangered epiphytic lichen-forming fungus *Lobaria pulmonaria* is that lower levels of sexual reproduction may hinder populations from responding to dynamic environmental changes. This necessitates a consideration of mating systems and their potential implications for conservation strategy. With the combination of reproductive strategy and genetic diversity understanding, translocation operations aimed at conservation can be planned to increase genetic heterogeneity as well as resilience in diminishing populations (Cornejo et al., 2012). Fungi and plant conservation also needs a sophisticated strategy aimed at habitat preservation, policy development, and public awareness. Biodiversity threats in the way of extinction due to habitat loss, climate change, and pollution need specialized responses involving local communities and stakeholders. Policies aimed at the conservation of habitats and the encouragement of sustainable land-use strategies are critical in alleviating these threats. At the same time, educational programs can help build respect and appreciation for plants and fungi, helping to garner public support for conservation efforts. Highlighting how these organisms and their respective habitats are interconnected can strengthen cooperative efforts at the local as well as global scales. Last but not least, successful conservation of fungi and plants will not only be founded upon scientific research but also upon societal participatory efforts to value and conserve our natural heritage (Van der Putten et al., 2010) (Figure 5.21).

Figure 5.21. Conservation of fungi and plants.

Source: https://cdn.prod.website-files.com/620a6e6edc2558058657da07/6548fa30114ca3eae5d3f567_fungi111.webp.

5.6.1. Importance of Preserving Fungal Diversity

Under ecological resilience and health, the preservation of fungal diversity is essential. Fungi are ecologically vital components of terrestrial ecosystems in that they act as decomposers, nutrient recyclers, and plant symbionts. They contribute to root-to-nutrient uptake improvement, which enhances plant growth and overall ecosystem productivity. Without them, with a diverse population, these roles would be lost, and the fertility of the soil and plant diversity could be threatened. Specifically, the role of ectomycorrhizal fungi (EMF) is critical, as they form symbiotic relationships with many tree species. Through the preservation of a varied fungal community, we promote ecosystem function and stability and, in turn, plant diversity and resistance to environmental change. Designation of reserved habitats, like ancient-growth and mixed-wood forests, is significant in the conservation of these critical fungal populations, as evident from the positive impacts of set-aside plantations on old EMF species (Bidartondo et al., 2015). Moreover, the genetic diversity of fungal species is a significant factor in their adaptive potential, and hence very much crucial for conservation. Sexually reproducing fungi are generally more genetically diverse than their asexually reproducing relatives and thus more resilient to fluctuating environmental conditions. Genetically diverse populations of fungi can facilitate more robust ecosystem connections and nutrient cycling. For instance, frayed epiphytic lichen-forming fungus *Lobaria pulmonaria* in the endangered condition has limited sexual reproduction due to skewed distributions of mating-type genes that promote genetic restrictiveness and thus discourage frayed adaptive capacity to stress factors (Cornejo et al., 2012). Based on such reproductive patterns and genetic structure, conservation strategies can be developed for enhancing gene heterogeneity in fungal populations to sustain their critical roles in ecosystems and to achieve increased biodiversity for plant societies. Conservation of fungal diversity is also necessarily interlinked with the battle against other large-scale environmental problems, , such as climate change and habitat loss. When ecosystems are under more pressure, highly diverse fungal assemblages can be crucial for maintaining ecosystem resilience and flexibility. If fungi like EMF can grow in a variety of tree species, this means that they have the potential to dynamically react to changing climatic factors and assist in forest recovery and carbon sequestration. Furthermore, with our ongoing expansion in knowledge of fungi's contribution to ecosystem services, there is no doubt that conservation of the fungi has to be prioritized as part of classical plant conservation. Conservation of fungal diversity has to be integrated into land use and conservation policy to promote establishment of durable ecosystems resistant to environmental stresses but capable of sustaining high biodiversity, ultimately sustaining the well-being of natural and man-dominated environments (Figure 5.22).

Figure 5.22. Close-up view of bracket fungi growing on decaying wood, illustrating their ecological role in decomposition and nutrient cycling.

Source: https://www.tigersafariindia.com/blog/wp-content/uploads/2024/07/Fungi-and-its-role-in-environment.jpg.

5.6.2. Conservation Strategies for Endangered Plant Species

Threatened plant species conservation is an urgent affair that encompasses numerous strategies aimed at limiting threats and ensuring survival. Habitat restoration is one primary strategy that aims at restoring habitats where these species had been before. Effective restoration involves not only the reintroduction of native plants but also the management of invasive plants that vie for resources. This strategy is particularly crucial in areas where there has been urbanization and resulting habitat loss. Through restoring such ecosystems, conservationists can create environments conducive to associations with other organisms and fungi that are essential for plant health and thus promote biodiversity as a whole. Collaboration with local communities would also enhance the success of such projects since they are a source of invaluable information regarding traditional land use systems that can prove useful for restoration initiatives (Kiik, 2017). Another required approach is the establishment of seed banks, which are crucial to preserving genetic diversity among endangered plant species. Seed banks not only preserve seeds under conditions that mimic their natural habitats but also act as a research facility for understanding plant resilience and adaptability when subjected to changed climate conditions. It is a method of extinction protection because researchers can reintroduce genetically diverse populations back into the native habitats. In addition, through the identification of the viability of seeds stored, researchers can ascertain the best candidates for rewilding initiatives, hence critical to ensure healthy plant societies. Cooperation between seed banks and conservation authorities can thus further enhance such efforts to ensure that a gene material repository is created to support conservation procedures (Knapp et al., 2019). The inclusion of public awareness and education initiatives is also an adjunct approach that contributes significantly to the conservation of endangered plant species. Through enlightening people on the importance of biodiversity, conservationists can enlist people to work toward the conservation of indigenous plants. Activities that can be engaged in include workshops, walks in nature, and school programs through which people learn to appreciate the ecological roles of plants and their associated fungi. Increased public interest will generally result in volunteerism and support for neighborhood conservation programs. Furthermore, the promotion of sustainable gardening keeps homeowners motivated to create conditions favoring native vegetation, thus sustaining the diversity of local ecosystems. This type of grass-roots conservation system places greater focus on the interconnectedness of all species, upholding the role of each in maintaining ecosystem stability (Kiik, 2017).

5.6.3. Role of Biodiversity in Ecosystem Resilience

Biodiversity is also important in increasing the resilience of the ecosystem, which is a concept that is important in ensuring the stability of the ecology despite the environmental stressors. These diverse species interactions also help give a systems capacity to withstand disturbances, whether caused by climate variability, invasive species, or habitat degradation. The richer the community of plants and fungi, the stronger the network of interactions, the more adaptive the ecosystems will be and the more they will be able to resist the impending dangers of anthropogenic changes. As an example, a diverse ecosystem may provide a variety of plant species that may provide different resources to support diverse animal populations to maintain ecological stability and recovery processes following disturbances. This resilience is particularly important as ecosystems are increasingly subjected to stressors, necessitating a greater understanding of how biodiversity influences adaptive capacities within community structures (Meredith et al., 2015). Besides creating resilience, biodiversity acts as a cushion against loss of species, which may lead to secondary effects on ecosystem functions and services. Functional redundancy is possible with the existence of different species; whereby different organisms may perform similar functions in an ecosystem. This redundancy means that should one species be struck down by disease or environmental stress, then others will be able to take its place, and thus maintain ecosystem processes of nutrient cycling, pollination, and provision of habitat.

Furthermore, fungi-plant interactions play the most significant role in this discussion, where mycorrhizal networks may promote nutrient exchange and lead to the well-being and prosperity of plants. Enhanced species interactions result in more resilient ecosystems that can better cope with variation and uncertainty, suggesting that conservation efforts should prioritize maintaining and restoring biodiversity (University of Montana-Missoula, 2023). Moreover, the current loss of biodiversity in habitat erosion and climatic change highlights the importance of deciphering its contribution to ecosystem resilience as a matter of urgency. The implications are even more salient to resilience as human activities proceed to fragment landscapes and strain ecological systems. This interdependence among species implies that the disappearance of a single living being can be the first step towards a chain of adverse events, which worsen the overall state of ecosystems and their stability. Conservation measures should thus not be species-centered but should also take into consideration the whole ecological picture , such as maintenance of genetic diversity and ecological interactions. By fostering diverse communities, particularly those involving key players like fungi and plants, we can enhance ecosystem resilience against future challenges, ultimately securing the vital functions that these ecosystems provide to humanity and the planet (Meredith et al., 2015).

Test Your Knowledge

1. What are the conservation strategies for endangered plant species?
2. What is the role of biodiversity in ecosystem resilience?

SUMMARY

The intricate relationship between the plant and the fungi is a basis for the understanding of ecosystems. Fungal associations, particularly AMF, are key agents of nutrient uptake, water regulation, and plant disease. The findings in the research conducted on strawberry plants under water stress demonstrate the way AMF, particularly fungi , such as *F. geosporus* and *F. mosseae*, have the ability to significantly enhance root colonization and growth, even under severe environmental stress conditions. This identifies the importance of considering mycorrhizal colonization as an essential factor that will influence plant resilience and, as a result, greater plant diversity in ecosystems. Since agriculture is increasingly being threatened by climatic variability, the application of AMF benefits can promote more sustainable agriculture (van der Putten et al., 2010). Besides, the relative colonization ability and differential abundance among different AMF species reaffirm the delicacy of soil environment interaction with fungal processes. That a rehabilitated forest had significantly higher AMF density compared to a logged-over forest is an example of how chemical soil properties can affect fungal diversity in a direct way. It is particularly significant that enhanced organic matter in rehabilitated soils provides conducive conditions for AMF sporulation and thus promotes more extensive root colonization. These findings suggest that effective forest management strategies must focus on soil health, as this is inextricably linked with the creation of complex and diverse assemblages of fungi that support plant growth and resilience. This relationship points to the possibility for restoration treatments to enhance biodiversity by targeted action in soil chemistry and microbiome control (Chubo et al., 2012). Lastly, establishing an appreciation for fungi's integral role in plant diversity can inform conservation and agricultural policy in a way that conserves ecosystem integrity. In a time when anthropogenic activity grows more intense, it is necessary to learn about the symbiotic interactions of fungal and plant species to develop more resilient ecosystems. Combining microbial management with conservation efforts, especially in forest and agriculture ecosystems, can make increasing plant diversity and ecosystem services possible. The overall weight of evidence in AMF studies of their interaction with soil factors and plant productivity emphasizes their central function in the preservation of ecological health. Thus, while we grapple with climate deterioration and loss of habitat, an integrated emphasis on fungi in the larger context of plant diversity will prove essential to preserving ecosystems capable of managing future stress.

The intricate relationship between plants and fungi is underscored by the role of mycobiomes in regulating community dynamics in various ecosystems. Research indicates that fungal community assembly is highly regulated by deterministic selection by the host plant, particularly in considering the varied physiological requirements of the different plant species. Notably, sorghum studies have discovered stochastic forces , such as dispersal and drift, to play a primary role in early growth stages or under drought stress, but not thereafter. After recovery from growth to normal levels, the influence of the hosts on fungal communities comes into focus, proving the complex relationship between environmental stressors and biotic associations. These findings are significant in pointing out the significance of understanding the mechanisms underpinning the diversity of fungi to be able to fully appreciate the broader implications for plant well-being and ecosystem resilience, particularly within the context of pressures around climate change and variability of resources (Coleman-Derr et al., 2020). Additionally, the significance of mycorrhizal fungi, AM (fungi) specifically, should not be overstated in terms of eliciting plant health. While high soil phosphorus (P) levels have been shown to inhibit root colonization by AM fungi, the mechanisms of diversity of these symbiotic organisms are not yet fully understood. Research has shown that colonization between plant species , such as maize, soybean, and field violet is extremely variable, demonstrating how host identity traditionally outweighs soil nutrient status in determining AM community composition, especially at low P. This is also underscored by the fact that despite diversity generally declining when P concentrations are high, these trends are negligible and context-specific. The intricate interaction of different plant roots and their mycorrhizal symbionts depicts a critical aspect of ecological equilibrium that tends to be regulated by both biotic as well as abiotic conditions (Abdi et al., 2013). In summary, the interactions between fungi and plants reveal a multifaceted world of ecological interaction that greatly affects both organism diversities. Plant host species and even their specific growth phases determine the organization of the accompanying fungal communities in their roots. As demonstrated by current studies, while deterministic processes like the plant genotype and environmental stresses govern such relationships largely, stochastic processes can also operate on community structure under certain conditions temporarily. Further, the decreased diversity of the AM fungi at high soil phosphorus levels suggests the need for more work on the adaptive options of mycorrhizal fungi against fluctuating nutrient availability. Lastly, a clear comprehension of these dynamics is imperative to advancing agriculture as well as enhancing ecosystem resilience in the presence of environmental variability.

KEY TERMS FROM THIS CHAPTER

Angiosperms	Bryophytes	Chitin
Fungi	Gymnosperms	Hyphae
Lichen	Mushroom	Mycelium
Mycorrhizae	Parasite	Phloem
Photosynthesis	Pollination	Saprophyte
Seedless vascular plants	Spores	Tracheophytes

CRITICAL THINKING QUESTIONS

1. **Fungi obtain nutrients primarily by:**
   a. Photosynthesis
   b. Absorbing nutrients from dead or decaying organic matter
   c. Ingesting food like animals
   d. Nitrogen fixation
2. **Which structure is characteristic of fungi?**
   a. Roots
   b. Hyphae
   c. Chloroplasts
   d. Stomata
3. **Which group of plants is seedless and relies on spores for reproduction?**
   a. Angiosperms
   b. Gymnosperms
   c. Ferns
   d. Flowering plants
4. **What is the dominant generation in most plants?**
   a. Gametophyte
   b. Sporophyte
   c. Zygote
   d. Spore
5. **Which plants produce flowers and fruits?**
   a. Gymnosperms
   b. Mosses
   c. Angiosperms
   d. Ferns

Short Answer Questions

1. What are the main characteristics of fungi that distinguish them from plants and animals?
2. How do fungi obtain their food, and what term describes this method of nutrition?
3. What is the structure of a typical fungus, including the terms hyphae and mycelium?
4. What are the four main groups of plants, and how are they different in terms of structure and reproduction?
5. What adaptation allowed plants to move from water to land?
6. What is the difference between vascular and nonvascular plants?

Answers to Critical Thinking Questions

1. (b); 2. (b); 3. (c); 4. (b); 5. (c)

REFERENCES

1. Abdi, H., Mead, A., Clarke, K. R., Bending, G. D., Hazard, C., Hammond, J. P., et al. (2013). Contrasting arbuscular mycorrhizal communities colonizing different host plants show a similar response to a soil phosphorus concentration gradient. https://core.ac.uk/download/9847390.pdf (accessed on 13th June 2025).
2. Ahonen, S. H. K., Markkola, A., Männistö, M. K., Ruotsalainen, A. L., et al. (2021). Reindeer grazing history determines the responses of subarctic soil fungal communities to warming and fertilization. https://core.ac.uk/download/482672403.pdf (accessed on 13th June 2025).
3. Ambrogi, J. (2024). *Fungi's Contributions to People in Glaciered Environments.* https://core.ac.uk/download/617932810.pdf (accessed on 13th June 2025).
4. Amselem, J., Bonfante, P., Ghignone, S., Grigoriev, I. V., et al. (2020). At the nexus of three kingdoms: The genome of the mycorrhizal fungus *Gigaspora margarita* provides insights into plant, endobacterial and fungal interactions. https://core.ac.uk/download/323071981.pdf (accessed on 13th June 2025).
5. Bahram, M., Cooper, E. J., Eidesen, P. B., Elberling, B., et al. (2016). Ectomycorrhizal and saprotrophic fungi respond differently to long-term experimentally increased snow depth in the High Arctic. https://core.ac.uk/download/392170607.pdf (accessed on 13th June 2025).
6. Barberán, A., Ender, C., Parsons, L. S., Rodrigues, J. L. M., et al. (2020). Soil microbial communities in restored and unrestored coastal dune ecosystems in California. https://core.ac.uk/download/287622311.pdf (accessed on 13th June 2025).
7. Barrett, D. J., Beijerinck, M. W., Brookes, P. C., Dunger, W., Finlay, B. J., Fitter, A. H., et al. (2005). Darkness visible: Reflections on underground ecology. https://core.ac.uk/download/59119.pdf (accessed on 13th June 2025).
8. Barry, K., Daum, C., Erba, L., Grigoriev, I. V., et al. (2019). Shallow genome sequencing for phylogenomic of mycorrhizal fungi from endangered orchids. https://core.ac.uk/download/323082387.pdf (accessed on 13th June 2025).
9. Bever, J. D., Cavalier-Smith, T., Clapp, J. P., et al. (2002). Molecular diversity of arbuscular mycorrhizal fungi and patterns of host association over time and space in a tropical forest. https://core.ac.uk/download/59301.pdf (accessed on 13th June 2025).
10. Bidartondo, M. I., Doncaster, C. P., Newton, A. C., Spake, R., et al. (2015). Similar biodiversity of ectomycorrhizal fungi in set-aside plantations and ancient old-growth broadleaved forests. https://core.ac.uk/download/77019015.pdf (accessed on 13th June 2025).
11. Bongard, C. (2012). A review of the influence of root-associating fungi and root exudates on the success of invasive plants. https://core.ac.uk/download/18325938.pdf (accessed on 13th June 2025).
12. Boucher, D. H., James, S., & Keeler, K. H. (1982). The ecology of mutualism. https://core.ac.uk/download/323061299.pdf (accessed on 13th June 2025).
13. Broussard, K., Chaturvedi, S., Chaturvedi, V., Davis, T., et al. (2014). Mycobiome of the bat White Nose Syndrome (WNS) affected caves and mines reveals high diversity of fungi and local adaptation by the fungal pathogen Pseudogymnoascus (Geomyces) destructans. http://arxiv.org/abs/1403.0631 (accessed on 13th June 2025).

14. Bueno de Mesquita, C. P., Farrer, E. C., King, A. J., Porazinska, D. L., et al. (2019). Soil microbial networks shift across a high-elevation successional gradient. https://core.ac.uk/download/286361320.pdf *(accessed on 13th June 2025).*

15. Buscot, F., Charcosset, J. Y., Gardes, M., Hampp, R., et al. (2000). Recent advances in exploring physiology and biodiversity of ectomycorrhizas highlight the functioning of these symbioses in ecosystems. https://core.ac.uk/download/12041797.pdf (accessed on 13th June 2025).

16. Cabello, M. N., Schalamuk, S., Simon, M. R., Velázquez, S. J., et al. (2014). Effect of Septoria leaf blotch and its control with commercial fungicides on arbuscular-mycorrhizal-fungal colonization, spore numbers, and morphotype diversity. https://core.ac.uk/download/159294458.pdf (accessed on 13th June 2025).

17. Chubo, J. K., King, J. H., Ong, K. H., Sipen, P., et al. (2012). Influence of soil chemical properties on relative abundance of arbuscular mycorrhiza in forested soils in Malaysia. https://core.ac.uk/download/12233173.pdf (accessed on 13th June 2025).

18. Coleman-Derr, D., Dahlberg, J. A., Gao, C., Hollingsworth, J. A., et al. (2020). Fungal community assembly in drought-stressed sorghum shows stochasticity, selection, and universal ecological dynamics. https://core.ac.uk/download/323081747.pdf (accessed on 13th June 2025).

19. Cornejo, C., Dal Grande, F., Scheidegger, C., Schmitt, I., et al. (2012). Genetic basis of self-incompatibility in the lichen-forming fungus Lobaria pulmonaria and skewed frequency distribution of mating-type idiomorphs: Implications for conservation. https://core.ac.uk/download/14527481.pdf (accessed on 13th June 2025).

20. Cornejo, P. S., Fernández, N. V., Fontenla, S., Meie, M. A., et al. (2018). Nothofagus pumilio forest affected by recent tephra deposition in northern Patagonia: II—Shifts in diversity and structure of rhizosphere fungal communities. https://core.ac.uk/download/268916678.pdf (accessed on 13th June 2025).

21. Cárdenas-Flores, A., Roger, A., Gonzalez, A. V., Rinkevich, *B., Santelices, B.*, et al. (2019). Considering intra-individual genetic heterogeneity to understand biodiversity. https://core.ac.uk/download/286357369.pdf (accessed on 13th June 2025).

22. Dobson, A., Donnelly, A., Fitter, A., MacLeod, A., Rytkönen, A., Woods, *A., Zocca, A., et* al. (2012). Impacts of climate change on plant diseases–opinions and trends. https://core.ac.uk/download/10932891.pdf (accessed on 13th June 2025).

23. East Malling Research, & Robinson Boyer, L. (2014). Population biology and interactions of arbuscular mycorrhizal fungi and their benefits in strawberry cultivation. https://core.ac.uk/download/30707690.pdf (accessed on 13th June 2025).

24. Field, K. J., Pastok, D., & Thirkell, T. J. (2020). Carbon for nutrient exchange between arbuscular mycorrhizal fungi and wheat varies according to cultivar and changes in atmospheric carbon dioxide concentration. https://core.ac.uk/download/324070546.pdf (accessed on 13th June 2025).

25. Finlay, B. J., Groombridge, B., Kowalchuk, G. A., Liu, W. T., Manefield, M., Meselson, M., et al. (2005). Biodiversity and ecosystem function in soil. https://core.ac.uk/download/58969.pdf (accessed on 13th June 2025).

26. Folwell, B. D., McGenity, T. J., McKew, B. A., Sanni, G. O., et al. (2012). Marine crude-oil biodegradation: A central role for interspecies interactions. https://core.ac.uk/download/9995659.pdf (accessed on 13th June 2025).

27. Francini, G., Hui, N., Jumpponen, A., Kotze, D. J., et al. (2017). Ectomycorrhizal fungal communities in urban parks are similar to those in natural forests but shaped by vegetation and park age. https://core.ac.uk/download/157587562.pdf (accessed on 13th June 2025).

28. Fransson, P. M. A. (2001). Responses of ectomycorrhizal fungi to changes in carbon and nutrient availability. https://core.ac.uk/download/11694502.pdf (accessed on 13th June 2025).

29. Gollotte, A., Heggo, A., Hodge, A., Liu, A., Manjunath, A., Mollier, A., Mozafar, A., et al. (2008). Agronomic management of indigenous mycorrhizas. https://core.ac.uk/download/62455716.pdf (accessed on 13th June 2025).

30. Gopalakrishnan, M., Kavitha, M., Maragatham, S., Rajeswari, R., et al. (2024). Resilient soils for a changing climate: Navigating the future of sustainable agriculture. https://core.ac.uk/download/641609462.pdf (accessed on 13th June 2025).

31. Hodge, A., Aylott, M., Azcon-Aguilar, C., Bardgett, R. D., Baum, C., et al. (2009). Mycorrhizas and biomass crops: Opportunities for future sustainable development. https://core.ac.uk/download/47152.pdf (accessed on 13th June 2025).

32. Ianson, D., & Smeenk, J. (2010). Mycorrhizae in the Alaska landscape. https://core.ac.uk/download/162573706.pdf (accessed on 13th June 2025).

33. Kiik, K. (2017). Ohustatud Euroopa Naaritsa (Mustela lutreola) Sigimine Ja Käitumine Tehiskeskkonnas. https://core.ac.uk/download/143642945.pdf (accessed on 13th June 2025).

34. Knapp, S., Schmauck, *S., & Zehnsdorf, A. (2019). Biodi*versity impact of green roofs and constructed wetlands as progressive eco-technologies in u*rban areas. https://core.ac.uk/download/287885433.pdf (accessed on 13th June 2025).*

35. Lumbsch, T., & Rikkinen, J. (2017). Evolution of lichens. https://core.ac.uk/download/286389681.pdf (accessed on 13th June 2025).

36. Meredith, H. M. R., & Zoological Society of London. (2015). Improving the Impact of Amphibian conservation. https://core.ac.uk/download/30710673.pdf (accessed on 13th June 2025).

37. Murray, P. J., Smeding, F. W., & van Eekeren, N. (2007). Soil biota in grassland, its ecosystem services and the impact of management. https://core.ac.uk/download/10925819.pdf (acces*sed on 13th June 2025).*

38. Oliva, J., Stenlid, J., & Martínez-Vilalta, J. (2014). The effect of fungal pathogens on the water and carbon economy of trees: Implications for drought-induced mortality. New Phytologist, 203, 1028–1035. https://doi.org/10.1111/nph.12857 (accessed on 13th June 2025).

39. O'Leary, N. A., Wright, M. W., Brister, J. R., Ciufo, S., Haddad, D., McVeigh, R., Rajput, B., et al. (2015). Reference sequence (RefSeq) database at NCBI: Current status, taxonomic expansion, and functional annotation. Nucleic Acids Research, 44, D733–*D745. https://doi.org/10.*1093/nar/gkv1189 (accessed on 13th June 2025).

40. Phillips, L., Stockdale, E. A., & Watson, C. A. (2006). Impacts of farming practice within organic farming systems on below-ground ecology and ecosystem function. https://core.ac.uk/download/10925258.pdf *(accessed on 13th Ju*ne 2025).

41. Rahmann, G., & Seip, H. (2006). Alternative strategies to prevent and control endoparasite diseases in organic sheep and goat farming systems–A review of current scientific knowledge. https://core.ac.uk/download/10925183.pdf (accessed on 13th June 2025).

42. Rajala, T. (2008). Responses of soil microbial communities to clonal variation of Norway spruce. https://core.ac.uk/download/14916422.pdf (accessed on 13th June 2025).
43. Ruiz-González, M. X., & Vicente, O. (2022). The microbially extended phenotype of plants, a keystone against abiotic stress. https://riunet.upv.es/bitstream/10251/195427/1/Ruiz-GonzalezVicente%20-%20The%20Microbially%20Extended%20Phenotype%20of%20Plants%20a%20Keystone%20against%20Abiotic%20St....pdf (accessed on 13th June 2025).
44. Sjöberg, J. (2005). Arbuscular Mycorrhizal Fungi. https://core.ac.uk/download/11694918.pdf (accessed on 13th June 2025).
45. Stewart, J. R., 2010. Understanding evolutionary processes during past Quaternary climatic cycles: Can it be applied to the future? In: Grant, F., Mergeay, J., Santamaria, L., Young, J. and Watt, A.D., *eds. Evolution and Biodivers*ity: The evolutionary basis of biodiversity and its potential for adaptation to global change. Report of an e. Mallorca, Spain: Centre for Ecology & Hydrology, INBO, IMEDA, CSIC-UIB.(accessed on 13th June 2025).
46. University of Montana–Missoula. Office of the Registrar. (2023). 2023–2024 Course Catalog. https://core.ac.uk/download/621684373.pdf (accessed on 13th June 2025).
47. Van Der Merwe, M. (2020). The influence of cone age and urbanization on the diversity and community composition of seed fungal endophytes within native Australian Banksia ericifolia L.f. subsp. ericifoli*a. https://core.ac.uk/do*wnload/534127319.pdf (accessed on 13th June 2025).
48. van der Putten, W. H., Mudgal, S., Turbé, A., de Toni, A., Lavelle, P., Benito, P., & Ruiz, N. (2010). Soil biodiversity: functions, threats and tools for policy makers. Bio Intelligenc*e Service. http:/*/ec.europa.eu/*environmen*t/archives/soil/pdf/biodiversity_report.pdf (accessed on 13th June 2025).

ANIMAL DIVERSITY AND BODY PLANS

6

LEARNING OBJECTIVES

After studying this chapter, you will be able to:

1. List the classification of animals;
2. Define body symmetry;
3. Explain body plans and development;
4. Explore adaptations and evolution;
5. Identify the ecological roles of diverse body plans.

Introduction

Understanding the sheer diversity of animal form and body plans is one of the basic domains of evolutionary biology. Evolutionary biology examines how various species have attempted to modulate various body forms in order to survive on Earth. The study of body plans, or the overall organization and plan of an organism's tissues and organs, does more than simply reveal the functional diversifications of animals; it gives some insight into their evolutionary history. The complexity that emerges from such diversity is not just the consequence of genetic changes but a reflection of the intricate dynamics of gene regulation and developmental events. In this broader exploration, researchers seek to account for the interplay between body plan expression and genetic conservation across diverse taxa, ultimately culminating in a more nuanced appreciation of animal evolution. The interplay between genetic mechanisms and body-plan diversity precipitates pressing concerns about the evolution of body plans. Importantly, evo-devo has illuminated the reason why elevated genetic complexity does not necessarily equate to perceived organismal complexity. This implies that the variation in form seen across the animal kingdom could arise from the modification of gene regulation, rather than the incorporation of new genetic material. Described to be tissue-level simplicity can actually hide a profound genetic origin, thereby implying evolutionary adaptations to have more complex developmental coding than previously understood. Such an awareness underscores the need to distinguish conserved genes and their functional versatility in morphology, ultimately to further our understanding in animal phylogeny and the evolutionary history to current diversity (Becker et al., 2005).

Figure 6.1. Animal diversity.

Source: https://upload.wikimedia.org/wikipedia/commons/thumb/c/cd/Animal_diversity_October_2007.jpg/500px-Animal_diversity_October_2007.jpg.

Moreover, reconstructions of ancestral body plans are revised as more information is gleaned from both phylogenetic analysis and fossil evidence. The current consensus suggests a protostome-deuterostome ancestor with sophisticated body organization, but other ideas propose that a less complex arrangement may more accurately depict ancestral states. This revision calls for reserve when interpreting phylogenetic association and homologies in developmental organization across animals. It challenges a more nuanced interpretation of the fossil record, provoking scientists to seek evidence that can reveal the paths of bilaterian evolution with greater clarity. By putting conventional suppositions about the sophistication of early ancestors under challenge, scientists can better comprehend the evolution processes that have led to modern-day animal form, proposing new hypotheses and research directions into animal diversity and body plans (Erwin & Davidson, 2002) (Figure 6.1).

Definition of Animal Diversity

Animal diversity encompasses the thousands of species and taxa that represent the staggering variety of body shapes and ecological specializations in the animal world. The diversity is not merely a function of the species number that exceeds 8.7 million, as quantified by recent estimates, but also represents profound differences in morphology, behavior, and physiology among different environments. The evolutionary processes that have resulted in such diversity are complex, as they involve complex interactions between genetic variation, environmental stress, and ecological niches. As highlighted in evolutionary developmental biology research, although body plan diversity may appear humongous, the genomic blueprints that underlie them are very conserved within phyla that are widely separated. This paradox suggests that much of observed diversity is created by differences in gene regulation rather than extreme variations in genetic code and can result in a rich diversity of animal life from fairly few genetic components (Angelos et al., 2017). Understanding animal diversity goes beyond taxonomy and identification and involves exploring the evolutionary and ecological implications of the diversity. The word "biodiversity" does not just apply to diversity of species but to the way species interact within ecosystems to affect their evolution. The One Health concept identifies the interconnectedness of animal, human, and environmental health and how biodiversity loss may lead to more disease transmission and disruption of ecology. Here, diversity of animals is critical in informing conservation activities and maintaining ecological equilibrium. The great diversity of adaptation mechanisms exhibited by different species to endure their habitat also highlights the significance of diversity in maintaining resilience to change in the environment, hence promoting sustainability in ecosystems (Angelos et al., 2017). In order to fully understand the scope of animal diversity, there is a need to comprehend how body plans operate as critical signs of evolutionary lineages and adaptive functions. Body plans are not only reflective of how an organism's form contributes to existence but also of how forces of evolution have operated on these forms over time. Conservation of genetic components among species, as explained earlier, raises fascinating questions about the origins of complex traits and the roles of regulatory networks during body plan development. Definition and categorization of the different body plans within groups allow scientists to trace lineage diversification and adaptive radiation and reveal patterns that can anticipate the potential for future evolutionary changes. These observations not only come in handy for classifying current-day biodiversity but also to understand the evolutionary affinities that bind diverse forms of life under a greater comprehension of life's history and its ongoing dynamics (Becker et al., 2005).

Importance of Studying Body Plans

The importance of studying body plans is that they can reveal information about the evolutionary history of the various groups of animals. Body plans are templates of how organisms achieved adaptability to an environment of varying conditions over millions of years. By the study of these structural blueprints, scientists can make new discoveries about patterns of homology and evolutionary constraints that have shaped organism development. For example, the genetic correspondence seen in the shared invertebrate and vertebrate genes, even across their disparate evolutionary histories, points to a shared ancestral origin that has conserved vital developmental genes. This evo-devo perspective highlights the potential to investigate body plan research to reveal fundamental biological principles and mechanisms in organismal diversity that lead to new insights relating comparative morphology and genetics and finally tell us about life's complexity (Becker et al., 2005). Body plan research also has significant implications for the understanding of organisms' reaction and interaction with the environment. As the environment undergoes ecological transformations, including those caused by climate change, body plans affect not only survival and reproduction but also the dynamics of ecosystems at large. In the Quaternary, for example, climate change was responsible for changes in ecological community composition, mirroring intricate interactions between environmental variables and evolutionary processes. Body plan investigations can reveal how specific morphological characteristics offer advantages under various conditions and thus shed light on evolutionary adaptability. This information is crucial for the prediction of species response to ongoing problems, like climate fluctuations and habitat loss, as it allows scientists to determine the evolutionary paths that can enable resilience to ongoing environmental change (Stewart, 2010). Moreover, comparative anatomy relies upon the study of body plan in order to highlight the importance of evolution of morphological adaptations. By examining anatomical structure components responsible for distinguishing various taxonomic groups, scientists can reconstruct the evolutionary innovations that have evolved over time. For instance, the origin of specialized limbs in terrestrial vertebrates is an example of how body plans have adapted to specific ecological niches, indicative of functionality and efficiency of locomotion or foraging strategy. Researching these adaptations enhances our knowledge of evolutionary biology, which supports concepts , such as natural selection and adaptive radiation. Also, the understanding of the genetic basis of the body plans can illuminate the processes by which new forms develop, locating the investigation of body plans at the boundary of ecology, evolution, and genetics and rendering them essential for analyzing animal diversity (Becker et al., 2005) (Figure 6.2).

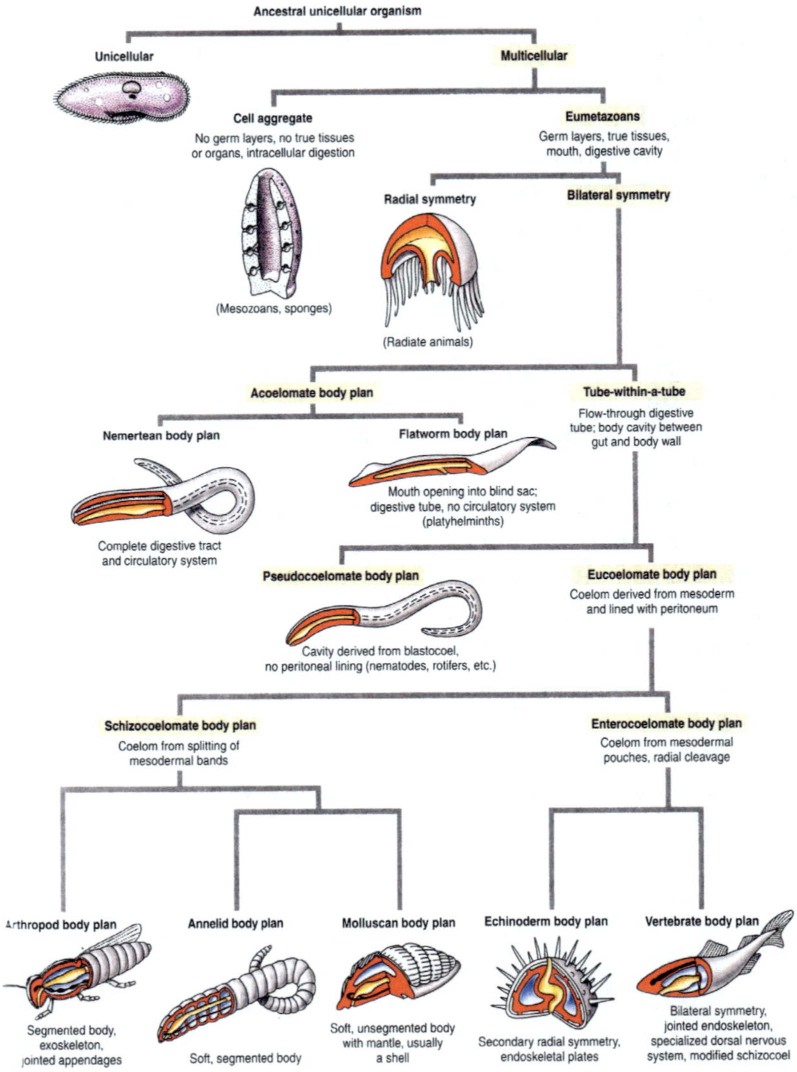

Figure 6.2. Animal body plans.

Source: https://biocyclopedia.com/index/general_zoology/images/images11/fig009.jpg.

6.1. Classification of Animals

Learning Outcome

1. Taxonomic hierarchy;
2. Major animal phyla;
3. Criteria for classification.

Animal classification is a complex phenomenon that requires a thorough understanding in order to embark on the animal diversity richness and the diversity of body plans that are represented in the terrestrial and aquatic ecosystems. A hierarchical classification system can be used to classify animals with domains, kingdoms, phyla, classes, orders, families, genus, and species. This scientific method enables the scientists to classify the great diversities of life forms into groups with similar features and evolutionary connections. Of these categories, the difference between invertebrates and vertebrates forms a fundamental division in the animal kingdom describing significant differences in body plans. Lack of a backbone (invertebrates) comprises a large percentage of animal diversity, and can further be subdivided into several phyla, including Arthropoda and Mollusca. This complicated division shows the evolution changes which various groups have obtained due to the environment. Ecological and evolutionary implications of animal classification are not restricted to taxonomy; they are complex genetic and developmental processes. Recent research in evolutionary developmental biology (evo-devo) suggests that the observed diversity of body plans across various animal phyla may not necessarily parallel significant changes in genetic makeup.

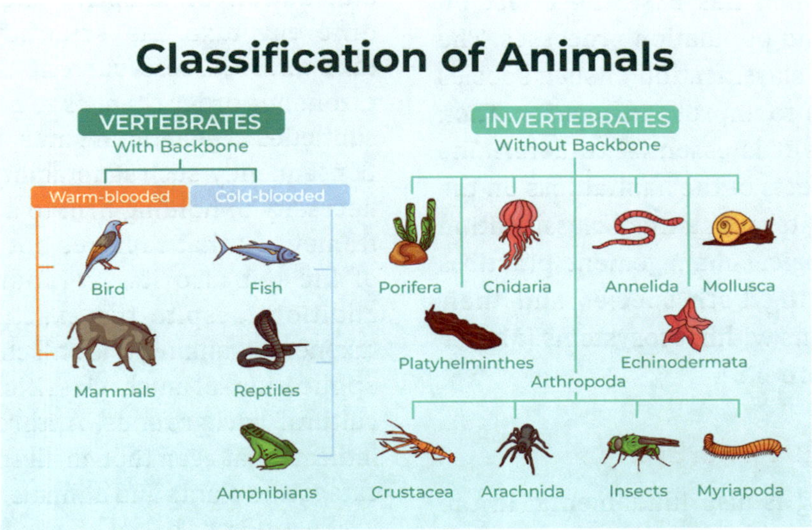

Figure 6.3. Classification of animals.

Source: https://media.geeksforgeeks.org/wp-content/uploads/20220919162335/Classificationofanimals.png.

An example is that the tissue level organization of certain organisms may be quite simple, yet their genetic blueprints may be highly complex, suggesting that evolutionary novelty is frequently generated by changes in gene regulation, not by radical modification of gene sequences. This preservation of genetic machinery is a key to the statement that phylogenetic relationships are important to be studied in order to get a vision of the homology of developmental genes in various species. Therefore, the study of animal classification serves as a window into the evolutionary history of life, highlighting the role of genetic conservation and modification in shaping diverse body plans across the animal kingdom (Becker et al., 2005). Further, the ecological functions of different groups of animals further demonstrate the importance of classification in the study of biodiversity. For example, members of the Mustelidae family, including the European pine marten (*Martes martes*), exhibit diverse feeding behaviors that influence the ecological balance within their habitats. Being included in the group of carnivores, they emphasize their influence on the food web, as they can act both as predators and scavengers. Studies on the ecology of the European pine marten suggest that the species adjusts its resource utilization to habitat fragmentation, which has a strong effect on prey abundance and population processes. The importance of the classification of such species is in the possibility to improve the conservation strategies because it is essential to determine and reduce the effects of the habitat loss on the biodiversity. Insights from animal classification thus inform ecological management practices aimed at preserving both species and their complex interactions within ecosystems (Mestre et al., 2012) (Figure 6.3).

6.1.1. Taxonomic Hierarchy

Taxonomic ranking is also fundamental in the achievement of biological diversity organization, particularly in the classification of animals. The hierarchy ranks organisms into progressively broad categories , such as domain, kingdom, phylum, class, order, family, genus, and species. Each level of hierarchy positions closely related organisms into groups that may be related by having similar characters or a common origin. With this systematic method, biologists have a better chance of understanding relationships between various animal species and their evolutionary history. The relevance of taxonomic ranking is shown by its ability to facilitate the exchange of information among scientists since a standard system of classification allows researchers to easily exchange information regarding specific organisms, thus facilitating collaborative research in various branches of biology and ecology. Further, taxonomic hierarchy extends beyond classification but also provides insight into evolutionary processes that account for diversity in animals. Subsequently, as recent work has highlighted, this hierarchical framework not only depicts relationships between organisms but also determines evolutionary features describing different groups (Lake et al., 2009). For instance, phyletic differences can be shown to define dramatic variations in body organization and developmental patterns, important for the elucidation of adaptive mechanisms among different taxa. As scientists discover new taxa and reassess current classifications, the taxonomic order changes to better represent the subtleties of genetic and morphological diversity. Consequently, such adaptability is in line with the necessity of holding firm to a robust taxonomic framework that captures not only the diversity of life but also its evolutionary histories. In addition, despite the exactitude afforded by taxonomic ranking, it is still challenging in actual application of such classifications in varying cultural backgrounds. Anthropological studies indicate that even though all societies worldwide categorize plants and animals, their concepts are vastly different based on experience and culture (Medin et al., 2025). This diversity indicates that local knowledge can differ immensely from science-mandated concepts with a resulting complex interplay between folk categorization and taxonomic arrangements sanctioned by science. An awareness of these classifications' differences can enhance our conception of diversity among animals in the context of local environments and cultural significance. Consequently, the integration of traditional and scientific models becomes essential towards a holistic understanding of diversity among animals, which calls for inter-disciplinary research in ecology and taxonomy (Figure 6.4).

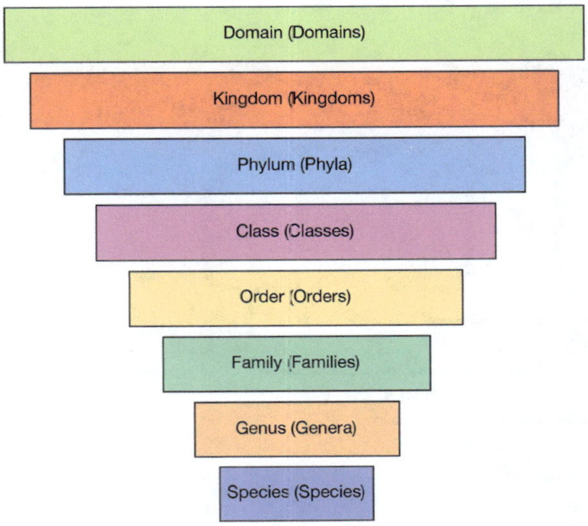

Figure 6.4. Taxonomic hierarchy of animals.

Source: https://cdn.britannica.com/78/103778-050-D797CF4F/Animals-groups-organisms-succession-general-particular.jpg.

6.1.2. Major Animal Phyla

The diversity of the animal kingdom rests largely upon major phyla that are characterized by discrete body plans and evolutionary adaptations to their niches. Such phyla include Arthropoda, Mollusca, and Chordata, which show the diversity through dramatic differences in morphology, physiology, and mode of reproduction. Arthropods, for instance, have a segmented body plan, jointed appendages, and an exoskeleton that sustain diverse forms like insects and crustaceans to coexist in different environments. Mollusca, on the other hand, displays a soft body shape typically sustained by a hard shell, with attributes , such as a muscular foot and a mantle to facilitate movement and defense. These structural differences among the phyla maximize survival in various environments but determine the intricate species relationships, demonstrating how body plans coalesce with ecological niches in the broader tapestry of biodiversity. The phyla's evolutionary history reveals a macroevolutionary trend that is determined by a dynamic interplay between body plan complexity and environmental adaptability. Early multicellular animals probably had a simple body structure, but gradually phyla like Cnidaria and Echinodermata evolved specialized systems that reflect their success through millions of years. They both have a radial symmetry and stinging organ to catch prey in the case of cnidarians, whereas echinoderms have a pentaradial symmetry as well as a water vascular system, which helps them in locomotion and feeding. Second, this evolutionary trend is spurred by various forces, including predator interactions and parasitic encounters, as seen in current studies.

Most importantly, the immunological phenomena across broad phyla reveal shared features and lineage-specific trends that drive diversity; for instance, immune system innovations are essential to both host adaptation and parasite specificity, illustrating how body plan developmental stresses mold over time (Loker et al., 2012). The study of regulatory genes through animal phyla has yielded greater understanding in body plan development and evolution. Studies of organisms like Drosophila and chordates indicate a homology in the genetic processes that govern body part development, suggesting functional homologies independent of lineage. This perspective dissuades the conventional conception of a complex last common protostome-deuterostome ancestor to one of a simpler early body plan that acted as a precursor to the sophisticated forms of modern bilaterians. This new perspective compels scientists to reinterpret the fossil record and biological data to better discern paths of evolution and body plan origin (Erwin & Davidson, 2002). By taking into account the genetic origins of such adaptations, scientists can better understand the evolutionary explanations that mapped out the complicated zoological terrain we observe today (Figure 6.5).

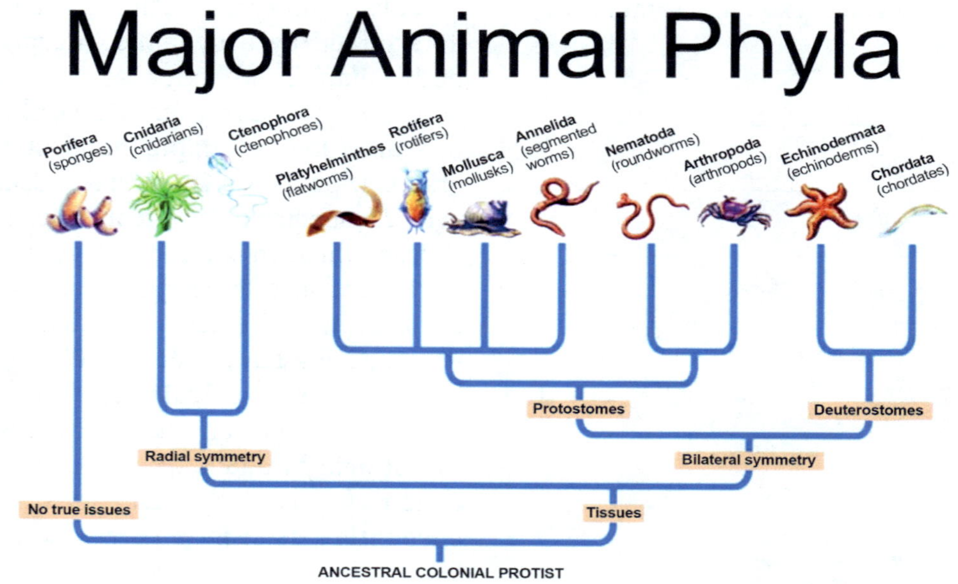

Figure 6.5. Major animal phyla.

Source: https://i.ytimg.com/vi/pCb4XnY2xl4/maxresdefault.jpg.

6.1.3. Criteria for Classification

The animal taxonomy is at its base in the various criteria that distinguish one from the other, emphasizing the importance of a structured method in biodiversity studies. The conventional taxonomic system is based primarily on morphological characteristics, , such as body plan symmetry, segmentation, and organ pattern formation. These tangible attributes are used as a basis on which to begin classifying organisms, but their limitations sometimes result in further investigation. As an illustration, morphology's use can be , such as to conceal interrelations between species, particularly between taxa that exhibit convergent evolution. For this reason, alternative methods, like molecular taxonomy, have gained popularity because they use DNA sequence information in describing evolutionary relationships that morphological techniques may overlook. This paradigm is concerned with the importance of holistic methods in species identification and highlights the sophistication involved in understanding animal diversity, and also the intricacies of classification (Day et al., 2009). With the development of the study of animal diversity, it is increasingly important to include technology development in classification criteria. Such new technologies as DNA barcoding are a prime example, offering a speedy and accurate means of species identification that bypasses some of the inherent drawbacks of morphological taxonomy. The molecular approach not only speeds up classification and makes it more accurate but also improves the accuracy of the detection of cryptic species that may present common morphological characteristics but be genetically distinct. Also, developments in non-destructive sampling methods allow the preservation of samples for morphological identification to allow scientists to maintain whole records even as they utilize genetic data. These developments are critical to solving long-standing issues embedded in taxonomic research, particularly with diverse groups where traditional methods may fail, thus underlining the need for adaptive responses in ongoing discovery of animal diversity (Atanacio et al., 2010). In the broad context of taxonomic classification, the interaction between classical and molecular approaches gives a clearer broader picture of animal diversity. While morphological needs provide the basis for classifying organisms, the incorporation of molecular data fills in and elaborates the existing classifications, giving a clearer picture of evolutionary linkage. The use of computerized dichotomous keys and identification systems , such as FishBase, shows the prospect that technology can automate the process of classification and make it accessible for use by additional researchers and conservationists. These technologies not only enable monitoring and management of biodiversity in an effective manner but also encourage the calling for a new generation of taxonomists with old skills and new technological skills. By taking a multi-dimensional approach to the criteria of classification, the scientific community can better respond to the complexity that pervades animal diversity and the necessity of its conservation (Day et al., 2009).

Test Your Knowledge

1. What is taxonomic hierarchy?
2. What are the major animal phyla?

6.2. Body Symmetry

Learning Outcome

1. Types of symmetry: bilateral, radial, and asymmetry;
2. Evolutionary significance of symmetry;
3. Examples of symmetrical and asymmetrical animals.

Body symmetry is a primary phenomenon in the study of animal diversity and body strategies since it not only acts as a significant feature of different species but also as an indication of their evolutionary adjustments. There are generally two kinds of symmetry in animals, bilateral and radial. One side of the organism being a mirror of the other, bilateral symmetry, is more common in the case of more complex organisms, including mammals and insects, and enables directed movement and more complex nervous systems. Radial symmetry, on the other hand, as seen in coelenterates , such as jellyfish, enables environmental interactions on many sides and this can increase survival in water environments. Morphological consequences of these symmetrical patterns have led to queries concerning the genetic and developmental processes behind such diversity, giving an example of a profound correlation between symmetry and the ecological niche of organisms. Research in evolutionary developmental biology (evo-devo) has shed light on the genetic underpinnings that influence body symmetry and complexity across animal taxa. Interestingly, numerous species share a large number of similarities on the genomic level regardless of having considerable differences in the body plan. As an example, the preservation of a few genes in widely divergent taxa has led to the idea that evolutionary novelties can be produced more by changes in gene regulation than by changes in the coding sequences of the genes themselves. This observation highlights the significance of the mechanism of how genetic pathways allow different body plans and their symmetries. It poses compelling questions regarding the ancestral roles of these genetic mechanisms, particularly whether they can be traced back to common ancestors or if they have adapted to serve new functions independently in distinct evolutionary branches (Becker et al., 2005). Body symmetry is also explored in terms of the investigation of the process of development and the way it can tell us about the phylogenetic relationship between animal groups. The differences in symmetry may reflect the adaptation to specific ecological niches and show the evolutionary forces which operate on the species. Researchers need to examine gene expression patterns and to evaluate their correlation with developmental outcomes in various organisms to ascertain the homological importance of such characteristics. This method can be used to understand whether homologous genes have constant functions in the evolutionary history, or they have acquired new functions independently. Future research endeavors that expand the diversity of studied organisms, beyond the conventional model species, will be essential in enriching our knowledge of body symmetry and its evolutionary implications (Gruhl et al., 2016). This type of research is bound to make us appreciate the complexity of the interaction between genetic build and morphological diversity throughout the animal kingdom and eventually make us understand the complexity of life (Figure 6.6).

Figure 6.6. Animal body symmetry.

Source: https://cdn.britannica.com/50/195350-050-B5046285/moon-jellyfish.jpg.

6.2.1. Types of Symmetry: Bilateral, Radial, and Asymmetry

Symmetry is fundamental to the learning and comprehension of animal body plan evolution and diversity. Of the main types of symmetry, bilateral symmetry is likely to be the most salient among animal morphology. Bilateral symmetry works by having a body plan which can be bisected into mirror-image halves along a single plane, typically into well-defined left and right halves. Bilateral symmetry is universal in most advanced organisms, including vertebrates, with high levels of organization and specialization along their body axes. The advantages of this arrangement during evolution would be efficient locomotion and increased sensory input, enabling them to explore their worlds more effectively. Thus, the bilateral body plan must have facilitated survival and adaptability of most species to demonstrate how symmetry can influence the direction of evolution for organisms as hypothetical in present genomic studies investigating the origin of Bilateria's body axes (Baxevanis et al., 2007). Radial symmetry utilizes another form of organizing template whereby members of the body manifest around an axis. This kind of symmetry is well exemplified in animals , such as cnidarians and echinoderms, which are divisible into similar sections from the middle to the exterior. These organisms can effectively interact with their surroundings in more than one direction by using radial symmetry, which is helpful in their feeding and protection strategies. The pattern of having body parts arranged in radiating or circular form can provide enhancement in the capture or avoidance of prey or predators, depending on the situation. This is especially useful in habitats with open water. Radial symmetry also allows for a life form that might not require sophisticated locomotion. This adaptability uncovers the different strategies of evolution employed by different organisms in response to their ecological niches further underlining the significance of symmetry in evolutionary biology (Korenić et al., 2024). Asymmetry, the absence of any regular symmetry, violates traditional notions of organization within biological structures. Such symmetry appears in certain organisms, specifically in the majority of sponge species, certain fish, and mollusks. Asymmetrical body plans can be caused by various evolutionary pressures, including adaptation to specific environmental conditions or dissimilar lifestyles. In some species of flatfish, for example, asymmetry, where the body side of a specimen is significantly different from the opposite side, is the outcome of their flat, bottom-dwelling lifestyle, allowing these fish to camouflage effectively. Asymmetry's evolutionary importance reminds us that successful adaptation may be asymmetrical, expanding our understanding of body plan diversity even more. By tracing the characteristics and evolutionary roles of bilateral, radial, and asymmetrical designs, researchers can have a clearer understanding of the complex dynamics of function and structure in animals as we examine ecological and evolutionary pressures that shape these diversities (Baxevanis et al., 2007) (Figure 6.7).

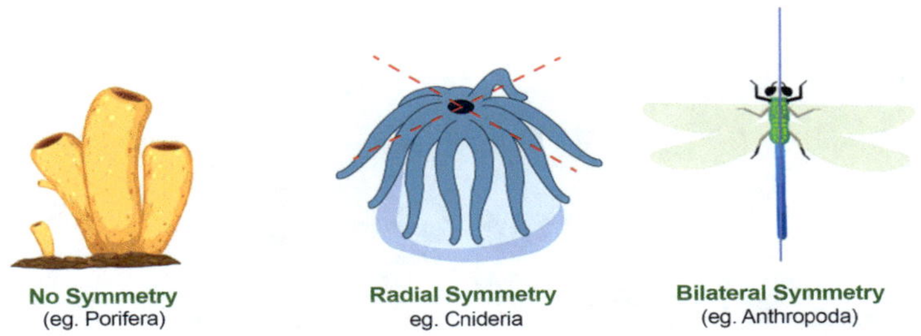

Figure 6.7. Types of symmetry.

Source: https://media.geeksforgeeks.org/wp-content/uploads/20220728114754/symmetry.jpg.

6.2.2. Evolutionary Significance of Symmetry

Evolution of symmetry is a significant component of animal diversity, offering basic information regarding the evolution of body plans across various taxa. Symmetry, particularly bilateral symmetry, is characteristic of a highly evolved form of organization that supports locomotion and coordination more effectively. The bilateral arrangement enables the organisms to create rapid movements towards the stimuli of the surroundings, allowing them to navigate through complex environments more efficiently. Moreover, symmetry development is not merely a question of taste; it is linked to evolutionary pressures tending to favor certain body plans. Here, symmetry is an intrinsic aspect of morphogenesis in demonstrating how genetic regulation is capable of producing diversification of form without detracting from functional integrity. Recent advances in evolutionary developmental biology (evo-devo) have shed light on the more basic genetic processes underlying the evolution of symmetry. These studies propose that body plan variation between animal phyla is not always mirrored by corresponding genetic transformation. For example, the genetic codes of different organisms, be they invertebrates or vertebrates, have a remarkable degree of conservation even when morphology varies extensively. This implies that the processes of symmetrical form evolution are possibly more than just single genetic mutations; instead, they occur due to intricate regulatory changes and re-employment of proven genetic networks. These developmental processes display developmental fluidity, where symmetrical forms can evolve from conservative genetic frameworks implementing a long-conserved developmental tool (Becker et al., 2005). Moreover, comprehension of symmetries evolutionary significance requires careful attention to the phylogenetic relationships. The serial conservation of genetic tools in diversified species raises outstanding issues of homology and functional evolution. Determining whether the symmetrical features evolved from a common ancestor or through independent evolutionary pathways is reliant on a delicate understanding of gene expression patterns and developmental operations. This sophistication is further enriched by the necessity of careful study outside of model organisms to account for the whole range of symmetry functions in animal diversity (Gruhl et al., 2016). As researchers go on to examine these relations, the evolutionary significance of symmetry not only enriches our understanding of developmental biology but also underscores the dynamic interaction between genetics, environment, and the evolution of body plans (Figure 6.8).

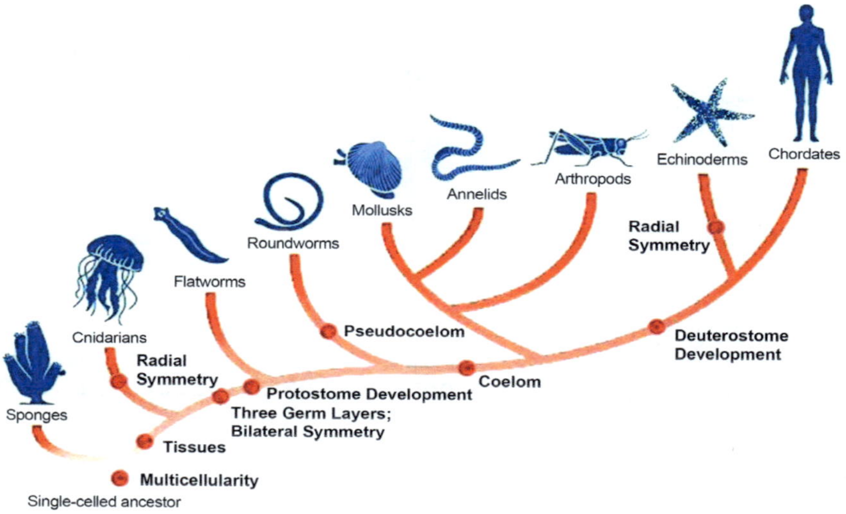

Figure 6.8. Evolutionary significance of symmetry.

Source: https://i.sstatic.net/QFXVb.png.

6.2.3. Examples of Symmetrical and Asymmetrical Animals

The symmetry principle in animals is a reflection of the elementary nature of the body plan and organization. Symmetrical animals come in two categories: bilateral and radial symmetry. Bilateral symmetry occurs in animals like human beings, dogs, and most terrestrial animals, whose body can be divided into two equal parts in one plane. This organizational structure has profound evolutionary importance because it provides for directional movement and enhanced spatial awareness within the organism. Radial symmetry is characteristic of organisms , such as jellyfish and sea anemones, however, which allows them to interact with their environment in a variety of directions, highly useful in aquatic settings. The study of such symmetrical arrangements provides a window into the evolutionary history of body plans and how different shapes fit into their niches. Similar to symmetrical patterns are asymmetrical animals, which exhibit a more complex and varied body plan. Of particular note is that most sponges are a prime example of asymmetry, without any detectable body plan to categorize them in symmetrical forms. Their asymmetrical forms allow them to effectively filter water and extract nutrients, demonstrating the capability of asymmetrical arrangements to confer significant adaptational advantages. Similarly, certain animals like the flatfish adult exhibit an odd asymmetrical development in that one aspect of their body becomes flattened and adapted for a benthic existence. These variations challenge traditional notions of symmetry in biological structure, highlighting the vast adaptability exhibited within the animal kingdom. These advances in body shape highlight the concept that symmetry, while ubiquitous, represents just one of many pathways through which animals move along their evolutionary trajectory. The development and evolution of symmetry in animals can be further examined through embryological development, illustrating the close coordination between development and adult body shape. For instance, the spiral cleavage observed in snails and earthworms, though producing opposite adult body forms, suggests a shared evolutionary history among different phyla. Such embryonic development provides the foundation for the observation of how complex body forms arise from simpler ones, suggesting a shared origin of allegedly disparate animals. As scientists extend their investigations into processes of symmetry and asymmetry, for example, breakthroughs in genome editing and phylogenetics, they are discovering the underlying rules that govern these developmental processes. Knowledge of this sort is necessary, as it gives insight into how evolution builds the incredible variety of animal life, whether symmetrical or asymmetrical (Marletaz et al., 2020; Korenić et al., 2024).

Test Your Knowledge

1. What are the types of symmetry?
2. What is the evolutionary significance of symmetry?

6.3. Body Cavities

Learning Outcome

1. Types of body cavities: acoelomate, pseudocoelomate, and coelomate;
2. Functions of body cavities in animal physiology;
3. Examples of animals with different body cavities.

The anatomical variation in the body cavity of animals is an important factor in the complexity of the entire body plan, a factor that is determinant in the physiological performance and evolutionary capacity of animals. Body cavities, which can be coelomate, pseudocoelomate, and acoelomate, in general, affect the development of organs, nutrients transportation, and the entire work of biological systems. Coelomates, with a proper body cavity (coelom) of mesodermal origin, e.g., annelids and vertebrates, enable more mobility and specialization of organs, facilitating complex organisms. By comparison, the pseudocoelomates (, such as nematodes) have a partially lined cavity that is beneficial in hydrostatic pressure, but do not have the complete complexity of organ systems that is found in the true coelomates. Acoelomates, which include flatworms, lack a coelomic cavity altogether and must rely on diffusion for internal processes, underscoring the trade-offs associated with bodily design (Gruhl et al., 2016). The existence of a body cavity also influences the reproductive strategies in various taxa, which shows the evolutionary importance of such an anatomical structure. As an example, coelomates can have more efficient reproductive systems, and in many cases have internal fertilization and viviparity because of the protection of the body cavities. The coelom in such organisms does not only accommodate the reproductive organs, but it also aids in embryo development, which offers protection to the embryos against environmental risks. But external fertilization is characteristic of pseudocoelomates and acoelomates which depend upon water currents to distribute their gametes.

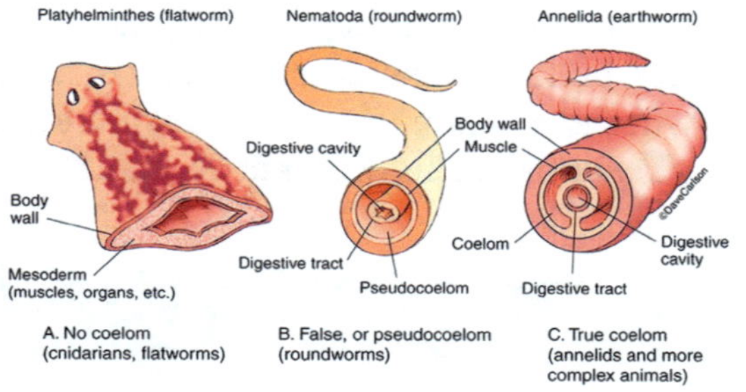

Figure 6.9. Animals' body cavities.

Source: https://i.pinimg.com/736x/0b/58/02/0b58025f6ff2d3a457116c491d173888.jpg.

This divergence emphasizes how body cavity structure directly influences reproductive success, adaptability, and survival strategies among diverse animal groups (Mestre et al., 2012). Besides, the evolutionary shift in body plans, acoelomate to coelomate, is a critical stage in the evolution of animals, which adds to biodiversity. The appearance of a proper coelom allowed the formation of more complicated organ systems and the separation of body functions. This permitted the further development of complex systems like the circulation and digestive systems of higher animals. These innovations have resulted in efficiency in metabolism and better homeostasis. Consequently, the exploration of body cavity evolution reveals critical insights into the adaptive radiation of diverse life forms, underlining the unity and versatility of biological organization across the animal kingdom (Gruhl et al., 2016) (Figure 6.9).

6.3.1. Types of Body Cavities: Acoelomate, Pseudocoelomate, and Coelomate

The separation of body cavities in animals to a great extent defines their general structural and functional diversity. Acoelomates, , such as flatworms, lack an actual coelomic cavity and instead possess a compact body that is packed with mesenchyme, a type of gelatinous substance. The absence of a body cavity means that acoelomates rely on diffusion for the delivery of nutrients and the expulsion of waste, therefore limiting their complexity and size. Their body organization is relatively simple, with bilateral symmetry and centralized digestive system. Due to this, the lack of a coelom impacts their ability to develop specialized organs and systems, and the relationship between forms of body cavities and evolutionary adaptation is obviously seen in structure differences among various taxa (Baguñá et al., 2008). Conversely, pseudocoelomates, , such as nematodes, contain a body cavity known as a pseudocoelom that is only partially lined with mesodermal tissue. The hydrostatic skeleton, the pseudocoelom, provides structural support and capability for locomotion. The pseudocoelom enables more complex organ development than in the acoelomates and enables the formation of specialized systems, , such as a more developed nervous and reproductive system. The presence of a pseudocoelom has some benefits for such organisms, for instance, the ability to possess an elaborate body plan and more developed physiological processes. Thus, the pseudocoelom is significant as an evolutionary development permitting greater complexity in the body (Passamaneck, 2003). The coelomates, including annelids and chordates, have a true coelom, or a body cavity completely lined with mesoderm. This developmental organization facilitates the development of sophisticated organ systems and allows for better movement and the transportation of nutrients. The coelom acts as a shield protecting internal organs and aids in suspending and organizing them, thereby enhancing the function of organs in their respective systems. Coelomates can also develop more diverse types of body plans, leading to amazing morphological diversity of animal phyla. The contrast between these three body cavity types underscores one of the most characteristic aspects of evolutionary biology, as it illustrates how crucial body plans are in determining the differential potential for survival strategy and adaptation between different environmental contexts (Baguñá et al., 2008) (Figure 6.10).

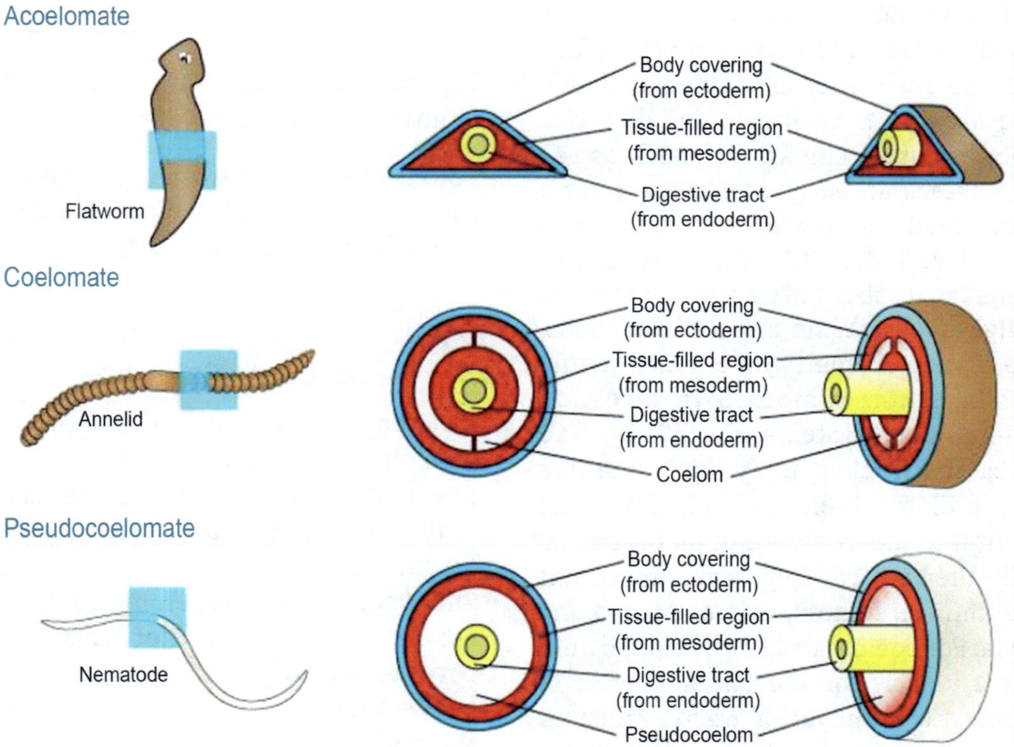

Figure 6.10. Classification of body cavities in animals.

Source: https://bioclass.cos.ncsu.edu/bio402_315/Classificaton/body%20cavities/types-of-coeloms.gif.

6.3.2. Functions of Body Cavities in Animal Physiology

Body cavities, or coeloms, form a significant aspect of the physiology of animal groups, and they exist as compartments for housing the internal organs. They are categorized into two groups: fluid-filled coeloms and pseudocoeloms, depending on the evolutionary history of the organism. In bilaterians, the presence of a true coelom enables complex organ development and functional differentiation, which enhances the organism's capacity for the regulation of homeostasis. For instance, the coelomic fluid provides a hydrostatic skeleton for successful movement and support in soft-bodied organisms, , such as annelids and mollusks. Second, the separation of the body wall from visceral organs prevents damage to important systems during the movement of the organism, which equates to overall fitness in the organism's environmental niche. The evolution of the body cavities has immediate implications on digestive processes in animals, particularly in relation to extracellular digestion. The primary function of the coelom is not merely the support structure; it plays a key function in the absorption of nutrients, , such as in those organisms with complex digestive systems. For instance, cnidarians , such as sea anemones possess an early evolutionary form of a digestive cavity, both where food is broken down and where nutrients are absorbed. Not just does this tell us about the evolution of more complex digestive systems among bilaterians, but it also shows dramatic divergence in the process of ingestion compared to digestion across animal phyla (Steinmetz, 2019). The coelomic cavity is thus a dynamic volume that accommodates important physiological processes vital to the existence of an organism. Additionally, the body cavity plays an important function within circulatory systems of many animal phyla. For coelomates, the coelomic fluid may act as a vehicle for the dissemination of nutrients and the removal of waste, much as blood does in more complex circulatory systems. This organization facilitates efficient movement of substances within the body, supplementing metabolic processes. Furthermore, the presence of a coelom might also affect the development of specialized structures for circulatory function, , such as hearts or aortic arches. With growing diversification of body plans among animals, the evolutionary history of body cavity origins and their associated functions continues to reveal the intricacies of animal physiology and how structural modifications contribute to the success of many species in their environments (Machamer et al., 2002) (Figure 6.11).

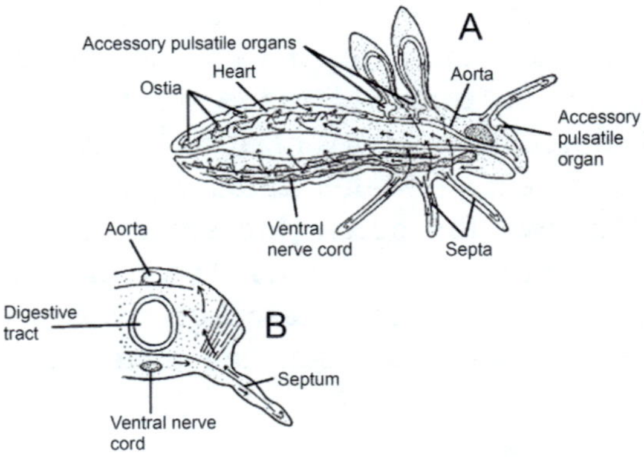

Figure 6.11. Functions of body cavities in animal physiology.

Source: https://ars.els-cdn.com/content/image/3-s2.0-B9780128203590000104-f07-07-9780128203590.jpg.

6.3.3. Examples of Animals with Different Body Cavities

An animal's diversity can be partially divided by their body cavities, which play a role in their physiology and evolutionary characteristics. A few of the animals with diversified body cavities are acoelomates, pseudocoelomates, and coelomates. Acoelomates, for example, are flatworms that do not have a body cavity between the body wall and the digestive system, hence their body cavity is solid. This limits their size and the intricacy of their internal organization but allows them to occupy specific ecological niches where less complex systems of organization are advantageous. On the other hand, pseudocoelomates, like roundworms, possess a fluid-filled cavity within the body known as a pseudocoelom that provides space for the organs and may assist in locomotion and transportation of nutrients. These body cavity variations, thus, illustrate the diverse animal evolutionary designs that render them more resilient and adaptable. The coelomates are another improvement in the evolution of body cavities, which occur in organisms , such as earthworms and vertebrates. The former possesses a true coelom, a fluid-filled cavity lined completely by mesodermal tissue, which allows them to have complex organ systems and also facilitates ease of movement. With a coelom, more advanced physiological activities can take place, including the development of a circulatory system and differential development of the gut. The advantage in evolution of the coelom is also manifested in the increased capacity for organs to grow and specialize, the basis for more complex interaction with their surroundings. This seeming complexity within coelomate body plans is paralleled by a particular evolutionary transition which allows these animals to occupy more than one ecological niche, maximizing their survival probabilities under disparate environmental conditions. Embryonic developmental mechanisms relevant to the different body plans play an important role within the evolutionary trajectory of different taxa. As dealt with in metazoan research, embryonic diversification displays special mechanisms in charge of body cavity and organ formation. For instance, the coelomate development pattern is more reliant on spatial organization and cell differentiation compared to acoelomates, which lack this complexity. The exploration of developmental strategies in these groups of organisms provides insights regarding how evolutionary forces have shaped the incredible diversity of animal form that we find today. By looking at the embryo developmental processes underlying the development of body cavities, we gain a better understanding of principles governing animal diversity. This comparative approach not only highlights the significance of body cavities in interpreting anatomical and functional variation but also has broader applications in understanding evolutionary biology as a whole (Davidson, 1991; Gruhl et al., 2016).

Test Your Knowledge

1. What are the types of body cavities?
2. What are the functions of body cavities in animal physiology?

6.4. Body Plans and Development

Learning Outcome

1. Embryonic development and body plan formation;
2. Role of Hox genes in body plan development;
3. Variations in body plans across different species.

Understanding body plans and development is important to understand the evolutionary paths of animal diversity. Body plans provide the basis of the morphology of organisms, the way organs and structures are laid out. Researchers have discovered a somewhat shocking paradox in a number of animal phyla: genetic conservation is frequently accompanied by a reduction in morphological variability. This effect implies that, although animals display a huge variety of shapes and capabilities, the GMs required to develop them might not be as dramatic as it might seem. The consequences of this difference are profound, and it has cast doubt on how the mechanism of evolution, especially the aspects that deal with gene regulation, contributes to the variation seen among animal species. Investigating these relationships can yield insights into the evolution of developmental pathways and the role of genetic networks in shaping body forms, further enriching our understanding of evolutionary biology (Becker et al., 2005). Evolutionary developmental biology or evo-devo is a concept that has a central role in the study of the evolution of body plans. Comparative studies have enabled researchers to uncover important similarities in the regulation of gene action across a wide range of taxa, including the model organisms , such as Drosophila and vertebrates. Such similarities indicate that the formation of similar structures may be controlled by common genetic mechanisms, and point to the possibility of functional homologies between distantly related species. But this view is not unblemished by complications; the presuppositions of ancestral complexity have been modified by the emergence of new discoveries. For instance, an alternative model proposes that the last common protostome-deuterostome ancestor may have exhibited a simpler morphology than traditionally inferred, which reshapes our interpretations of the fossil record regarding early bilaterian evolution (Erwin & Davidson, 2002). These and other insights highlight the complexity of the interactions between genetic legacy and morphological variation and stretch the limits of what we know about developmental evolution. The complexity of body plan evolution can only be untangled by intensively researching the gene regulatory networks underlying development. Recent studies have stressed the importance of a great deal of phylogenetic analysis outside the usual model systems. With a greater diversity of organisms examined, scientists can learn more about the evolution and diversification of developmental processes. The procedures used in the research of gene homology and functionality need to be streamlined to include the inconsistent appearance of homology among different species. Such a method can, in principle, help to resolve whether superficial similarities are the result of common ancestral features or of convergent evolution. As evolutionary models advance, they increasingly inform our understanding of the dynamic nature of body plans and development, spotlighting how evolutionary pressures and genetic regulation converge to produce the remarkable diversity observed in the animal kingdom (Becker et al., 2005) (Figure 6.12).

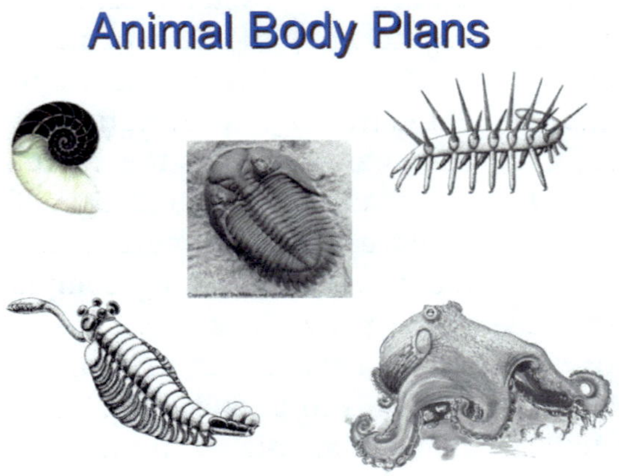

Figure 6.12. Body plans and development in animals.

Source: https://image5.slideserve.com/9593045/animal-body-plans-l.jpg.

6.4.1. Embryonic Development and Body Plan Formation

The intricate embryo development process sits at the core of animal body plan evolution as an indicator of evolutionary change in different groups of taxa. At its core, embryogenesis is a chain of tightly regulated events where individual cells proliferate, differentiate, and spatially organize to generate the body plan. Vertebrate early developmental stages reveal a fascinating divergence from the invertebrates, an indicator of the complexity and diversification of developmental processes. Different regions of the embryo impose different paths influencing cellular destiny, highlighting both the action of lineage and positional information. This flexibility early in development not only determines the final organization of the organism but also holds the clue to phyletic evolutionary history, giving a glimpse into how disparate body plans have evolved and proliferated over millions of years. In the overall metazoan body plan establishment, the regulatory genes play determinant roles in the developmental pathways. The genes have been reported to have conserved and specialized functions across different taxa and to orchestrate the spatio-temporal gene expression patterns that are required for proper organism organization. Significantly, gene homologies that appear in Drosophila and in chordates imply an underlying commonality that occurs beyond contemporary classes, necessitating discussion of the ancestral traits common to protostomes and deuterostomes. It is this information that prompts the reconsideration of the complexity of the last common protostome-deuterostome ancestor as potentially less complicated than traditionally has been assumed. This shift in perspective invites more research into evolutionary routes to current animal morphology, emphasizing the role of regulatory genes in body plan establishment (Erwin & Davidson, 2002). The three major types of embryogenesis, Type 1 as exemplifying invertebrates, Type 2 as exemplifying vertebrate development, and Type 3 as exemplifying insects, disclose the different tactics employed in embryogenesis between different lineages. Type 1, for instance, exemplifies the significance of cell lineage in early embryonic organization, where cell fate is determined by positional information. Type 2 is a more intricate mechanism in which cell specification is reduced to lineage independence with emphasis on morphogen and cell migration control. Type 3, however, represents a specialized regulatory system with sharp spatial expression of development genes before cellularization. Understanding of these embryogenic processes teaches us about the evolutionary significance of animal diversity and body plans and how numerous mechanisms can lead to the fine complexity observed in the animal kingdom (Davidson, 1991) (Figure 6.13).

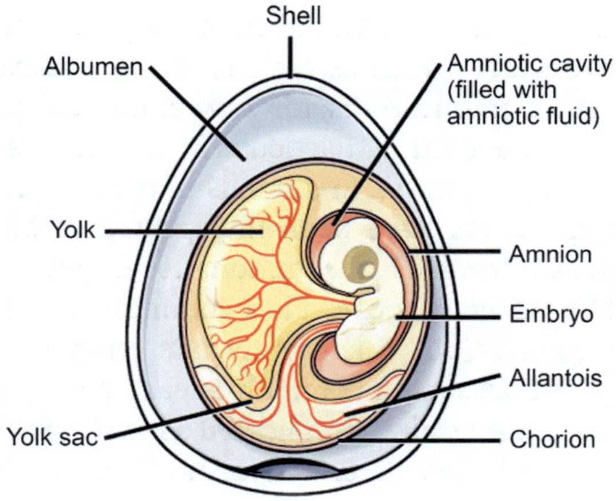

Figure 6.13. Embryonic development and body plan formation in animal.

Source: https://bpb-us-e1.wpmucdn.com/sites.gatech.edu/dist/6/1810/files/2017/05/Figure_29_04_01.png.

6.4.2. Role of Hox Genes in Body Plan Development

Their central role in body plan organization has been one of evolutionary developmental biology's (evo-devo) central interests. The genes, members of an extremely conserved family, specify the body segment identity along the anterior-posterior axis in a variety of animal taxonomic groups, both invertebrates and vertebrates. Regulation of Hox genes is what brings together the subtle mechanisms of an organism's anatomy, influencing aspects , such as limb growth, organ location, and body shape. The conservation of these genetic components in so many species suggests that there has been a shared evolutionary history, where complex body plans emerged from a relatively simple genetic plan. This conservation generates a fascinating paradox: in spite of wide morphological diversity, genomic architecture is fantastically stable, prompting further exploration of how differences in gene regulation could permit evolutionary innovation in body plans (Becker et al., 2005). In addition, studies examining the functional homology of genes that regulate, , such as the Hox genes, indicate profound implications for our understanding of animal evolution. Particularly, the Hox gene expression patterns in animals like Drosophila and chordates have been interpreted to offer evidence towards a common developmental toolkit, which suggests a last common ancestor with complex body plans. Alternate perspectives suggest that such a common ancestor was not as morphologically complicated as traditionally envisioned (Erwin & Davidson, 2002). This model proposes that while Hox genes perform similar regulatory roles between taxon groups, their specific functions and interactions have been redirected by evolutionary processes. Thus, the dynamics of Hox gene expression can illuminate the evolutionary mechanisms behind the staggering diversity of body plans within the phylum of animals. The intricate dance of Hox genes and morphological transformation underscores the significance of gene co-option and regulatory evolution in creating body plan diversity.

As newer studies suggest, the emergence of new traits could often be attained through the hijacking of existing genetic machinery rather than significant changes in the gene sequences themselves. This mechanism shows how the spatial and temporal expression of genes is crucial in guiding the developmental fate and, in the process, generating the vast number of body forms existing in modern animals. Expression studies of Hox genes across phyla are increasingly revealing important aspects of body plan architecture, so that deciphering these genetic modules may be the key to understanding the mysteries of the evolutionary past and life form diversification. Thereby, Hox genes are an integral aspect of unravelling the evolutionary history of animal body plans and diversity (Becker et al., 2005) (Figure 6.14).

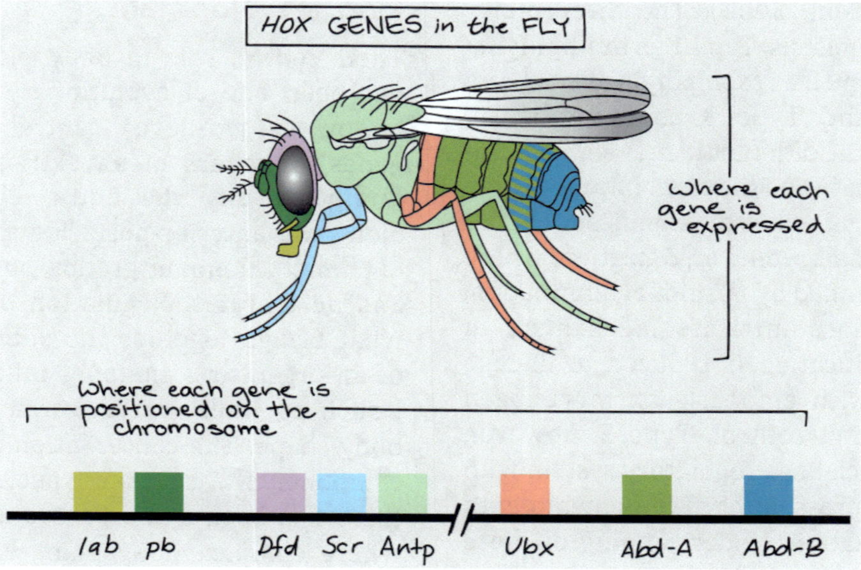

Figure 6.14. Role of Hox genes in body plan development in animals.

Source: https://bpb-us-e1.wpmucdn.com/sites.gatech.edu/dist/6/1810/files/2017/05/3a2e43d25e68e05e524fbc34b01db3d8390b534a.png.

6.4.3. Variations in Body Plans Across Different Species

The anatomical diversity in the animal kingdom is largely due to body plan differences, which are the fundamental template for organismal structure and function. Body plans occur in a range of ways from uncomplicated bilateral symmetry in worms to more elaborate radial symmetry in cnidarians , such as jellyfish. These differences are resultant from evolutionary adaptations to environmental niches allowing for survival and reproduction. For instance, the streamlined bodies of water-dwelling animals like fish are hydrodynamically shaped to reduce resistance when they move through water. These morphological developments are not superficial; they have effects on physiological performance, reproductive life, and even behavioral characteristics, contending that body plans play a crucial role in understanding the big picture of animal diversity and ecological processes in various ecosystems. Besides, the genetic and developmental processes that condition body plan variations are intricate and heterogeneous among groups. Recent research has found that developmental variation of regulatory genes has profound influences on morphology, enabling species to adapt to their environments for millions of years. The action of such genetic determinants can be particularly observed in the examination of model systems, where mutations in some developmental genes cause drastic alterations in body morphology. For instance, research on the gene networks that regulate arthropod segmentation sheds light on the mechanisms of evolution of different body architectures in this phylum. Such genetic knowledge not only enhances our understanding of phylogenetic connection among species but also highlights the remarkable plasticity in body plans, as shown by evolution's competence to create diverse forms from common ancestral blueprints (Tice et al., 2016). The importance of various body plans also transcends species to influence entire communities. For instance, the evolution of varying body shapes can influence predator-prey interactions, the use of resources, and space occupation. Aquatic animals, for instance, exhibit a variety of adaptations, ranging from long, flexible predatory fish to flat bottom-dwellers, each occupying unique ecological niches. They enable different species to live together and thrive in shared environments, thus raising overall biodiversity. Such complex interactions are only comprehensible with an interdisciplinary framework integrating genetics, ecology, and evolutionary biology. Since scientists are still learning about the complexities of how body plans have evolved and the ecological significance of the same, it is certain that learning about variations in body plans is the only way to grasp the total network of diversity in animals and its impact on ecosystems (Gupta et al., 2004).

Test Your Knowledge

1. What is embryonic development?
2. What is the role of Hox genes in body plan development?

6.5. Adaptations and Evolution

Learning Outcome

1. Influence of environment on body plan adaptations;
2. Evolutionary trends in body plans;
3. Case studies of specific adaptations in various species.

Adaptations are the most important concept in the study of the evolution process that leads to animal diversity and body strategies. We can define an adaptation as any characteristic that increases the capacity of the organisms to survive and reproduce in a particular environment. In the course of time, such characteristics may become common in a population because individuals who have adaptations that help them survive have better chances to prosper and leave their genes. A driving force in this sense is the process of natural selection, in which pressures of the environment select traits that make organisms more fit. In this light, it is possible to see that adaptations are not fixed; they are, on the contrary, dynamic features that change in accordance with the changing environmental factors. This constant process of adaptation highlights the beauty of animal diversity, with different species developing distinct features to enable survival in diverse environments and ecological niches. The issue of adaptations and evolution is even complicated by the fact that there are what is known as evolutionary convergences whereby different species tend to independently evolve similar features due to similar environmental pressures. The phenomenon points to the flexibility of life and how selective pressures can influence not just individual species, but whole ecosystems. The fossil record is a good source of evidence to this notion, as it shows how some adaptations have reoccurred many times in the history of evolution, indicating some sort of directionality attributed to adaptations that allow organisms to survive successfully. However, although these convergent adaptations can give the impression of predictability in the evolutionary results, it does not imply progress in a straight line towards complexity or progress. As research continues to unveil the mechanisms that drive these adaptations, it becomes evident that evolution operates within a complex interplay of chance and environmental demands, producing a myriad of body plans that embody the diversity of life itself (Vermeij, 2019). Recent insights from evolutionary developmental biology (evo-devo) have elucidated the underlying genetic mechanisms responsible for the diversity in animal body plans. It has been found that, although morphological diversity is spectacular among phyla, the genetic code is usually remarkably preserved. , such as, numerous genes, even those participating in the developmental processes, are unexpectedly similar among a wide variety of species, suggesting that evolutionary novelties are frequently produced not by modifications in the genetic make-up but by variations in the regulation of these genes. This implies that the complex sequence of body plan changes might not involve the recruitment of whole new genes so much as the hijacking and regulatory refinement of the existing genetic circuits. As researchers continue to explore these patterns, the distinction between homologous structures and their evolutionary origins remains a critical area of inquiry, highlighting the necessity of expanding research beyond traditional model organisms to truly grasp the full spectrum of evolutionary adaptations (Becker et al., 2005).

6.5.1. Influence of Environment on Body Plan Adaptations

The environmental impact on the variation of the body plan is observed in all animal clades, and increasingly in marine environments. The case of darters, benthic fish *Percidae*, demonstrates the impact of prior climatic history, , such as Pleistocene glaciations, on morphological and genetic variation within the species. The glacial periods' habitat fragmentation favored speciation, but favored interbreeding and hence could not be used to distinguish interrelated species due to hybridization and introgression. In the province of the Ozark Highlands, researchers separated 12 distinct species of *Etheostoma* using extensive sampling and phylogenetic reconstruction. The findings point out how environmental stress factors promote diversity but also complicate tracing the evolutionary path of organisms. These intricacies highlight the habitat and genetic variation interaction in the specification of body plans, and they point out the necessity of large sampling across varied habitats to enhance adaptation mechanisms of darters (George et al., 2018). Adaptations in African great apes, , such as chimpanzees and gorillas, similarly indicate how habitat could influence body plan specifications. By constructing a time budget model from data at multiple study sites, researchers had predicted how the species allocate time on essential activities , such as feeding, resting, and socializing. Surprisingly, body mass and climatic variables were predictors of site-specific behavior, with a direct correlation between environmental factors and survival strategies. The model also revealed that movement time was a controlling factor in the abundance of the apes in different habitats, revealing that more energy expenditure in more stressful environments could lower population viability. The link between body mass, activity allocation, and habitat supports the co-evolution of physical traits and environmental niches, also illustrating the impact of environmental factors on body plan adaptations of primates (Dunbar et al., 2008). The significance of environmental forces as determinants of adaptations in body plan extends beyond species to larger ecological networks. As they evolve, the morphological features of animals have a tendency to reflect not only their survival mechanisms but also their interactions with other organisms within their ecosystems. For instance, predator-prey interactions can lead to evolutionary arms races that precipitate adaptations in body shape, locomotion, and sensory systems. Further, ecological changes, , such as habitat destruction or global warming, can exert new selective pressures, which can lead to rapid adaptive evolution or even extinction. Therefore, knowledge of how environmental drivers influence body plans is essential for informing conservation efforts, so that species can evolve in response to changing environments and endure more transformed ecosystems. As such, the research on these adaptive trends not only improves our knowledge of animal diversity but also facilitates the conservation of biodiversity in light of global environmental changes.

6.5.2. Evolutionary Trends in Body Plans

The exploration of evolutionary trends in body plans reveals a complex interplay among morphological disparity and genetic conservativeness among the animal phyla. The evo-devo research has demonstrated that expanding disparity in the animal body plans is not invariably associated with drastic changes in genomic composition. For example, while complex body plans possess a vast repertoire of shapes, the genetic code underlying them is relatively conserved, particularly when invertebrates and vertebrates are compared. This suggests that body plan diversity is derived less from variation within coding regions of genes and more from combinatorial mechanisms, like variations in gene regulation and co-option of genes into new developmental pathways. This type of understanding challenges traditional views of evolution, emphasizing the significance of deciphering phylogenetic patterns and gene homology to establish how new body plans are created in an apparently constraint-ridden genetic environment (Becker et al., 2005). In assessing the genetic roots for body plans, attention must also be drawn to the function of regulatory genes in development within taxonomic groups. The most important is the similarity in deployment of such genes in Drosophila and chordates, which has precipitated making deductions on functional homologies at the process level during development. Early interpretations suggested that the most recent common ancestor of the protostomes and deuterostomes possessed a much more intricate body plan than extant bilaterians. This is currently being reversed, with the suggestion that the most recent common ancestor possessed a less complex morphology than was previously assumed. This reconstruction offers us a revised expectation of the Neoproterozoic fossil record, necessitating reevaluation of early bilaterian complexity as well as evolutionary trends. To further enhance our understanding of these trends, further analysis of ancient and extant organisms must be undertaken (Erwin & Davidson, 2002). Along with this, the relevance of studying the patterns of evolution in body forms is to lead us toward greater insight into the nature of how evolutionary processes operate within a context of genetic continuity and adaptation to the environment. The frequent conservation of genetic elements across broad taxa suggests a shared developmental inheritance, where evolutionary innovations may occur more through the reorganization of existing genetic tools rather than through the generation of entirely new sequences. This revolution of paradigm not only impacts our knowledge of body plan evolution but also highlights the intricate relationship between environmental and genetic factors that mold morphology. Synthesizing data from several disciplines, including paleontology and genetics, enables scientists to sweep a holistic map of how body plans evolved across geological timescales, gaining more information on the tale of animal diversity, which is continuing to unfold in contemporary ecosystems (Figure 6.15).

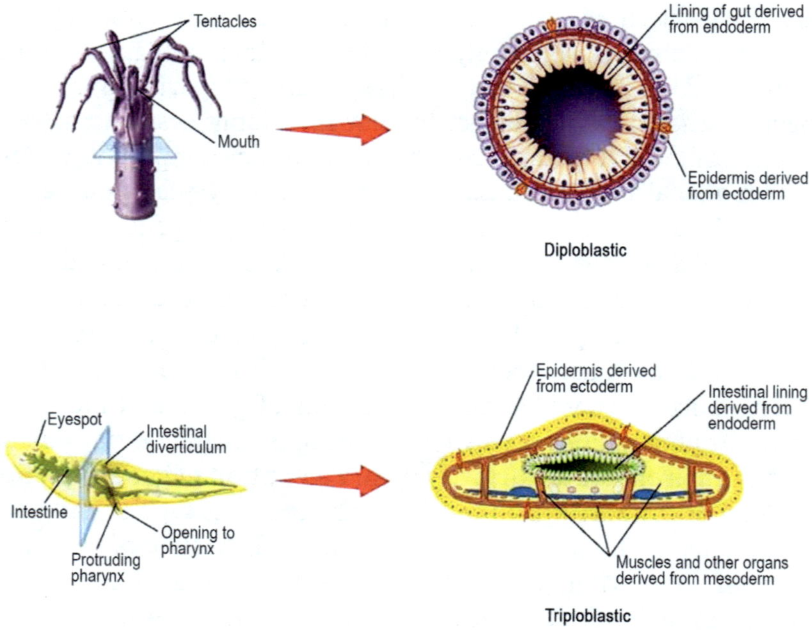

Figure 6.15. Evolutionary trends in body plans in animals.

Source: https://schoolbag.info/biology/concepts/concepts.files/image633.jpg.

6.5.3. Case Studies of Specific Adaptations in Various Species

The very diversity of animal adaptations is a witness to the creative powers of evolutionary processes. A good case study is the African cichlid fish, which is adaptive radiation within one environment. The fish have developed a wide range of feeding structures and morphological characters, enabling them to inhabit various ecological niches in freshwater lakes. For instance, there are forms with specialized herbivorous jaws, while others possess more robust teeth to consume smaller fish. This adaptability not only guarantees enhanced survival within competitive ecosystems but also the evolutionary forces that determine the body plans and behaviors of a species. The cichlids, then, emphasize the crucial role of environmental factors in determining specific adaptations and emphasize a close relationship between ecological interaction and bodily variation in the animal kingdom. Terrestrial habitats, as opposed to aquatic ones, also provide instances of spectacular adaptation. One of the most traditional instances of natural selection against the background of ethnological modification is the peppered moth (*Biston betularia*).

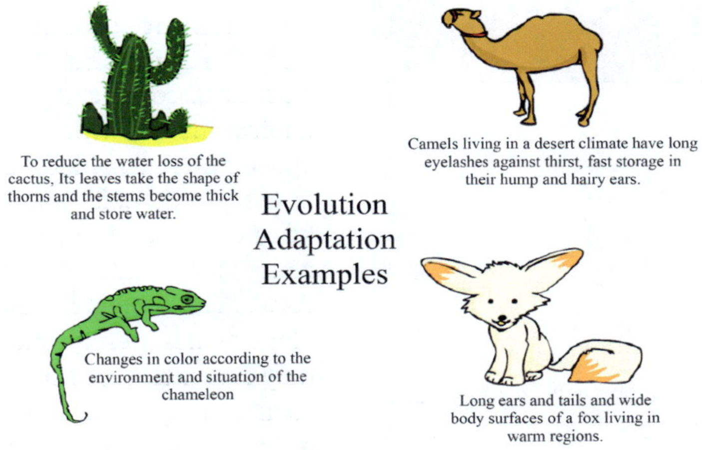

Figure 6.16. Specific adaptations in various species.

Source: https://synopsisias.com/uploads/blog/posts/tinymce/Synopsis-IAS-1722677093.png.

More dark moths became common during the Industrial Revolution as more pollution rendered the bark on city trees darker. Darkening in this case was an adaptive benefit since darker moths were less observable to predators on the soot-covered surfaces. This shift is a prime example of how drastic environmental changes may affect the genetic composition of populations directly, where the dynamic mutual interaction between organisms and the environment plays a role. Such instances not only illuminate the adaptation process but also illustrate why ecological wholes are crucial in maintaining diversity. The peppered moth is therefore a very significant example of how adaptations came about from immediate reaction to changing environments. Amphibians also possess unique adaptations that illustrate the complexities of survival mechanisms within different ecosystems. The poison dart frog, particularly in Central and South America, has chemical defense mechanisms in the way of predator deterrence. The small, brightly colored frogs acquire their toxins from the same insects that they consume, which is a large portion of their survival. The bright color is an aposematic warning, discouraging predators from approaching since they are poisonous. This adaptation is not only for defense against self but also influences the dynamics of predator-prey relationships in their natural habitat. Understanding the intricacy of such adaptations, as the ecological interactions leading to such attributes, provides insight into the larger world of evolution and biodiversity. The poison dart frog example clearly illustrates how adaptations can be both an evolutionary strategy and an ecological balance mechanism in intricate environments simultaneously (Figure 6.16).

Test Your Knowledge

1. How influence of environment on body plan adaptations?
2. What are the evolutionary trends in body plans?

6.6. Ecological Roles of Diverse Body Plans

Learning Outcome

1. Impact of body plans on ecological niches;
2. Interactions between different body plans in ecosystems;
3. Importance of body diversity for ecosystem stability.

The connection between the ecological roles and the animal body plans is very complex and is central to the operation of the ecosystems. Different body plans help the different species to have a different niche thereby bringing an ecological balance. As an illustration, radial symmetrical organisms (e.g., jellyfish) live in the pelagic zone and use their body shape to efficiently hunt. On the other hand, bilaterally symmetrical organisms (, such as mammals) are specialized towards directional locomotion and intricate behavioral patterns required during hunting or foraging. Such a variety of body strategies is an adaptation to the environment and resource distribution history, which further demonstrates how structural diversifications enable ecological interactions to support species persistence and ecosystem stability. Besides, the different body shapes are important in ecosystem services, including nutrient cycling and habitat creation. Mineralized shells and hard exoskeletons of organisms add structural complexity to marine and terrestrial environments, which serves as refuge to other species increasing local biodiversity. An example is the coral reefs which provide a home to a huge diversity of marine organisms as they are composed of the calcareous organisms of the corals. The ecological roles of these diverse body plans do not merely enhance habitat complexity; they also intersect with evolutionary processes, , such as adaptation and extinction events, which have shaped ecological community structure over time (Stewart, 2010). Therefore, it is important to know what role different body plans play in an ecological setup so as to conserve them in a bid to maintain biodiversity. The ecological research is still innovating to highlight the diverse purposes the animal body plans can have in the agricultural systems. Sustainable farming strategies are becoming aware of the potential of animal diversity to increase the resilience and productivity of agriculture. To take some examples, incorporation of livestock with varied body plans into agro-ecosystems can provide better nutrient cycling and resistance to diseases since various species serve ecological functions that bring about general health of the farm. This strategy is consistent with the thinking of eco-functional intensification, which aims to use the ecosystem services embodied in biodiversity to achieve greater agricultural productivity. By understanding these interconnections, as elucidated by Cuoco et al. (2010), researchers can create more effective policies aimed at fostering both biodiversity and sustainable agricultural practices, emphasizing the importance of diverse body plans in both natural and managed ecosystems (Figure 6.17).

Figure 6.17. Ecological roles of diverse body plans.

Source: https://assets.labroots.com/_public/_files/system/content-articles/images/body/63c2a77f652b8ee0d1fa7d4cdda8fb829bde004a_600x532.jpg.

6.6.1. Impact of Body Plans on Ecological Niches

The interface between body plans and ecological niches is central to animal diversity and its evolutionary significance. Body plans determine the morphological and physiological characteristics of organisms, which in turn determine how species live in their worlds and exploit available resources. For example, aquatic organisms' diverse body plans, , such as the fish's streamlined shape and amphibians' more body weight, provide for different food acquisition and water locomotion strategies. Specialization leads to niche differentiation where species evolve to occupy specific niches in their environment, thereby lessening competition and optimizing the use of resources. Therefore, the body plan of an organism not only indicates evolutionary adaptations to environmental demands but also prescribes its ecological roles, establishing a tangible link between body forms and functional dynamics in ecological systems. Evolutionary pressures exerted on body plans also affect ecological niches, which are vastly different across various environments. For instance, research findings show that urban environments may pose certain challenges to free-living animals, as evidenced by research on blue tit populations. The animals are less fertile in cities primarily due to reduced food availability, particularly caterpillars, which are important for their offspring during development. These findings illustrate how body plans and the resultant foraging tactics must modify according to diverse ecological conditions. The influence of urbanization on the presence of niches and food habits of species indicates the interconnectedness of body form, environmental pressures, and survival, and how anthropogenic forces can interfere with the natural process of evolution in various organisms. The study of these processes teaches us about the greater implications of differences in body plans on species fitness and ecological relationships. The body plan research also provides interesting insights into speciation and diversity, particularly in communities with high levels of competition. In environments , such as the Nicaraguan lakes, the adaptive radiation of the Midas cichlids illustrates how various body forms evolved as adaptations to selective pressures in specialized ecological niches. Genomic and morphological studies have shown population differentiation differences among such cichlids that highlight ecological specialization by means of body plans. Interestingly, more intense genomic divergence in benthic species than in their pelagic counterparts suggests that unique body features are the reasons for inhabiting specific niches and maintaining biodiversity. As body plans make possible different feeding strategies and interactions within the environment, they become critical to the process of recognizing the evolutionary forces that sustain ecological diversity and emphasizing the need for rigorous studies , such as those mentioned in species groups , such as Midas cichlids, where ecological and morphological characteristics co-evolve with subtle detail with conditions of the habitat (Capilla-Lasheras et al., 2017; Card Lín et al., 2024).

6.6.2. Interactions Between Different Body Plans in Ecosystems

The interactions between different body plans in communities are crucial factors influencing ecological dynamics and community structures. The different animal body plans characterized by different anatomical and physiological attributes determine how species compete for resources, defend territory, and engage in symbiotic associations. For instance, the existence of diverse body plans results in intricate food webs under which predators, prey, and decomposers interact in intricate networks. This engagement is what will be able to bring stability in ecosystems, e.g., in multi-trophic systems involving large carnivores that control prey numbers to facilitate a balance that allows different species, like plants and herbivores, to live together. In the long term, the understanding of these connections enhances our ability to grasp the nuances of biodiversity and functional roles that give body plans ecosystem resilience. Ecological interactions also emphasize the significance of evolutionary adaptations within various body plans. The evolutionary history of life unfolds emergent directionality, indicating competition and cooperation among species with altered ecosystems over time, hence influencing the capacity of these systems to trap power and energy efficiency (Vermeij, 2019). For example, body plans that enable efficient foraging, , such as the extended appendages of certain fish or the proboscis-feeding organs of proboscis-bearing insects, can alter nutrient flow within ecosystems. As such organisms adapt to optimize energy transfer through food chains, they become participants in the net power status of their environments. In this way, body plan diversity-driven interactions not only document evolutionary trends but also facilitate energy flow, which is vital for life sustenance despite varied habitats. Lastly, the ecological threshold concept must be considered during the study of interactions among species with varying body plans. These tipping points are thresholds beyond which an ecosystem may dramatically shift in the face of alteration of significant environmental drivers, , such as climate change or habitat alteration (Holyoak et al., 2020). Depending on how different body plans react to these stressors, the new associations between organisms can amplify or suppress these changes, leading to cascading responses throughout the ecosystem. For instance, large predators with specific body architecture specializing in highly preyed habitats can determine vegetation patterns, and their extinction can lead to herbivore overpopulation and consequently ecosystem decline. Understanding how two or more body plans interact along these thresholds is significant to conservation strategies, allowing for the creation of targeted interventions to maintain ecological equilibrium and protect high-risk species in an ever-changing landscape (Figure 6.18).

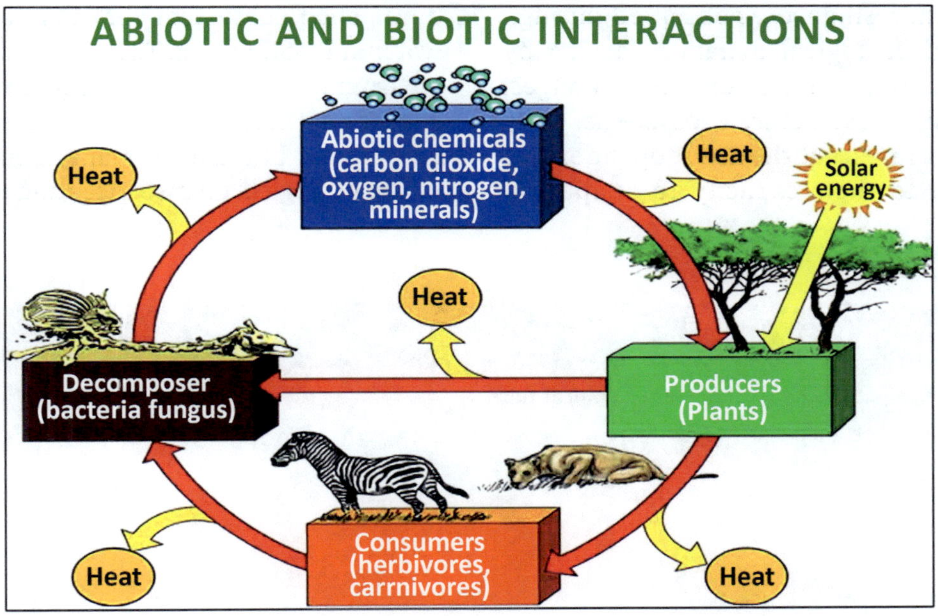

Figure 6.18. Interactions between different body plans in ecosystems.

Source: https://www.nextias.com/blog/wp-content/uploads/2024/08/abiotic-and-biotic-interactions.png.

6.6.3. Importance of Body Diversity for Ecosystem Stability

Body diversity within ecosystems has a critical role to play in guaranteeing stability and resistance to change in the environment. Body dimensions, form, and physiological variation among species form a complex network of interactions, enriching ecosystem processes , such as nutrient cycling, energy transfer, and habitat provision. Diversity in body plans allows for specialized niches, where species occupy substitute roles from predation to decomposition. Thus, this dependence ensures that when one group of species is disrupted, others can take over similar ecological roles, thereby supporting ecosystem recovery. Additionally, this variability becomes increasingly important in the context of global climate change and habitat destruction, where those ecosystems lacking such variation will struggle to adapt and, therefore, collapse. Further, body diversity allows for other forms of species' interaction, including competition, predation, and symbiosis, all of which come under ecosystem dynamics. Various bodies are likely to correspond to various methods of exploiting resources, which minimizes direct competition and promotes coexistence among species. Herbivores with differing digestive tracts, for instance, can exploit different types of plants, utilizing resources to the maximum without exhausting shared food sources. This availability of diverse resources has a direct correlation with ecosystem resilience, as observed through the long-term ecological dynamics driven by species populations and interaction changes. As climate shifts restructure ecosystems, body diversity generates subtle interdependencies that are vital to their capacity to adapt and persevere under shifting challenges (Stewart, 2010). In addition, species extinction and accompanying body plans can have negative effects, such that conserving biodiversity is crucial. Reduced body diversity has a tendency to make ecosystems vulnerable to instability, as realized via previous extinction events and their ecological contribution. Such cases have already reshaped ecological communities, asserting that both biodiversity and richness of body plans are significant in contributing towards stability (Davies et al., 2016). Their demise can distort food webs, disrupt nutrient cycling, and cause diminished ecosystem services , such as clean water and pollination. Therefore, establishing an environment that accepts different body forms is essential not only to help maintain current ecosystems but also to make them resilient to future ecological disturbances and uncertainties.

Test Your Knowledge

1. What is the impact of body plans on ecological niches?
2. What is the importance of body diversity for ecosystem stability?

SUMMARY

Investigation of animal body designs and diversity reveals that the diversification of life forms is supported by common genetic blueprints among species. Such developmental processes indicate that distantly related organisms like Drosophila and chordates evolved from a common protostome-deuterostome ancestor, which probably possessed a more rudimentary body plan than previously speculated. This model, though demanding, necessitates a reevaluation of our assumptions about evolutionary paths and morphological diversification. Understanding the genealogical basis that controls body plan variability not only sheds more light on the subject of evolutionary biology but also aids in uncovering adaptation processes that made it possible for the tremendously wide variety of animal forms we observe today. The credibility of this study is extended to the examination of Neoproterozoic fossils for potential disclosures of ancient pathways in bilaterian evolution, hence firming up links between diverse forms of life (Erwin & Davidson, 2002). Also, the controversy of animal welfare and health, particularly in organic agriculture, influences our understanding of animal diversity in relation to ecological processes. Projects , such as the ANIPLAN project demonstrate the applied welfare and health planning for organic dairy cows. The projects' emphasis on minimizing the use of medicine with active health control indicates a trend towards sustainable agriculture that respects animal autonomy while providing optimal outcomes in terms of health. This confluence of animal well-being and biodiversity is essential; as farmers implement improved animal health measures, they not only promote the well-being of individual animals but also guard against the genetic diversity upon which healthy populations rely. Such measures bring valuable information to bear on how ethical concerns in agriculture can inform our knowledge of animal diversity as it fits into larger ecological systems (Atkinson et al., 2008). Last, the understanding achieved from the study of animal body plans and diversity makes us adopt evolutionary, ecological, and ethical perspectives in our approach to save biodiversity. Appreciation of the intricate interactions between adaptive mechanisms and genetic bases provides an expanded understanding of ecological sustainability and robustness of animal populations. As we struggle with questions , such as climate change, habitat loss, and biodiversity decline, creating an appreciation for the fundamental biological principles that govern animal diversity rises to the top of our agendas. Future scientists will have to navigate back to interdisciplinary thinking, blending genetics, ecology, and animal welfare science, and then proceed to establish a unified picture of what influences diversity and adaptation. By doing so, we not only pay homage to the rich brocade of life on Earth but also enhance our capacity to preserve it for the generations to come.

Knowledge of animal diversity involves analyzing various body plans that have been developed over time as being adaptive to their respective environments. Body plans function as a blueprint of structural and functional properties of animals, including symmetry, segmentation, and the development of specialized organs. A specimen of various body plans of invertebrates and vertebrates displays the wide range of evolutionary strategies employed for survival and reproduction in a variety of ecosystems. An examination of these body plans not only provides a greater understanding of biological diversity but also calls attention to the complex relationships between organisms and their environments. As scientists continue to investigate these topics, an in-depth understanding of the principles that rule animal form and function becomes increasingly important within the context of environmental fluctuations and conservation of biodiversity. The intricacies of animal welfare and health are significantly informed by the research principles derived from various study programs, , such as those outlined in ANIPLAN. This project especially focuses on the importance of meticulous animal health and welfare planning, particularly in the organic dairy farming sector. With an emphasis on minimizing medicine use, the project pinpoints the pragmatics of implementing health plans that advance animal welfare under the specific conditions of various farms. By developing guidelines that promote effective communication between farmer groups and advisory organizations, , such as the Danish Stable School, the project aims to establish a culture for continuous improvement. Such a pioneering approach focuses on the necessity to bring together scientific information and functional applications, leading eventually to healthier animal populations and sustainable agriculture in the broader context of animal diversity and preservation of biological integrity. Moreover, the concentration on sustainable development, as represented by the U.S. Partnership for Education for Sustainable Development, is consonant with the pillars of animal diversity and well-being. Educational materials categorized according to various groups show a commitment towards encouraging complete understanding of ecological balance. In its promotion of sustainable approaches at various levels of education, ranging from basic education to university education, the initiative emphasizes the value of animal care and health in achieving strong ecosystems. The synthesis of animal diversity, body plan, and sustainable development creates a vivid image of high interconnectedness that not only informs conservation and management practice at the individual level but also inspires comprehensive conservation and management approaches. Interconnectedness here symbolizes the importance of collaboration in conservation of animals' diversity and encouraging sustainable practice for future generations.

KEY TERMS FROM THIS CHAPTER

Acoelomate	Asymmetry	Bilateral symmetry
Cephalization	Coelom	Coelomate
Deuterostome	Ectoderm	Endoderm
Germ layers	Invertebrates	Mesoderm
Notochord	Phylum	Protostome
Pseudocoelomate	Radial symmetry	Segmentation

CRITICAL THINKING QUESTIONS

1. **Which of the following is a characteristic of all animals?**
   a. Cell walls
   b. Autotrophic nutrition
   c. Multicellularity and heterotrophy
   d. Photosynthesis
2. **What type of body symmetry do most animals exhibit?**
   a. Asymmetry
   b. Radial symmetry
   c. Bilateral symmetry
   d. Circular symmetry
3. **Which animal group has no true tissues and is asymmetrical?**
   a. Cnidarians
   b. Flatworms
   c. Sponges
   d. Mollusks
4. **What is the function of a coelom in animals?**
   a. To perform photosynthesis
   b. To serve as a digestive gland
   c. To act as a body cavity that cushions internal organs
   d. To transport oxygen
5. **Which of the following groups includes animals with a segmented body plan?**
   a. Cnidarians
   b. Arthropods
   c. Echinoderms
   d. Sponges

Short Answer Questions

1. What are the key characteristics that all animals share?
2. What is the difference between radial and bilateral symmetry?
3. What are the three germ layers in animal embryos, and what do they develop into?
4. What is cephalization, and in which type of body symmetry is it usually found?
5. What is the difference between protostomes and deuterostomes?

Answers to Critical Thinking Questions

1. (c); 2. (c); 3. (c); 4. (c); 5. (b)

REFERENCES

1. Angelos, J. A., Arens, A. L., Cadriel, J. L., et al. (2017). One Health in food safety and security education: Subject matter outline for a curricular framework. https://core.ac.uk/download/323068605.pdf (accessed on 13th June 2025).
2. Atanacio, R., Bailly, N., Froese, R., et al. (2010). Simple identification tools in Fish Base. https://core.ac.uk/download/11900256.pdf (accessed on 13th June 2025).
3. Atkinson, C., Brinkmann, J., Gratzer, E., et al. (2008). Planning for better animal health and welfare, Report from the 1st ANIPLAN project workshop, Hellevad, October 2007. https://core.ac.uk/download/10930606.pdf (accessed on 13th June 2025).
4. Baguñá, J., Martinez, P., Paps, J., et al. (2008). Unravelling body plan and axial evolution in the Bilateria with molecular phylogenetic markers. https://core.ac.uk/download/84158269.pdf (accessed on 13th June 2025).
5. Baxevanis, A. D., & Ryan, J. F. (2007). Hox, Wnt, and the evolution of the primary body axis: Insights from the early-divergent phyla. https://core.ac.uk/download/pdf/7912101.pdf (accessed on 13th June 2025).
6. Becker, M. B., Begemann, G., Meyer, A., et al. (2005). Conservation and co-option in developmental programmed: The importance of homology relationships. https://core.ac.uk/download/pdf/3800618.pdf (accessed on 13th June 2025).
7. Burt, P., Chancellor, T., Morton, J. F., et al. (2010). Climate change, agriculture and Fairtrade: Identifying the challenges and opportunities. https://core.ac.uk/download/316279.pdf (accessed on 13th June 2025).
8. Capilla-Lasheras, P., Dominoni, D. M., Helm, B., et al. (2017). Integrated behavioral and stable isotope data reveal altered diet linked to low breeding success in urban-dwelling blue tits (Cyanistes caeruleus). https://core.ac.uk/download/84148271.pdf (accessed on 13th June 2025).
9. Cuoco, E., Halberg, N., Huber, M., et al. (2010). Implementation Action Plan for organic food and farming research. https://core.ac.uk/download/10931243.*pdf (accessed on 1*3th June 2025).
10. Davidson E. H. (1991). Spatial mechanisms of gene regulation in metazoan embryos. Development (Cambridge, England), 113(1), 1–26. https://doi.org/10.1242/dev.113.1.1(accessed on 13th June 2025).
11. Davies, I. M., Gubbins, M., Irvine, K. N., et al. (2016). The Cooperative Participatory Evaluation of Renewable Technologies on Eco*system Services (CORPORATES). https:*//core.ac.uk/download/42130390.pdf (accessed on 13th June 2025).
12. Day, J. C. (2009). Using molecular tools to differentiate closely related blackfly species *of the genus Simulism. https://core.*ac.uk/download/14512133.pdf (accessed on 13th June 2025).
13. Dunbar, R. I. M., Korstjens, A., & Lehmann, J. (2008). Time and distribution: A model of ape biogeography. https://core.ac.uk/download/75042.pdf (accessed on 13th June 2025).
14. Erwin, D. H., & Davidson, E. H. (2002). The last common bilaterian ancestor. Development (Cambridge, England), 129(13), 3021–3032. https://doi.org/10.1242/dev.129.13.3021(accessed on 13th J*une 2025*).

15. George, M. (2018). Phylogeny of the Orange throat Darter (Etheostoma spectabile) species complex in the Ozark Highlands of Arkansas. https://core.ac.uk/download/158200288.pdf (accessed on 13th June 2025).

16. Gruhl, A., & Okamura, B. (2016). Development and myogenesis of the vermiform Buddenbrockia (Myxoz*oa) and implications* for cnidarian body plan evolution. https://core.ac.uk/download/42046060.pdf (accessed on 13th June 2025).

17. Gupta, M.V.; Bartley, D.M.; Acosta, B.O. (eds.). (2004). Use of genetically improved and alien spe*cies for aqua*culture and conservation of aquatic biodiversity in Africa. WorldFish Center Conf. Proc. (68): 107 p. (accessed on 13th June 2025).

18. Holyoak, M., Jiang, G., Ning, Y., Qi, et al. (2020). Ecological thresholds and large carnivores' conservation: Implications for the Amur tiger and leopard in China. https://core.ac.uk/download/287622211.pdf (accessed on 13th June 2025).

19. Korenić, A., Perović, S., & Tadić, J. M. (2024). Symmetry and historicity in the Game of Life: Modifying rules to simulate evolutionary dynamics. https://core.ac.uk/download/617033123.pdf (accessed on 13th June 2025).

20. Lake, M. W., & Venti, J. (2009). Quantitative analysis of macroevolutionary patterning in technological evolution: Bicycle design from 1800 to 2000. https://core.ac.uk/download/1670989.pdf (accessed on 13th June 2025).

21. Leal Cardín, M. (2024). Patrones fenotípicos y de expresión génica convergentes y divergentes en cíclidos neotropicales. https://core.ac.uk/download/617193399.pdf (accessed on 13th June 2025).

22. Loker, E. S. (2012). Macroevolutionary immunology: A role for immunity in the diversification of animal life. https://core.ac.uk/download/pdf/8720845.pdf (accessed on 13th June 2025).

23. Machamer, P., Quartz, S., Scarantino, A., et al. (2002). Session 5: Development, neuroscience and evolutionary psychology. https://core.ac.uk/download/11920468.pdf (accessed on 13th June 2025).

24. Marletaz, F., & Martin-Duran, J. M. (2020). Unravelling spiral cleavage. https://core.ac.uk/download/286400025.pdf (accessed on 13th June 2025).

25. Medin, D., Ross, N., & Atran, S. (2025). Evolution and devolution of knowledge: A tale of two biologist. https://core.ac.uk/download/pdf/7308744.pdf (accessed on 13th June 2025).

26. Mestre, H. M. (2012). The effect of habitat fragmentation in the diet of pine marten. https://core.ac.uk/download/12426352.pdf (accessed on 13th June 2025).

27. Passamaneck, Y. J. (2003). Molecular phylogenetics of the Metazoan clade Lophotrochozoan. https://core.ac.uk/download/4166949.pdf (accessed on 13th June 2025).

28. Stein, A. L. (2010). State fish stocking programs at risk: Takings under the Endangered Species Act. https://core.ac.uk/download/62546805.pdf (accessed on 13th June 2025).

29. Steinmetz, P. (2019). A non-bilaterian perspective on the development and evolution of animal digestive systems. https://core.ac.uk/download/479090059.pdf (accessed on 13th June 2025).

30. Stewart, J. R., 2010. Understanding evolutionary processes during past Quaternary climatic cycles: Can it be applied to the future? *In:* Grant, F., Mergeay, J., Santamaria, L., Young, J. and Watt, A.D., eds. *Evolution and Biodiversity: The evolutionary basis of biodiversity and its*

potential for adaptation to global change. Report of an e. Mallorca, Spain: Centre for Ecology & Hydrology, INBO, IMEDA, CSIC-UIB.(accessed on 13th June 2025).

31. Tice, A., Fiore-Donno, A., Lahr, D., Voelcker, E., Spiegel, F., Schuler, G., Zhang, J., et al. (2016). Including food systems, biodiversity, nutrition and health in the Post-2020 Global Biodiversity Framework: A submission from the Alliance of Bioversity International and the International Center for Tropical Agriculture. https://core.ac.uk/download/288633284.pdf (accessed on 13th June 2025).

32. Vermeij, G. J. (2019). Power, competition, and the nature of history. https://core.ac.uk/download/287622827.pdf (accessed on 13th June 2025).

PLANT STRUCTURE AND FUNCTION

7

LEARNING OBJECTIVES

After studying this chapter, you will be able to:

1. Describe the basic plant cell structure;
2. Explore the tissues in plants;
3. Discuss roots and their functions;
4. Explain about stems and their functions;
5. Identify the leaves and photosynthesis;
6. Illustrate the reproductive structures in plants.

Introduction

An appreciation of the intricate relationship between structure and function in plants lies at the heart of appreciating how plants survive and accommodate different surroundings. The chief structures of a plant, including roots, stems, leaves, and reproductive bodies, possess inherent functions basic to realizing the overall function of the plants and influencing such critical processes as photosynthesis, nutrient acquisition, and reproduction. This interconnectedness of structure and function not only maintains the development and growth of plants but also enables them to thrive well even in changing environments, validating their viability and sophistication as living organisms. These relationships become increasingly relevant to the solution of contemporary issues in agriculture, ecology, and sustainability, and thus to studying the structures of plants not only as stationary components but as key parts of a complex system upon which life on Earth is founded. The focal point of plants functional potentiality is the detailed structure of their cells and tissues that maintain important processes, like photosynthesis and transpiration. For instance, chloroplasts present in the cells of leaves are specially adapted to trap maximum sunlight, illustrating the connection between cell structure and functional efficiency. Similarly, the vascular tissues consisting of xylem and phloem illustrate how plant structure enables water, nutrients, and organic substances to be transported through varying distances. These intricate adaptations show direction towards evolutionary forces acting upon the morphology and physiology of plants for thousands of years.

As research continues to deconstruct the complexities of these relationships, it opens up new avenues for understanding how the performance of plants can be maximized under changing situations, hence providing insight into crop improvement and ecological stability. Moreover, recent advances in molecular biology are shedding light on the genes and biochemistry controlling plant form and function. The investigation into RNA-binding proteins, , such as the Mei2-like family, has demonstrated the complex regulatory mechanisms that control cellular processes critical to plant development and response to the environment. For example, it has been demonstrated how particular proteins interact with DNA in terms of its localization and functionality in the cell (Benaroya et al., 1990). This molecular strategy refines our understanding of plant systems through the marriage of cellular functions with phenomenological traits, breaking down the micro-macro gap between microlevel interactions and macrolevel patterns of growth. It is important to consider these interdisciplinary results as we continue with further research of plant structure and function because they have a significant contribution to our fundamental information and field utilization in the fields of botany, agriculture, and environmental science (Leung et al., 2003) (Figure 7.1).

Figure 7.1. Plant structure.

Source: https://image.pbs.org/poster_images/assets/thnkgard_vid_structure.jpg.

Definition of Plant Structure and Function

The meaning of plant structure and function is all about the intricate combination of various parts which collaborate to provide a plant with a general performance and adaptability in the environment. Plant structures like roots, stems, leaves, and reproductive parts all share specialized functions for survival. For instance, roots anchor the plant and absorb water and nutrients, which are required for growth. Stems are channels through which transportation of these resources is facilitated from leaves to roots and structural support. Leaves, apart from intercepting light for photosynthesis, have the critical function of gas exchange. Plant structural specialization helps to enhance plant functions and make adequate responses to varied environmental stimuli and stresses. The integration of advanced modeling techniques in plant biology gives us useful information in understanding plant traits and adaptations. Mathematical models of plants can simulate physiological processes occurring in plants and predict the impact of genetic variations on plant performance under a variety of environmental regimes. By employing optimization methods, researchers can suggest the best plant characters that result in optimal performance parameters , such as yield and light capture. This iterative modeling approach not only explains the complex interactions between traits but also assists in determining optimal solutions for trait conversions, thus providing a groundwork for future advances in plant breeding and cultivation methods. This is particularly pertinent in a time of high emphasis on agricultural efficiency and sustainability, and it illustrates the necessity of ongoing research into plant structure and function (Casadebaig et al., 2017). Moreover, the interaction between plants and their fluid surroundings is an intrinsic aspect of their structural and functional dynamics. Plants are under persistent mechanical loads from wind, water, and other environmental factors, which cause extensive deformations essential in optimizing biomass production. Information on fluid-structure interactions (FSI) plays a central role in understanding how plants react to external loads, which directly affects their growth patterns and ecological success. FSI mechanics, in addition to illuminating resilience and adaptation mechanisms, also present pioneering fields for enhancing biomass yield for food and material manufacturing. Investigating how plants reorganize under fluid forces requires interdisciplinarity between physics and biology and hence broadening our knowledge of plant functional morphology and intricate mechanics of plant systems (Gosselin et al., 2019).

Importance of Studying Plant Biology

Plant biology study is critical towards understanding not only the general mechanism of life but also the intricate relationship between plants and their ecology. Observation of the plant structure and function allows scientists to learn about the physiological processes of growth, reproduction, and adaptation. This information is critical in the fight against environmental degradation, , such as climate change and logging. Further, the relationship between plants and soil organisms, highlighted by recent research, demonstrates the complexities of ecosystem processes. For instance, the role played by microbial soil diversity in nutrient cycling is central to both plant health and ecosystem stability generally. Resolving these complexities requires a comprehensive understanding of plant biology, with a push for specialist research in this area to address actual ecological issues (Becker et al., 2005). Climate change poses one of the biggest challenges to plant biology, making research into it even more relevant. The climatic variability has a great impact on species evolution, adaptability, and extinction, as seen during the Quaternary. Knowledge of how plants would react to these changes is crucial for conservation and sustainable agriculture. Plant biology studies enable researchers to identify long-lasting species, understand how they evolve, and come up with ways to enhance crop yields in a changing climate. Particularly, plant physical adaptations to different environmental conditions inform agricultural practices and conservation strategies, hence assuring food security and maintaining the ecosystem. By developing a solid foundation in plant biology, researchers can better anticipate and manage the impacts of climate change on plant communities and their associated ecosystems (Stewart, 2010). Moreover, plant biology research is also of significant importance to the health and well-being of humans. Plants are the basis of the food chain and are among the most crucial sources of nutrients necessary for the maintenance of life. Knowledge about plant structure and function not only aids in the improvement of crop yield and quality but is also pivotal in identifying medicinal substances. Most medicines are derived from plants, and one can enhance new treatment regimen identification using information on plant biology. Also, as industry and urbanization advance and penetrate natural habitats, conservation of plant diversity becomes necessary. Studying plant biology gives society the background it needs to implement effective conservation methods while engendering citizen appreciation of the plant community, ultimately increasing sustainability and health in a changing world (Figure 7.2).

Figure 7.2. Plant biology.

Source: https://iimtu.edu.in/blog/wp-content/uploads/2023/08/Botany-21.jpg.

Overview of Key Concepts to Be Discussed

In keeping with the examination of plant structure and function, a strong comprehension of such principles is important. One of the first areas to explore is the differentiation of the different classes of plant cells, specialized to fulfill particular functions within the organism. One case in point is the parenchyma, collenchyma, and sclerenchyma cells, each with specialized structural characteristics for the execution of various functions, , such as storage, support, and transport. Parenchyma, the most general of cell types in most cases, is attributed to photosynthesis, storage, and tissue repair, while collenchyma is attributed to structural support, especially in young stems and leaves. Sclerenchyma, with thickened cell walls, is attributed to providing rigidity to hardened plant tissues. It is significant to understand these cellular foundations fully to comprehend the role they play in contributing to the overall physiological processes and development of plants. Aside from cellular construction, tissue organization into larger systems even better explains plant function. The differentiation of plant tissues into meristematic and permanent tissue highlights the dynamic nature of plant development. Meristematic tissue, which is made up of undifferentiated cells, is responsible for the continuous growth of plants via primary and secondary growth processes. Permanent tissues , such as epidermal, ground, and vascular tissues form the foundation of mature plant organs. Vascular tissue, which consists of xylem and phloem, is essential for the transport of water, nutrients, and photosynthates within the plant and therefore forms a vital link between plant parts. The interaction among these kinds of tissues not only underscores structural strength but also makes the plants adaptable to varying conditions in their environments. Additionally, the understanding of plant physiology is crucial to comprehend how plants utilize their structures in an effort to interact with their environments. Significant themes , such as photosynthesis, respiration, and water regulation demonstrate the functional significance of plant structures. For instance, chloroplasts, which are membrane-enclosed organelles found in plant cells, facilitate photosynthesis through the use of light to convert CO_2 and water into glucose. It is very well associated with plant health and growth. Similarly, stomata play a very vital role in gas exchange as well as transpiration and regulate water evaporation to allow uptake of CO_2. Observations of these physiological processes validate the structural-functional relationship, illustrating how adaptations in a particular form enable a plant's survival and reproductive success in different ecological niches. Overall, these simple components form a general framework to explore the intricate relationship between plant structure and function in the remainder of this essay.

7.1. Basic Plant Cell Structure

Learning Outcome

1. Components of plant cells;
2. Differences between plant and animal cells;
3. Role of cell walls in plant integrity.

The fundamental organization of plant cells is made up of a number of discrete but incorporated components which enable them to undertake the critical biological functions. At the pinnacle of plant cell structure is the firm cell wall, primarily made up of cellulose, and this provides strength and form to the cell. This ECM not only protects the cell from mechanical stress but also plays a critical role in regulating growth and structural support. Inside the cell wall is the plasma membrane, which is selectively permeable and through which material is passed into and out of the cell. All these structures underscore the importance of plant cell compartmentalization, laying the groundwork for a highly intricate web of cellular processes necessary to plant development and adaptation to the outside world. In addition to the cell wall and plasma membrane, plant cells contain several specialized organelles that each serve a particular and critical purpose. The chloroplasts, for instance, carry out photosynthesis, in which they harness the sunlight and convert its energy into chemical form via glucose molecules.

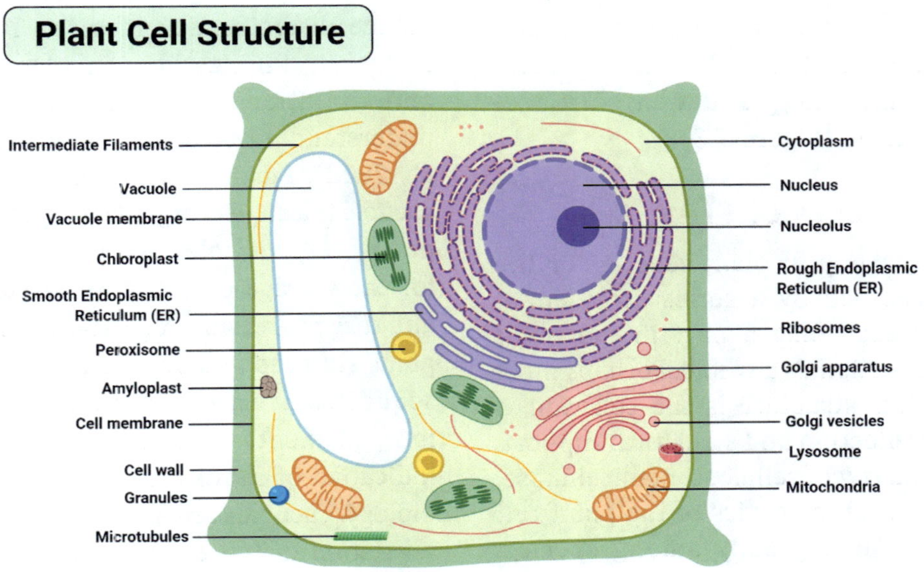

Figure 7.3. Basic plant cell structure.

Source: https://microbenotes.com/wp-content/uploads/2020/02/Plant-Cell.jpeg.

Moreover, the organelles contain the green pigment chlorophyll, which plays a pivotal role in absorbing light. Similarly, the vacuole performs different functions, including holding waste products and nutrients and also maintaining turgor pressure that rigidifies the cell. It is worth understanding how these organelles are coordinated since their function relies on each other, influencing the efficiency of the plant metabolism and ensuring stability to environmental alterations as the last progress in imaging technologies that offer real-time cellular dynamics (Aller et al., 2014). The contribution of genetic regulation to the plant cell organization is also worthy of consideration since it is directly associated with cell organization and function. The discovery of the *Arabidopsis thaliana* genome sequence has provided us with a model of plant transcriptional regulation, emphasizing the complexity of gene expression interactions in the cellular context (Riechmann, 2002). TFs, a significant proportion of plant genes, control these processes by activating or repressing target genes according to internal and external stimuli. Regulation of this kind allows for adaptive processes to take place in relation to environmental stimuli and, accordingly, informs how basic cell structure in plants forms plant resilience. Therefore, an integrated interpretation of structural and genetic analysis of plant cells illuminates general principles in plant biology and their functional roles in ecosystems (Figure 7.3).

7.1.1. Components of Plant Cells

The intricate makeup of plant cells is, for the most part, described by a series of essential elements that each play a part in the plant structure and function as a whole. Perhaps most significant to this structure is the cell wall, a hard layer of protection and structural support. Comprised primarily of cellulose, hemicellulose, and pectin, the characteristics of the plant cell walls can vary quite greatly among species. Current research highlights the biochemical interrelations among these components and their susceptibility to enzymatic modification, particularly in such functions as fruit ripening. As the chapter continues to elucidate, physical alterations responsible for fruit softening are correlated with loss of the neutral sugars, solubilization and depolymerization of the cell wall polysaccharides (Goulao & Oliveira, 2008). It is necessary to understand these dynamics in fruit marketing strategies and agricultural practices due to the fact that different cultivars possess different characteristics. Another interesting component of plant cells is the chloroplast, which plays an important role in photosynthesis, which is the process of generating chemical energy from light. Chloroplasts contain their own DNA and have a highly developed protein import system to be operational. This import of protein is facilitated by sophisticated equipment in the chloroplast envelope, i.e., Translocon at the outer chloroplast membrane (TOC) and Translocon at the inner chloroplast membrane (TIC).

The TOC complex of key players , such as Toc75 and the other receptors, organizes the import of approximately 3,000 nucleus-encoded proteins, differing to meet the specialized demands of various plant species. Recent studies focus on the diversification of these TOC complexes and their implications in plant adaptability and evolutionary biology, which builds our understanding of plant productivity (Bölter et al., 2012). Further, in the cell architecture, the vacuole is a superb organelle, at times filling most of the cell. Vacuoles are primarily responsible for the storage of nutrients, waste compounds, and turgor pressure regulation, which maintains plant stiffness and support. vacuole's ability to sequester a variety of compounds, including ions, pigments, and secondary metabolites, is indicative of its diversity of functions in plant biology. Vacuoles are also involved in cellular pH regulation and the detoxification of metabolites in a more defined view. This functional adaptability is an homage to the evolutionary tendencies that allow plants to develop within a diverse assortment of environments. By examining the functions of the various components , such as the vacuole, researchers can disentangle the complex interdependencies characteristic of plant cell function, which in turn guide plant development and survival processes (Figure 7.4).

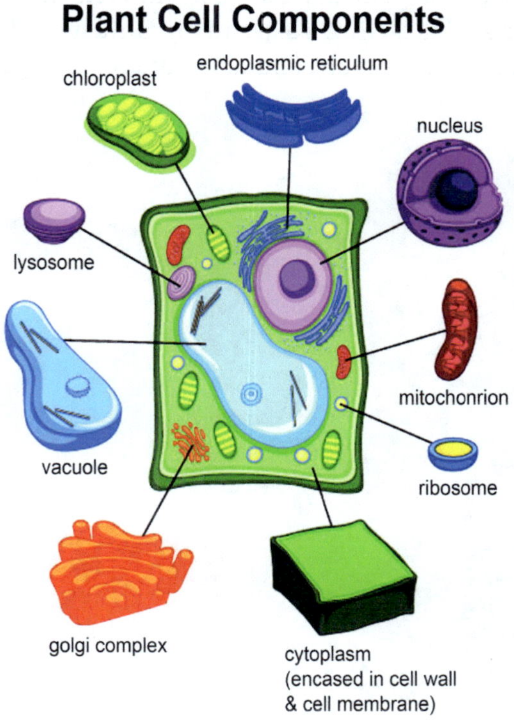

Figure 7.4. Plant cell components.

Source: https://ogden_images.s3.amazonaws.com/www.motherearthgardener.com/images/2017/05/19160243/E00EB5ED0F5F4ECE86F6E17B62F0259E.jpg.

7.1.2. Differences Between Plant and Animal Cells

The structural differences between plant cells and animal cells are the foundation of their varied roles in ecosystems and biological systems. Perhaps one of the most prominent features of plant cells is the presence of a stiff cell wall that provides structural support and defines the cell's shape, allowing plants to develop as tall structures and provide resistance against exterior pressure. This cell wall is largely made of cellulose, an adaptive advancement that, besides being strong, facilitates inter-cellular communication through plasmodesmata, aiding nutrient and signal transfer. Animal cells lack a cell wall but have a flexible plasma membrane, enabling cells to take on many varied forms and the formation of elaborate tissues and organs. This inherent plasticity is of the greatest significance in movements, signaling, and general cell interaction in multicellular organisms, focusing attention on the significance of cellular construction in numerous biological processes. Furthermore, the presence of specialized organelles works to distinguish plant cells from animal cells to a large degree. The chloroplast, an organelle for photosynthesis, is perhaps the most typical feature of plant cells. It is one of the most specialized organelles that transform light energy into chemical energy in the form of glucose. This ability makes the plants the principal producers within ecosystems, generating organic compounds that sustain other organisms. Animal cells do not contain chloroplasts and obtain their energy by utilizing ingested organic matter. Animal cells do possess lysosomes, however, which are necessary for intracellular digestion and waste removal, suggesting their diverse array of metabolic needs. These organelles illustrate the contrasting patterns of energy acquisition and utilization in plants and animals, highlighting their distinctive evolutionary adaptations and niches. Lastly, the plant and animal cell function contrasts are illustrated in their modes of growth and differentiation. Plant cells can exhibit a form of growth known as indeterminate growth, whereby they can continue to grow and differentiate throughout their lifetime. This capacity for regeneration is in part due to the presence of meristematic tissue, which consists of undifferentiated cells with the ability to become a variety of cell types. Plant-specific sophistication exists in how gene expression is regulated in this differentiation, as found in studies on TFs like AS1 in *Arabidopsis* (Phelps-Durr et al., 2024). Contrary to plant cells, animal cells generally display determinate growth, where the majority of the cells stop dividing once, they have reached a size or level of development. This early divergence of growth patterns not only impacts the physical composition of the organisms but also impacts the way they respond to environmental change, as in flowering plants that respond by adjusting their growth according to seasons, further exemplifying plant and animal cellular strategy divergence (Bernardello et al., 2001) (Figure 7.5).

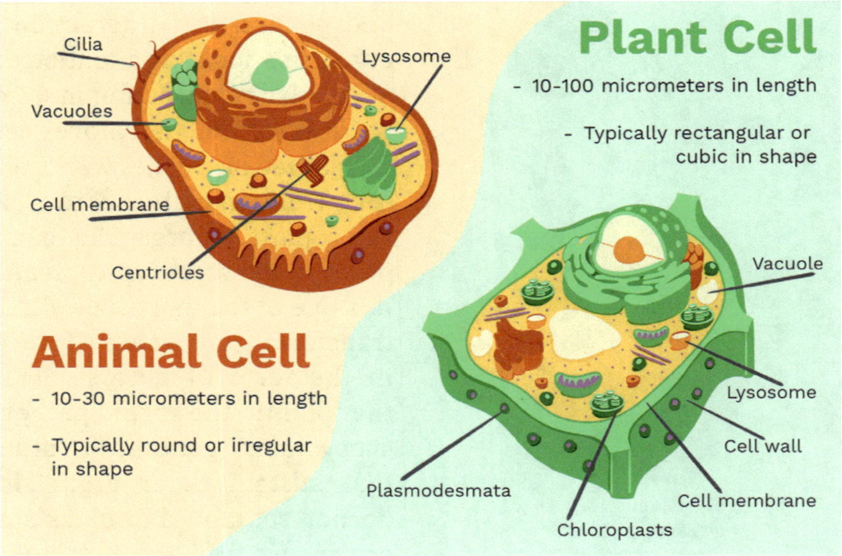

Figure 7.5. Differences between plant and animal cells.

Source: https://www.thoughtco.com/thmb/KC4a3LJ9Lvarsw6QaZ69z-tsKEU=/1500x0/filters:no_upscale():max_bytes(150000):strip_icc()/animal-cells-vs-plant-cells-373375_final-5b462d7fc9e77c00375014f1.png.

7.1.3. Role of Cell Walls in Plant Integrity

The structural stability of plant cells is mostly provided by the cell wall, which is a tough outer layer mostly comprised of polysaccharides. This structural element is vital in maintaining turgor pressure in the cell, which is a vital factor in plant growth and overall stability. The primary cell wall is largely made up of cellulose, hemicellulose, and pectin, which together provide mechanical strength and flexibility. As the plant matures, the ability to regulate cell wall composition is critical in adapting to various environmental conditions. For example, ferulic acid metabolism, which forms a significant constituent that is involved with lignification, plays a significant role in reinforcing cell walls and thus enhancing the plant's resistance to biotic stress , such as pathogen invasion. Studies have shown that cell wall property changes affect plant defense immediately, highlighting the central role of cell walls in maintaining structural integrity and responsive functionality to environmental stress (Bellincampi et al., 2016). In addition to their structural role, plant cell walls are dynamic systems that undergo considerable remodeling during growth and development. Extensins, O-glycoproteins in one subgroup, are important here since they contribute to the formation of a three-dimensional network that gives structural support to the cell wall architecture. These proteins must be post-translationally modified so that they can perform their function, illustrating a multiplicity of regulation in the integrity of the plant cell wall. There has been a renewed focus on how extensin assembly and processing must be strictly controlled for proper growth, particularly in rapidly growing tissues like root hairs and pollen tubes. The intricate interaction between extensin modifications and cell wall composition therefore, does not just influence mechanical strength but has a key function in plant developmental processes, enabling cells to become pliable to maximize growth while responding to outer stresses (Borassi et al., 2018). The interaction of cell walls with other environmental influences therefore, underscores their importance in plant integrity. Cell wall composition modification might trigger defense mechanisms that are important to ensure plant survival. For instance, reduced levels of monomeric ferulic acid in plants suggest enhanced susceptibility to pathogens and thus that cell wall composition indirectly affects disease resistance. Further, plant defense mechanisms, like the induction of defense gene expression when infected with a pathogen, are organically linked with cell wall integrity. The ability to modify cell wall components, thus, is a strategic weapon that enables the plants to strengthen their defense effectively. Therefore, the multifunctional nature of cell walls not only plays a vital role in providing structural support but also in enabling adaptive responses that ensure the survival of the plants under various environmental stresses (Bellincampi et al., 2016; Borassi et al., 2018).

Test Your Knowledge

1. What are the components of plant cells?
2. What is the role of cell walls in plant integrity?

7.2. Tissues in Plants

Learning Outcome

1. Types of plant tissues: meristematic and permanent;
2. Functions of vascular tissues: xylem and phloem;
3. Ground tissue and its role in photosynthesis.

While learning about the organization and function of plant tissues, one should know the variation and specialization of tissues that contribute to the health and functioning of plants. Plant tissues are primarily classified into two broad groups: meristematic and permanent tissues. Meristematic tissues, present in areas of growth like the shoots and roots, are marked by undifferentiated cells capable of dividing and forming new cells. Permanent tissues, on the other hand, are made up of differentiated cells that carry out specialized functions like transport, support, and photosynthesis. This division underscores the dynamic pattern of plant development wherein the meristematic tissues are responsible for growth and permanent tissues for structure and function during the life cycle of a plant. Understanding the role of permanent tissues is accompanied by an analysis of their subgroups, i.e., parenchyma, collenchyma, and sclerenchyma. Parenchyma tissues are varied in function and are involved in a number of activities like storage, photosynthesis, and tissue repair. Sclerenchyma, made up of thick-walled cells, provides rigidity and support, while collenchyma offers flexibility so that young, growing plants can endure a high number of mechanical stresses. These permanent tissues combined allow for effective resource management in the plant, so nutrients can be transported effectively and structural support ensured. In addition, the interaction between such tissue types exhibits the evolutionary adaptations that have led land plants to thrive in different environments, thus illustrating the complexities of plant tissue systems in relation to success as autotrophic plants. The genetic control is imperative in plant tissue structuring and functioning, as seen in significant research on gene identification and expression. For instance, the *Arabidopsis* JAGGED (JAG) locus is crucial in determining the morphology of lateral organs and influencing their sizes and shapes. Loss-of-function mutations in this locus lead to irregularities , such as thin floral organs and serrated leaves, indicating the importance of the gene in order to maintaining the morphological features of most plant organs. These studies not only chart the genetic basis of tissue development but also describe how specific genes may specify coordination between different tissues during development. As a productive field of study, understanding these genetic processes opens gates to potential future agricultural development and knowledge about plant adaptability, therefore emphasizing the intrinsic role of tissues in the broader picture of plant form and function (Heisler et al., 2004; Ningrum et al., 2013) (Figure 7.6).

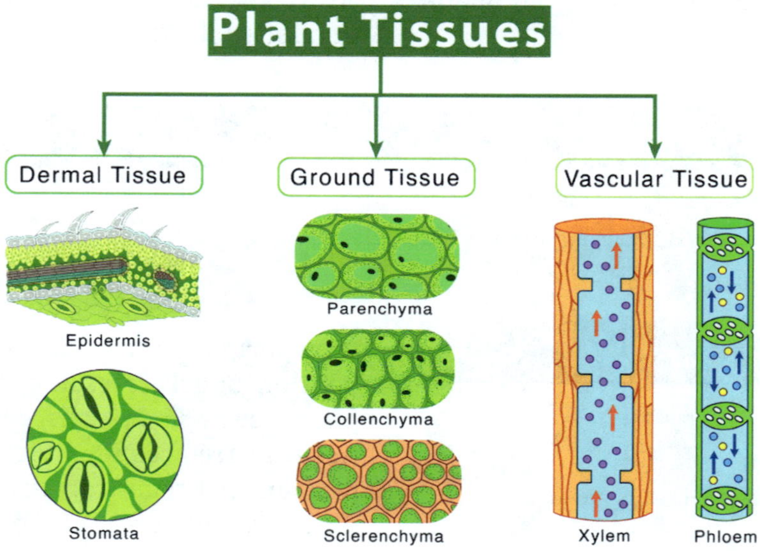

Figure 7.6. Tissues in plants.

Source: https://www.sciencefacts.net/wp-content/uploads/2022/02/Types-of-Plant-Tissue.jpg.

7.2.1. Types of Plant Tissues: Meristematic and Permanent

Plant tissues are categorized into two groups: meristematic and permanent tissues, both with vital roles in the life cycle of plants. Meristematic tissue consists of undifferentiated cells that have a constant capability to divide and differentiate, thereby bringing growth and regeneration. This tissue is present predominantly in regions of growth, i.e., root and shoot apices, where it gives rise to various cell types needed for plant formation. Meristematic tissues can be divided into three main categories: apical meristems, lateral meristems, and intercalary meristems, each contributing a special attribute to plant development. For instance, apical meristems promote growth vertically, whereas lateral meristems increase stem and root thickening. Ongoing cell division in such cells is accountable for how a plant can evolve and continue to exist within an ever-changing environment, establishing an integral function in form and function in plants. Permanent tissues, on the other hand, are developed due to the differentiation of the meristematic tissues and are liable for numerous functions that help to contribute towards the structural integrity of the plants and support their physiological processes. Permanent tissues are also divided into simple and complex types. Simple tissues consist of uniform cell types like parenchyma, collenchyma, and sclerenchyma, each of which is liable for certain functions like storage, support, and protection. Composite tissues, , such as xylem and phloem, are more than a single kind of cell; integrated, mainly responsible for the conduction of minerals, water, and nutrients within and between the plant. Structural specialization of the permanent tissues enables plants to effectively manage resources and respond to environmental stimuli, contributing significantly to overall adaptability and survival of the plants. The structural complexity of such tissues highlights the elaborate architecture of plants, which is responsible for their environmental success. In addition, temporally and spatially dynamic character of plant tissues can also be determined by traits like epigenetic variation, which is, in turn, vital for phenotypic plasticity. Research suggests that reversible heritable epigenetic marks are potentially assisting the adaptation of plants in changing environments, particularly in long-lived species , such as forest trees. These epigenetic alterations can turn out to be crucial in stress or hybridization stages, influencing how the meristematic tissues exhibit growth and differentiation. Therefore, the interaction between meristematic and permanent tissues and epigenetic processes together constitutes a sensitive network enhancing the plants' ability to respond to environmental changes effectively. An understanding of these processes sheds light on the way structure and function in a plant may vary over time, with emphasis placed upon the function of both tissue types in aiding a plant's development and potential to adapt in an ever-changing ecosystem (Lima et al., 2009; Akimoto et al., 2013) (Figure 7.7).

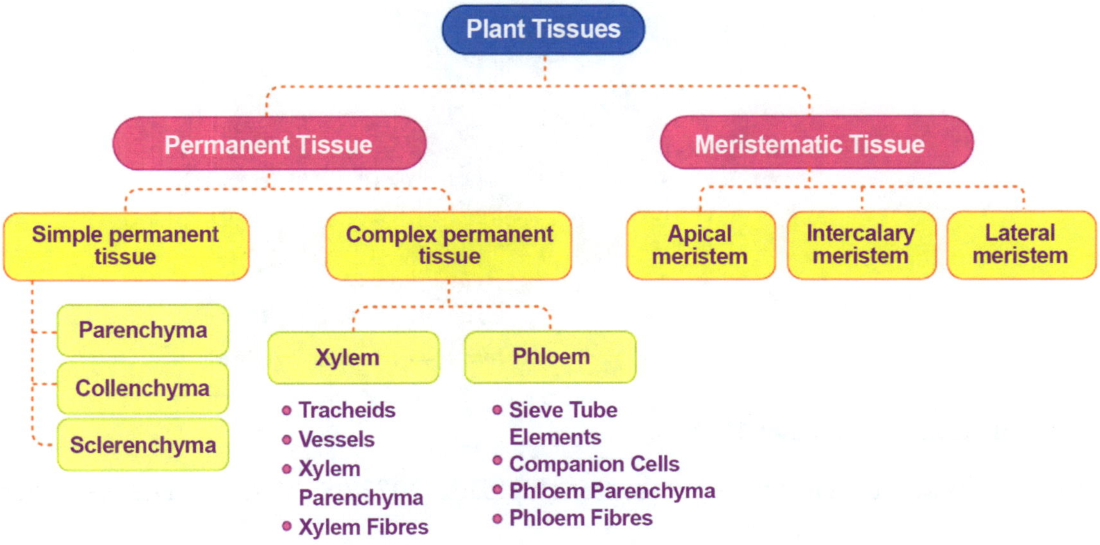

Figure 7.7. Types of plant tissue.

Source: https://cdn1.byjus.com/wp-content/uploads/2022/06/Types-of-Plant-Tissues.png.

7.2.2. Functions of Vascular Tissues: Xylem and Phloem

The vascular tissues of a plant, including xylem and phloem, play a vital role in the nourishment of physiological processes necessary for plant life. Xylem is primarily the water and dissolved minerals conveyor from the roots to the aerial portions of a plant. Such a transport not only benefits nutrient uptake via photosynthesis by providing sufficient hydration to leaves but also facilitates the uptake of nutrients from the soil. The mechanical rigidity of xylem, composed of lignified cells, allows a plant to withstand the negative pressure generated by transpiration. It has been discovered that there are significant molecular mechanisms that regulate xylem differentiation and the formation of secondary walls, and these new discoveries indicate that specific proteins , such as HipI-SOD target lignified structures to make these processes possible and hence raise the net efficiency of the function of xylem and plant health (Karlsson, 2003). On the other hand, phloem is committed to transporting organic compounds, that is, photosynthesis products from the leaves to non-photosynthetic organs , such as roots and developing fruits. Translocation is essential for energy resource distribution in the plant for proper development and reproductive achievement. Phloem consists of companion cells and sieve tubes, which are responsible for regulating nutrient transport and maintaining metabolic processes.

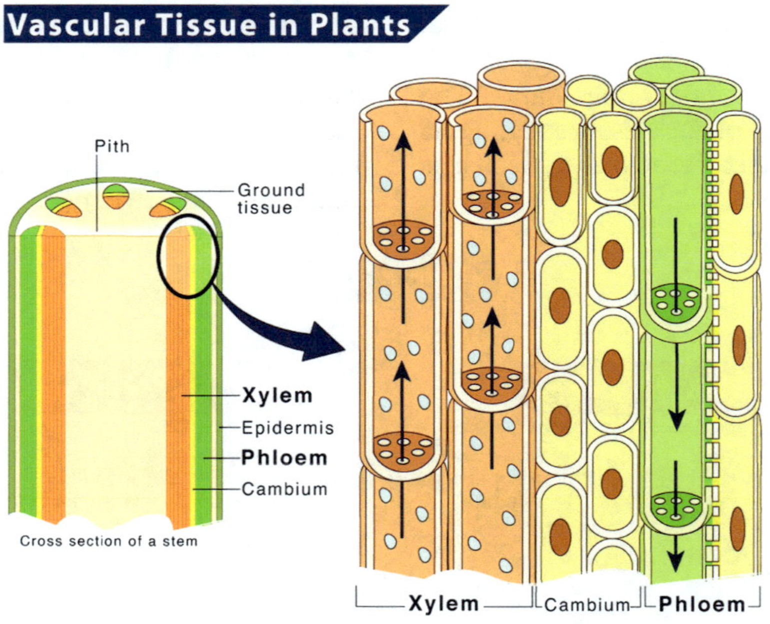

Figure 7.8. Functions of vascular tissues.

Source: https://www.sciencefacts.net/wp-content/uploads/2022/07/Vascular-Tissue-in-Plants.jpg.

The complex regulatory interaction between hormones and the environment, i.e., auxin levels, indicates the active condition of phloem tissue. This observation is significant since recent studies have indicated that auxin level changes induce profound changes in gene expression for phloem function, thus supporting interdependency between plant growth and environmental response (Schrader, 2003). In addition, coordination of interaction between xylem and phloem within the vascular tissue is important to overall plant health. Efficient water transportation by the xylem provides conditions for transpiration, which in turn creates negative pressure to facilitate phloem loading and unloading procedures. The interdependence relationship identifies the effective functioning of vessels in order to sustain plant physiology under various conditions. Experiments have shown that the year-to-year variation in cambial activity as a result of environmental stress has the potential to affect xylem and phloem development, causing changes in biomass production and plant fitness. The extremely intricate control of these vascular tissues not only reflects the plasticity of plants toward their environment but also the response of plants to climatic change, and therefore it is a field of significant research interest (Karlsson, 2003; Schrader, 2003) (Figure 7.8).

7.2.3. Ground Tissue and Its Role in Photosynthesis

Ground tissue is crucial in plant physiology, with extensive influences on photosynthesis, storage, and support. Parenchyma, collenchyma, and sclerenchyma all constitute ground tissue as the site of important metabolic activity and supporting function. Parenchyma cells are distinguished by thin walls and large vacuoles, which are especially significant since they facilitate photosynthesis because of the presence of chloroplasts in green tissues , such as leaves. These cells are actively involved in the photoreduction of light energy to chemical energy, rendering them vital for the overall growth and energy production of the plants. Through the provision of the major locus of photosynthetic reactions, the ground tissue cells provide grounds for plants to be able to capture solar energy effectively, adding substantially to their growth success and environmental sustainability. In addition to the photosynthetic function, the role of ground tissues goes further to encompass critical storage functions. Ground tissue functions as a storage reservoir for critical compounds , such as starches, sugars, and proteins in most plants, including those with high-productivity profiles like cassava. Parenchyma cell storage allows the plant to ration energy resources efficiently, especially where there is low light or poor growing conditions. For instance, root bulking photosynthesis has been found to influence yield traits based on the metabolic processes in the cells, which are directly associated with storage capacity. Synchronization of these functions ultimately allows for maximum energy use during growth and reproduction, illustrating the complex relationship between ground tissue, photosynthesis, and a plant's general health and productivity, as seen through practices informed by crop research (Fernie et al., 2020). Moreover, studies indicate that plant photosynthesis efficiency is very much related to the characteristics of ground tissue, including enzyme activities involved in major metabolic processes. High levels of enzyme activities in the Calvin-Benson cycle contribute to high rates of carbon assimilation, and this enhances photosynthetic productivity. The correlation illustrates the contribution of ground tissue to physiological adaptation in plants to the environment. Alterations in ground tissue composition could also be the cause of differences in photosynthetic efficiency among cultivars, and hence, the possible avenues for genetic enhancement and an increase in yield. Research on tropical crops , such as cassava, points to the importance of metabolic characteristics in optimizing photosynthesis and thus underscores the central place of ground tissue in the general equation of plant structure and function. Understanding these interactions makes it easier to develop policy for maximizing plant growth under diverse environmental conditions, highlighting the important role of ground tissue in crop control and ecological habitats (Aber et al., 2002) (Figure 7.9).

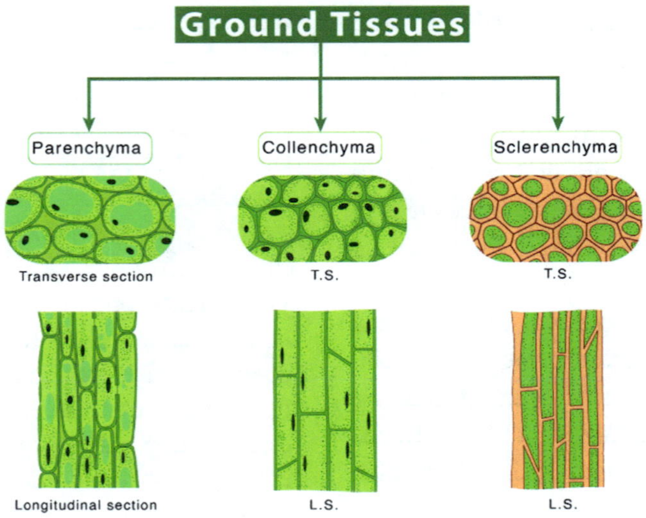

Figure 7.9. Ground tissue.

Source: https://www.sciencefacts.net/wp-content/uploads/2022/05/Ground-Tissue.jpg.

Test Your Knowledge

1. What are the types of plant tissues?
2. What are the functions of vascular tissues?

7.3. Roots and Their Functions

Learning Outcome

1. Structure of root systems;
2. Functions of roots in nutrient absorption;
3. Adaptations of roots for different environments.

Roots are an essential plant structure, playing essential functions that contribute to the overall health and viability of the plant. Anchorage is among the most essential functions of roots, allowing plants to stay in one place in their growing habitat. The efficient root system supports the intake of nutrients and water from the soil, which are essential for photosynthesis and metabolism. The root system, including primary roots and lateral branches, offers maximum surface area exposure to the soil for enhancing the plant's ability to obtain required resources. Furthermore, roots establish symbiotic relationships with soil microorganisms, which can further improve the availability of nutrients. It is crucial to comprehend the complexities of root architecture because it influences nutrient uptake efficiency along with the plant's response towards environmental stress, an area increasingly crucial with climate change and soil erosion. In addition to the role of water and nutrient absorption, roots have a critical role to play in maintaining the health of the plant through storage and transport. Roots serve as energy reservoirs of carbohydrates, enabling the plant to survive during periods of dormancy or adverse conditions. For perennial plants, these reserve units are utilized to sustain new growth the following season. Roots are also directly engaged in water and nutrient transport towards the aerial part of the plant through specialized vascular tissues. The efficiency of this transport system is characterized by root morphology and can impact the plant's general growth and yield. Learning about how root systems adapt to differences in soil and water can provide insight into ways to improve agricultural practices, drawing attention to the invaluable role roots play in natural environments as well as agriculture. Recent studies have unmasked the complex interactions between plant roots and their corresponding affiliated microbial communities, which regulate root function to a great degree. Experiments on Salicornia europaea, for instance, have demonstrated how certain microbial populations host the roots under salinity and seasonal regimes (Furtado et al., 2019). Such microbial populations play a crucial role in enhancing the uptake of nutrients and plant growth, emphasizing symbiotic associations underpinning root function. Beyond this, bacteria associated with root systems help recycle nutrients and can impact the health of the whole plant through its effects on physiological processes. As our understanding of these interactions deepens, it becomes increasingly clear that roots are not merely structural components but dynamic systems that engage in complex ecological relationships, emphasizing the integral role they play in the broader context of plant structure and function (Figure 7.10).

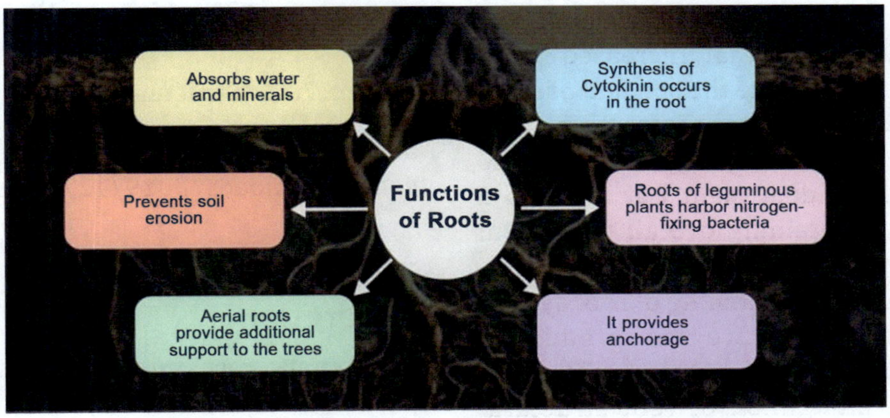

Figure 7.10. Roots and their functions.

Source: https://media.geeksforgeeks.org/wp-content/uploads/20240105130349/Function-of-Roots.jpg.

7.3.1. Structure of Root Systems

Root system morphology is a fundamental element of plant structure and function, regulating a range of important processes in terrestrial environments. Root system morphology varies greatly in architecture and design and is divided into fibrous and taproot systems. Fibrous root systems, which are composed of numerous small roots, allow for extensive exploratory potential in the soil and enhance water and nutrient uptake. On the other hand, the taproot develops deeper into the soil, providing support and accessing water resources unavailable to shallow-rooted plants. It is this root structure diversity that one has to get the hang of because it directly relates to plant performance and environmental stress tolerance for drought and starvation. Also, simulation of such root systems has become much better and has provided tools like the DigR model, which uses a number of parameters to simulate root growth and architecture under different environmental conditions, thus enhancing knowledge about plant-root interaction in various ecosystems (Barczi et al., 2011). Root systems also play a strong role in influencing soil traits, and therefore, they affect ecosystem functioning. Experiments have shown that root architecture, as well as root characteristics, are primary factors controlling soil aggregate stability, which is the critical parameter governing water holding and nutrient supply in soils. Inclusive root systems can improve the physical structure of soils, which offers better aeration and drainage, resulting in healthy plants and increased biomass production. The relationship between plant diversity and root structure is particularly important, since studies have indicated that increased plant diversity serves to stabilize the soil and prevent erosion. This being the case, structural root characteristics are not only important for the success of an individual plant but are also a component of maintaining overall ecosystem stability and health intact (Abiven et al., 2016). Thus, an intimate understanding of root system architecture is imperative for productive ecosystem management and restoration strategies. The integration of root structural analysis into larger plant functional models presents a path to expanding our understanding of how plants adapt to environmental change. Utilizing powerful modeling tools , such as DigR, enables researchers to model root systems in a manner that captures both structural and functional features. This multi-faceted method incorporates a number of parameters, including root length, patterns of branching, and spatial pattern, which enable studies to make projections regarding how various species react to environmental stress. Furthermore, these models enable researchers to examine root plasticity, which enables scientists to ascertain how plants may alter their root morphology in response to changes in soil conditions or competition. It is this knowledge that informs the design of farming practices and conservation strategies for maximizing plant growth and provision of ecosystem services, highlighting the root architecture-function relationship with its profound implications in a dynamic environment (Barczi et al., 2011) (Figure 7.11).

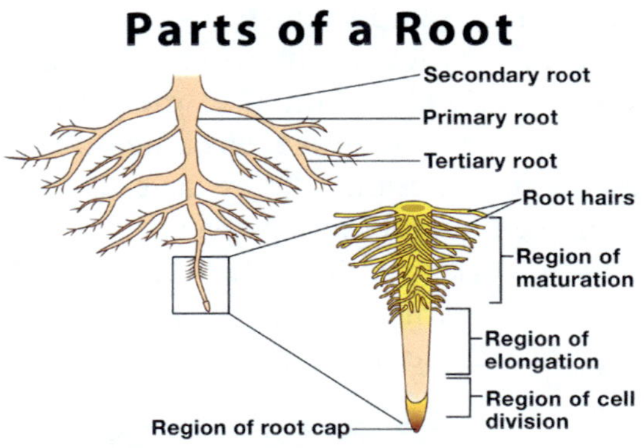

Figure 7.11. Structure of root systems.

Source: https://www.sciencefacts.net/wp-content/uploads/2019/12/Parts-of-a-Root.jpg.

7.3.2. Functions of Roots in Nutrient Absorption

The complex root system of plants plays a central part in the uptake of nutrients, facilitating the acquisition of essential minerals from the soil. Root systems are developed to ensure maximum contact with the soil environment. The primary roots, lateral roots, and root hairs all work in conjunction to search through vast amounts of soil, thereby enhancing the availability of nutrients. As roots grow and branch out, they exchange with the soil matrix, which allows them to absorb nutrients through diffusion and mass flow. Research points out that the exchange of the roots with the surrounding soil is essential for effective uptake of nutrients. With a model suggested by Nye and Tinker, the quantitative description of nutrient uptake is possible, pointing out the influence of root anatomy on nutrient acquisition dynamics in plants (Darrah et al., 2001). This kind of model lends support to the role of root structure in optimizing nutrient uptake, which is essential to plant well-being and productivity. In addition to playing structural functions, roots also establish symbiotic relationships with mycorrhizal fungi that significantly enhance nutrient acquisition potential. Symbionts increase the surface area of the absorptive root system on the soil to facilitate greater access to water and primary nutrients, primarily phosphorus, that often exist in immobilized form. Sieverding's studies indicate that mycorrhizal networks can sample much greater volumes of soil than roots by themselves and thus are central to nutrient capture (Gollotte et al., 2008). Not only does this symbiotic association supply the advantages of nutrient capture, but it also benefits , such as improved drought tolerance and disease resistance, thus highlighting the various facets of the functions of roots in plant nutrition. The symbiotic nature of such interactions is an indication of the co-evolution of plant and fungal organisms in nutrient-poor habitats. The effectiveness of nutrient acquisition by roots also testifies to the plasticity of plants in responding to variations in soil texture. Root architecture is influenced by abiotic factors , such as soil type, water content, and nutrient concentration, promoting changes in root growth patterns and form. For instance, in poor soils, plants may develop deeper or thicker root systems to access maximum nutrients. This adaptability is central to the survival of species in a broad variety of ecological niches and reflects the dynamic relationship between soil environments and root systems. Healthy root systems also contribute to the overall health of the soil ecosystem through the promotion of soil structure and nutrient cycling. When roots decay, they release organic matter that acts as fertilizer to the soil, which supports a thriving microbiotic community that also assists in mobilizing and absorbing nutrients by plant roots (Darrah et al., 2001). Thus, knowledge regarding root role in nutrient uptake is a central requirement for understanding plant structure and its ecological importance (Figure 7.12).

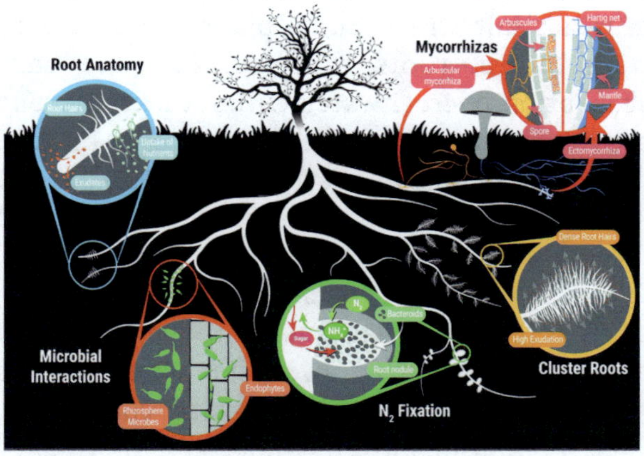

Figure 7.12. Functions of roots in nutrient absorption.

Source: https://rseco.org/sites/plantsinaction.science.uq.edu.au/files/4.0-Ch-Fig-4.0-NoNum%20v2%20resized.jpg.

7.3.3. Adaptations of Roots for Different Environments

Roots play an important role in guaranteeing the survival and acclimation of plants to a variety of habitats. Under water-stressed conditions, , such as arid deserts, many plant species possess deep taproots capable of penetrating underground water reservoirs. Such a characteristic enables these species to endure lengthy periods of drought, effectively keeping water-dependent physiological processes active. In addition, some plants develop extensive fibrous root systems that spread laterally rather than deeply, aiming to maximize water intake from surface moisture in rare rain showers. These adaptations enable such plants to draw hydration from diverse layers of the soil, representing their ability to accommodate fluctuations in the surroundings. Understanding these root systems is significant in terms of the potential to forecast how plants will respond to changing climates, particularly in cases where climate change can be predicted to lead to higher water deficiency in many regions, as addressed by Lukac et al. (2011). Alternatively, for saline soils found in coastal regions, among others, plants have evolved special mechanisms to resist salt stress. Some plants possess very wide lateral roots that facilitate the easy absorption of freshwater without salt, thereby providing an internal balance of ions. In such roots, there are excretion systems for salt wherein specific glands excrete excess salt, and the plant is able to thrive in salty conditions. Some halophytes possess thickened root tissue, not just as a water storage component but also to aid osmoregulation, a very important process required for withstanding salty conditions. Such adaptations are a profound evolutionary adjustment to specific environmental stresses, allowing plants to occupy ecological niches that would otherwise be inhospitable, and show how important root adaptations are to plant survival, particularly under the circumstances of climatic change and salinity stress highlighted by Lukac et al. (2011). In disturbed ecosystems , such as floodplains, root adaptations are equally important for plant survival. Such environments support many plants to develop shallow yet robust root systems that anchor the plants firmly in loose and saturated soils. Such roots are most often adapted to withstand the mechanical stress of water movements without being uprooted when floods occur. Some plants have aerenchyma growth in roots, allowing for gas exchange and buoyancy to sustain oxygen supply even when waterlogged. These adjustments not only enhance plant stability but also optimize nutrient uptake with changing water regimes. The resilience that accompanies these root architectures is becoming increasingly significant as climate models predict more instances of flooding across the globe, and focus is being placed on further studying root adaptation amidst a context of changing environmental regimes (Figure 7.13).

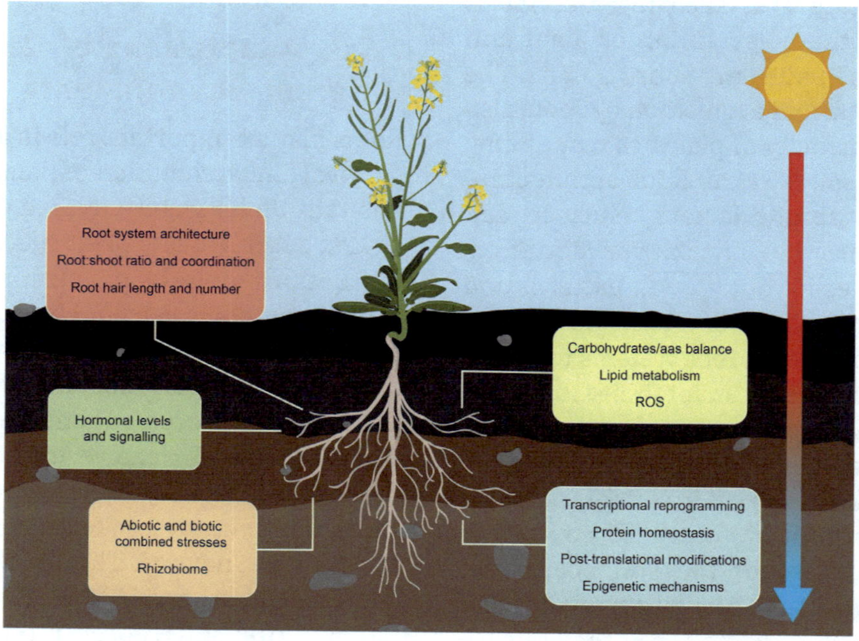

Figure 7.13. Adaptations of roots for different environments.

Source: Calleja-Cabrera J, Boter M, Oñate-Sánchez L and Pernas M (2020) Root Growth Adaptation to Climate Change in Crops. Front. Plant Sci. 11:544. doi: 10.3389/fpls.2020.00544

Test Your Knowledge

1. Illustrate the structure of root systems.
2. What are the functions of roots in nutrient absorption?

7.4. Stems and Their Functions

Learning Outcome

1. Structure and types of stems;
2. Role of stems in support and transport;
3. Modifications of stems for storage and reproduction.

Stems of plants are a significant aspect of plant architecture, serving several crucial functions in ensuring the survival and development of the organisms. Foremost, stems offer tubes through which water, nutrients, and photosynthesis yield move from the roots to the leaves. This vascular process is supported by two main types of tissue: xylem, which transports water and dissolved minerals from roots, and phloem, through which sugars produced via photosynthesis are channeled to other areas of the plant. This twin vascular system is vital in maintaining plant homeostasis and enables plants to compensate for changes in their environment, , such as differential water supply and deficiencies in nutrients. The efficiency in such functions highlights the importance of stems in the overall physiological functioning of plants with reference to supporting photosynthesis and nutrient transport. Moreover, stems are also responsible for structural support and plant growth in a vertical direction, enabling plants to reach light sources for photosynthesis. By elevating leaves above the ground plane, stems enable optimal light capture that is necessary for energy production. In others, stems have remarkable adaptations , such as thorns to protect against herbivores, or organs , such as tendrils that facilitate climbing. These adaptations enable the plants to occupy diverse ecological niches. Climbing plants, for instance, use their stems to ascend towards light, enhancing their competitiveness in dense vegetation. Besides, other research suggests an association between stem structure and environmental conditions, whereby adaptations may occur based on soil salinity or nutrient availability, influencing stem integrity and functioning (Blumwald et al., 2017).

The emphasis is on the adaptive capacity of stems in responding to environmental stresses. Stems not only provide structural support but also store and grow. Stems in many plants serve to store carbohydrates and water, thus serving as reservoirs that can be utilized during periods of drought. This is particularly significant in plants that are succulent, whereby stem adaptation facilitates them to live in desert regions. Stem growth habits, also controlled through hormone stimuli , such as auxins and gibberellins, shape their development and branching activities. A current nodule function study has emphasized the importance of symbiotic relationships to nutrient acquisition, with the suggestion that specific stem adaptations have the potential to maximize efficiency in nitrogen acquisition, which is vital to plant health and yield (Chiurazzi et al., 2017). As a result, learning the complex roles played by stems not only broadens our understanding of plant morphology but also provides insights into farming methods for enhancing crop resistance and production (Figure 7.14).

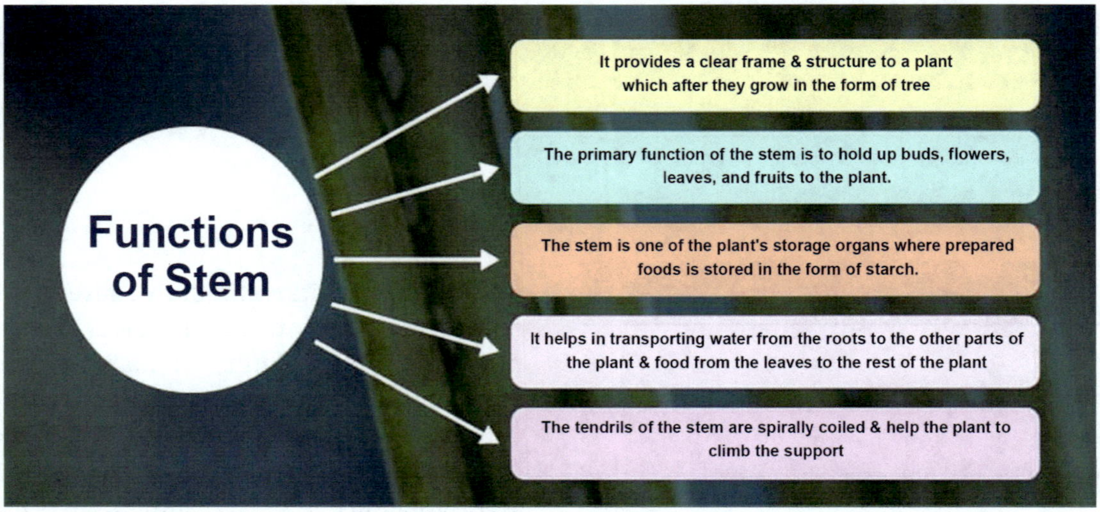

Figure 7.14. Stems and their functions.

Source: https://media.geeksforgeeks.org/wp-content/uploads/20240105130423/Function-of-stem.jpg.

7.4.1. Structure and Types of Stems

Stem structure is important in the general functioning of plants as an essential component of their growth and development. Stems consist mainly of three tissue types, namely the epidermis, vascular tissue, and ground tissue. The epidermis protects the plant from water loss and herbivory. Under the epidermis, vascular tissues , such as xylem and phloem transport water, nutrients, and sugars within the plant. Xylem, which is made up of tracheids and vessel elements, enables the upward movement of water from the roots, while phloem transports organic materials produced through photosynthesis. These specialized tissues are augmented by ground tissue, which contributes to photosynthesis, storage, and support, thereby making stems intricate structures that contribute to the survivability and health of plants. There are different forms of stems, each evolved for the purpose of performing in various environmental conditions and functional requirements. Herbaceous stems, prevalent in the majority of annuals, are mostly soft and green, allowing for straightforward growth and flexibility but not structural hardness to support strong biomass. Woody stems, as evidenced in shrubs and trees, possess a tough outer layer composed of secondary growth, which is mostly formed by the vascular cambium. This evolutionary step enables woody plants to rise tall, achieving greater sunlight exposure and facilitating reproduction. Also apparent in modified stems, , such as rhizomes, tubers, and runners, are differences in stem organization, which allow for asexual reproduction and resource storage. These structures not only protect the plant but also enhance the efficiency of its reproduction as well as its resistance to environmental change. External processes, , such as frequency of fires and environment heterogeneity, could also affect the spatial arrangement of stems and, in turn, affect plant community processes. For instance, Gandiwa et al. (2009) shows how different frequencies of fire produce different structural changes in woodland habitats, which produce differences in species composition and stem density in different habitats. In addition, studies like those quoted by Beaudrot et al. (2011) show how microhabitats, , such as termite mounds, can significantly influence stem density and species distribution. These interactions emphasize the importance of stem structure in both the survival of an individual plant and the ecological interactions within changing environments. Thus, it is understood that the study of stems encompasses a wide array of biological and environmental aspects that play important roles in plant structure and functioning (Figure 7.15).

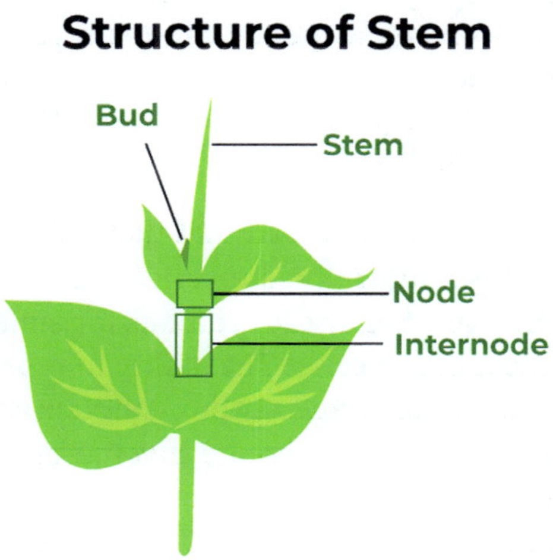

Figure 7.15. Structure of stems.

Source: https://media.geeksforgeeks.org/wp-content/uploads/20220922155857/StructureofStem.png.

7.4.2. Role of Stems in Support and Transport

The function of stems in plant architecture is more than the provision of mere support; stems are important conduits for the transportation of water, nutrients, and hormones throughout the organism. Constituted mainly of vascular tissues, or xylem and phloem, the stems provide passage for the movement of needed substances from leaves to roots and from roots to leaves. The xylem transports water and dissolved minerals absorbed from the soil, while the phloem transports organic compounds, , such as sugars produced during photosynthesis. This intricate transport system enables plants to maintain homeostasis and react to environmental stress. For instance, during water shortages, the efficiency of nutrient transport can make a significant difference to a plant's overall health and development. The structural support provided by stems is necessary for the preservation of this transport as well and allows plants to exist within various settings through mechanical rigidity and adaptive patterns of growth. In addition to transport, stems are significant in structural support within plants as well. They must withstand various forces, including wind, gravity, and their own weight, while not compromising the vascular system's integrity. This mechanical support relies on the structure and organization of the stem's tissues, , such as lignin in the xylem, which stiffens the structure but makes it flexible. Divergences between plant species have large variations in the structure of the stem; for example, lianas have greater maximum stem-specific hydraulic conductivity than trees but also exhibit a trade-off between efficiency and safety of water transport. These adaptations point to the importance of stem structure in several species, indicating that stem characteristics are acclimated to optimize support as well as transport functions against the requirements of the environment (Guzman et al., 2016). Also, recent studies have elucidated the multifaceted evolution and function of transporters in woody species, attesting to their utility in stem physiology. For example, studies on auxin transport proteins in trees have revealed complexities in how these signaling molecules influence stem development and growth. Populus species have demonstrated diverse auxin transport protein families that are expressed differently within tissues, which implies specialized function in woody stem development (Carraro et al., 2012). This understanding of variations is able to enhance knowledge in relation to stem functions in hormonal transport and overall plant development. With continued advances in research, clarification of the functional interactions among stems, transport networks, and environmental relationships will shed vital information on plant resilience and adaptability in shifting climates (Figure 7.16).

Figure 7.16. Close-up of a stem node of Polygonum amphibium showing the development of adventitious roots.

Source: https://upload.wikimedia.org/wikipedia/commons/thumb/f/fb/Polygonum_amphibium_%284973671699%29.jpg/250px-Polygonum_amphibium_%284973671699%29.jpg.

7.4.3. Modifications of Stems for Storage and Reproduction

Plant structural adaptations in stems play an important role in enhancing storage and reproductive success. Many angiosperms possess specialized stem modifications like bulbs and tubers that enable them to store nutrients and water. Bulbs like those in onions (*Allium cepa*) consist of fleshy leaves covering a central bud, which maintains the plant alive under adverse seasonal conditions by storing starches. Similarly, tubers, , such as those of the potato (*Solanum tuberosum*), serve dual roles: they act as reservoir organs and concurrently provide a means of asexual reproduction by the establishment of eyes or buds. Such characteristics not only ensure plant survival during unfavorable seasons but also allow for vegetative dissemination, promoting population resistance to environmental stressors. In addition to nutrient storage, stems can evolve to enable reproduction between species of plants. A case in point is stolons, a type of horizontal stem that forms on the ground surface, which facilitate strawberry plants (*Fragaria* spp.) to reproduce sexually by giving rise to new clones that root and grow independently. This vegetative reproduction increases genetic continuity and is able to rapidly expand a plant's range under optimal conditions.

It should be stressed that these alterations are not structurally superficial but rather complex evolutionary adaptations for enhancing the fitness of a species. The ability to reproduce both sexually and asexually through stem changes has the capability to create diverse ecological strategies and ultimately affect plant community dynamics. Moreover, interaction between stem changes and the environment provides further support for the significance of these changes in plant function. As amply illustrated in recent studies, recruitment of plant species is based on the interaction between their reproduction mechanisms and the environment. For example, developments like those described in terms of the CLE family of peptides help optimize root architecture under optimum conditions, facilitating general plant health and efficiency of reproduction (Van et al., 2015). This encounter assumes particular significance in changing climates when the ability to modulate growth or reproduction in response to environmental cues can determine the fate of a species. Adaptations of stems thus demonstrate how plant structure development and function are integrally linked with ecological responsiveness, equating storage potential and reproductive strategies to survival under changing conditions (Figure 7.17).

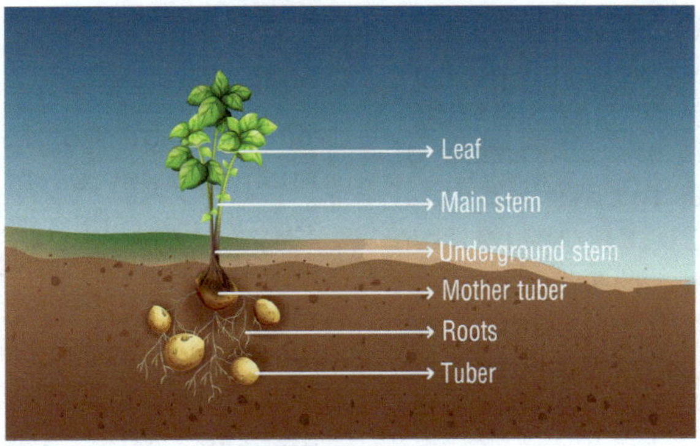

Figure 7.17. Modifications of stems for storage and reproduction.

Source: https://www.olabs.edu.in/userfiles/1/1702288826_POTATO.jpg.

Test Your Knowledge

1. What is the role of stems in support and transport?
2. What are the modifications of stems for storage and reproduction?

7.5. Leaves and Photosynthesis

Learning Outcome

1. Structure of leaves and their adaptations;
2. Process of photosynthesis in leaves;
3. Factors affecting photosynthesis efficiency.

Leaves also bear the responsibility of the process of photosynthesis since they are the primary locations for the absorption of sunlight and the transformation of it into chemical energy. The structural design of the leaves, based on their large surface area and specialized cells, ensures optimal light absorption and gas exchange. Mesophyll cells, which exist within the leaf, contain chloroplasts filled with chlorophyll pigments required for the process of photosynthesis. Furthermore, the arrangement of these chloroplasts inside the mesophyll cells is not random; it is specially made to enhance the capture of light. Photosynthetic efficiency depends on various factors like the intensity of light, CO_2 concentration, as well as anatomy of the leaf itself. Whenever environmental conditions are favorable, leaves can potentially produce a great amount of sugars, which are necessary for the maintenance of plant growth and metabolic activities, thus emphasizing the necessary interface between the structure of the leaf and the photosynthetic process. The impact of herbivory on leaf physiology can give us a hint at the fine balance between plant defense and energy production. Experimental studies have found that while herbivory can influence the physical traits of leaves, , such as mass per area and nitrogen content, the source of photosynthesis rates could be from natural variation between trees as individuals rather than damage induced by the herbivores themselves. Specifically, one study found leaf nitrogen content correlated with photosynthetic rates, indicating the importance of nutrient conditions in what a leaf is capable of doing in relation to solar capture. This complexity also means that leaves possess some level of resistance to herbivorous attack, and they can continue to photosynthesize proficiently even when under stress, exhibiting the dynamic character of plant physiological responses to environmental stress (Gripenberg et al., 2020). Apart from structural adaptations, leaves are also important sites for communication in the plant and are governed by nutrient availability. For example, the presence of sucrose is a requirement for leaf growth and can alter growth patterns and metabolic processes. Sucrose treatment induces shifts in cell proliferation and chloroplast development in *Arabidopsis* leaves. High sucrose levels trigger glucose-6-phosphate transporter (GPT2) expression, inducing the uptake of essential energy substrates into chloroplasts. As such, it affects chloroplast differentiation and contributes to the transition from expansion to proliferation of cells, and thus leaf growth in general. This sucrose sensitivity not only highlights the prominence of leaf metabolism in regulating growth but also illustrates how leaves integrate extrinsic signals to optimize photosynthetic performance as well as overall plant health (Bruyne et al., 2016) (Figure 7.18).

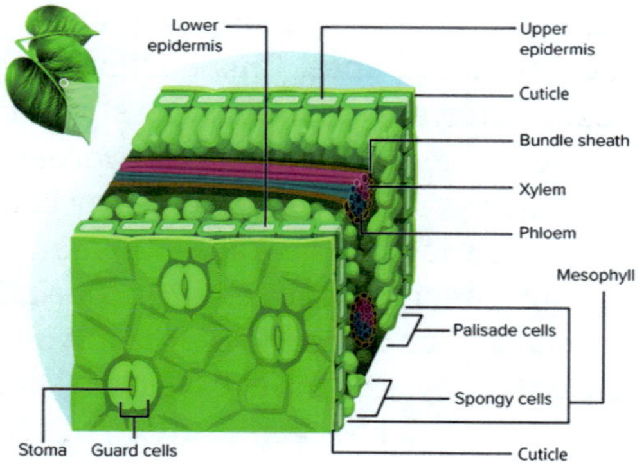

Figure 7.18. Leaves and photosynthesis.

Source: https://dr282zn36sxxg.cloudfront.net/datastreams/f-d%3Ad2801dc161ea9e91474694974cc84e35020d9caa1e47c8fd7ff9bcbd%2BIMAGE_TINY%2BIMAGE_TINY.1.

7.5.1. Structure of Leaves and Their Adaptations

The leaf morphology is well specialized to maximize photosynthesis at the cost of as much water as possible, indicating several adaptations to make it more functional. On a basic level, a leaf consists of different layers like the epidermis, mesophyll, and vascular tissues. The epidermis is a shield against environmental stresses and disease, while specialized cells like guard cells regulate gas exchange through stomata. Furthermore, the mesophyll, which is composed of palisade and spongy parenchyma, is organized to optimize light capture and gas diffusion. The tissue structural organization allows leaves to exchange light absorption and CO_2 entry for loss of water, demonstrating their accommodations for endurance in a broad range of environments. These structural complexities are essential to the plant's survival across various conditions and therefore illustrate the importance of leaf structure to plant ecology and physiology. Leaf adaptations extend beyond structural features to involve complex biochemical processes that enhance the survival and reproduction of plants. Some plants, for instance, have evolved leaf traits that allow them to better withstand herbivory, a significant problem in their ecosystems. In the Brassicales order, the plants produce glucosinolates, chemical compounds that serve as herbivore deterrents. Meanwhile, some herbivorous insects, e.g., Pierid butterflies, have become sensitive by developing special gut proteins that degrade these chemical defenses and thereby allow them to make use of such plants as a food source effectively. This sophisticated interaction is an excellent example of coevolution wherein the adaptation of one species becomes the driving force in shaping another species' evolutionary path. Understanding these adaptations provides us with fundamental information regarding the dynamic processes of ecosystems and the part played by leaf structure in facilitating plant-herbivore relationships, as is illustrated by evidence of distinct performance differences in Pieris species according to their host plant glucosinolate chemistries (Hirai et al., 2019). In addition, the phyllosphere, the outer layer of leaves, exhibits yet another fundamental part of leaf adaptations necessary for plant well-being and productivity. This microhabitat serves not only for gas exchange but also sustains intricate microbial communities that play critical roles in plant health. The structure of these microbial communities is the result of a complex interaction between environmental conditions and the leaves' physico-chemical properties, which has a significant impact on plant fitness as well as agricultural yield. Research has indicated that plant genotype may act upon these microbial communities to influence the overall health of the plant as well as its tolerance to environmental stress. With increasing knowledge of the complexities of these dynamics, it is apparent that leaf structures are crucial not only for gas exchange and photosynthesis but also in determining and sustaining symbiotic microbial interactions within the phyllosphere (Adams et al., 2008). These specializations reflect the diverse roles of leaf structures within the general context of plant structure and function (Figure 7.19).

Figure 7.19. The structure of a leaf.

Source: https://www.passmyexams.co.uk/GCSE/biology/images/leaf-structure.jpg.

7.5.2. Process of Photosynthesis in Leaves

The process of photosynthesis is essential to the function and structure of plants in that it allows energy from light to be transferred to chemical energy. Photosynthesis within leaves occurs primarily in chloroplasts, which contain pigments and chlorophyll essential for absorbing sunlight. The light reactions initially occur in thylakoid membranes, whereby sunlight is absorbed and used to split water molecules, with oxygen being a byproduct. This stage generates ATP and NADPH, both of which are used in the subsequent light-independent reactions, or the Calvin cycle. During this process, CO_2 is fixed into organic molecules, ultimately leading to the synthesis of glucose and other carbohydrates. The efficiency of these processes is crucial in plant growth and energy generation and thus emphasizes the dependent relationship between leaf anatomy and photosynthetic process. Leaf structure performs a key role in optimizing the photosynthetic process so that plants become acclimatized to variations in environmental conditions. For instance, the arrangement of chloroplasts in leaf cells optimizes light gathering, while stomatal openings regulate gas exchange and minimize water loss. Monocotyledons specifically show advanced adaptations , such as the ability to regulate photosynthesis on various surfaces of leaves. At water-stressed times, such plants can close metabolic activity on the adaxial surface but still carry on photosynthesis on the abaxial surface, which serves a survival advantage during unfavorable times. Such morphological adaptations not only enhance photosynthetic efficiency but also highlight plant evolutionary strategies in responding to varying environments as they strive for optimal growth and energy production for survival (Ana et al., 2008). Moreover, the exchange of leaf longevity with photosynthetic efficiency is pivotal to probing ecosystem processes. Models , such as the University of Sheffield Conifer Model (USCM) illustrate such an interaction by simulating carbon, nitrogen, and water fluxes in coniferous forests. By focusing on leaf longevity as a significant trait, researchers can predict forest performance in a variety of climatic regimes. This model stresses the importance of leaf architecture, thicker leaves will live longer and hence enhance their photosynthetic capacity and storage of resources over time. This is crucial in the study of plant management and conservation strategy design because it reveals the intricate interaction between leaf structure, function, and plant environmental interaction that governs plant productivity (Aerts et al., 2002) (Figure 7.20).

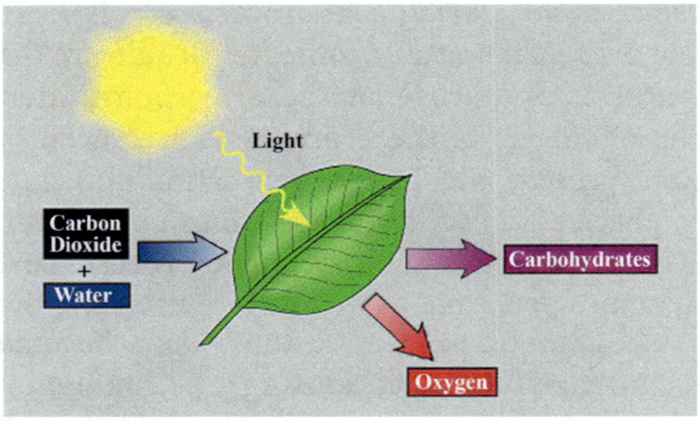

Figure 7.20. Process of photosynthesis in leaves.

Source: https://leavingbio.net/wp-content/uploads/2017/11/phtosynthesis.gif.

7.5.3. Factors Affecting Photosynthesis Efficiency

Several factors influence the efficiency of photosynthesis, starting with light availability. Light intensity itself influences photosynthetic rate, with insufficient light being a source of reduced photosynthetic activity. Chloroplasts' photosystems employ photons for energy to facilitate the transformation of water and CO_2 into glucose and oxygen. However, too low or too high light conditions can inhibit photosynthesis; at elevated light intensities, photoinhibition causes photochemical damage to photosynthetic equipment. Adding to this complexity, leaf morphology is an important factor in the absorption and utilization of light. Changes in leaf anatomy, including thickness and chlorophyll content, can have major effects on the light-harvesting capacity of a plant and the photosynthetic efficiency overall. As such, then, studying the morphology of a leaf becomes integral to predicting a plant's potential for absorbing and utilizing light energy to fuel growth and development, thus connecting structural features with photosynthetic function. Availability of water is also another major factor that influences photosynthesis efficiency directly. Stomatal conductance, which is responsible for gas exchange into and out of the leaves, is directly associated with the availability of water; decreased content of soil water content will close the stomata in order to maintain water from being lost, leading to decreased CO_2 intake. This is a trade-off as water stress can lead to decreased assimilation of carbon and hence lower photosynthesis rates. Research highlights that water use efficiency (WUE) differs significantly between plant species, revealing genetic and anatomical disparity that modulates this aspect of photosynthesis (Easlon et al., 2013). On a broader landscape scale, drought can heavily limit net primary productivity in ecosystems, with implications for growth and carbon sequestration. This water supply-photosynthetic rate interaction emphasizes the significance of moisture in maintaining proper plant functioning and demonstrates how structural mechanisms can ward off water stress. Another key factor determining the efficiency of photosynthesis is environmental pollution, primarily ground-level ozone. Elevated ozone levels can lead to reduced carbon gain by leaves due to a direct impact on stomatal conductance that limits CO_2 uptake (Aber et al., 1997). The combination of interior leaf concentration and ozone flux is of primary importance since it characterizes how pollution impacts directly on photosynthetic physiological mechanisms. In addition, mature forests have exhibited varied sensitivity to the pollutants, which highlights the necessity of a holistic comprehension of how environmental conditions, coupled with structural characteristics, regulate photosynthetic capacity over time. The impacts of ozone exposure can have significant declines in forest productivity, demonstrating the means by which external environmental and internal plant architecture factors both influence photosynthesis efficiency. Understanding how they interact is crucial to developing ways of preventing the impacts of pollution on plant health and productivity.

Test Your Knowledge

1. What is the process of photosynthesis in leaves?
2. Write the factors affecting photosynthesis efficiency.

7.6. Reproductive Structures in Plants

Learning Outcome

1. Overview of plant reproductive systems;
2. Structure and function of flowers;
3. Role of seeds and fruits in plant reproduction.

The plant's reproductive organs play a vital role in the plant's life cycle, performing sexual reproduction and continuation. The flowers are central to these organs, which are male and female reproductive organs. The male organs are the stamens, and they produce pollen. The female organs or carpels, hold ovules. The flower morphology is intricate, with features , such as striking coloration of the petals and pleasant fragrance serving to attract pollinators. An understanding of these structures requires integrating facts about their form and function; for instance, the pollen distribution relies on successful pollination operations, leading to fertilization needed to establish seeds. Research has indicated that students have a weakened perception of plants focused largely on flowers, reflecting poor understanding of the broader implications of reproductive organs on plant structure and ecology (Oesterlin, 2022). Investment in reproductive organs can contribute significantly towards affecting a plant's overall development and future performance, something that is more nuanced than in the past. Latest evidence shows that the relationship between reproductive resource investment and subsequent plant vitality is not linear. Low levels of investment, for instance, can have a negligible impact on future performance, whereas increased reproductive effort can lead to extensive resource depletion with adverse effects. This curvilinear trade-off is used to highlight the necessity for plants to temper their reproductive strategy relative to survival and growth requirements. Results of experiments in Plantago major reveal that vegetative tissues are more resource-intensive than reproductive tissues at some stages of growth, and this implies that a strategic reproductive strategy can maximize fitness in the long term (Baltzer et al., 2002). This knowledge is vital in understanding how plants will evolve to suit the environment while promoting species survival. In school learning environments, knowledge about plant reproduction and structure has been credited to experience-based learning activities, especially within garden learning environments. A case study with fifth-grade students indicated that students who became more engaged with garden activities constructed a stronger concept of plant structure and function, indeed in relation to reproduction. However, although these results were anticipated, the correlation of time spent in the garden to total comprehension was not as strong as was predicted and raises questions regarding what types of activities foster the best comprehension. Because students consistently labeled flowers in their sketches, this indicates an initial focus on reproductive structures. Therefore, it is important to conduct further research into how various garden-related activities can enhance diversity in learners' knowledge and perception towards plant reproduction to form an inclusive educational plan (Oesterlin, 2022) (Figure 7.21).

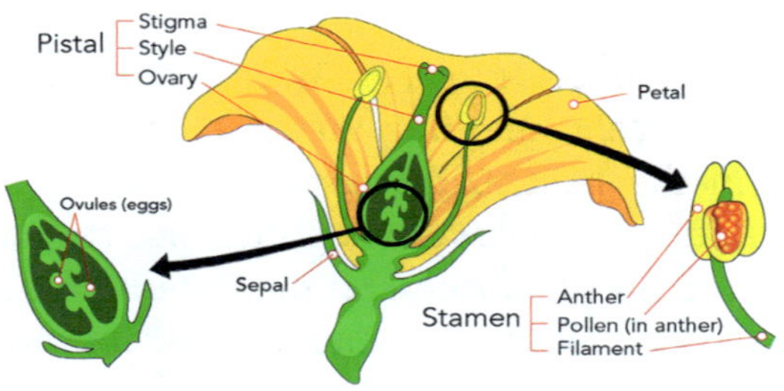

Figure 7.21. Reproductive structures in plants.

Source: https://letstalkscience.ca/sites/default/files/2020-06/parts_of_a_flower.png.

7.6.1. Overview of Plant Reproductive Systems

The plant reproductive mechanisms are highly complicated in their design to allow species continuation regardless of various environmental conditions. At the center of the mechanisms is sexual versus asexual reproduction. Sexual reproduction involves gamete union—sperm and ova—leading to genetic diversity, while asexual reproduction allows quick population establishment by such mechanisms as vegetative propagation. Each of these mechanisms has specific advantages; sexual reproduction maintains genetic diversity, which can result in greater adaptability, while asexual reproduction ensures survival in persistent environments with optimally fitted features (Wyse, 2014). Understanding these reproduction mechanisms is critical in order to understand how plants operate within the ecosystem and respond to changes in habitat conditions. The morphology of the reproductive structures of a plant, , such as inflorescences and flowers, is crucial in establishing mating systems and reproductive strategies. Flower morphology, e.g., size and configuration, affects pollination efficacy and success rates, with direct implications for pollen and ovule allocation. Current research indicates that the pollen:ovule (P:O) ratio is indicative of this allocation strategy, and differences are seen across various mating systems. For example, xenogamous species tend to have higher P:O ratios compared to autogamous species, indicating a strategic investment in male reproductive effort to attract generalized pollinators (Wyse, 2014).

This interaction indicates the importance of inflorescence size and structure in optimizing reproductive success in different ecological environments, thereby revealing sophisticated evolutionary specializations in plant reproductive systems. Moreover, environmental effects on plant reproductive strategy cannot be overemphasized, especially under climatic change and landscape alteration. Plants demonstrate remarkable plasticity of reproductive strategies, and they can adjust to changing conditions. For instance, in urban areas and altered landscapes, plants adapt by modifying investment in reproductive traits to maximize chances at successful pollination and seed dispersal. Research committed to exploring the interplay between plant morphology, functional traits, and abiotic stressors is crucial to the demystification of such complex dynamics. In light of continued advancements in the comprehension of plant reproductive systems, insights derived from diverse ecological environments will facilitate better conservation and management practices, which foster the persistence of plant communities amid continued global change (Figure 7.22).

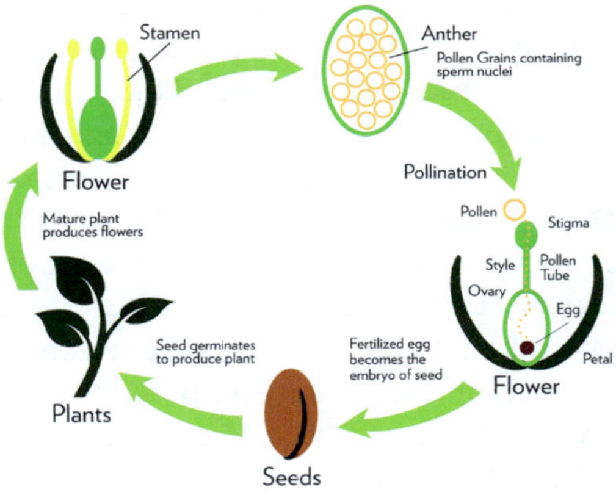

Figure 7.22. Plant reproductive systems.

Source: https://classnotes.ng/wp-content/uploads/2020/04/plant-reproduction-biology-classnotesng.png.

7.6.2. Structure and Function of Flowers

The structure and functioning of flowers are a perfect example of evolutionary adaptation in angiosperms, which are vital for the successful completion of flowering plant reproduction. Flowers consist of disparate structures, each with its specific role in the reproductive process. The primary components are the petals, the stamens, and the carpels. Petals attract pollinators by their resplendent color and arousing fragrances, facilitating cross-pollination, thereby increasing genetic diversity. Stamens, the male organs, create pollen with sperm cells, whereas carpels contain the ovules, where fertilization takes place. Such a complex arrangement highlights not only the beauty of flowers but also their very purpose in perpetuating plant species by the complex processes of pollination and fertilization. Studies on what students know about plant structures show significant information on how flowers are perceived and understood. In research involving fifth-grade pupils from a school garden, the overwhelming majority (64%) of the students identified flowers as being at the center of their mental representations of plants (Oesterlin, 2022). This bias suggests flowers are more readily grasped by the imagination than other elements of plants, , such as leaves or roots. Furthermore, the research also identified that students who drew flowers had a better understanding of plant reproductive systems compared to students who drew trees or other plant parts.

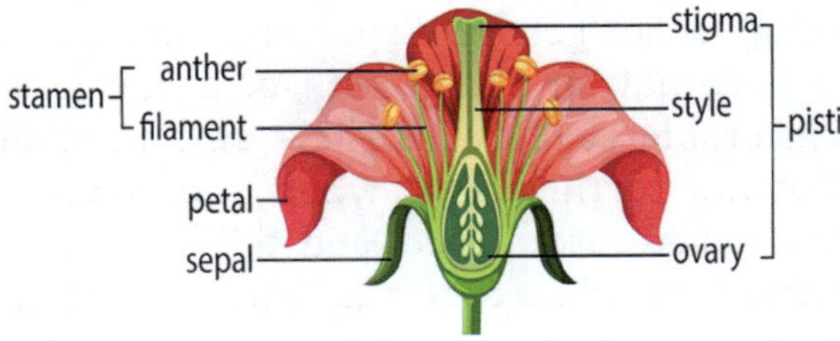

Figure 7.23. Structure and function of flowers.

Source: https://media.geeksforgeeks.org/wp-content/uploads/20240520101649/Flower.webp.

This result indicates the value of flowers not only as reproductive functional structures but also as effective teaching aids for studying plant biology and stimulating botany interest. The intricate flower development is partially controlled by gene mechanisms that guide cellular interaction among different layers within the plant. In more detail, studies have shown that the AGAMOUS gene is critical in floral structure development through the interaction among the plant meristem layers (Drews et al., 1998). Function of this gene in the L2 layer is essential to reproductive structure formation, i.e., stamens and carpels, and to influencing meristem determinacy as a whole. Their consequences are pronounced, offering a model of how structural variation arises in flowers and how these adaptations influence reproductive strategy. Therefore, flower structure and function complexity not only display the remarkable efficiency of angiosperms but also provide a fruitful research field for plant biology, integrating genetics and morphology (Figure 7.23).

7.6.3. Role of Seeds and Fruits in Plant Reproduction

The reproductive strategies of plants are highly coordinated with the structural specializations of the plants, the most prominent being in the growth of seeds and fruits. Seeds are the agency through which genetic dispersal occurs, having the potential for the growth of a new plant in favorable conditions. They grow normally following fertilization, when the ovule is transformed into a seed, and this is typically followed by storage of nutrients to facilitate early germination. The nutrients found in seeds ensure the seedling can accommodate in a variety of environments, highlighting structure and function in plant reproductive success. Seeds are therefore not just reproductive units, but also carriers of genetic variation and evolutionary potential, which ensure adaptation to new environments and plant species survival in general within their ecosystems. Fruits are not essential for seed dispersal and protection, though, but are a demonstration of incredible evolutionary success. Fruit formation from the ovary of a flower not only protects seeds but also disperses them through mechanisms , such as wind, water, or animal touch. Fruits are tender and visually appealing and attract animals that consume the fruit and then disperse the seeds away from the parental plant, a function beneficial to the species' growth and genetic diversity. The relationship between animals and plants is a mutualism in which both partners gain something, and as a result, there is increased reproductive success. As has been observed in different ecological environments, these interactions are significant in plant species distribution and the ability of new habitats to be colonized, depicting the dynamic interdependence of structure and function in plant reproduction. Seed and fruit morphology also reflects the sensitive correlation between environment and reproduction strategy. For example, seeds with wind-dispersal features , such as light structure or wing-like projections, show how physical characteristics may enable spreading and colonization of suitable habitats. Similarly, floating, or buoyant fruits enable water dispersal in water-based ecosystems. Such adaptability is critical where traditional growth may be hindered in ecological systems. Research on seed and fruit structures in various species establish the importance of understanding these structures in predicting the response of plants under shifting ecological and anthropogenic pressures. These are essential for conservation since they inform strategies to preserve plant diversity and ensure the persistence of various communities of plants under shifting conditions (Konuma et al., 2011; Choirina et al., 2019).

Test Your Knowledge

1. Give an overview of plant reproductive systems.
2. What is the role of seeds and fruits in plant reproduction?

SUMMARY

In bringing together the intricate relationship between plant function and structure, it is apparent that these elements are inextricably linked, informing the evolution and adaptability of plants across a multitude of ecosystems. The convergence of photosynthetic efficiency, structural integrity, and metabolic plasticity reveals the phenomenal adaptability of plants to changing environmental conditions. For instance, the polar adaptations indicate the resistance of plants to extreme temperatures and fluctuating light regimes, as indicated by the physiological mechanisms developed by photosynthetic autotrophs. These are not just vital for existence but also to maintain general ecological equilibrium, which affects global biogeochemical cycles by participating actively in carbon fixation and nutrient cycling, as indicated by recent studies (Arrigo et al., 2014). Thus, information about plant structure and function gives us not only information about their biology, but also information about environmental processes at a larger scale. In addition, the complexity of plant responses to the environment emphasizes the interplay of several structural and functional traits to facilitate their survival and reproduction. This is best evidenced in plant genera that have particular adaptations to their environment, , such as those in agricultural systems or those under thick secondary metabolites in the environment. The ruminant feeding behavior studies indicate that organisms, , such as goats, choose what to eat according to the availability or unavailability of such metabolites, which, even though they are possibly toxic, also express biochemical significance in nutrient intake and metabolism. Understanding this selectivity can help explain the way that plants and herbivores coexist and reciprocally, reveal the underlying ecology of the process (Lamy et al., 2012). In addition to merely validating the concept of an equilibrated ecosystem, these processes also speak to the importance of plant structures in facilitating these processes. Last, the deduction from plant structure and function reveals to us a systems perspective of plant ecological relationships and the processes of evolution taking place in plant environments. In the face of continued climate change posing the specter of plant extinction, further information regarding their adaptational properties, , such as polar microalgae and their molecular process, is vital. These findings not only guide conservation action but also suggest probable biotechnological applications based on the understanding of such resilient organisms. Through recognizing the dominant position of plant form, we could more highly regard their contribution to ecosystem health and resilience. In summary, the study of plant structure and function is not a mystical pursuit but a vital endeavor with implications for the maintenance of world biodiversity in a new world.

Plant structure and function is a multidisciplinary field with a great variety of valuable topics, including the morphological features of flowers and developmental patterns. Exceptionally reduced floral morphology is found in perianthless families of the Piperales, i.e., Piperaceae and Saururaceae, when compared to other plant families. Recent molecular phylogenetics isolated *Verhuellia* as the first diverging branch in Piperaceae, indicating its uniqueness. Detailed examination of *Verhuellia lunaria* has yielded data about its inflorescence, which is in the form of an indeterminate spike with sessile flowers. The flower appears in a helical pattern and possesses typical morphological traits in the form of two to three stamens and a rudimentary gynoecium, thereby further developing our knowledge of group evolutionary associations (Goetghebeur et al., 2010). In addition, the anatomical floral characters are also vital to their total function and fertilization success. From the research, *Verhuellia lunaria* inflorescence lacks such common traits as a style but has three to four stigma branches instead. This structural simplicity goes against conventional propositions of floral sophistication and reveals that evolutionary adaptation is likely to prefer simplified structures in certain ecological scenarios. Furthermore, the *Verhuellia* fruit is a drupe with multicellular epidermal protuberances of large size, which can further promote dispersal mechanisms. Data from pollen ultrastructure analysis show inaperturate pollen grains, a significant evolutionary feature that the species shares with a number of Peperomia species, which correlate morphological features with reproductive strategy (Goetghebeur et al., 2010). Briefly, the integration of molecular and morphological data proves to be a useful means in studying plant relationships and adaptations. Examination of *Verhuellia* and its relatives not only adds to knowledge of flower structure but also indicates the evolution of reproductive strategy in angiosperms. Consequently, the combination of extensive anatomical study and phylogenetic research allows more subtle insight into plant evolution and confirms the argument that structural adaptation is inseparable from functional consequence in diverse environments. Together, these findings highlight the merit of cross-disciplinary approaches to plant biology since they enhance the insights into plant diversity and evolutionary biology (Goetghebeur et al., 2010).

KEY TERMS FROM THIS CHAPTER

Apical meristem	Cambium	Chloroplast
Cuticle	Guard cells	Internode
Lateral meristem	Leaf	Meristem
Node	Petiole	Phloem
Photosynthesis	Root	Stem
Stomata	Transpiration	Turgor pressure

CRITICAL THINKING QUESTIONS

1. **Which plant tissue is responsible for transporting water from the roots to the leaves?**
   a. Phloem
   b. Xylem
   c. Cambium
   d. Cortex
2. **What is the primary function of the stomata in plant leaves?**
   a. Absorb nutrients
   b. Support the leaf structure
   c. Regulate gas exchange and water loss
   d. Store starch
3. **Which part of the plant anchors it in the soil and absorbs water and minerals?**
   a. Stem
   b. Leaf
   c. Flower
   d. Root
4. **The process by which plants make their own food using sunlight is called:**
   a. Respiration
   b. Transpiration
   c. Photosynthesis
   d. Germination
5. **Which plant structure is mainly responsible for reproduction?**
   a. Leaf
   b. Stem
   c. Flower
   d. Root

Short Answer Questions

1. What are the three main organs of a plant, and what are their primary functions?
2. What are the two main types of vascular tissue in plants, and what do each transport?
3. Describe the structure and function of a leaf, including the roles of the cuticle, stomata, and guard cells.
4. How do roots help a plant besides anchoring it in the soil?
5. What is transpiration, and how does it help move water through the plant?
6. Explain how photosynthesis and cellular respiration are connected in plant cells.

Answers to Critical Thinking Questions

1. (b); 2. (c); 3. (d); 4. (c); 5. (c)

REFERENCES

1. Aber, J. D., Alm, J., Anisimov, O., Apps, M., Bartlett, P., Bartsch, I., Belyea, L., et al. (2002). *Modeling Seasonal to Annual Carbon Balance of Mer Blaue Bog, Ontario, Canada.* https://core.ac.uk/download/72052578.pdf (accessed on 13th June 2025).
2. Abiven, M., Allan, E., Amundson, R., Bardgett, R., Bello, F., Berendse, F., Boivin, P., et al. (2016). *Plant Diversity and Root Traits Benefit Physical Properties Key to Soil Function in Grasslands.* https://core.ac.uk/download/76999242.pdf (accessed on 13th June 2025).
3. Adams, J., Andrews, M., Arnold, M. L., Arnold, S., Asins, M. J., Beattie, G. A., Beattie, A. J., et al. (2008). *Phyllo Sphere Microbiology with Special Reference to Diversity and Plant Genotype.* https://core.ac.uk/download/49037.pdf (accessed on 13th June 2025).
4. Aerts, R., Amthor, J., Amthor, J., Atkin, O. K., Axelrod, D. I., Axelrod, D. I., Beerling, D. J., et al. (2002). *A Process-Based Model of Conifer Forest Structure and Function with Special Emphasis on Leaf Lifespan.* https://core.ac.uk/download/60129.pdf (accessed on 13th June 2025).
5. Aguilar-Lasserre, A., Azzaro-Pantel, C., Dietz, A., Domenech, S., et al. (2006). *Multiobjective Multiproduct Batch Plant Design Under Uncertainty. https://core.ac.uk/download/120*39657.pdf (accessed on 13th June 2025).
6. Aguilar-Lasserre, A., Azzaro-Pantel, C., Dietz, A., Domenech, S., et al. (2008). A Fuzzy Multiobjective Algorithm *for Multiproduct Batch Plant: Application to Protein Production.* https://core.ac.uk/download/12039646.pdf (accessed on 13th June 2025).
7. Akimoto, H., Bastow, R., Baurens, F. C., Besnard, G., Birchler, J. A., Block, M. A., Bogeat-Triboulot, M. B., et al. (2013). *Epigenetic Regulation of Adaptive Responses of Forest Tree Species to The Environment.* https://core.ac.uk/download/16664504.pdf (accessed on 13th June 2025).
8. Aller, I., Conejero, G., Gjetting, S. K., Ivakov, A., Littlejohn, G., Martinière, A., Müller, K., et al. (2014). *Functional Imaging in Living Plants–Cell Biology Meets Physiology.* https://core.ac.uk/download/148791365.pdf (accessed on 13th June 2025).
9. Arrigo, K. R., Comeau, A. M., Comiso, J. C., Devos, N., DiTullio, G. R., Ferrara, S., Gabric, A. J., et al. (2014). *Polar Microalgae: New Approaches Towards Understanding Adaptations to An Extreme and Changing Environment.* https://core.ac.uk/download/41989130.pdf (accessed on 13th June 2025).
10. Atanacio, R., Bailly, N., Froese, R., Reyes Jr., R. B., et al. (2010). *Simple Identification Tools in Fish Base.* https://core.ac.uk/download/11900256.pdf (accessed on 13th June 2025).
11. Baltzer, J. L., Budge, S., & Reekie, E. G. (2002). The shape of the trade-off function between reproduction and future performance in *Plantago major* and *Plantago rugelii.* https://core.ac.uk/download/143679905.pdf (accessed on 13th June 2025).
12. Barczi, J. F., Jourdan, C., & Rey, H. (2011). DigR: How to Model Root System *in Its Environment? 1–The Model. http://agritrop.cirad.fr/561964/1/document_561964.pdf (accessed* on 13th June 2025).
13. Barrett, D. J., Beijerinck, M. W., Brookes, P. C., Dunger, W., Finlay, B. J., Finlay, B. J., Fitter, A. H., et al. (2005). Darkness Visible: Reflections On Underground Ecology. https://core.ac.uk/download/59119.pdf (accessed on 13th June 2025).
14. Beaudrot, L., Du, Y., Harrison, R., *Rahman Kassim, A., et al. (2011). Do Epigeal Termite Mounds*

Increase the Diversity of Plant Habitats in A Tropical Rain Forest in Peninsular Malaysia? https://core.ac.uk/download/323070091.pdf (accessed on 13th June 2025).

15. Bellincampi, D., Chambers, L., Held, M. A., Lionetti, V., et al. (2016). Decreased Polysaccharide Feruloylation Compromises Plant Cell Wall Integrity and Increases Susceptibility to Necrotrophic Fungal Pathogens. https://core.ac.uk/download/54532579.pdf (accessed on 13th June 2025).

16. Benaroya, H., & Nagurka, M. L. (1990). Space Structures: Issues In Dynamics and Control. https://core.ac.uk/download/213088418.pdf (accessed on 13th June 2025).

17. Bernardello, G. L. M., Freitas, L., Galetto, L., Paoli, A. A., et al. (2001). Nectarie's and reproductive biology of Croton sarcopetalus (Euphorbiaceae). https://core.ac.uk/download/159284549.pdf (accessed on 13th June 2025).

18. Blumwald, E., Du, J., Feng, H., Li, H., et al. (2017). Two NHX-type transporters from Helianthus tuberosus improve the tolerance of rice to salinity and nutrient deficiency stress. https://core.ac.uk/download/323069413.pdf (accessed on 13th June 2025).

19. Borassi, C., Bringas, M., Capece, L., Estevez, J. M., et al. (2018). Filling the gaps to solve the extension puzzle. https://core.ac.uk/download/224999470.pdf (accessed on 13th June 2025).

20. Bölter, B., Chang, W. L., & Soll, J. (2012). The gateway to chloroplast: Re-defining the function of chloroplast receptor proteins. https://core.ac.uk/download/19418339.pdf (accessed on 13th June 2025).

21. Carraro, N., Clouse, R. M., Knöller, A. S., Spicer, R., et al. (2012). Diversification and expression of the PIN, AUX/LAX, and ABCB families of putative auxin transporters in Populus. https://core.ac.uk/download/46704094.pdf (accessed on 13th June 2025).

22. Casadebaig, P., Costes, E., Da Silva, D., Faivre, R., et al. (2017). Using numerical plant models and phenotypic correlation space to design achievable ideotypes. http://arxiv.org/abs/1603.03238 (accessed on 13th June 2025).

23. Chiurazzi, M., Lacombe, B., Leran, S., Martins Alves, P., et al. (2017). The nitrate transporter family protein LjNPF8.6 controls the N-fixing nodule activity. https://core.ac.uk/download/288641678.pdf (accessed on 13th June 2025).

24. Choirina, Y., Ramli, M., & Rinanto, Y. (2019). Conceptual understanding of high school students on Plantae. Journal of Education and Learning (EduLearn). ISSN: 2089-9823; e-ISSN: 2302-9277

25. Clark, D. F., Jorritsma, I. T. M., Lasch, P., Leadley, P., et al. (2001). Regeneration in gap models: Priority issues for studying forest responses to climate change. https://core.ac.uk/download/pdf/29302888.pdf (accessed on 13th June 2025).

26. Curwiel, P. H., & Wieringa, R. J. (1986). Final report on the farmer's aid in plant disease diagnoses. https://core.ac.uk/download/pdf/11487250.pdf (accessed on 13th June 2025).

27. Darrah, P. R., Fowler, A. C., & Roose, T. (2001). A mathematical model of plant nutrient uptake. https://core.ac.uk/download/pdf/1568332.pdf (accessed on 13th June 2025).

28. De Bruyne, M., De Milde, L., De Rycke, R., Dhondt, S., et al. (2016). Chloroplasts are central players in sugar-induced leaf growth. https://core.ac.uk/download/55849641.pdf (accessed on 13th June 2025).

29. De Guzman, M. E., Santiago, L. S., Schnitzer, S. A., Álvarez-Cansino, L., et al. (2016). Trade-offs between water transport capacity and drought resistance in Neotropical canopy liana and

tree species. https://core.ac.uk/download/213089133.pdf (accessed on 13th June 2025).

30. Drews, G. N., Meyerowitz, E. M., & Sieburth, L. E. (1998). Non-autonomy of AGAMOUS function in flower development: Use of a Cre/loxP method for mosaic analysis in Arabidopsis. https://core.ac.uk/download/4880764.pdf (accessed on 13th June 2025).
31. Easlon, H. M., Hanson, D. T., Juenger, T. E., McKay, J. K., et al. (2013). The physiological basis for genetic variation in water use efficiency and carbon isotope composition in Arabidopsis thaliana. https://core.ac.uk/download/323068410.pdf (accessed on 13th June 2025).
32. Fernie, A. R., Gisel, A., Klemens, P. A. W., Ludewig, F., et al. (2020). Metabolic profiles of six African cultivars of cassava (Manihot esculenta Crantz) highlight bottlenecks of root yield. https://core.ac.uk/download/288634044.*pdf (access*ed on 13th June 2025).
33. Furtado, B. U., Gołebiewski, M., Hrynkiewicz, K., Hulisz, P., et al. (2019). Bacterial and fungal endophytic microbiomes of Salicornia europaea. https://core.ac.uk/download/232128542.pdf (accessed on 13th June 2025).
34. Gandiwa, E., *& Kativu, S. (2009)*. Influence of fire frequency on Colophospermum mopane and Combretum apiculatum woodland structure and composition in northern Gonarezhou National Park, Zimbabwe. https://core.ac.uk/download/pdf/29246901.*pdf (accessed on* 13th June 2025).
35. Goetghebeur, P., Hesse, M., Jiménez Rodríguez, F., Neinhuis, C., et al. (2010). Verhuellia is a segregate lineage in Piperaceae: More evidence from flower, fruit and pollen morphology, anatomy and development. https://core.*ac.uk/download/5583*8677.pdf (accessed on 13th June 2025).
36. Gollotte, A., Heggo, A., Hodge, A., Liu, A., Manjunath, A., Mollier, A., Mozafar, A., et al. (2008). *Agronomic Management* of I*ndigenous Mycorrhiza*s. https://core.ac.uk/download/62455716.pdf (accessed on 13th June 2025).
37. Gosselin, F. P. (2019). Mechanics of a plant in fluid flow. http://arxiv.org/abs/1905.08055 (accessed on 13th June 2025).
38. Goulao, L. F., & Oliveira, C. M. (2008). Ce*ll wall mo*difications during fruit ripening: when a fruit is not the fruit. Trends in Food Science and Technology, 19(1), 4-25. https://doi.org/10.1016/j.tifs.2007.07.002(accessed on 13th June 2025).
39. Gripenberg, S., Malhi, Y., Riutta, T., Salinas, N., et al. (2020). Changes in oak (Quercus robur) photo*synthesis after winter moth (Operophtera bruma*ta) herbivory are not explained by changes in chemical or structural leaf traits. https://core.ac.uk/download/286712731.pdf (accessed on 13th June 2025).
40. Heisler, M. G. B., Meyerowitz, E. M., Ohno, C. K., Reddy, G. V., et al. (2004). The Arabidopsis JAGGED gene encodes a zinc finger protein that promotes leaf tissue development. https://core.ac.uk/download/4879564.pdf (accessed on 13th June 2025).
41. Hirai, M., Kuroda, S., Murakami, M., Okamura, Y., et al. (2019). Interspecific differences in the larval performance of *Pieris butter*flies (Lepidoptera: Pieridae) are as*sociated with diffe*rences in the glucosinolate profiles of host plants. (accessed on 13th June 2025).
42. Karlsson, M. (2003). Molecular factors involved in the formation of secondary vascular tissues and lignification in higher plants. https://core.ac.uk/dow*nload/11694*622.pdf (accessed on 13th June 2025).

43. Konuma, A., Purvis, A., Worley, A. C., Worley, A. C., Worley, A. C., de Lacerda, A. E. B., Blundell, A. G., et al. (2011). Ecological Implications of a Flower Size/Number Trade-Off in Tropical Forest Trees. https://core.ac.uk/download/*177417*.pdf (accessed on 13th June 2025).

44. Kusuma Ningrum, F. (2013). The Theoretical Feasibility of Biology Worksheet Oriented by Active Learning with Mind Mapping Strategies. https://core.ac.uk/download/482199558.pdf (accessed on 13th June 2025).

45. Lamy, E., Baptista, E. S., Capela e Silva, F., Ferreira, A., et al. (2012). The Influence of Oral Environment on Diet Choices in Goats: A Focus on Saliva Protein Composition. In book: Goats: Habitat, Breeding and ManagementChapt*er: The Influence of oral environment on diet choices in goats: a focus on saliva* protein compositionPublisher: Nova Science Publishers IncEditors: Garrote DE, Arede GJ accessed on *13th June 2025).*

46. *Leung, S. C. S. (2003). Functional Analysis of Plant Mei2-Like Proteins: A Thesis Presen*ted in Partial Fulfilment of The Requirements for The Degree of Master of Science in Biochemistry *at Massey University, Palmerston North, New Zealand. https://core.ac.uk/download/148645771. pdf (accessed on 13th June 2025).*

47. *Lima, V. F. G. A. P., & Menezes, N. L. de. (2009). Morpho-anatomical analysis of the rhiz*ome in species of Scleria Berg. (Cyperaceae) from Serra do Cipó (MG). https://core.ac.uk/download/pdf/37458519.pdf (accessed on 13th June 2025).

48. Lukac, M., Pensa, M., & Schiller, *G. (201*1). Tree species' tolerance to water stress, salinity and fire. https://core.ac.uk/download/107821.pdf (accessed on 13th June 2025).

49. N/A. (2021). Genetic and Morphological Variation in Tropical and Temperate Plant Species. https://core.ac.uk/download/478139479.pdf (accessed on 13th June 2025).

50. Oesterlin, E*milie, "Botanically Inclined: Elementary Students' Knowledge of Plant Struct*ures and Their Functions Captured in Drawings" (2022). Electronic Theses and Dissertations. 3620. https://digitalcommons.library.umaine.edu/etd/3620(accessed on 13th June 2025).

51. Phelps-Durr, T., & Rueschhoff, G. J. (2024). Generating CRISPR mutations in the ASYMMETRIC LEAVES1 gene of Arabidopsis. https://core.ac.uk/download/616440375.pdf (accessed on 13th June 2025).

52. Published by Intelektual Pustaka Media Utama (IPMU) in collaboration with th*e Institute* of Advanced Engineering and Science (IAES).(accessed on 13th June 2025).

53. Riechmann, J. L. (2002). Transcriptional regulation: A genomic overview. https://core.ac.uk/download/4869897.pdf (accessed on 13th June 2025).

54. Schrader, J. (2003). *Developmental Biology of Wood Formation.* Publisher: Swedish University of Agricultural Sciences. https://core.ac.uk/download/11694666.pdf (accessed on 13th June 2025).

55. Soares, A. S., Foyer, C., Olmos, E., Arrabaç, M., Driscoll, S. P., & Pellny, T. K. (2008). *Incident Light Orientation Lets C4 Monocotyledonous Leaves Make Light Work Differently.* https://core.ac.uk/download/pdf/287717.pdf (accessed on 13th June 2025).

56. Stewart, J. R., 2010. Understanding evolutionary processes during past Quaternary climatic cycles: Can it be applied to the future? In: Grant, F., Mergeay, J., Santamaria, L., Young, J. and Watt, A.D., eds. Evolution and Biodiversity: The evolutionary basis of biodiversity and its potential for adaptation to global change. Report of an e. Mallorca, Spain: Centre for Ecology & Hydrology, INBO, IMEDA, CSIC-UIB.(accessed on 13th June 2025).

57. Van, B., Hodgman, C., Shi, C. L., De Cotte, B., De Smet, I., Vu, L. D., Butenko, M. A., et al. (2015). Modulation of *Arabidopsis* and monocot root architecture by CLAVATA3/EMBRYO SURROUNDING REGION 26 peptide. https://core.ac.uk/download/55843397.pdf (accessed on 13th June 2025).

58. Wyse, Lisa, "Meta-analysis of male and female function in plants with contrasting architecture and mating systems" (2014). Senior Honors Theses & Projects. 403. https://commons.emich.edu/honors/403

ANIMAL STRUCTURE AND FUNCTION

8

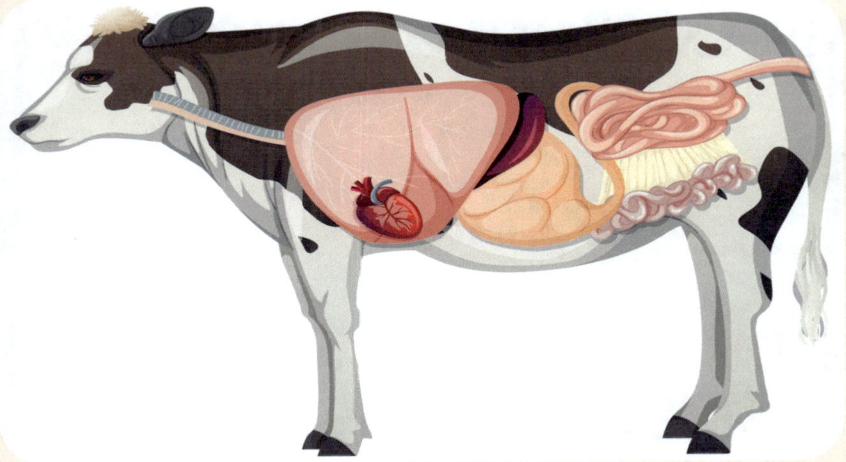

LEARNING OBJECTIVES

After studying this chapter, you will be able to:

1. List the levels of biological organization;
2. Explain about skeletal system;
3. Discuss about muscular system;
4. Describe the nervous system;
5. Explore the circulatory system;
6. Focus on the respiratory system.

Introduction

The study of animal structure and function reveals crucial details regarding the diverse adaptations required for survival in a range of disparate environments. Animals have a diverse set of anatomical and physiological features that allow for survival in specific ecological niches. From the minute details of the bird's wing to the complex systems regulating mammalian homeostasis, each aspect of animal biology reveals the intimate association between form and function. Understanding these interactions not only enhances our appreciation of biodiversity but also highlights the evolutionary pressures that have led to these traits. In addition, the examination of structures across organizational levels—cellular to systemic—is informative regarding how intricate mechanisms interact to support life. Such studies highlight comparative analysis among disparate species as a means to reveal patterns that cross the individual taxa, strengthening further the interconnectedness of life on this earth. In comparative biology, the kidney is a time-honored example of structure equaling function across disparate groups of animals. Each vertebrate group has special accommodations in renal functioning, based on environment, way of life, and evolutionary relationships. For instance, investigative studies of reabsorption of proteins in amphibian renal proximal tubule (PT) is fascinating to contrast with related events that take place in mammals and man. Such research provides valuable information regarding the efficacy and processes of renal function, , such as traits , such as epithelial cell absorption capacities and molecular mechanisms that control protein endocytosis. Through the use of sophisticated methods, , such as immunohistochemistry and confocal microscopy, researchers can see these mechanisms in operation, further explaining the physiological reactions that are essential to the maintenance of homeostasis in a dynamic environment.

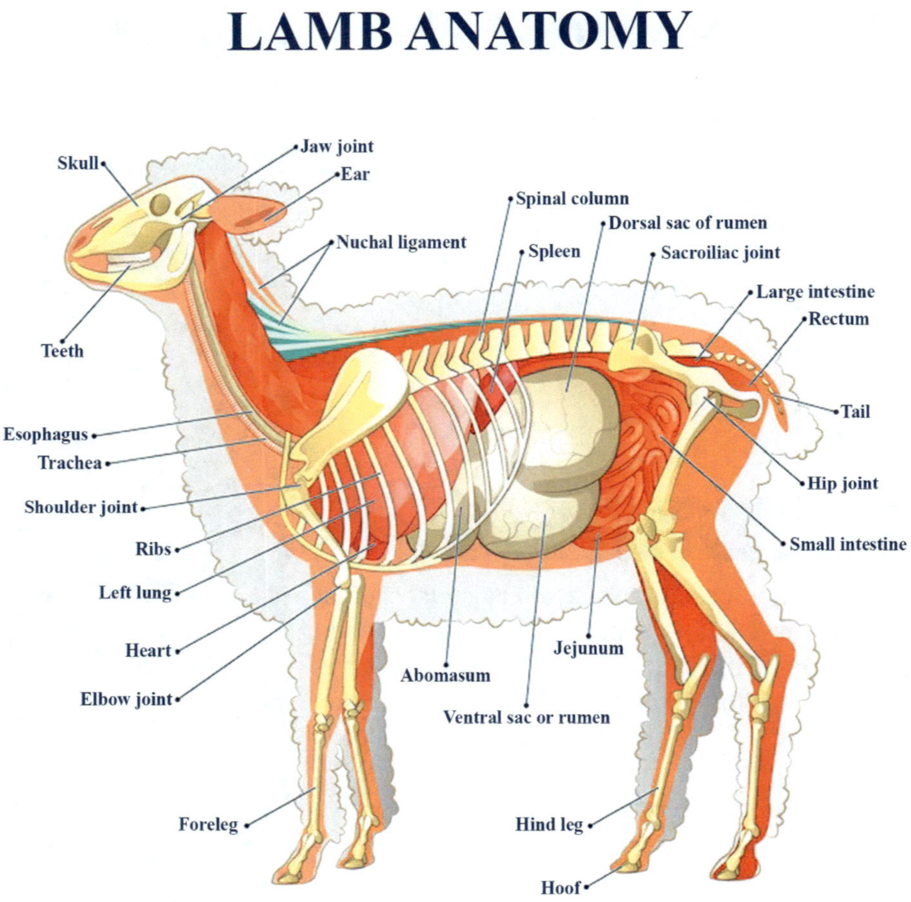

Figure 8.1. Animal structure and function.

Source: https://agristudies.co.za/wp-content/uploads/2023/06/lamb-anatomy-illustration-labeled-educational-inner-organ-structure_1995-533.jpg.

This comparative approach demonstrates not just functional congruence but renal physiological evolutionary adaptations. The application of knowledge about animal structure and function extends beyond tidy biology; it has seminal implications for applied fields , such as agriculture and environmental stewardship. In agricultural systems, an understanding of the structural and functional nuances of animal physiology can inform sustainable practice that optimizes productivity while minimizing ecological footprint. Research emphasizes the interaction of different systems of production, be it crop-based, livestock-based, or integrated, in a broad web of environmental drivers. Every system contains different degrees of potential for sustainability, as observed through modeling attempts that simulate these intricate interactions. The outcomes of such research emphasize the need to adopt a systems-based approach in tackling the concerns of modern-day agriculture. By integrating such knowledge of animal structure and function in these situations, we may open the door to intelligent practice and its facilitation of ecological and economic sustainability in the future (Figure 8.1).

Definition of Animal Structure and Function

Understanding the term animal structure and function lies at the foundation of the field of biology. Animal structure refers to the structured arrangement of anatomical structures, tissues, organs, and systems, which together constitute an organized organism. Animal function, conversely, pertains to the biological activities that these structures perform to maintain life and reproduce, , such as moving, respiration, and digestion. This relationship between form and function is governed by the laws of evolution because anatomical adaptation will normally have a definite physiological purpose. For instance, the flowing form of a fish allows for good movement within water, a classic case of how structural alteration has direct counterparts in functional efficiency. This fine balance between both of these elements creates a dynamic matrix that allows organisms to successfully relate to their surroundings. The study of animal structure and function is also aided through an examination of individual biological processes and their underlying mechanisms. In diseases , such as cancer cachexia, there is a definite disturbance of normal muscular function that illustrates how structural abnormality can lead to systemic dysfunction. It has been illustrated in studies that the disturbance of signaling pathways, e.g., the NF-kB and MAPK pathways, directly impacts muscle mass and functional integrity. In cachectic mice, there are drastic elevations in protein oxidation and ubiquitin-proteasome activity that cause large-scale muscle degradation and hence prevent movement and overall health. Through the study of these intricate connections, researchers can pioneer treatment methods for restoring normal structural and functional patterns in impaired individuals and hence bridging the gap between anatomy and physiology in a clinical scenario (Bal et al., 2014). Moreover, the philosophical meaning of structure and function, particularly in reference to a biological organism, makes it understandable how we conceptualize the nature of living organisms. Aristotle's hylomorphic conception, where he believed that living organisms are made of matter and form, forms a formidable theoretical framework for the study of the biology of animals through development. This perspective emphasizes that animal shape is not a mere collection of body pieces but is ontologically connected with the processes that result in the course of the organism's development and interaction with its environment. By employing such an ontological model, scientists are better able to grasp the complexity and interdependence of biological objects. As more progress is made in the field of developmental biology, these observations serve to strengthen the conceptual richness required to model the nature of organisms effectively, thereby increasing our understanding in theoretical and practical terms of animal structure and function (Austin et al., 2017).

Importance of Studying Animal Biology

Animal biology knowledge is fundamental to comprehend the mechanisms controlling the structure and function of multicellular organisms. In animal biology, scientists uncover the intricate interactions between anatomical features and physiological processes. The comprehension of interactions gives scientists the capacity to explore the survival strategies used by different species in living in their environment, and guides fields , such as ecology, conservation, and veterinary sciences.

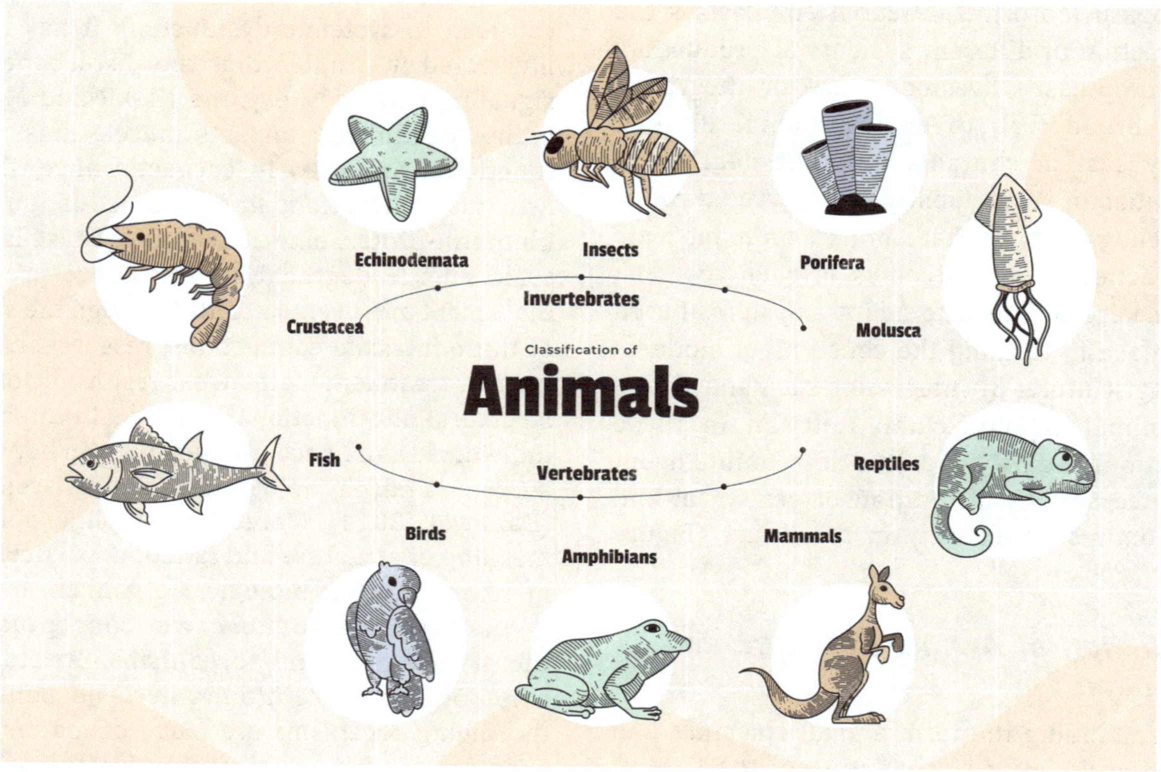

Figure 8.2. Animal biology.

Source: https://img.freepik.com/free-vector/hand-drawn-classification-animals-infographic_23-2149414674.jpg?semt=ais_hybrid&w=740.

For instance, the conclusions that can be drawn from the study of evolutionary adaptations can account for the impacts of climate change on species survival and community composition. The Quaternary period alteration in ecological processes, according to climatic influences research, referred to the two-way role of evolutionary pressures in the development of biodiversity and species extinction (Stewart, 2010). By being aware of such interactions, we can increasingly appreciate the worth of preserving biological diversity in current ecosystems. Furthermore, being informed about animal biology informs our understanding of health and disease in a more complete way, making significant contributions to developments in medicine and agriculture. The parallels in human and other animals' physiology allow researchers to develop medical therapies and interventions tailored to the treatment of most diseases. For instance, research on animal models has broad impacts on biology studies, particularly in genetics, immunology, and disease spread. This study also extends to marine ecosystems, where marine biotechnology is increasingly being studied, which promotes new applications in health and environmental stewardship. Collaborative efforts between industries and research organizations can bring about innovation in this area, providing access to research and discovery funds (Seaman et al., 2001). It is from these discoveries that the field of animal biology continues to be relevant for scientific research as well as practical application in numerous areas. The study of animal biology also fosters appreciation of the ethical and moral implications of our association with other living organisms. As people increasingly face problems concerning biodiversity loss and habitat degradation, knowledge of animals' biological and ecological functions is essential. Not only will this information guide conservation, but it also enhances the ethical use of animals in research and agriculture. Training in animal biology gives future decision-makers and leaders the necessary tools to promote accountable stewardship of natural resources. By building respect for the complex web of life that animals represent, society can move toward more sustainable ways that pay homage to the interdependency of all living beings. Lastly, the study of animal biology is a beginning point for the establishment of viable relationships between humans and the natural world, which guides the preservation and protection of the planet's varied ecosystems (Figure 8.2).

Overview of Key Concepts

A detailed description of the structure and function of animals necessitates an examination of the prevailing ideas that shape the nature of biological systems. Chief among these is the idea of interdependence among various physiological systems as the basis for the integrated operation of an organism. For instance, the coordination between the skeletal and muscular systems shows structure supporting function; the organization of bones and muscles to allow movement and support reflects sophisticated design created to meet specific environmental needs. Furthermore, the interplay among these systems is complex and requires sound neural control, reflecting the complexity within biological structures. The study of these relationships and the evolutionary significance thereof reveals not just the intricacies of anatomy but also the mechanisms of adaptation that have developed over time. Homeostasis, or the biological mechanism employed by animals to preserve internal stability in the face of changing environments, is yet another concept of supreme significance. This controlling process plays a vital role in sustaining life because it facilitates the sustenance of desirable conditions required for cellular activity. For instance, thermoregulation makes it possible to maintain body temperature within a relatively narrow range, which is a necessity for metabolic functions. In this context, organs like mammals' integumentary system play a significant role. The skin provides a barrier and facilitates the exchange of heat through various physiological adaptations. Also, understanding homeostasis is directly related to the evolutionary perspective in animal physiology because over generations, organisms' adaptations to environmental pressures have been developing such regulatory processes, indicating the dynamic interaction of structure and function in an attempt at survival. Lastly, the incorporation of evolutionary biology as a branch under the paradigm of animal structure and function creates an even greater appreciation for biodiversity. Evolutionary theory accounts for how structural properties arise through mechanisms , such as natural selection, allowing species to adapt to their ecological niches. An example of this is the evolution of specialized respiratory systems for aquatic animals , such as gills, which demonstrates the interconnectedness of structure with an organism's environment and lifestyle. Such an interaction is not only functional but also reflects the genetic processes that are involved, emphasizing the role of heritable variation in shaping characters in populations. Integrating evolutionary theory with the study of animal form and function, researchers can achieve a better appreciation of biological diversity and the intricate tapestry of life on Earth (Arbib et al., 2018; Hanisch & Eirdosh, 2020).

8.1. Levels of Biological Organization

Learning Outcome

1. Cellular level and its significance;
2. Tissue types and their functions;
3. Organ systems and their interconnections.

In thinking about the hierarchy of biological organization, one starts at the most general unit: the cell. Cells are the smallest structural and functional units of life and show a remarkable degree of complexity and specialization. Different cells, , such as muscle, nerve, and epithelial cells, all perform different functions that are directed towards the welfare and efficiency of an organism. This cellular complexity highlights the intricate interaction between structure and function, e.g., muscle cells' extendable shape permits contraction, and nerve cells' branching shape enhances their signal transmission. When cells aggregate, they form tissues, which function together to carry out special functions, with an enhanced level of organization required for the more sophisticated processes of multicellular organisms.

Appreciation of this first order of organization serves as the basis for understanding higher biological organization. As the organization progresses from tissues to organs, more and more apparent becomes the hierarchical organization. Organs composed of varying types of tissue work together to perform specific functions required for the survival of the organism. For example, the heart, which is of muscle, connective, and epithelial tissue origin, successfully circulates blood all over the circulatory system, illustrating how specialized tissues integrate to achieve a shared goal. Additionally, the coordination of multiple organs into organ systems, , such as the cardiovascular system, illustrates dynamic interrelation among structures.

Each organ system assists in homeostasis and overall organism function, illustrating the complex scheme present within biological organization. By appreciating these relationships, scientists can more easily be impressed with the evolutionary and developmental mechanisms that have built the anatomy and physiology of animals in various taxa. The study of levels of biological organization also teaches us about genetic and evolutionary conservation. Evolutionary developmental biology (evo-devo) studies determine how morphological variation in animals is not necessarily coordinated with genetic variation, suggesting that core genetic building blocks are retained across species while allowing for great variation in phenotype (Becker et al., 2005).

Conservation suggests that evolutionary adaptations tend to arise chiefly due to gene regulation modifications rather than the genetic code itself. Understanding these transformations is at the heart of deciphering the organization across species and the survival advantages that have emerged through time. The character of biological organization is thus more than anatomical structure to encompass gene blueprints directing developmental pathways. They disclose the intimate interconnection of structure, function, and evolution in the realm of animal biology and enhance our understanding of life's diverse manifestations (Figure 8.3).

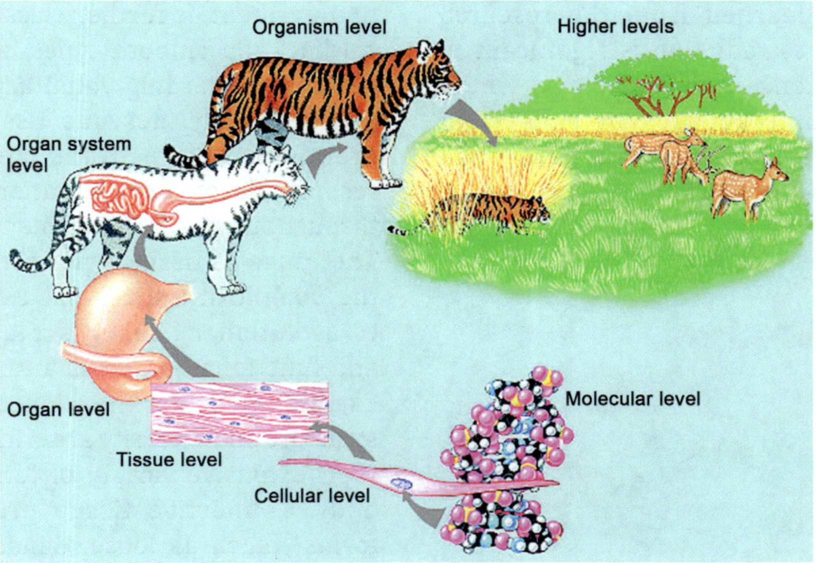

Figure 8.3. Levels of organization.

Source: https://michelleburden.weebly.com/uploads/1/7/0/0/17006738/4654434_orig.jpg?367.

8.1.1. Cellular Level and Its Significance

Understanding at the cellular level is most critical to learning about the overall structure and function of animals. Cells are the structural and functional units of life, and they collaborate to manage the complex biological systems to facilitate growth, development, and homeostasis. All cell types possess unique structures and functions that affect the general physiology of the organism. For instance, eukaryotic cells, to which animal cells belong, are characterized by the presence of organelles , such as the nucleus and mitochondria. These enable basic processes , such as the production of energy and control of genetics. The significance of the cellular level is even more emphasized when one is looking at specialized cells that specialize in specific functions in tissues, thus making the organism function at its best and respond accordingly to external stimuli. Cellular response to injury and stress is also a critical field of study that examines the need for the cellular level of animal biology. In models of spinal cord injury, for example, there are significant molecular responses that indicate the versatility and adaptability of cells. Findings from the aneurysm clip impact-compression spinal cord injury model show discrete global changes in mRNA levels over time that allow for a variety of biological processes , such as blood coagulation and inflammatory reactions (Boutros et al., 2013). These reactions engage cellular mechanisms that control healing and recovery in the damaged tissues and demonstrate an awe-inspiring coordination of cellular behavior both during acute and chronic phases of injury. This flexibility is a clear illustration of the cellular levels important role in guaranteeing the body's functional integrity irrespective of environmental conditions. Additionally, the cellular level is also significantly important in understanding the impact of radiation exposure on animals, clearly distinguishing between the various forms of radiation and their influences on living organisms. Studies have indicated that high LET radiations possess a greater potential to produce more extensive biochemical damage at cellular and organismal levels than low LET radiations (Wiley et al., 1972). These differ in the extent of cellular damage and repair capabilities, emphasizing the significance of cellular structure in modulating biological response.

The information learned from the research on cell responses to radiation is significant in environmental science and toxicology, where hardness and vulnerabilities of cells form the basis of understanding animal health and stability analysis of ecosystems (Figure 8.4).

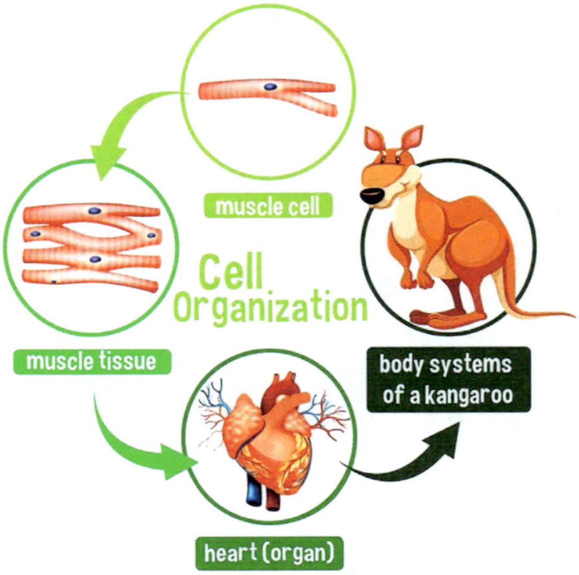

Figure 8.4. Cellular level.

Source: https://media.geeksforgeeks.org/wp-content/uploads/20221223151937/Structural-organization-in-Animals.webp.

8.1.2. Tissue Types and Their Functions

Forms of tissue are the fundamental building blocks of animal structure, each involved in a specialized function towards the overall physiological harmony of the organism. Tissues are generally classified into four major types: epithelial, connective, muscle, and nervous tissues. Epithelial tissue is responsible for the development of protective linings and boundaries and controls permeability and selective absorption. Connective tissue provides structural support, holding tissues together, and aids in the transportation of nutrients. Muscle tissue, which is significant to movement, is further classified into skeletal, cardiac, and smooth types, each with distinct functions regarding voluntary and involuntary action. Finally, nervous tissue, consisting of predominantly neurons and glial cells, is responsible for communication and coordination throughout the body through sending signals. The range of tissue types serves to emphasize the complexity of animal existence and refer to evolutionary changes that ensure survival and functionality across a range of conditions. Examining further the connective tissues, their structure and diversity are critical in determining their extensive variety of functions within the body. Connective tissue presents numerous forms , such as loose, dense, adipose, blood, and bone, and all have a unique contribution towards the structural integrity and metabolic processes of the body. For instance, loose connective tissue gives support and elasticity to nearby structures, while dense connective tissue is responsible for tensile strength and elasticity, helpful in structures like tendons and ligaments. In addition, specialized connective tissues like blood are not only responsible for transport but also for immune responses and homeostasis. The conservation of genetic coding in various phyla also supports the notion that these differentiated tissues have evolved to perform sophisticated functions, testifying to the refined choreography of form and function in animal biology. The knowledge of tissue types and functions may also provide insight into evolutionary pressures that sculpt animal bodies. Research in evolutionary developmental biology (evo-devo) shows that the vast disparity in animal form has not always been accompanied by extensive genetic change but rather is the outcome of alterations in gene expression and regulation (Bode et al., 1992). This perspective highlights the mechanism through which tissues evolve and develop to acquire new functions and stresses the co-option idea, according to which pre-existing regulatory networks are being repurposed. For example, the diminished tissue organization of Hydra animals illustrates an intriguing interaction between genomic complexity and functional differentiation in that specific proteins are restricted to specific cell types and represent a tailor-made biological function above general activity (Bode et al., 1992). Empirical explanation for these mechanisms not only enhances our understanding of tissue function but also illuminates evolutionary strategies that form the foundation of morphological diversity seen within the animal kingdom (Figure 8.5).

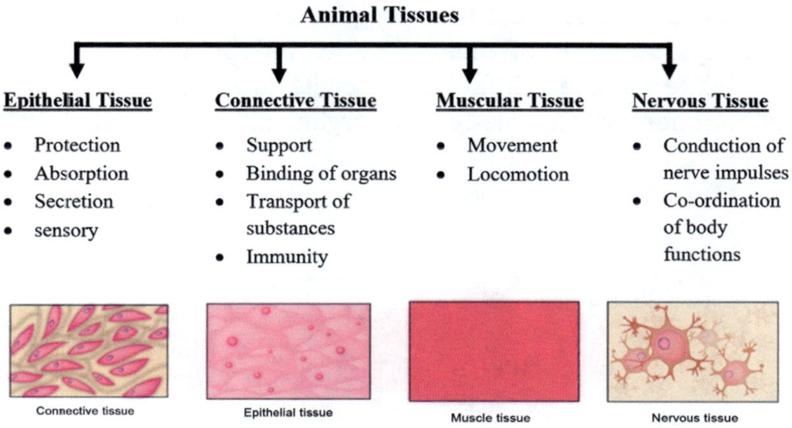

Figure 8.5. Tissue types and their functions.

Source: https://www.w3schools.blog/wp-content/uploads/2020/01/Pasted-into-Animal-Tissues-8-1.png.

8.1.3. Organ Systems and Their Interconnections

The complexity of animal physiology is in the intricate interaction between various organ systems, each functioning in a specialized manner but, in the process, relying on one another in a bid to attain homeostasis and overall health. The interdependence of these systems—, such as the muscular, circulatory, and nervous systems—illustrates the concept of networked functionality. For instance, the muscular system needs to be constantly supplied with oxygenated blood from the circulatory system regulated by control signals from the nervous system. This reciprocal relationship is a classic instance of disturbance in one system flowing into others and impacting the capacity of organisms to operate effectively. Identification of these interactions is significant to fields , such as medicine and biology, as it encompasses an entire assessment of how physiological health may be optimized through the cooperative interaction of related systems. There has been a great emphasis placed on understanding these organ system relationships from both evolutionary and developmental perspectives. Current studies of animals , such as *Neanthes arenaceodentata* indicate that their nervous system is complex very early in development, suggesting a deep evolutionary history with the rest of the annelids. Complexity not only benefits locomotion and foraging behavior but also points to the significance of the nervous system when working together with other body systems. Moreover, research shows that the variations in neural anatomy are the cause of variability in the functional implications across different species within a single taxonomic group. By comparative analysis, biologists can acquire knowledge on the development of organ systems, which have evolved to perform adaptive functions, thus increasing understanding of animal structure and function in general (Jacobs et al., 2010).

Furthermore, pedagogic studies on students' perceptions of their internal anatomical features highlight major aspects of the knowledge about organ systems and how they are interlinked. Through requesting students to sketch what they believe is contained within their bodies, researchers understand the knowledge and misconceptions held about anatomical features among various age groups. Analysis of these drawings reveals that while younger children are likely to possess rudimentary knowledge, older students are more likely to exhibit a wider understanding of the organ system and function. This pattern of developmental learning emphasizes the importance of instruction in increasing consciousness about the human body as a system. These discoveries enhance pedagogical models to ensure that coming generations have a sound, firm foundation of how independent organ systems contribute to the total functioning and well-being of the organism (Reiss et al., 2001) (Figure 8.6).

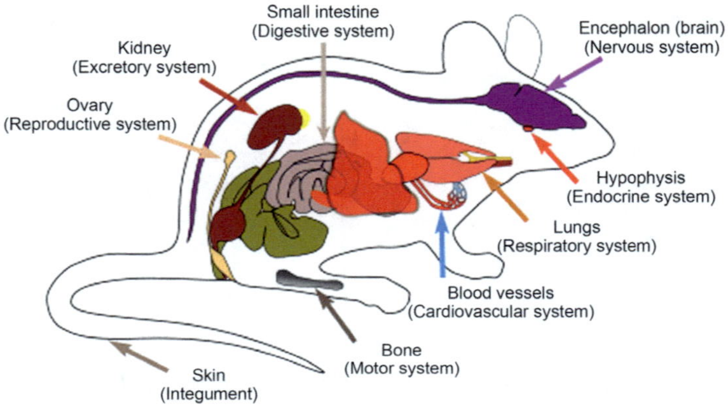

Figure 8.6. Animal organ systems and their interconnections.

Source: https://mmegias.webs.uvigo.es/02-english/2-organos-a/imagenes/organos.png.

Test Your Knowledge

1. What is the cellular level and its significance?
2. What are the tissue types and their functions?

8.2. Skeletal System

Learning Outcome

1. Structure of bones and cartilage;
2. Functions of the skeletal system;
3. Common disorders affecting the skeletal system.

The skeletal system is a simple framework that not only gives rise to support for the structure of an organism but also impacts many physiological functions. Comprised mainly of bone, cartilage, and ligaments, this complex system is responsible for the protection of vital organs, aiding in movement, and in the storage of minerals and blood cell formation. Bones, in their dynamic nature, remodel themselves continuously, and this process is regulated by various pathways of signaling and hormonal control. This remodeling can be pathologically compromised by states , such as cancer-induced cachexia, in which muscle wasting and systemic inflammation compromise the normal physiological homeostasis. Under such conditions, the intricate interplay between skeletal integrity and muscle function reflects the significance of this system within the grand scheme of animal structure and function (Bal et al., 2014). This interaction between the musculoskeletal system and muscle tissue is also observed in how they both cooperate to facilitate locomotion. The muscles are linked to the bones via tendons such that movement results from the contraction and relaxation of muscle fibers. This collaboration of biomechanics allows for numerous activities ranging from simple reflexes to coordinated compound movements to ensure survival in various environmental contexts. Additionally, the skeletal make-up varies between species, and this is indicative of adaptations to occupy different ecological niches. In vertebrates, bone tissue arrangement and density are variable in response to environmental stress and lifestyle requirements, and this implies an evolutionary component to skeletal formation that is essential to the understanding of animal function.

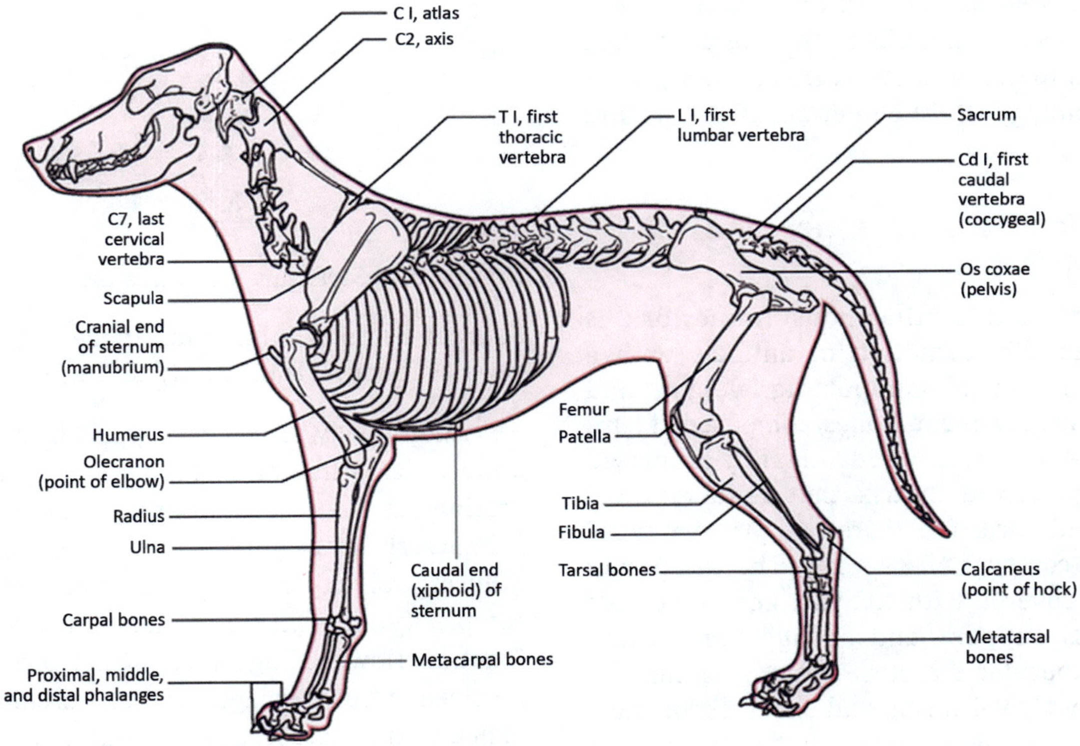

Figure 8.7. Animal skeletal system.

Source: https://veteriankey.com/wp-content/uploads/2016/07/B9780702029387000038_gr1.jpg.

Consequently, the skeletal system not only provides structural support but also assumes a substantial functional role in the facilitation of movement and interaction with the environment. Furthermore, advances in recent molecular biology have enhanced the understanding of skeletal muscle diseases and their implications for the skeleton's health. Studies have defined important proteins , such as amphiphysin 2, which is critical to the muscle's integrity and in the pathology of centronuclear myopathy and myotonic dystrophy. These conditions tend to emphasize structural and functional defects of skeletal muscle, pointing out the tenuous balance necessary for muscular and bony health. Disruptions at the molecular level were known to have serious consequences for muscle and bone tissue as well, affecting the fitness and life expectancy of animals. Investigating such relationships improves understanding of the physical support systems provided by the skeletal system and also for the maintenance of systemic health, emphasizing the interdependent relationship between structure and function in animal biology (Buj-Bello et al., 2013) (Figure 8.7).

8.2.1. Structure of Bones and Cartilage

The bony and cartilaginous framework is crucial to the formation of animals with a permanent shape, supporting weight, and facilitating movement. Bones, composed of hard connective tissue, are predominantly composed of a mineralized matrix that renders them strong and hard. There are two primary types of bone: compact and spongy. The exterior of bones is covered with compact bone that adds structural support, and spongy bone inside has a trabecular structure system within that helps in weight-bearing and shock absorption. This intricate arrangement not only protects vital organs but is also a reservoir of minerals , such as calcium and phosphorus, whose myriad metabolic functions in the organism are brought out by the dynamic interplay between the bone's structure and its various functions in animal physiology. Cartilage, however, possesses a unique but equally vital function in the skeletal arrangement. Composed of a tough matrix with elastin and collagen fibers, cartilage provides shock absorption as well as stress resistance in mechanically stressed joints, ribcage, and the ear. Cartilage is avascular and has no blood supply and relies on diffusion to get its nutrients. Some of these are hyaline, elastic, and fibrocartilage cartilages, and they are special functional adaptations. For instance, hyaline cartilage found in growth plates of long bones facilitates frictionless articulation at joints.

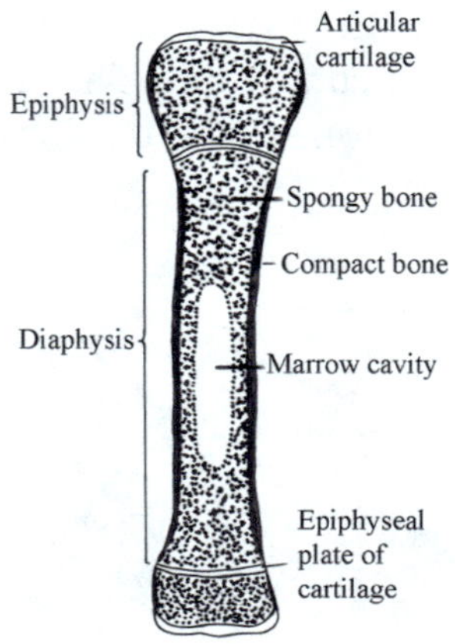

Figure 8.8. Structure of bones and cartilage.

Source: https://animalbiosciences.uoguelph.ca/~swatland/HTML10234/LEC26/PS109.gif.

Additionally, structural support by cartilage allows flexibility and elasticity required in motion, as well as stability of the skeletal framework as a whole. The distinction and differing roles of cartilage and bone show the complexity and efficiency of animal structure and function. New findings have further proved the involvement of cartilage and bone architecture in health and disease. Osteochondromas, benign bone tumors often seen to afflict antlers and zygomatic bones of deer, are a perfect example of how modification in disease results in the altered functioning of the skeleton (Flohr et al., 2017). Furthermore, impacts of cartilage faulty structure are witnessed in poultry, as the environmental conditions produce endemic lameness due to compromised cartilage health

(Stark et al., 2013). These findings emphasize that abnormal structural manifestations in bone and cartilage not only limit the physical functioning of the animal but also reflect broader physiological well-being, ultimately shaping longevity and quality of life. Hence, researching the structure and function of these components is invaluable for enhancing animal welfare and improving management practices across different species (Figure 8.8).

8.2.2. Functions of the Skeletal System

The skeletal system is directly involved in the overall functioning of animals, playing several important functions in addition to just structural support. One of its significant functions is to provide a framework that defines an organism's shape and allows for locomotion. Bones anchor muscles, allowing movement through their interaction with the muscular system. This interaction is particularly pronounced in higher vertebrates, where a more advanced muscular anatomy has evolved to enable various forms of movement. The skeletal system also protects vital internal organs by enclosing organs , such as the brain, heart, and lungs within bony protective cavities. This protective function becomes important for the survival of an organism because it delays the risk of damage from external harm and predation. In addition to providing structural support and protection, the skeletal system also plays an important role in hematopoiesis, the production of blood cells. This happens mainly within the red marrow of certain bones, which generates erythrocytes, leukocytes, and thrombocytes employed for transporting oxygen, fighting off infections, and clotting blood. The presence of this bone marrow reinforces the unification between the circulatory and skeletal systems. Aside from that, the skeletal system plays a vital role in controlling the balance of minerals within an organism, particularly calcium and phosphorus. The bones also serve as depots for these important minerals, releasing or taking them up to maintain vital physiological processes, including contraction of the muscles and nerve impulse conduction. Lastly, the skeleton is involved in body mechanics regulation and homeostasis maintenance through the regulation of metabolic processes. Various signaling molecules, including hormones parathyroid hormone and calcitonin, signal to bones to regulate mineral metabolism, hence keeping calcium in narrow ranges that are critical for physiology. In addition, the design of the skeletal system gives it the flexibility to conform to environmental demands; for instance, bones may be strengthened and densified according to mechanical load, as postulated by Wolff's law. This dynamic regulation is particularly relevant in pathologic states, where skeletal changes can affect overall health and cause such disorders as cancer cachexia, which can affect the strength of muscle and bone as seen in cachectic mouse models (Bal et al., 2014). The skeletal system, therefore, is not just a static structure but a dynamic structure that significantly contributes to the organism's functional capacity and overall health. Furthermore, understanding the genetic and molecular interactions involved in contributing to skeletal health, , such as those involving amphiphysin 2 in muscle disease, provides insight into the complexities of the skeletal system's function (Buj-Bello et al., 2013).

8.2.3. Common Disorders Affecting the Skeletal System

The individual and collective form and function of animals in bony form are open-ended to many potential disorders that can afflict mobility and health. A common disorder of this system involves osteoarthritis (OA), which consists of cartilage loss with subsequent bone remodeling. It is defined as a degenerative joint disease caused by such factors as progressing age, trauma, or excessive weight, causing inflammation and pain, thus limiting motion. The consequences of OA are not only restricted to individual joints but also extend to the overall health, lifestyle, and well-being of animals. Moreover, prevention and control measures, including weight control and physical therapy, are integral components in minimizing the impact of this long-term disease on the skeletal system. Another significant disorder that affects the skeletal framework is osteoporosis, which is defined by the loss of bone density and strength. It most commonly affects mature animals, leading to an increased susceptibility for fractures as well as other skeletal injuries. Osteoporosis is caused by a variety of factors, including bad diet, hormonal imbalance, and disease. While it most commonly affects elderly humans, there are a few breeds that can be predisposed to its development through genetics. Osteoporosis treatment is regularly coupled with nutritional therapy, , such as supplementation of vitamin D and calcium, and regular exercise for healthy bone density. Awareness of the necessity of a healthy skeletal system is the most important point to enhance the lifespan and well-being of the affected animals. Apart from degenerative disorders, inherited myopathy is one of the genetic diseases that emphasizes another essential aspect of the skeletal system's health. Canine Inherited Myopathy of Great Danes (IMGD) is a case in point, where mutations in genes like BIN1 lead to structural abnormalities in skeletal muscle that extend to skeletal function (Buj-Bello et al., 2013). Although presently classified as a muscle disease, its effects on the skeletal system are important because the muscle and skeleton are linked. These genetic conditions need ongoing research to explore targeted treatment, particularly through drug delivery technology directed at bone and muscle tissue (Alméciga-Díaz et al., 2017). Researchers can create improved treatments and improve results for affected species by discovering interrelated disorders within the skeletal system (Figure 8.9).

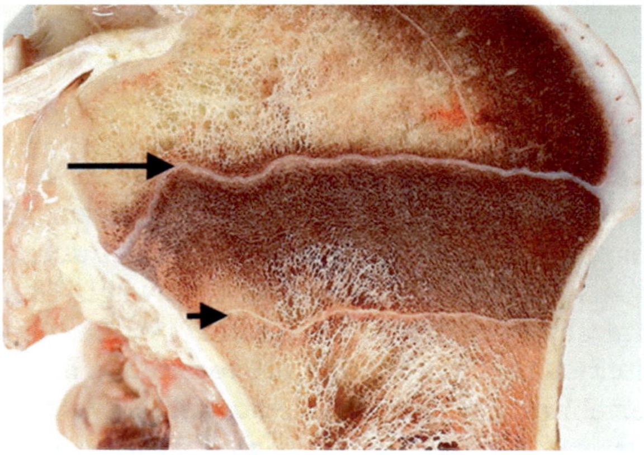

Figure 8.9. Animals can suffer from a wide variety of bone diseases.

Source: https://static.producer.com/wp-content/uploads/2023/01/04143040/39-1-col_animal-health-growth-arrest-lines-annotated-copy_Jen-Davies-Photo.jpg.

Test Your Knowledge

1. Illustrate the structure of bones and cartilage.
2. What are the functions of the skeletal system?

8.3. Muscular System

Learning Outcome

1. Types of muscle tissue: skeletal, cardiac, and smooth;
2. Mechanisms of muscle contraction;
3. Role of the muscular system in movement.

The muscular system has a pivotal role in the overall design and function of animal bodies to provide movement, aid in physiological functions, and assist with homeostasis. Composed of three large groups of muscle tissue, skeletal, smooth, and cardiac, the system demonstrates extreme specialization to fit multiple functions. Skeletal muscles, under voluntary control, give locomotion and posture support, while smooth muscles, found in internal organs, regulate involuntary movements like digestion and circulation. Cardiac muscle is exclusive to the heart, autonomously creating contractions required for circulating blood within the organism. The complex interaction between these muscle types demonstrates the complexity of animal structure and function, showing their significance to maintaining life and facilitating adaptation to environmental conditions. Adaptation of the muscular system is most clearly demonstrated through response to stimuli, as evidenced in the research of microgravity effects on musculoskeletal conditions. Extended periods of weightlessness, , such as those experienced through space travel, drastically alter the mechanical loading of the muscles and bones. This is a stimulus that produces adaptive responses and resultant functional impairments with time as the system adapts from normal use to a situation of disuse (NASA-SP-7011, 1974). These studies emphasize the importance of acknowledging the plasticity of the muscular systems because not only does it push the rehabilitation processes, but it also gives the researchers clues on what underlies the processes that determine muscle adaptation. The study of the effects of altered gravitational conditions helps researchers know how muscle physiology can guide medical interventions and improve muscle health in general. Compounding the intricacy of muscular disease and adaptation is the inherited basis of diseases , such as laminin-α2 deficient congenital muscular dystrophy (MDC1A). This disabling muscle-wasting disorder is the result of the deficiency of specific proteins central to the stability and function of muscle fibers (Meinen et al., 2005). The importance of such genetic conditions demonstrates the crossroads of animal structure and function and necessitates studies into possible therapies for improving muscle function. New methods, , such as the administration of mini-agrin to improve muscle integrity, hold potential in lessening the toxic consequences of dystrophy. As research persists, it continues to become apparent that understanding the genetic origins of muscular disorders is essential to the development of successful treatments and achieving optimal quality of life for individuals who have them, further demonstrating the interconnectedness of the muscular system with overall physiological wellness and resiliency (Figure 8.10).

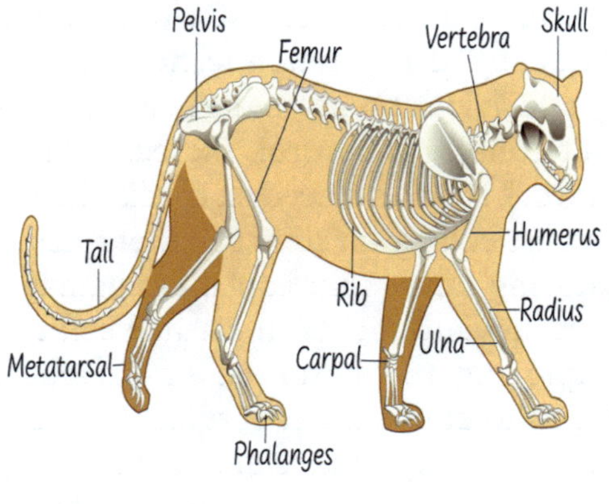

Figure 8.10. Canine and feline musculoskeletal system.

Source: https://vetwebsites.co.za/articles/images/Canine%20and%20feline%20musculoskeletal%20system.jpg.

8.3.1. Types of Muscle Tissue: Skeletal, Cardiac, and Smooth

The diversity of tissue types in animal muscles is important for various physiological functions, and they are classified into three main groups: skeletal, cardiac, and smooth muscle. Striated muscle with voluntary control enables the movements by attaching to bones using tendons. This muscle is extremely pliable, demonstrating the ability to grow with hypertrophy through exercise and strengthen the fibers while continuing to enhance the functioning of the overall system. This elasticity is necessary for locomotion and other bodily functions and enables organisms to be able to interact with their environment effectively. As skeletal muscle fibers can contract at a forceful yet speedy manner, they play an important role in actions , such as running, jumping, and lifting, and reflect a strong relationship between the structure of a muscle and an organism's functional requirements. Cardiac muscle, however, is only found in the heart and is responsible for the involuntary contractions that pump the blood around the body. Its characteristic striated appearance, similar to that of skeletal muscle, is accompanied by intercalated discs, which facilitate communication and regulated contractions among cells. Cardiac muscle is regulated by the autonomic nervous system (ANS), which lets it react to the needs of the body without conscious control. Increasing focus in recent work is being placed on cardiac health and repair, particularly on stem cell therapies aimed at restoring heart function after damage. For instance, advancements in cardiac cell therapy are exploring the potential of various stem cell types to supplement cardiac regeneration, indicating the importance of understanding this type of muscle in sickness and health (Barile et al., 2009). Smooth muscle is the third type, which plays a critical role in the function of inner organs , such as the gastrointestinal tract, blood vessels, and respiratory tract. Unlike skeletal and cardiac muscle, smooth muscle is unstriated and functions involuntarily, allowing for automatic regulation of body functions. Smooth muscle is characterized by slow, sustained contractions, which sustain processes like peristalsis during digestion and blood flow control via vasodilation and vasoconstriction. The function of smooth muscle in homeostasis is revealed in its responsiveness to hormones and neural stimuli, which integrate its contractions. Emerging treatment strategies aim at utilizing advances in smooth muscle biology, particularly for diseases affecting muscle function, highlighting its central role in overall animal structure and function (Brockington et al., 2010) (Figure 8.11).

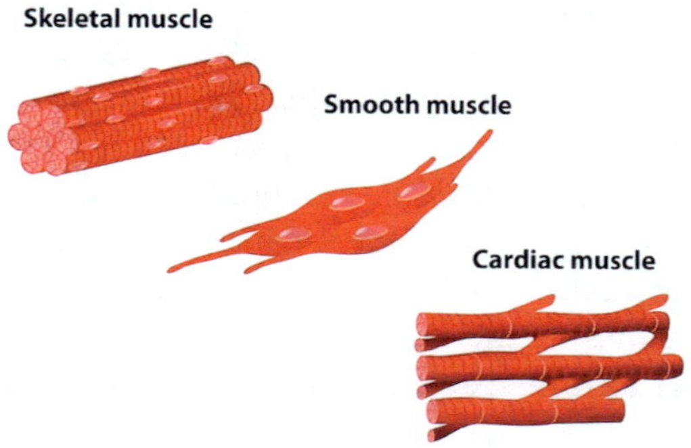

Figure 8.11. Types of muscle tissue.

Source: https://www.sciencelearn.org.nz/_next/image?url=https%3A%2F%2Fwww.datocms-assets.com%2F117510%2F1722387917-types-of-muscle-tissue20161111-16640-1c2x74o.jpg&w=3840&q=75&dpl=dpl_4n9pVfrF4H6vbtWLHqt8okXU1ahu.

8.3.2. Mechanisms of Muscle Contraction

Muscle contraction is a highly organized process that involves the cooperative action of numerous molecular and mechanical components. At the cellular level in muscle fibers, there are myofibrils composed of sarcomeres—the fundamental building units that produce muscle tension and contract. Contraction initiation begins when an action potential travels along a motor neuron, causing the excitation of the release of acetylcholine at the neuromuscular junction. This neurotransmitter, on receptors located on the muscle cell membrane, causes depolarization and ultimately leads to the release of calcium ions from the sarcoplasmic reticulum. This increase in calcium concentration leads to the binding of actin and myosin filaments and results in the sliding filament model of contraction, as the myosin heads bind with actin and pull it towards themselves, leading to the shortening of the muscle fiber. Knowledge of muscle contraction mechanisms also involves taking into account factors that affect the production of force, e.g., temperature and muscle architecture. Findings demonstrate that temperature can have a major impact on muscle functioning, as shown by studies on the stomatogastric system of lobsters, which reveal that muscle force production can decline with an increase in temperature but can be regulated by the presence of dopamine. This hormone amplifies the generation of muscle force with alterations in temperature, and it displays an adaptive mechanism that allows for organisms to maintain muscle function under varying conditions in the environment. Therefore, muscle contraction is not only dependent on internal characteristics but also on extrinsic biochemical factors that can amplify or suppress its efficiency, reflecting the complexity of muscle physiology under different environments (Burstein et al., 2013). Furthermore, muscle fiber architecture matters to muscle force-generating capacity. Length-force relations are dynamic, not fixed, and are a function of a variety of interacting factors like activation levels and history of the muscle. Potentiation and fatigue have impacts that vary their overall functioning during muscle fiber contractions. These effects make the appreciation of muscle contraction as a dynamic relationship between structure design and functional demands a necessity. In addition, the spatial patterning of muscle fibers, whether multi-articular or mono-articular, contributes to the distribution of forces throughout the muscle. The results of existing studies on muscle structure shed light on the fact that effective motor control relies on this intricate interaction, which showcases the character of muscular systems in generating precise movements and maintaining locomotor capabilities in varied animal species (Huijing et al., 1995) (Figure 8.12).

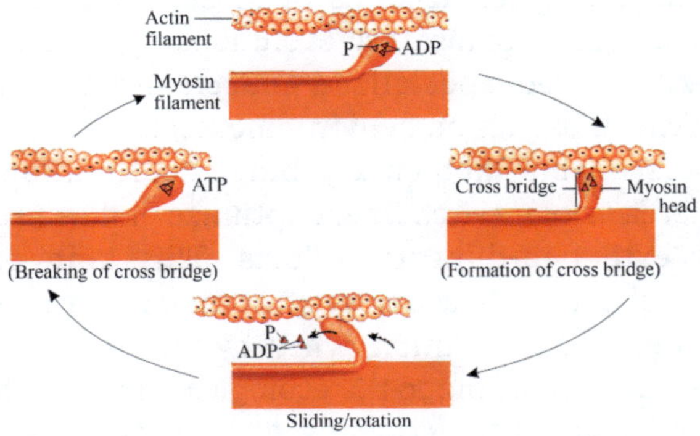

Figure 8.12. Mechanisms of muscle contraction.

Source: https://blogger.googleusercontent.com/img/b/R29vZ2xl/AVvXsEjCjtffbqzJ7qEBc8DtvL7RQWd90ZujApXNpHMbrVXFOe6hSHOFXuUnN3AUUwSM41NoDbjaN9lEsN8UQxrx80OJUmPH2wgeCg8rP_qnHp1lQtDjCo_VXw5BkTGJrZIWX59G7XF3Sgj9LdI/s757/capture.png.

8.3.3. Role of the Muscular System in Movement

The most crucial mechanism making movement possible in the animal kingdom, the muscular system is the primary pathway for the conversion of neurological stimuli into actual motion. Made up of three types of muscle tissue—skeletal, cardiac, and smooth—the muscles specialize in particular functions that collectively make each body capable of movement and physiological function. Skeletal muscles that are responsible for voluntary movement are attached to bones by tendons and regulated by the CNS. Such intricate coordination enables organisms to perform a wide range of movements from simple reflex actions to complex motor activity, pointing towards adaptability that aids survival in different environments. Besides contributing to movement, the muscular system also provides postural stability and balance, indicating its sophisticated role in the organism's dynamic interaction with the environment. Besides locomotion, the muscular system enhances the effectiveness and speed of movement through numerous specializations of muscle structure and function. For instance, evolutionarily, animals have evolved specialized muscle fibers for specific lifestyles and environments; fast-twitch fibers enable momentary burst acceleration, whereas slow-twitch fibers optimize endurance for extended-duration efforts. This kind of specialization is a demonstration of a trade-off in muscle evolution, whereby the organism develops according to the ecological niche, such that its mobility according to survival needs. Moreover, coordination of the muscular system with other body systems, , such as the skeletal and nervous systems, increases the effectiveness of movement and sensitivity to outside inputs. Such an integrated function highlights the importance of the muscular system to the overall organization of animal structure and function. Secondly, the musculoskeletal system is not just involved in the motion but also plays a role in functions other than the motion; it plays a central role in the overall homeostasis of the organism. Muscles generate heat as a byproduct of metabolic reactions with muscle contraction, keeping the body temperature, an essential factor for endothermic animals. Besides, muscle contractions make it possible to move fluids from one place to another, say blood in the circulatory system or food in the digestive system, to facilitate vital survival processes. The coming together of these varied functions is proof that the muscular system is not just standalone; rather, it works in liaison with other systems to develop a collective working model of an organism. The interaction of physiological systems and muscle function reflects a complicated balance that allows animals to maintain life and react to environmental change, demonstrating the complexity of animal structure and function.

Test Your Knowledge

1. What are the types of muscle tissue?
2. What is the mechanism of muscle contraction?

8.4. Nervous System

Learning Outcome

1. Structure of neurons and glial cells;
2. Central and peripheral nervous systems (PNS);
3. Functions of the nervous system in response to stimuli.

The nervous system is a very intricate web that helps to coordinate and control various body functions among various animal species. It comprises neurons, glial cells, and intricate connections for facilitating communication in the body. Specifically, nerve impulses are transmitted by neurons to facilitate rapid responses to external and internal changes. The organization of the nervous system varies between animals, with some being centralized (for instance, vertebrates) while others display a decentralized system (e.g., in cnidarians). These differences are an expression of evolutionary adaptation that optimizes the survival of an organism in the environment. For instance, the process of complex neural circuit formation is crucial for the processing of sensory input and the execution of motor function, ultimately leading to behavior and homeostasis maintenance in the organism. Current research has disclosed the functions of sleep concerning neural associations and synaptic regulation. Sleep, a generalized phenomenon across animal species, appears to have critical restorative roles in the nervous system. Disruption of the sleep pattern leads to significant impairment of cognitive function, suggesting that sleep plays a role in maintaining synaptic integrity. Experiments suggest that there are multiple molecular processes that occur in synapses during sleep, and this emphasizes how fluid synaptic connections are. Specifically, de Vivo et al. and Diering et al. have revealed synaptic structural and molecular machinery alterations, critical to understanding the biological mechanisms of sleep and its significance to cognitive wellness. These findings emphasize the intricate interaction between sleep neural activity and the overall processes of brain activity, concretizing the nervous system's involvement in overall wellness (Acsády et al., 2017). Also, the development and maintenance of the CNS are largely regulated by ECM molecules that function as guidance molecules in neural development. ECM receptors and their ligand interactions play a pivotal role in directing axonal and dendritic projection formation that later form successful synaptic connections. Once stable synapses have been formed, these receptors play a vital role in controlling synaptic plasticity, which plays a significant role in learning and memory. Plasticity versus connection stability is vital to adaptive behavior. The research highlights how ECM composition and receptor activation control cytoskeletal dynamics, which in turn change synaptic activity and, indirectly, behavioral outcome. Understanding these mechanisms extends our knowledge of the complexity of the nervous system and its ability to respond to internal and external environments, continuing to enlarge our knowledge of animal structure and function (Abrahams et al., 2014) (Figure 8.13).

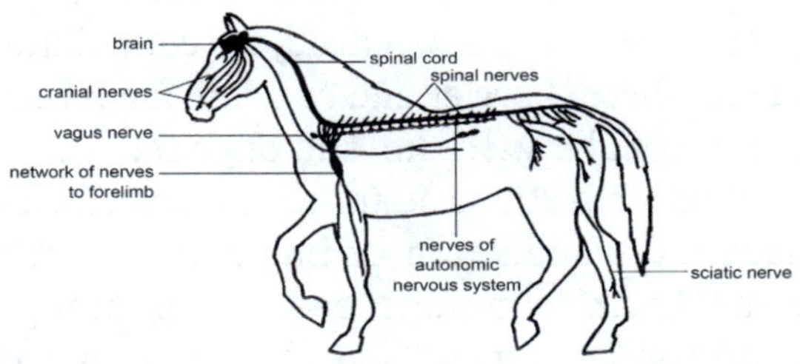

Figure 8.13. Animal nervous system.

Source: https://files.mtstatic.com/site_7338/3259/0?Expires=1748408612&Signature=ZJu3ENmT6SKM168Luw8Y7oQGCbWCT1RZzCjctg337k0Vho4uCEB0-CaWg3rBdVEyMI3XM6ZE3cif34yN~6lMb-ILa0PYily-7aokcDNTelQLZ~3xbpUQgx8Gi7~1Ekx8NlWsmg9eKBaS~Ep6lmdTJ~083JvX7ZGMOKZmdxmifqM_&Key-Pair-Id=APKAJ5Y6AV4GI7A555NA.

8.4.1. Structure of Neurons and Glial Cells

Morphology of the neurons and glial cells is at the forefront of nervous system function as a whole. Neurons, the primary communicative units, are characterized by a characteristic morphology of dendrites, cell body, and axon. Dendrites relay the signals from other neurons to reach the cell body, where critical metabolic processes occur. The axon, at times myelinated, transmits electrical impulses from the cell body to the synaptic terminals, where neurotransmitter release conveys information to adjacent cells. Such a complex structure makes signal transmission and processing rapid, highlighting the nervous system's responsiveness and efficiency to internal and external stimuli, a significant consideration in investigating animal vulnerability and adaptation. In contrast to neurons, glial cells that were previously considered to be merely support cells are now recognized to have an active role in homeostasis, structural support, and assistance in neuronal function. Astrocytes, oligodendrocytes, and microglia are various forms of glial cells that serve unique purposes. Astrocytes contribute to the blood-brain barrier and regulate ion levels in the extracellular medium, and oligodendrocytes form myelin sheaths around axons within the CNS to facilitate conduction of impulses. Microglia are the brain's immune cells, responding to infection and damage. Their physical properties allow them to fit varied environments within the nervous system, highlighting the dynamic and multifunctional nature of animal glial cells in function and structure. Recent developments in electrophysiological techniques, , such as whole-cell patch recording, have allowed us to understand the functional roles of animal glial cells in the whole brain. Touch and Zap, for example, have enabled scientists to clarify the biophysical properties of glial cells and neurons without exposing them to injury.

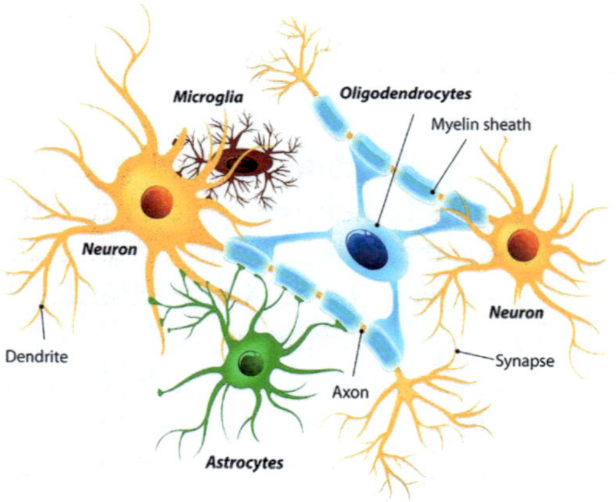

Figure 8.14. Structure of neurons and glial cells.

Source: https://www.ukdri.ac.uk/sites/default/files/styles/wide/public/2024-11/neurons_glia_diagram_designua_shutterstock_235015666.jpg?itok=IT_4Iwkr.

This method, exclusive of *in vivo* recordings, enables gentle and direct entry into cells, revealing intrinsic and visual responses that would otherwise be difficult to quantify. By giving us less disruptive and more precise measurements, this method gives us greater insight into how glial and neuronal structures interact with each other in various states of function, thus giving a more insightful look into the complexity of the nervous system and its function in animal physiology and behavior (Gener et al., 2014; Gordleeva et al., 2016) (Figure 8.14).

8.4.2. Central and Peripheral Nervous Systems (PNS)

Its intricate architecture is largely comprised of two broad divisions: the CNS and the peripheral nervous system (PNS). The brain and spinal cord form the CNS, the main information-processing center, which conducts sensory input and motor output integration. The CNS processes emotions, thoughts, and regulates body functions. Conversely, the PNS makes communication between the CNS and the rest of the body, , such as sensory organs, muscles, and glands, which is a double structure that allows animals to play with their environment by the well-coordinated feedback mechanism that facilitates survival and adaptation, describing the significance of the two systems in maintaining organismal function in general. Dysfunction in such systems can lead to a variety of neurological diseases, indicating the susceptibility and complexity of nervous system processes. Neurotoxicity is of special interest, since it deals with chemicals which are capable of influencing both central and peripheral parts of the nervous system. Exposure to neurotoxic substances can lead to a variety of interferences, ranging from inflammation, maldevelopment to even permanent neural impairment. For instance, certain environmental toxins have already been proven to affect cognitive function and behavioral outcomes, further implying that the interaction between exposure and neurological health is of utmost significance. Research in these neurotoxic impacts is essential not only for the maintenance of animal and human health but also for the determination of how such perturbations evolve with time after repeated exposure or environmental change (Brust et al., 2014). Further, research on disorders , such as autism underscores the importance of an empathetic appreciation for CNS and PNS interactions. Differences in behavior in the autism population are grounded in physiological issues that engage both the central and the peripheral systems. Trouble with social cognition and communication, for example, tends to go along with sensory processing abnormality related to the PNS. Studies demonstrate that examination of micro-movements and motoric variation provides significant insight into these large-scale systemic interactions, illustrating how multiple physiological processes combine to influence behavior (Brincker et al., 2017). Such complexity is reflective of the necessity of interdisciplinarity in neuroscience, as individual behaviors often respond to complex underlying processes spanning both central and PNS, thus illuminating the deep interconnectedness of animal structure and function (Figure 8.15).

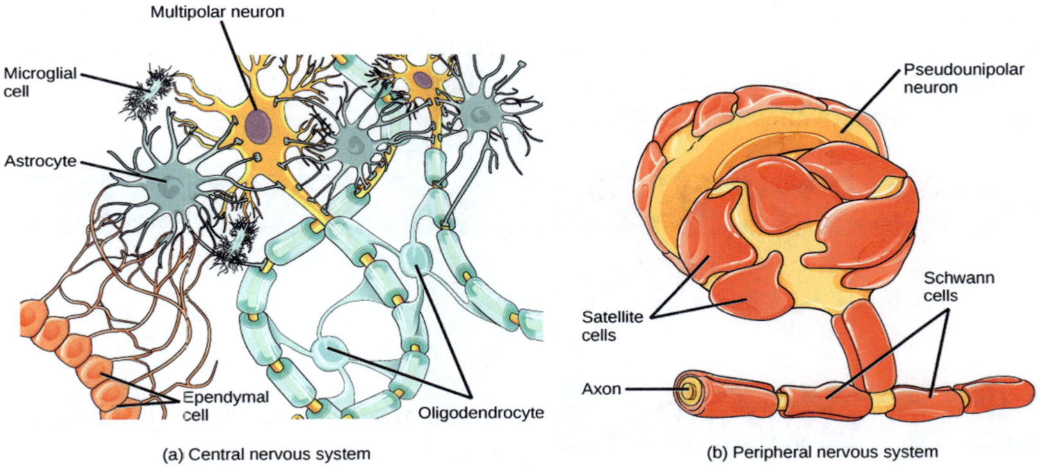

Figure 8.15. Central and peripheral nervous systems.

Source: https://opentextbc.ca/biology/wp-content/uploads/sites/96/2015/03/Figure_35_01_06.jpg.

8.4.3. Functions of the Nervous System in Response to Stimuli

The nervous system is a critical system that dictates how animals respond to stimuli because it functions as a complex network in which sensory input and motor output can be conveyed. The process uses specialized neurons that detect changes in the environment and transmit impulses to the brain for processing. For instance, visual stimuli are picked up by photoreceptors, specialized cells that sense light and transduce it into electrical signals, with tremendous effects on the behavior of an animal upon sensing various visual stimuli. The experimental results on the functional properties of movement-detector neurons reveal how various animals, , such as insects and vertebrates, use similar mechanisms to sense movement, initiating rapid behavioral changes. This coordination ensures that organisms can move appropriately in their worlds, showing an evolutionary advantage to persevere in complex ecosystems (Hateren et al., 1990). Additionally, the nervous system's ability to integrate sensory information extends beyond detection. The response of an animal to stimuli often relies on the context in which the stimuli are presented. As an example, research into the nematode *C. elegans* shows that its behavioral responses not only depend upon the amplitude of mechanosensory stimuli but also significantly on their temporal properties, indicating the depth of computation of the nervous system. Such responsiveness allows organisms to modulate their response according to prevailing environmental parameters in order to maximize their interaction with the environment. By processing not just the presence of stimuli but also their nuances, the nervous system enhances survival through allowing subtle and contextual responses (Leifer et al., 2018). As a result of these interactive dynamics, the processes of the nervous system's reaction to stimuli demonstrate an incredible evolutionary advancement of animal behavior. This complex interaction is present as animals usually need to integrate various sensory modalities in an attempt to receive a unifying perception of the environment. A good example is that of combining visual and somatic inputs to make more rational decisions, for instance, predator avoidance or tracking food sources. Sensory processing impacting the capacity to gauge relevance and import of stimuli significantly controls behavior, not just reactive functions, but also anticipatory based on sensory processing. Thus, the nervous system is a vital pillar of animal structure and function, providing animals with a proper response to remain alive in an ever-changing environment (Hateren et al., 1990).

Test Your Knowledge

1. Illustrate the structure of neurons and glial cells.
2. What are the functions of the nervous system in response to stimuli?

8.5. Circulatory System

Learning Outcome

1. Components of the circulatory system: heart, blood vessels, and blood;
2. Mechanisms of blood circulation;
3. Importance of the circulatory system in homeostasis.

The circulatory system is a fundamental concept in animal physiology, facilitating the delivery of vital materials , such as oxygen, nutrients, and hormones throughout the body. It operates through an intricate system of vessels, divided into arteries, veins, and capillaries, which work together in synchrony to deliver efficient transportation. Not only does it do the work of cellular respiration and metabolism, but it is also in charge of thermoregulation and immune response. In mammals, the heart is the pump at the center of a very efficient two-pump system for maintaining blood flow from oxygen-rich to oxygen-poor blood. Such cytological organization reflects the intrinsic compromise of the circulatory system to the varying energy demands of tissues and illustrates the intricately interwoven nature of structure and function of living systems. Structural subtleties of the vascular architecture are reflected through the study of mammalian vascular systems. It has also been cited that recent studies have found the correlation between aesthetic order and complexity among mammalian vascular systems, which would provide clues regarding their importance in the evolutionary process. Applying the Birkhoff aesthetic value in the study of vascular configurations, researchers have discovered that these structures are very well organized, which means highly efficient to achieve a streamlined look and function that is part of maximal operation (Nubiala et al., 2009). The mathematical analysis reveals that complexity and symmetry are not merely random but significant roles in enabling blood circulation efficiently based on the demand of various physiological activities. Understanding these structural principles will enhance our comprehension of how evolutionary forces shape circulatory systems to develop particular environmental and metabolic requirements. Beyond the structure, circulatory system functionality is dependent on dynamic regulatory processes. Circulatory system homeostasis is carefully controlled by neural and hormonal communications that can change heart rate, blood pressure, and vessel diameter according to different physiological demands. For instance, with intense exercise, the sympathetic nervous system is activated, which raises heart rate and diverts blood to muscles and decreases perfusion of the gastrointestinal tract. This adaptability is one example of the circulatory system engaging in actively establishing balance, which is vital to life. Furthermore, research explores how variations in vascular structuring among different species may be employed to account for ecological adaptations and evolutionary trends, as reflected through the research into the mammalian arterial systems (Astolfi et al., 2017). The strategy here points toward interaction between function and evolutionary emergence in the grand pattern of animal form and function, including the prominent role played by the circulatory systems in making life possible (Figure 8.16).

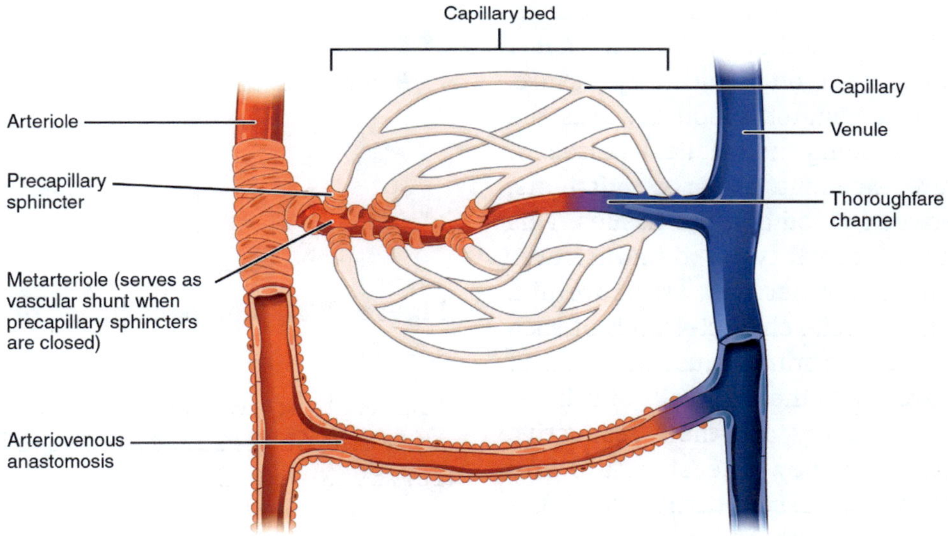

Figure 8.16. Animal circulatory system.

Source: https://bpb-us-e1.wpmucdn.com/sites.gatech.edu/dist/6/1810/files/2023/11/2105_Capillary_Bed-1024x605.jpg.

8.5.1. Components of the Circulatory System: Heart, Blood Vessels, and Blood

The circulatory system is an extremely complex system of organs that is responsible for the transportation of nutrients, gases, hormones, and waste products throughout the body. The heart, which sits at the center of this system, is a muscular organ that has the function of a strong pump in order to distribute the blood efficiently. The heart is composed of four chambers—the right atrium, right ventricle, left atrium, and left ventricle—each of which plays a critical role in regulating the movement of blood. The deoxygenated blood from the body is supplied to the right side of the heart and passed on to the lungs to get oxygenated, and the oxygenated blood is received by the left side and passed out to the rest of the body. In understanding the heart's function and structure lies the key to grasping the complexity of cardiovascular physiology, in that it defines how specialized organs perform as part of the broader context of animal function and structure. The blood vessels, which are pathways for the circulation of blood in the body, provide support to the heart's central position. There are three basic types of vessels: arteries, veins, and capillaries. Arteries carry oxygenated blood from the heart to other sections of the body under a pressure. Veins return deoxygenated blood towards the heart. Capillaries, being the smallest vessels of blood, facilitate exchange of gases, nutrients, and waste between blood and the tissues. This arrangement of vessel types demonstrates the dynamic interplay between structure and function, and the special characteristics of each type of vessel—i.e., the sturdy muscular walls of arteries compared with the thin walls of veins—enabling them to carry out their respective functions with efficiency. The structure and functioning of blood vessels demonstrate the circulatory system's sophistication, reflecting the need for a highly organized vascular system in order to sustain life. Blood, the vital fluid that circulates within this intricate system, is composed of plasma, red blood cells, white blood cells, and platelets. The fluid portion, plasma, acts as a transport medium for waste products, hormones, and nutrients. Red blood cells, or erythrocytes, function to transport oxygen, with hemoglobin being utilized to release and absorb oxygen within tissues. On the other hand, white blood cells have an important role to play in the immune system as a defense mechanism for the body against infection. Platelets or thrombocytes assist in clotting to prevent excess bleeding when injured. The composition and role of the blood are important in gaining homeostasis and supporting life processes and demonstrating how every component makes the circulatory system more effective in total. Advances in imaging tools , such as SPIM-μPIV have enabled scientists to understand blood flow dynamics better so that researchers can examine these elements in great detail, as observed in the intricate behaviors of zebrafish larvae (Taylor et al., 2018) (Figure 8.17).

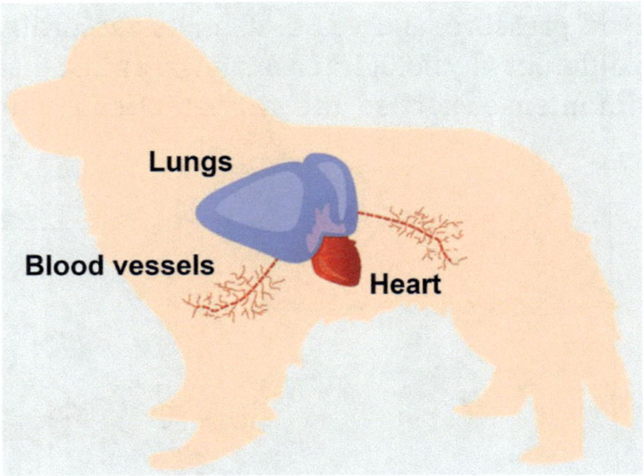

Figure 8.17. Components of the circulatory system.

Source: https://www.healthyheartsfordogs.com.au/static/6e9412aa8c2494924b7a342cdfae01e7/59139/circulatory1.png.

8.5.2. Mechanisms of Blood Circulation

The circulatory processes play a vital role in homeostatic upkeep of animal systems, by guiding the delivery of necessary nutrients and oxygen, and clearing of metabolic wastes. The highlight of this process is the heart, the blood vessels, and the interaction among these components. The heart is a muscular pump that pumps cyclically to propel blood through a circulation of arteries, veins, and capillaries. Arteries carry oxygenated blood from the heart out into smaller arterioles and ultimately to capillary beds, where substance exchange occurs. Veins carry deoxygenated blood to the heart, demonstrating a seemingly endless loop that represents the efficiency of circulatory systems. Through coordinative interactions of the components, circulation sustains essential body functions , such as thermoregulation and immune response, which reinforce the subtlety and significance of this physiological system. Variations in circulatory mechanisms among different animal taxonomic groups are adaptations for different environmental stressors and physiological demands. For instance, mammals and birds possess a double circulatory system with clear-cut pulmonary and systemic circuits that allow better oxygenation and metabolic regulation than the single circuit found in amphibians and reptiles. In mammals, high-pressure oxygenated blood is ejected from the left ventricle into the systemic circulation, enhancing exercise endurance potential. Also, the presence of such units as valves and special vessels guarantees unidirectional blood flow, thereby preventing of backflow and maintaining optimal functioning under various pressures. Such intricate form not only maximizes transportation efficiency but also demonstrates evolutionary trends in circulatory evolution in particular ecological niches. Current studies stress the importance of a realization of the pharmacological effects on circulatory mechanisms, particularly in clinical and experimental situations. For instance, the relationships between catecholamines, such as adrenaline and dopamine, and anesthetic drugs have an impact on cerebral blood flow and hemodynamic responses extensively during medical procedures. Cerebral blood flow is found to increase in a dose-dependent manner through dopamine, influencing intracranial pressure and cerebrovascular resistance, for instance, in propofol anesthesia, which skews normal physiological responses (Fortin et al., 2002). Additionally, the impact of stent placement in aortic coarctation clarifies potential long-term consequences, , such as hypertension caused by changes in arterial stiffness and baroreceptor sensitivity (Aresu et al., 2016). These findings serve to emphasize the dynamic nature of blood circulation and the importance of possessing an absolute comprehension of both pharmacological and physiological considerations that control circulatory health and effectiveness (Figure 8.18).

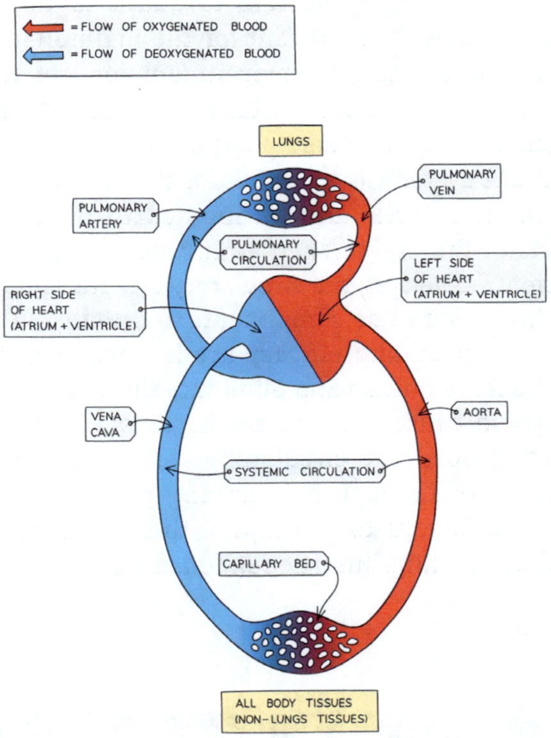

Figure 8.18. Mechanisms of blood circulation.

Source: https://cdn.savemyexams.com/cdn-cgi/image/w=1920,f=auto/uploads/2020/01/The-double-circulatory-system-in-mammals.png.

8.5.3. Importance of the Circulatory System in Homeostasis

The circulatory system lies at the heart of homeostasis in animal bodies by the transport of essential nutrients, gases, and wastes. In a ubiquitous network of blood vessels, the system ensures oxygen and glucose are supplied to the tissues and CO_2 and metabolic wastes are removed efficiently. This intricate process is vital in keeping cellular processes and total metabolic function in order. Furthermore, the circulatory system has a role in thermoregulation through the regulation of body temperature by transporting heat generated in the core to peripheral tissues and thus maintaining temperature. The dynamic response is important because animals encounter varying environmental conditions. The ability of the circulatory system to adapt to tissue needs makes it significant for the maintenance of a stable internal environment, one of the characteristics of homeostasis. Interestingly, the circulatory system also plays a focal role in the immune system, as a conduit for the distribution of immune cells and antibodies to different parts of the body. When an invader crosses an animal's defenses, the circulatory system rapidly mobilizes immune components to the site of the attack. Speed and efficacy are important here, as it acts to encase and eliminate threats before they can cause harm to the organism's health. Furthermore, the circulatory system supports healing of damage by carrying platelets and clotting factors to the injury, thus supporting wound healing. Interdependence between the circulatory system and the immune system is a testament to how physiological functions must be regulated in a way that supports homeostasis, cementing the fact that circulatory disorders can lead to compromised immunity and vulnerability to disease. The circulatory system influences pervades to hormonal regulation, serving as a transport medium for hormones, coordinating various body functions. Hormones released by endocrine glands are pumped into the blood, enabling system-wide dissemination within the organism. For example, insulin and glucagon play a crucial role in the management of blood glucose levels. Moreover, the hormone signal pathways are central to growth, stress response, and reproduction. Deviation in these signal transduction pathways can induce diabetes or thyroid disease, emphasizing the circulatory system's pivotal role in hormonal equilibrium. Understand of the circulatory system reveals the complexity of animal structure and function, showing how this system not only plays an essential role in transport of nutrients but is also essential for regulation among various physiological processes necessary for homeostasis, , such as with the complexity of human disease and health (Angelini et al., 2016; Klempin et al., 2020).

Test Your Knowledge

1. What are the components of the circulatory system?
2. What is the importance of the circulatory system in homeostasis?

8.6. Respiratory System

Learning Outcome

1. Anatomy of the respiratory system;
2. Gas exchange processes;
3. Impact of environmental factors on respiratory function.

The respiratory system is essential to the homeostasis of animal physiology by gas exchange, indispensable to cellular respiration. In mammals, this complex system is made up of multiple structures like the nasal passages, trachea, bronchi, and lungs, each one of them geared to maximize oxygen intake and the expulsion of CO_2. The very efficient composition of these structures allows for the efficient transfer of gases across the alveolar membrane into the blood. This is not just vital to maintaining metabolic processes but also has extremely important implications regarding overall health. For instance, impairments in adequate respiratory function can lead to severe health complications, , such as the fact that chronic respiratory conditions were among the leading causes of death in the United States in 2015, highlighting the systems' significant role in human health and disease (Miller et al., 2017). Examination of the contributors to respiratory health reveals how environmental contributors can have a significant impact on lung function. Exposure to toxins, allergens, and pathogens causes the integrity of the respiratory system to break down, leading to conditions , such as asthma, pneumonia, and chronic obstructive pulmonary disease. Vulnerability of the lungs to these external stressors proves the adaptive nature of the system and its vulnerabilities. Research with the model of nonhuman primates has provided valuable information on the mechanisms that are implicated in such diseases and revealed the points of similarity in lung function as well as structure between the nonhuman primates and human beings. All these are crucial in formulating proper therapies and prevention for respiratory diseases and enhancing our capability to sustain respiratory health within populations (Miller et al., 2017). Besides, the interconnection of respiratory function and systemic disease, , such as cancer, reveals a subtle balance of physiological reactions that complicate the administration of treatment and recuperation.

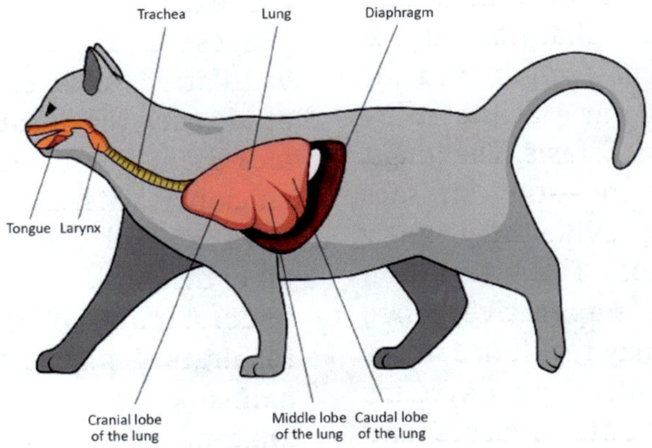

Figure 8.19. The respiratory system in animals.

Source: https://animalalmanacblog.wordpress.com/wp-content/uploads/2016/09/cat-respiratory-system-308067818-600x399.jpg?w=600.

More importantly, cancer cachexia, which is a syndrome of weight loss and wasting of muscles, profoundly compromises the respiratory muscles, hence destroying respiratory function during the illness. Recent studies have revealed that pharmacologic therapy designed to counteract the signaling pathways responsible for muscle breakdown can restore muscle function and mass by a significant degree. This finding points to the necessity of addressing respiratory health not merely as a discrete problem but also in conjunction with more universal systemic diseases of health. By understanding the intricate interdependencies between the respiratory system and the other body systems, researchers can begin whole systems-directed intervention that will strive to increase therapeutic outcomes for these patients with respiratory illnesses involving multifaceted underlying conditions like cancer (Figure 8.19).

8.6.1. Anatomy of the Respiratory System

The anatomy of the respiratory system plays a crucial role in how animals exchange gases, a most essential process for life sustenance. The respiratory system of mammals is comprised of a sophisticated system of structures that include the nasal passages, pharynx, larynx, trachea, bronchi, and lungs. Oxygen-rich air intake and expulsion of CO_2 are facilitated by the system to maintain homeostasis. The lungs, which have millions of alveoli—tiny air sacs effective in gas exchange—provide the primary site for this critical function. The integration of these components is imperative since alterations in any of them may result in serious respiratory complications, reflecting the systems susceptibility to environmental sources and biological pathogens that can affect overall well-being, as emphasized in studies concentrating on respiratory illness, especially in nonhuman primates, which are vital models for studying human respiratory illnesses (Miller et al., 2017). Besides, the avian respiratory system has unique adaptations to facilitate efficient gas exchange, which is an affirmation of evolutionary strategies designed to help satisfy the demands of flight. Birds have a structured lung-associated immune system designed to combat varied pathogens penetrating the respiratory system. The highlight of such an immunohierarchical framework is the bronchus-associated lymphoid tissue (BALT), which develops into discernible structures to facilitate immune responses as the birds mature. This lymphoid tissue is found at the intersection points of the primary and secondary bronchi and is marked by specialized spaces for B and T cells, maximizing the respiratory systems preparedness to be ready to react against infection and foreign matter. Understanding the anatomical intricacies of avian lungs, including where tissue macrophages are found and a unique organization of immune cells, is critical for host-pathogen interaction studies, particularly in avian biology (Dalamani et al., 2006). Anatomical studies of the respiratory system of various animal species reveal an impressive range of evolutionary adaptations adapted to their specific environmental context. For example, the anatomical differences among mammalian, avian, and other respiratory systems reflect the impact of evolutionary pressures on their associated gas exchange mechanisms. While mammals possess a diaphragm-driven mechanism of ventilation, birds contain a remarkable system of air sacs that ensures unidirectional airflow into the lungs for maximum oxygen extraction efficiency. This shows how the different species have solved similar biological challenges in a dissimilar manner, which provides an understanding of functional morphology and respiratory physiology in the entirety of the animal kingdom. Extensive knowledge of these anatomical variations is not only capable of aiding veterinary and wildlife health care but also informing us about respiratory diseases and their evolution across diverse species (Dalamani et al., 2006; Miller et al., 2017) (Figure 8.20).

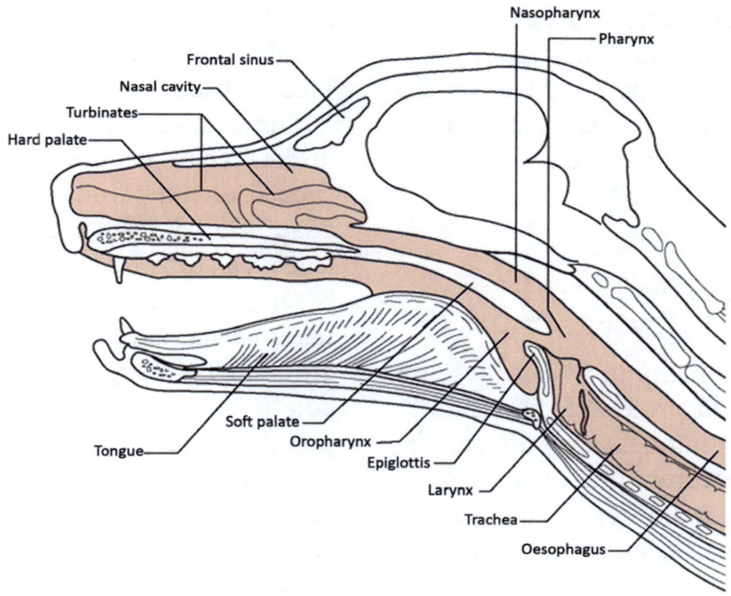

Figure 8.20. Animal anatomy of the respiratory system.

Source: https://veteriankey.com/wp-content/uploads/2016/07/B9780702029387000087_gr2.jpg.

8.6.2. Gas Exchange Processes

Gas exchange is a vital biological function that occurs over dissimilar structures in various animal phyla, specialized according to their lifestyle and ecology. In most terrestrial vertebrates, gas exchange is facilitated by organs that evolved to provide a large surface area for oxygen to diffuse into the bloodstream while CO_2 is released. The efficiency of gas exchange depends on several factors, including respiratory surface area, diffusion gradients, and the presence of transport molecules like hemoglobin. Aquatic animals, however, employ gills, from which they absorb oxygen from water. These anatomical structures mirror the versatility of respiratory mechanisms among animals in facilitating efficient gas exchange suitable for terrestrial and aquatic habitats, mirroring the sensitive balance between structure and function in animal biology. In mammals, the trachea is an airway well adapted for optimal flow of air rather than gas exchange. This stiff tube is reinforced in shape by cartilaginous rings so that there is unhampered flow of air to the lungs. While significant in respiratory mechanics, the area for gas exchange occurs at the level of alveoli, where thin walls facilitate the exchange of gases between air and blood. The adaptation of lung architecture in different species also generally reacts to their specific environmental needs and forces of adaptation. In the case of dolphins, the tracheal characteristics exclusive to them, , such as the presence of vascular lacunae, attest to their evolution for living underwater, with emphasis on structural uniqueness contributing to optimized gas exchange processes (Bagnoli et al., 2018). New studies in experimental modeling identified the potential of artificial systems for investigating gas exchange mechanisms more, for comparative physiology. An artificial animal/plant system experiment combined mathematical modeling and empirical evidence to uncover the operation of gas exchange characteristics in controlled systems. These systems allow scientists to regulate environmental conditions, , such as atmospheric oxygen and CO_2 levels, thus gaining a better insight into how various adaptations lead to efficient gas exchange (Averner et al., 1985). When such systems are studied, researchers can explore the limits of physiological flexibility and reactivity and provide new perspectives on the way gas exchange structures develop in animals, greatly advancing our understanding of animal structure and function.

8.6.3. Impact of Environmental Factors on Respiratory Function

The impact of environmental factors on respiratory function is profound, particularly because ecosystems are increasingly affected by climate change. Temperature is among the most significant factors that influence respiration. The physiological processes of ectothermic organisms, , such as fish and amphibians, rely on external temperatures. Heat can lead to reduced metabolic function and respiratory efficiency. For instance, research indicates that higher temperatures adversely alter physiological traits, and this results in lower oxygen content in water. Such alterations generally slow growth and increase mortality rates among sensitive organisms. Since higher temperatures raise the metabolic demands of animals, their respiratory systems need to acclimatize or collapse, thus exhibiting the intricate connection between climate conditions and animal respiratory performance. Availability of oxygen is yet another environmental determinant of importance, particularly in aquatic habitats where pollution and eutrophication can further exacerbate the condition. The degree to which animals can draw oxygen from their surroundings increases proportionally with respiratory efficiency. Where oxygen levels are diminished because of increased nutrient loading or thermal stratification, many aquatic species exhibit decreased aerobic capacity. This is then aggravated by species variation within individuals, where some organisms are discovered to possess greater aerobic adaptation capacities compared to others. For example, the adaptation of fish species to alternating oxygen levels can also be influenced by their ability to respire at the mitochondrial level. In laboratory tests, organisms with lower mitochondrial efficiency struggled to cope in low-oxygen conditions, with serious ramifications for population ecology and community structure in affected communities, again validating the significant link between oxygen content and respiratory capacity (Anderson et al., 2016). In addition, the interconnection between environmental poisons and respiratory capacity cannot be overstressed. Toxins , such as heavy metals, plastics, and other chemicals cause harm to the respiratory systems of most animals, leading to an increase in respiratory diseases or failure. For instance, particulate matter in terrestrial systems can be absorbed into the lungs of mammals, impairing gas exchange and overall respiratory functioning. The outcome of such exposure is heightened susceptibility to respiratory disease, which leads us to insights about the implications at the population health and biodiversity levels. Positioning respiratory pathophysiology within environmental stressors exhibits the interconnectedness of ecological health and animal physiology. This encounter calls for addressing environmental degradation with regard to respiratory health among animal populations, in a bid to justify further research on measures of mitigation in conservation processes (NASA-SP-7011, 1974).

Test Your Knowledge

1. What are the gas exchange processes?
2. What is the impact of environmental factors on respiratory function?

SUMMARY

The examination of animal form and function, the intricate correlation between an organism's structure and its functioning, truly establishes the foundations of biological knowledge. Organ systems and specialized tissues demonstrate how the structural characteristics of animals directly affect their capacity of animals to interact with and respond to their environment. This network allows for an intimate comprehension of the mechanisms driving evolutionary processes. The significance of examining these relations extends beyond mere anatomical description; it is interesting information about the physiological adaptation that allows species to exist in diverse ecological niches. It also entails a broader appreciation of biodiversity because structural differences reflect different functional strategies among different taxa. Besides, learning material like the ESQ booklet used for enhancing animal tissue comprehension underlines the importance of good communication of complex biological data. As seen in recent research, this learning material has shown great validity in various aspects, including content, language, and media presentation. The high rates of validity, 92.05% to 95.83%, indicate that such resources significantly enhance students' comprehension of structures and their functions in animal biology. Through the provision of learners with well-researched and readable content, educators can better facilitate the combining of theory and practice in learning animal biology so that future generations comprehend major biological concepts. Finally, the ongoing debate on whether or not ancient fallacies can be applied to scientific inquiry and the role they have played in epistemic progress during periods of uncertainty offers a compelling model of how knowledge is constructed that can be compared with the domain of animal structure and function, where understanding comes from recognizing the dynamic interplay between structural adaptation and relationship with environment. By doing a contextual learning and inquiry process, scholars can embrace the complexities that exist in biology exploration. Not only is this process encouraging scholarship at the scholarly level, but it also encourages broader discovery of the natural world and the great influence of animal structure on function and survival. The combination of different educational techniques ultimately enriches the field, creating a multidimensional understanding of life itself.

The intricate relationship between the structure and function of animals is ultimately accounted for by the lens of evolutionary biology. This perspective stresses how anatomical features developed to achieve specific functions over time. Organisms developed special physiological processes that enable them to thrive in a wide range of habitats, showing the extreme interdependence between the physical structure of an organism and its environment. For example, the development of modified respiratory structures in aquatic organisms , such as gills, shows how evolution has modified anatomical characteristics to optimize gas exchange in their respective water environments, enabling them to utilize available resources for optimal utilization. This concept is also supplemented by models that examine constructions of the brain and cognition, showing that knowledge about animal behavior is also directly connected to physical and mental structural adjustments (Gelepithis et al., 2009). This type of knowledge emphasizes the significance of evolutionary compatibility as a foundation to examine the functional capabilities of animals. In addition to anatomical specializations, environmental factors , such as climate play important roles in shaping animal size and shape. Bergmann's rule provides a prime example of temperature's effect on the size of various species, particularly along geographical gradients. The rule encapsulates the broader ecological processes at play and shows how such characteristics affect community interactions and trophic structure.

The size of an organism's body, particularly in primary marine food webs, could potentially have a significant impact on species interactions and their functions within such ecosystems. A study indicates that climate change has a profound effect on not only body sizes but also the scale of trophic cascades, and hence biotic community connectivity. This information contributes to our understanding of how ecological alterations can generate cascading effects across ecosystems, revealing the complexity of animal form and their functional consequences under varying climatic conditions (Lacerot et al., 2010). Furthermore, investigation of animal structure and function extends to understanding intra- and interspecific associations, which provides insight into the complexities of ecological interactions. These interactions illuminate the important roles animals play in their worlds, not just influencing their survival but that of other species within their realities. The behavioral adaptations that arise as a function of structural features, particularly in reaction to predation and competition, also exhibit this interdependence. For instance, the means of communication among different species, underpinned by the structural elements of their body, lead to social organization and survival mechanisms, and these further reflect an intricate design of relationships based on both structure and behavioral adaptabilities. Essentially, a holistic study of these components offers a more elaborate understanding of the generic principles underlying animal life, incorporating structure, function, and ecosystem behavior into a holistic context.

KEY TERMS FROM THIS CHAPTER

Alveoli	Capillaries	Circulatory system
Digestive system	Endocrine system	Excretory system
Homeostasis	Homeostatic regulation	Hormones
Immune system	Integumentary system	Muscular system
Nervous system	Neurons	Organ
Organ system	Reproductive system	Respiratory system

CRITICAL THINKING QUESTIONS

1. **Which system in animals is responsible for transporting nutrients, gases, and wastes?**
   a. Digestive system
   b. Respiratory system
   c. Circulatory system
   d. Nervous system
2. **The basic functional unit of the nervous system is the:**
   a. Neuron
   b. Alveolus
   c. Nephron
   d. Capillary
3. **Which organ is primarily responsible for filtering blood in vertebrates?**
   a. Liver
   b. Kidney
   c. Lung
   d. Heart
4. **Which system works closely with the skeletal system to produce movement?**
   a. Respiratory system
   b. Digestive system
   c. Muscular system
   d. Endocrine system
5. **The exchange of oxygen and carbon dioxide occurs in the:**
   a. Heart
   b. Kidneys
   c. Alveoli
   d. Stomach

Short Answer Questions

1. What are the four main types of animal tissue, and what is the function of each?
2. What is homeostasis, and why is it important in animals?
3. What is the function of the circulatory system, and how does it work with the respiratory system?
4. Describe how the digestive system breaks down food and absorbs nutrients.
5. What is the role of the excretory system in maintaining homeostasis?

Answers to Critical Thinking Questions

1. (c); 2. (a); 3. (b); 4. (c); 5. (c)

REFERENCES

1. Acsády, L., & Harris, K. D. (2017). Synaptic scaling in sleep. https://core.ac.uk/download/79556238.pdf (accessed on 13th June 2025).
2. Alarcon, V. J., Archer, D. W., Halloran, J. M., et al. (2016). Exploring agricultural production systems and their fundamental components with system dynamics modelling. https://core.ac.uk/download/pdf/82405339.pdf (accessed on 13th June 2025).
3. Alméciga-Díaz, C. J., Mackenzie, W. G., Mason, R. W., et al. (2017). Development of bone targeting drugs. https://core.ac.uk/download/129191458.pdf (accessed on 13th June 2025).
4. Anderson, G. J., Auer, S. K., Metcalfe, N. B., et al. (2016). Inadequate food intake at high temperatures is related to depressed mitochondrial respiratory capacity. https://core.ac.uk/download/42370220.pdf (accessed on 13th June 2025).
5. Angelini, A., Arcaro, A., Chimenti, C., et al. (2016). Novel perspectives in redox biology and pathophysiology of failing myocytes: Modulation of the intramyocardial redox milieu for therapeutic interventions – A review article from the Working Group of Cardiac Cell Biology, Italian Society of Cardiology. https://core.ac.uk/download/54526664.pdf (accessed on 13th June 2025).
6. Arbib, M. A., Bernstein, L., Carnap, R., Chomsky, N., et al. (2018). *Spring school on language, music, and cognition: Organizing events in time*. https://core.ac.uk/download/287606534.pdf (accessed on 13th June 2025).
7. Aresu, L., Basso, C., Bernardini, D., et al. (2016). Aortic stenting in the growing sheep causes aortic endothelial dysfunction but not hypertension: Clinical implications for coarctation repair. https://core.ac.uk/download/53531427.pdf (accessed on 13th June 2025).
8. Astolfi, L., Chicca, M., Danti, S., et al. (2017). Nanoparticle drug delivery systems for inner ear therapy: An overview. https://core.ac.uk/download/80157791.pdf (accessed on 13th June 2025).
9. Austin, C. J. (2017). A biologically informed hylomorphism. https://core.ac.uk/download/131216095.pdf (accessed on 13th June 2025).
10. Averner, M. M., Bartholomew, I., Moore, B. III, et al. (1985). Atmosphere behavior in gas-closed mouse-algal systems: An experimental and modelling study. https://core.ac.uk/download/pdf/10381201.pdf (accessed on 13th June 2025).
11. Bagnoli, P., Ballarin, C., Cozzi, B., et al. (2018). Vascularization of the trachea in the bottlenose dolphin: Comparison with bovine and evidence for evolutionary adaptations to diving. https://core.ac.uk/download/157767123.pdf (accessed on 13th June 2025).
12. Bal, E. D., Barreiro, E., Chacon Cabrera, A., et al. (2014). Pharmacological strategies in lung cancer-induced cachexia: Effects on muscle proteolysis, autophagy, structure, and weakness. https://core.ac.uk/download/159291085.pdf (accessed on 13th June 2025).
13. Barile, L., Chimenti, I., Di Consiglio, A., et al. (2009). New perspectives to repair a broken heart. https://core.ac.uk/download/54376905.pdf (accessed on 13th June 2025).
14. Becker, M. B., Begemann, G., Meyer, A., et al. (2005). Conservation and co-option in developmental programmed: The importance of homology relationships. https://core.ac.uk/download/pdf/3800618.pdf (accessed on 13th June 2025).

15. Bode, H. R., Haigler, H. T., & Schlaepfer, D. D. (1992). Distinct cellular expression pattern of annexins in *Hydra vulgaris*. https://core.ac.uk/download/323075552.pdf (accessed on 13th June 2025).

16. Boutros, P. C., Chamankhah, M., Eftekharpour, E., et al. (2013). Genome-wide gene expression profiling of stress response in a spinal cord clip compression injury model. https://core.ac.uk/download/323076209.pdf (accessed on 13th June 2025).

17. Brincker, M., & Torres, E. (2017). Why study movement variability in autism? https://core.ac.uk/download/131210732.pdf (accessed on 13th June 2025).

18. Brockington, M., Brown, S. C., Cirak, S., et al. (2010). Transgenic overexpression of LARGE induces alpha-dystroglycanopathy hyper glycosylation in skeletal and cardiac muscle. https://core.ac.uk/download/1730280.pdf (accessed on 13th June 2025).

19. Brust, V., Creeley, C., Feldman, R., Grandjean, P., Herken, H., Panzica, G., Parent, A., et al. (2014). Neurotoxicity. https://core.ac.uk/download/286729125.pdf (accessed on 13th June 2025).

20. Buj-Bello, A., Lujan Feliu-Pascual, A., Razzaq, A., Tjondrokoesoemo, A., Toussaint, A., Durieux, A. C., Beggs, A. H., et al. (2013). Altered splicing of the BIN1 muscle-specific exon in humans and dogs with highly progressive centronuclear myopathy. https://core.ac.uk/download/16460074.pdf (accessed on 13th June 2025).

21. Burstein, H. J., Hobbs, K. H., Hooper, S. L., et al. (2013). Temperature sensitivity of the pyloric neuromuscular system and its modulation by dopamine. https://core.ac.uk/download/16213803.pdf (accessed on 13th June 2025).

22. Cummings, L. (2002). Reasoning under uncertainty: The role of two informal fallacies in an emerging scientific inquiry. https://core.ac.uk/download/30639769.pdf (accessed on 13th June 2025).

23. Dalamani, G., Kaspers, B., & Reese, S. (2006). The avian lung-associated immune system. https://core.ac.uk/download/12174987.pdf (accessed on 13th June 2025).

24. Hanisch, S., & Eirdosh, D. (2020). Conceptual clarification of evolution as an interdisciplinary science. EdArXiv. https://doi.org/10.35542/osf.io/vr4t5Flohr, S., Gomez, S., Kierdorf, H., et al. (2017). Multiple osteochondromas of the antlers and cranium in a free-ranging white-tailed deer (*Odocoileus virginianus*). https://core.ac.uk/download/148067969.pdf (accessed on 13th June 2025).

25. Fortin, D., Motulsky, H., Runciman, W. B., Stephan, H., Teasdale, G., & Tuor, U. I. (2002). The cerebrovascular effects of adrenaline, noradrenaline and dopamine infusions under propofol and isoflurane anesthesia in sheep. https://core.ac.uk/download/12730631.pdf (accessed on 13th June 2025).

26. Gelepithis, P. A. M. (2009). Outline of a new approach to the nature of mind. https://core.ac.uk/download/pdf/87224.pdf (accessed on 13th June 2025).

27. Gener, T., Graham, L., Marinazzo, D., et al. (2014). The touch and zap method for *in vivo* whole-cell patch recording of intrinsic and visual responses of cortical neurons and glial cells. https://core.ac.uk/download/55844894.pdf (accessed on 13th June 2025).

28. Gordleeva, S., Ivanchenko, M., Kazantsev, V., et al. (2016). Mammalian brain as a network of networks. https://core.ac.uk/download/42130432.pdf (accessed on 13th June 2025).

29. Hateren, J. H. van. (1990). Directional tuning curves, elementary movement detectors, and the estimation of the direction of visual movement. https://core.ac.uk/download/pdf/12926216.pdf (accessed on 13th June 2025).

30. Huijing, P. A. (1995). Parameter interdependence and success of skeletal muscle modelling. https://core.ac.uk/download/pdf/11461276.pdf (accessed on 13th June 2025).

31. Hutchins, B. R., Brown, Buchwald, Clarke, D., D., D., et al. (2015). Descartes, corpuscles and reductionism: Mechanism and systems in Descartes' physiology. https://core.ac.uk/download/147061302.pdf (accessed on 13th June 2025).

32. Ikhtiarni, F., Ardi, A., & Darussyamsu, R. (2021, May). Validity of ESQ booklets about structure and function of animal tissue as biology learning supplement. *Edubiotik: Jurnal Pendidikan Biologi dan Terapan, 6*(1), 26–39. https://core.ac.uk/download/427132261.pdf (accessed on 13th June 2025).

33. Jacobs, D. K., Valencia, J. E., & Winchell, C. J. (2010). Confocal analysis of nervous system architecture in direct-developing juveniles of *Neanthes arenaceodentata* (Annelida, Nereididae). https://core.ac.uk/download/4886525.pdf (accessed on 13th June 2025).

34. Kerrisk, M. E., Cingolani, L. A., & Koleske, A. J. (2014). ECM receptors in neuronal structure, synaptic plasticity, and behavior. *Progress in Brain Research, 214*, 101–131. doi: 10.1016/B978-0-444-63486-3.00005-0 (accessed on 13th June 2025).

35. Klempin, F. (2020). Signaling pathways linking behavior to neurogenesis in healthy brain and disease. https://core.ac.uk/download/322846707.pdf (accessed on 13th June 2025).

36. Lacerot, G. (2010). Effects of climate on size structure and functioning of aquatic food webs. https://core.ac.uk/download/pdf/29245099.pdf (accessed on 13th June 2025).

37. Leifer, A. M., Liu, M., Shaevitz, J. W., et al. (2018). Temporal processing and context dependency in *C. elegans* mechanosensation. http://arxiv.org/abs/1803.04085 (accessed on 13th June 2025).

38. Meinen, S. (2005). Artificial restoration of the linkage between laminin and dystroglycanopathy ameliorates the disease progression of MDC1A muscular dystrophy at all stages. https://core.ac.uk/download/18233984.pdf (accessed on 13th June 2025).

39. Miller, L., Pinkerton, K., Royer, C., et al. (2017). Nonhuman primate models of respiratory disease: Past, present, and future. https://core.ac.uk/download/323069778.pdf (accessed on 13th June 2025).

40. (NASA-SP-7011(129)) AEROSPACE MEDICINE N74-29442 'AND BIOLOGY: A CONTINUING BIBLIOGRAPHY WITH INDEXES, SUPPLEMENT 129, JUNE 1974 *Aerospace Medicine and Biology: A Continuing Bibliography with Indexes, Supplement 144.* https://core.ac.uk/download/pdf/42887122.pdf (accessed on 13th June 2025).

41. Nakata, T. (2002). Generation of whole-body expressive movement based on somatically theories. https://core.ac.uk/download/pdf/85265.pdf (accessed on 13th June 2025).

42. Nikolić, D. (2015). Practopoiesis: Or how life fosters a mind. https://core.ac.uk/download/pdf/82680806.pdf (accessed on 13th June 2025).

43. Nubiala, J., Gil, J., Gimeno, M., & Lahoz-Beltra, R. (2009). The beauty of the mammalian vascular system. https://core.ac.uk/download/pdf/288349.pdf (accessed on 13th June 2025).

44. Prutskova, N. P., & Seliverstova, E. V. (2018). Protein reabsorption in the amphibian kidney: Comparative and evolutionary aspects. https://core.ac.uk/download/322434774.pdf (accessed on 13th June 2025).

45. Reiss, M., & Tunnicliffe, S. D. (2001). Students' understandings of human organs and organ systems. https://core.ac.uk/download/pdf/82146.pdf (accessed on 13th June 2025).

46. Seaman, W., & Whitehouse, J. H. (2001). Florida marine biotechnology: Research, development and training capabilities to advance science and commerce. https://core.ac.uk/download/11016354.pdf (accessed on 13th June 2025).

47. Stark, J. M. (2013). Efficacy of probiotics for reducing the incidence of lameness in broilers grown on wire flooring. https://core.ac.uk/download/72842771.pdf (accessed on 13th June 2025).

48. Stewart, J. R., 2010. Understanding evolutionary processes during past Quaternary climatic cycles: Can it be applied to the future? *In:* Grant, F., Mergeay, J., Santamaria, L., Young, J. and Watt, A.D., eds. *Evolution and Biodiversity: The evolutionary basis of biodiversity and its potential for adaptation to global change. Report of an e.* Mallorca, Spain: Centre for Ecology & Hydrology, INBO, IMEDA, CSIC-UIB.(accessed on 13th June 2025).

49. Taylor, J. M., & Zickus, V. (2018). 3D + time blood flow mapping using SPIM-microPIV in the developing zebrafish heart. https://core.ac.uk/download/157769456.pdf (accessed on 13th June 2025).

50. Wiley, A. (1972). A review: Some biological effects of high LET radiations. https://core.ac.uk/download/pdf/10297185.pdf (accessed on 13th June 2025).

THE NERVOUS AND ENDOCRINE SYSTEMS

9

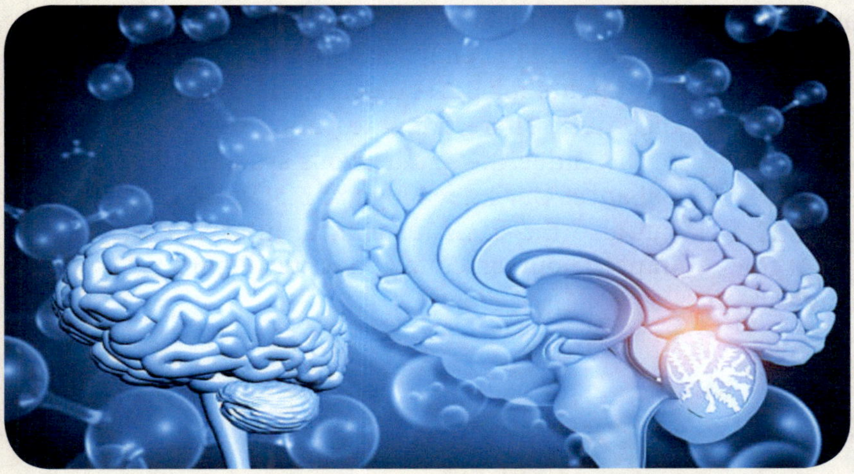

LEARNING OBJECTIVES

After studying this chapter, you will be able to:

1. Give an overview of the nervous system;
2. Illustrate the structure of the nervous system;
3. Give an overview of the endocrine system;
4. Explain the mechanisms of communication;
5. Focus on the interactions between the nervous and endocrine systems;
6. Explore disorders of the nervous and endocrine systems.

Introduction

The intricate relationship between the endocrine and nervous systems underlies knowledge of human physiology. The two systems, although dissimilar in method, work together to regulate the activities of the body, from metabolism to emotional response. The nervous system employs rapid electrical impulses in the form of neurons to pass information, hence providing an immediate response to stimuli. As opposed to this, the endocrine system releases hormones directly into the circulatory system, which causes slower but longer-lasting effects upon target tissues and organs. It is this regulatory alliance that allows for homeostasis, the provision of balance between physiological demand in a changing world. As we continue to describe the roles of these two important systems, an understanding of their interactions, particularly how they establish states of health and disease, becomes necessary in comprehending the complexities of biological processes that govern body function. Examination of the nervous and endocrine systems illustrates a very intricate web upon which homeostatic processes occur.

The communication between these two systems exists in the hypothalamus as it integrates information from both systems to control reactions to stress, hunger, and temperature. Corticotropin-releasing factor (CRF), a peptide that is produced in the hypothalamus, serves to be a significant modulator not only of the hypothalamic-pituitary-adrenal (HPA) axis but also to mediate responses between different systems, including the cardiovascular and immune system. The derangement of CRF control can result in disease states, underscoring the significance of this neuroendocrine interaction in health. Knowledge of these inter-relations provides some insight into the role of imbalances as causes of disorders like depression and anxiety and further underlines the requirement for studying both systems together, not separately. In the study of the therapeutic potential of the interaction between the nervous and endocrine systems, scientists have learned new lines of intervention. One is the recent evidence on the character of natriuretic peptides (NPs), emphasizing the endocrine function of the heart and its regulatory effects on blood pressure and fluid balance. These hormones, which control cardiovascular and renal functions, also possess traits capable of reversing overactivity of the sympathetic nervous system in heart failure (Volpe et al., 2022). As more such relationships are brought into the light by scientific studies, treatments that operate based on these hormonal properties are being developed to provide novel treatments for heart disease and hypertension disorders. Therefore, integration of knowledge of the endocrine and nervous systems not only enhances our knowledge of biological function but also the basis for clinical interventions to improve outcomes for patients (Figure 9.1).

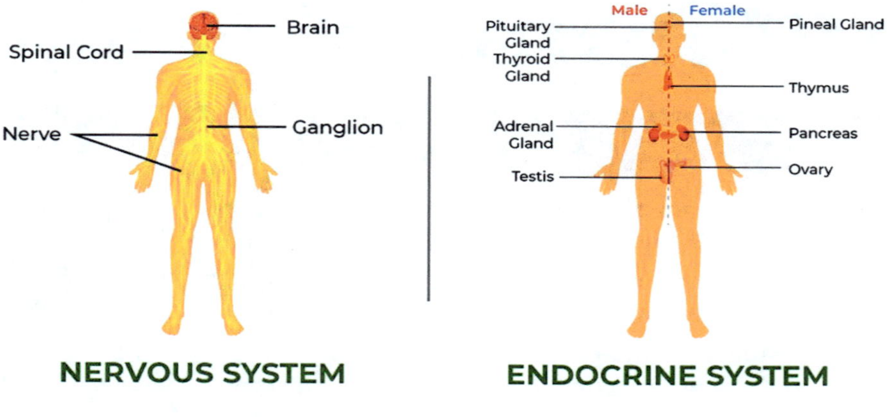

Figure 9.1. The nervous and endocrine systems.

Source: https://media.geeksforgeeks.org/wp-content/uploads/20230313102134/Difference-Between-Nervous-System-And-Endocrine-System.jpg.

Definition of the Nervous System

The nervous system is a sophisticated system that is crucial for the coordination and integration of body functions in order to allow organisms to react to their internal and external environment. The foundation of the nervous system is the CNS and the PNS. The CNS, which includes the brain and spinal cord, is tasked with interpreting sensory information and determining the response, accordingly, serving as the body's control center for all its activities. While the PNS links the CNS to limbs and organs, allowing for the exchange of signals around the body. This complex setup enables speedy communication between body parts, permitting effective responses to stimuli in time. As a principal regulatory system, the nervous system not only talks to other body systems but also contributes greatly to behavioral, cognitive, and emotional functions, making it the foundation of human physiology. Additional research indicates that the operation of the nervous system transcends traditional boundaries, being deeply integrated with other systems , such as the endocrine and immune systems. There is evidence that psychological processes, e.g., emotion and cognition, are linked to physiological activities throughout the body and suggest a more integrated model of human functioning. For instance, the idea that people supervene on nervous, endocrine, immune, and genetic networks' computational processes emphasizes the interconnectedness of such systems, meaning that mental states are not reduced to activities in brains but are constituted by broader biological networks (Steinhart et al., 2001). Hence, the nervous system must be imagined not as a separate system but as an integral part of a total physiological system that includes complex relationships among various systems in the body. Moreover, the nervous system is also accountable for how stress responses work in an individual, affecting overall wellness. It addresses the need for a holistic perspective when examining health and disease, moving beyond the purely cognitive or psychological approach. As implied, there has been a persistent lack of congruence between healthcare providers who adopt top-down (brain-focused) and bottom-up (body-focused) approaches (ILC Workshop ReportILC Workshop Report,2006). This indicates there is more imperative for a holistic view, one which recognizes the interaction between the nervous system and body responses in shaping health outcomes. Embracing the high-level operation of the nervous system in harmony with the endocrine system, scientists and clinicians can more readily engage the complexity of human health, ultimately to improve interventions and results (Figure 9.2).

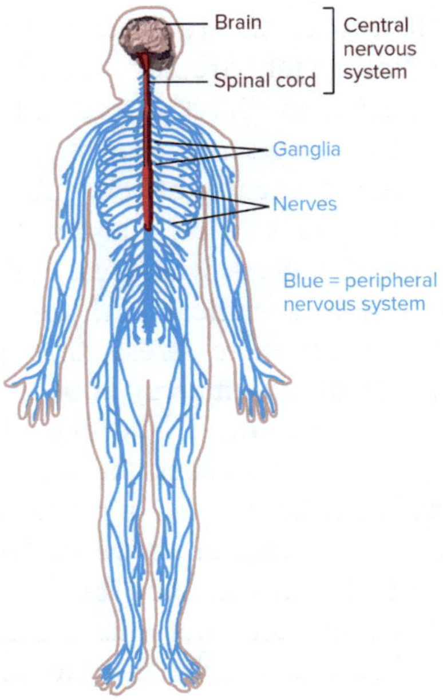

Figure 9.2. The nervous system.

Source: https://bpb-us-e1.wpmucdn.com/sites.gatech.edu/dist/6/1810/files/2017/06/f7fa1bdddc896c2c635a1 929c6578b745a6030c2.png.

Definition of the Endocrine System

It is important to be familiar with the definition of the endocrine system so that one can comprehend its role in the grand picture of human physiology. The endocrine system is a network of interconnected glands that secrete hormones directly into the circulatory system and thereby regulate various body functions. Unlike the nervous system, with its widespread network of neurons transmitting signals, the endocrine system transmits signals in the form of chemicals, controlling long-term adjustment in the metabolism, growth, and development of the body. The pituitary gland, thyroid gland, adrenal glands, and pancreas are the major glands of the endocrine system, each of which secretes specific hormones that influence target organs and tissues. The intricate relationship among these glands provides the regulation of homeostasis, depicting the system's important role in preserving health throughout a person's lifespan. Moreover, the regulatory functions of the endocrine system are not just the excretion of hormones; it is directly involved with the body's response to stress and environmental changes. Stress, acute or chronic, triggers hormonal responses that can fundamentally alter the physiological processes. For instance, the HPA axis is a critical pathway that gets activated in reaction to stress, through which the brain controls a vast array of body reactions by releasing hormones. The ILC Workshop Report (2006) emphasizes the importance of the awareness of how stress impinges on health and advocates a comprehensive understanding that takes into consideration both the effects of the endocrine and nervous systems. This interrelatedness also addresses the importance of health care professionals having a holistic perspective in assessing the patient's health because the relationship between these systems has long-term implications on overall welfare. Moreover, the review of the endocrine system involves its susceptibility to environmental toxins, whose impact on health is of significant concern. Some recent studies have applied network biology approaches to examine the effects of exposure to these pollutants on hormonal pathways and, subsequently, human health. This kind of research establishes that certain environmental contaminants are capable of disturbing endocrine function at a cost to health. Iida & Takemoto (2018) illustrate the complexity of this phenomenon by describing how environmental pollutants influence genetic and molecular pathways that play a central role in maintaining endocrine balance. Recognizing these dynamics both emphasizes the critical role of the endocrine system in contexts of health as well as underscores their urgent need to know more about environmental influence on hormonal control and disease expression (Figure 9.3).

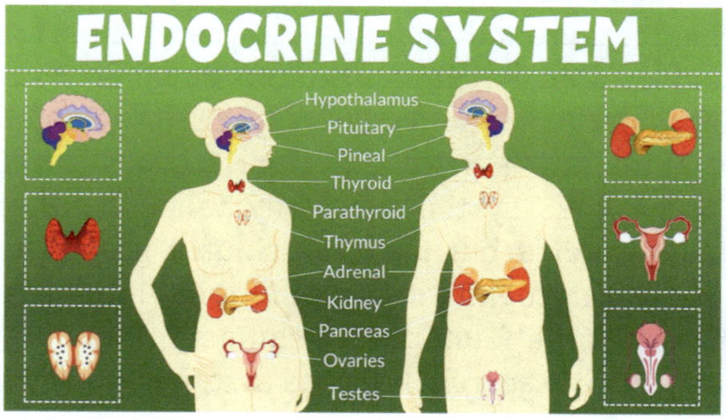

Figure 9.3. Endocrine systems.

Source: https://i.ytimg.com/vi/PvuHB9XYENQ/maxresdefault.jpg.

Importance of Studying Both Systems

A study of the relationship between the endocrine and nervous systems is essential for a holistic knowledge of human physiology. Both systems are complementary in their action, influencing a variety of functions of the body ranging from stress reactions, metabolic regulation, and homeostasis. For instance, while the nervous system sends rapid signals allowing rapid responses to stimuli, the endocrine system releases slow, residual effects through the delivery of hormones. The duality points to the need for a comprehensive study of both systems because dysfunctions in one system will lead to cascading dysfunction in the other. By integrating knowledge from both domains, researchers and clinicians can better understand complex disease states and develop more effective therapeutic strategies that are specific to the complex interaction between the systems. Also, stress research that looks at the psychophysiological effect is where the nervous and endocrine systems need to be discussed together. Stress triggers a series of reactions involving the hypothalamus, pituitary gland, and adrenal glands, that are basic components of the hormonal axis that are related to the nervous system. It is important to know how age and sex differences impact these responses because, as studies have shown, differences in cortisol levels and activity of the sympathetic nervous system can emerge from these differences. Expansion of such research into the fields of clinical practice and public health is where identification of such individual differences in stress responses can inform more personalized avenues of intervention in the management of stress-related health conditions, and ultimately improved health outcomes (Hidalgo et al., 2018). In the overall context of health care, recognizing interdependence among the nervous and endocrine systems has the promise to bridge the gap among disciplines that have traditionally been either neurological or based on the hormonal emphasis of health. As previously stated, stress plays a huge role in health throughout the life cycle, yet much of the health care continues to emanate from the dichotomous approach to care (ILC Workshop Report,2006). By promoting interdisciplinarity and the sharing of combined care models, professionals can more effectively react to the multifaceted complexity of health disorders. Interdisciplinarity not only enriches our understanding of the health experiences of individuals, but also advances a more inclusive paradigm for care, leading to richer prevention, therapeutic, and well-being outcomes.

9.1. Overview of the Nervous System

Learning Outcome

1. Major components: Brain, spinal cord, and nerves;
2. Functions of the nervous system;
3. Types of nervous systems: Central and peripheral.

The nervous system, which is a very intricate net, is basically tasked with the coordination of the countless body functions. Composed of two main components—the CNS and PNS—it operates to transfer messages from the brain, spinal cord, and the rest of the body. The CNS, composed of the brain and spinal cord, serves as the control center, processing information and making responses to environmental changes. Contrarily, the PNS links the CNS to limbs and organs, allowing for communication essential to voluntary movement and involuntary functions.

Through electrical impulses and neurotransmitters, the nervous system provides a timely and effective response to stimuli, which illustrates the extent of its regulatory function in maintaining homeostasis and underpinning health in general. Furthermore, the nervous system's interaction with the endocrine system illustrates its complexity and importance in human physiology. This exchange is particularly evident in the manner that which stress impacts bodily processes. Stressors activate the hypothalamus-pituitary-adrenal (HPA) axis, a functioning component of each system. This axis activates cortisol release, a hormone that assists in modulating stress.

Chronic stress, nonetheless, can unbalance this process, leading to a plethora of health ailments. According to one study, across the life course, stress contributes extensively to health and disease. Although it is well documented that the brain is responsible for coordinating the myriad means by which the body responds to these experiences, there exists a disconnect in the delivery of health care between head-up and head-down providers (ILC Workshop Report,2006). It is essential that this dynamic interaction be understood in order to appreciate how the nervous system influences global health and wellness. In addition to stress responses, the nervous system significantly affects immune response, demonstrating its wide scope. Recent research has identified exposure to endocrine-disrupting chemicals that can undermine immune response, potentially connecting the function of the nervous system to respiratory ailments or diseases. More specifically, such chemicals can suppress the secretion of pro-inflammatory cytokines from activated alveolar macrophages, suggesting a compromised immune system. The ability of the nervous system to communicate with both the immune and endocrine systems is a witness to its health maintenance function. From studies, it is noted that Cells exposed to PMA and PBDEs released significantly lower amounts of pro-inflammatory cytokines compared to controls. Suppression of pro-inflammatory cytokines, characteristic of a compromised immune system, is a sign that long-term exposure to PBDEs can increase the vulnerability to respiratory disease (Hennigar et al., 2010). Through this lens, one can appreciate the intricate connections between the nervous system, endocrine responses, and immune health (Figure 9.4).

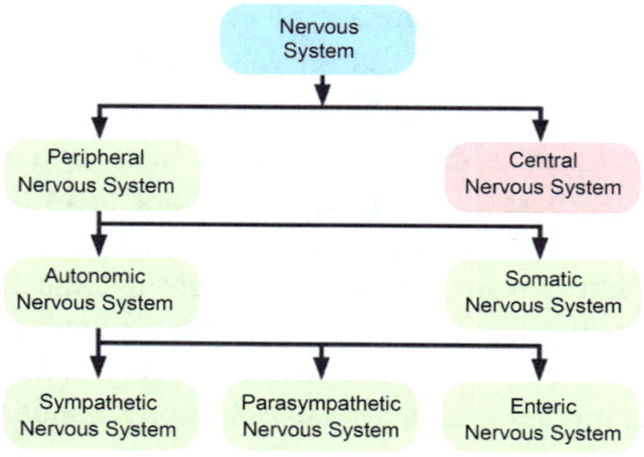

Figure 9.4. The nervous system.

Source: https://humanbiology.pressbooks.tru.ca/wp-content/uploads/sites/6/2019/06/Divisions-of-the-Nervous-System.png.

9.1.1. Major Components: Brain, Spinal Cord, and Nerves

Most of the nervous system, i.e., brain, spinal cord, and nerves, act in harmony with one another to produce complex functions required for living and adaptation to the environment. The brain, which is the command center, receives sensory information, sends motor commands, and regulates homeostasis. It is made up of various components, each of which is specialized for a specific function, i.e., the cerebral cortex for thinking and the cerebellum for movement regulation. Moreover, this central organ communicates with the body via the spinal cord, an important route of signals between the brain and peripheral nerves.

The heavy interconnection of these components allows for instant transmission of information, which enables voluntary action as well as reflex responses essential to everyday life. The spinal cord serves as a nerve impulse pathway in addition to being integral to reflex action. Segmented into individual segments, it gives rise to pairs of spinal nerves that go on to innervate different parts of the body, thereby facilitating communication from the CNS to the peripheral organs. Its segmentation facilitates reflexes independent of the brain, allowing for instant response to a stimulus, e.g., retraction of one's hand when placing it on a hot surface. Furthermore, the spinal cord also plays a role in coordinating the motor activities, such that it facilitates accurate movements required for actions as simple as walking to activities as specific as typing.

Reflex actions thus portray the integrated character of the nervous system, highlighting the spinal cord's role in ensuring operability and responsiveness within an ever-changing environment. Nerves are the basic building blocks extending the length of the body, transmitting signals between the CNS and peripheral tissues. Composed of groups of axons, nerves are divided into sensory, motor, and mixed nerves, each with particular roles in bodily functions. Sensory nerves convey feedback from the receptors to the brain, facilitating feelings of temperature, pain, and touch. Motor nerves, on the other hand, convey orders from the brain to the muscles of the skeleton, facilitating movement and coordination. The intricate relationship between the spinal cord, brain, and nerves serves to underscore the primary avenues employed in homeostasis and responses to external stresses. In addition, recent research on the gut-brain axis continues to explore the interaction between the enteric nervous system and the CNS, proposing intricate interactions with general well-being (Prins et al., 2011) and possibly new ways of treating nervous system disorders (Figure 9.5).

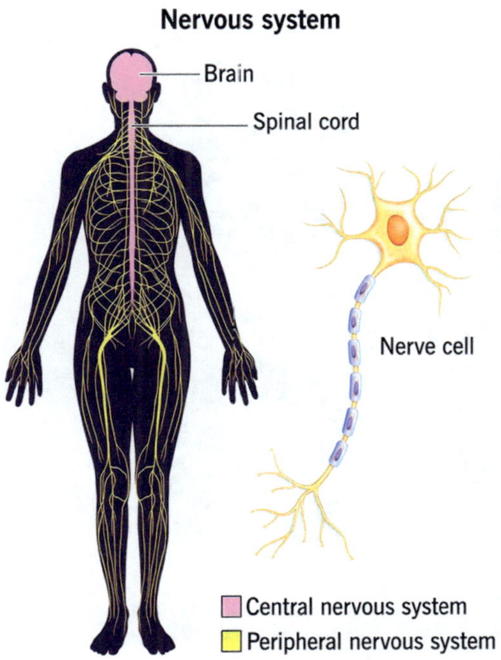

Figure 9.5. Major components of the nervous system.

Source: https://my.clevelandclinic.org/-/scassets/images/org/health/articles/21202-nervous-system.

9.1.2. Functions of the Nervous System

The nervous system carries out a variety of basic functions required for automatic as well as voluntary operations in the human body. Its key function is the rapid transmission of information within the organism to respond immediately to stimulus from the environment. This capability is made possible by a highly developed system of neurons and synapses that form intricate paths between the brain, spinal cord, and peripheral nerves.

Furthermore, the nervous system plays a significant role in sensation perception since it enables a person to sense and interpret various kinds of stimuli, ranging from vision and hearing to touch and smell. A perfect synchronization of sensory input and motor output is essential to ensuring homeostasis and to enabling effective interaction with the environment, thus affirming the place of nervous systems as a sophisticated command center for bodily functions. Another critical aspect of the nervous system is that it has a role in regulating body functions through interactions with other systems, specifically the endocrine system.

One such interaction is energy homeostasis regulation, where the CNS integrates information from endocrine determinants like leptin and adiponectin from adipose tissues, and osteocalcin from bone. These chemicals impact not only energy balance but also bone metabolism, forming a multifaceted axis that highlights the interrelatedness of the skeletal, adipose, and nervous systems. Such an interaction highlights how the nervous system acts as a coordinating center, collating information from various sources in order to control physiological responses, hence reflecting the sophistication of biological feedback loops that guarantee overall health.

Accordingly, Fica et al. (2021) assists in delineating the significant connections between different forms of tissue and how these influence the nervous system. Furthermore, the connection between stress and the way the nervous system operates is an increasingly relevant factor when considering outcomes of health. How the brain perceives and responds to stress stimuli has a profound influence on mental and physical health. Chronic stress response can lead to dysregulation of neuroendocrine systems, resulting in adverse outcomes , such as mood disorders, cardiovascular diseases, and metabolic syndromes.

The head-up vs. head-down approach to healthcare emphasizes the necessity of an integrative approach to understanding how mental states influence physiologic responses within the body. It is in an appreciation of how the nervous system regulates a variety of body responses to stress that its extensive role in health and disease is apparent, a valuable realization on our path to more holistic medical practices. These realizations have their foundation in the holistic approach provided by the ILC Workshop Report (2006), which insists on a unified method of studying the interrelations among body systems (Figure 9.6).

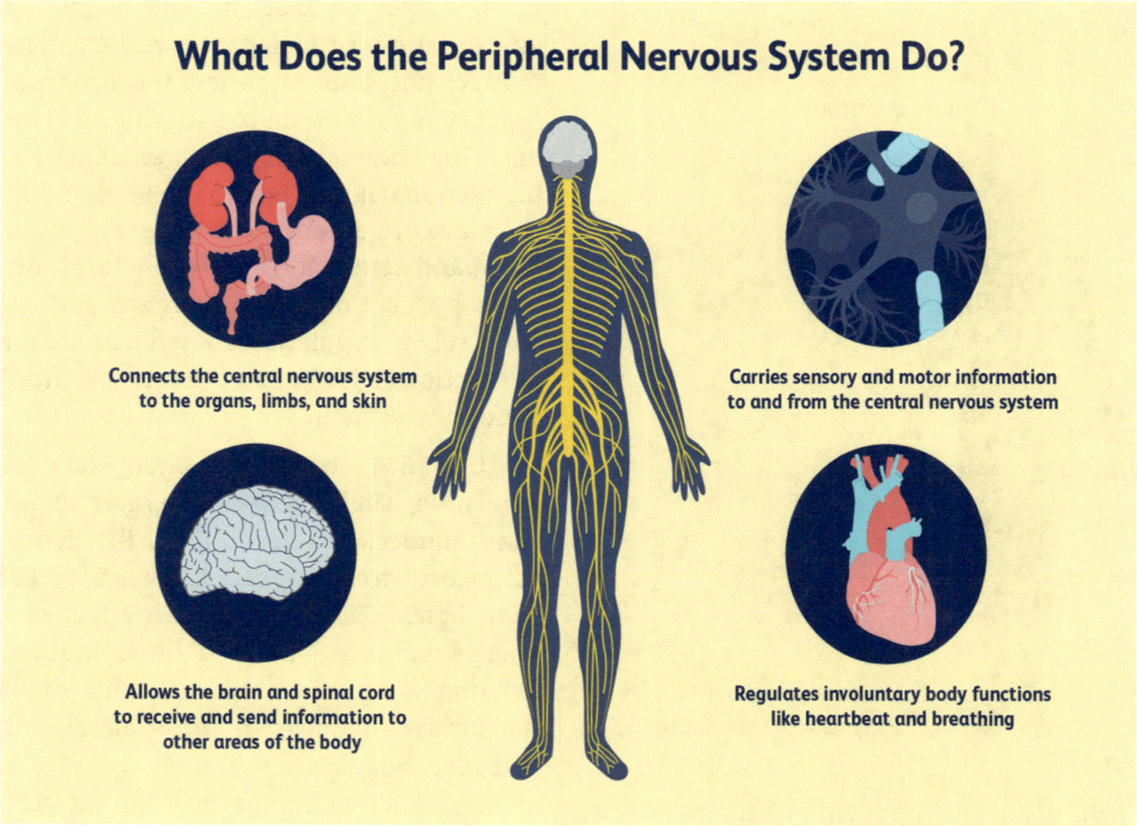

Figure 9.6. Functions of the nervous system.

Source: https://www.verywellmind.com/thmb/S6e2zuErsAfYKwHcrd1yKL2iLt8=/1500x0/filters:no_upscale():max_bytes(150000):strip_icc()/what-is-the-peripheral-nervous-system-2795465-FINAL-b69e1bb803654212a83d9e68eb4847d0.png.

9.1.3. Types of Nervous Systems: Central and Peripheral

The nervous system, a complex system that is responsible for the transmission of signals throughout the body, is primarily broken down into two distinct categories: the CNS and the PNS. The CNS encompasses the brain and spinal cord, the center hub for information processing and giving responses. The central hub integrates sensory information from our surrounding world and coordinates voluntary and involuntary responses. In contrast, the PNS contains all neural structures outside the CNS, including cranial and spinal nerves. It serves as a middle function linking organs and limbs to the brain, providing a two-way communication pathway so that the body may appropriately react to stimuli. These two systems, when integrated, work in harmony in the control of homeostasis and interaction with the external world, underlining their basic importance in all body functions. The autonomic division of the PNS also adds to the interdependence of physiological functions by explicitly differentiating involuntary control of muscles and control of the visceral organs into sympathetic and parasympathetic divisions. The sympathetic nervous system is primarily associated with the fight or flight response, calling on the body's resources upon sensing danger. The parasympathetic nervous system, however, invokes a state of rest and digest, facilitating bodily maintenance and energy conservation upon exposure to stress. Research has determined that both branches of the ANS respond dynamically to conditions , such as overfeeding and obesity. Specifically, alterations in sympathetic and parasympathetic function have been associated with alterations in autonomic outflow, implying a dynamic interaction of nervous system regulation and metabolic states, namely that of peripheral energy substrate homeostasis, as in recent evidence indicating heightened sympathetic tone during hyperphagic states (Prins et al., 1996). Understanding the intricate activities of the CNS and PNS reveals their interconnected relation to the endocrine system, most specifically through organs like the pituitary gland. Acting as an essential regulatory intermediary, the pituitary gland secretes and releases hormones that affect diverse physiological processes in the body, linking neural processes to endocrine responses. Gland responses have a close relation with various feedback mechanisms with various neurotrophic factors, which may influence not only normal gland function but also disorder formation , such as pituitary adenomas. Such a dependency indicates that hormonal and neurological responses serve crucial functions in maintaining homeostasis, regulating stress, and coping with developmental processes (Agostinelli et al., 2017). Thus, nervous system activities extend beyond reflex activity to encompass more diffuse metabolic and endocrine effects in a display of the ubiquitous richness of body regulation.

Test Your Knowledge

1. What are the functions of the nervous system?
2. What are the types of nervous systems?

9.2. Structure of the Nervous System

Learning Outcome

1. Neurons: Structure and function;
2. Glial cells: Support and protection;
3. Synapses and neurotransmitters.

Knowledge of the nervous system's structure is at the heart of understanding its role within various physiological and psychological processes. The nervous system is commonly divided into two broad divisions: the CNS and the PNS. The CNS, which consists of the brain and spinal cord, is the central command for information processing, while the PNS includes all the neural elements outside of the CNS, providing for communication with the outside world. Such a complex organization not only enables motor control and sensory perception but also supports higher-level functions , such as emotion and cognition. The interaction between the PNS and CNS also illustrates how the body receives internal and external stimuli, highlighting the importance of the structural division in its appreciation of general health and sickness. In the CNS, certain areas and structures consolidate a variety of functions that are essential to survival. The brain is a highly intricate organ with various divisions and every division with some specific function to carry out, , such as the cerebrum to carry out higher-order cognition and the cerebellum for movement coordination and motor response regulation. The spinal cord is also a major tract that transmits information from the brain to the peripheral nerves. A highly significant aspect of this structure is the influence of stress on health via neurobiological mechanisms. The integration of brain region responses together presents the capability to make comprehensive responses to stressors, thus demonstrating the CNS's coordinating function in the body's responses. As implied, there exists a mismatch between head up oriented health care practitioners and head down oriented ones, thereby necessitating that both structural and functional entities be considered in addressing health issues (ILC Workshop Report,2006). The PNS helps the CNS in the sense that it helps convey signals effectively around the body. It is further divided into the somatic nervous system, which controls voluntary movements, and the ANS, which controls involuntary actions , such as heartbeat and digestion. The complexity of the ANS is illustrated by the recently discovered superior cupric autonomic nervous system (SCANS) in *Drosophila larvae*, consisting of new neuro-enteroendocrine cells that are responsible for orchestrating important physiological processes. The cells play a vital role in peristalsis, illustrating the complexity of the PNS's role in many body processes (Johnson et al., 2009). This structuring of the organism highlights the central importance of the nervous system not only to reflex activity but also to guiding key functions , such as digestion and growth, showing its central place in the regulation of homeostasis and overall health.

9.2.1. Neurons: Structure and Function

Neurons are the components of the nervous system and the fundamental units to carry information throughout the body. The neuron's shape has three key elements: the cell body, dendrites, and the axon. The nucleus is found in the cell body and is crucial in maintaining the health of the neurons, while dendrites are branching-like projections that receive signals from other neurons. The axon, usually a slender, elongated projection, transmits electrical impulses from the cell body to other neurons, muscles, or glands. This complex structure enables neurons to communicate through synapses, where neurotransmitters transmit information. The interconnectivity and plasticity of these cells make complex information processing possible, adding to the concept that the neuron is more than just one individual but part of an integrated system that controls behavior, cognition, and physiological reaction in the organism. The neuron is not only concerned with communicating; it is necessary for environmental stimulus interpretation and response. This interaction of sensory receptors and neurons forms the foundation of sensation, allowing the nervous system to receive and process external and internal signals. Neurons are connected into vast networks that are responsible for parallel distributed processing, which is critical to voluntary and involuntary movements. This perspective highlights how all aspects of biological functions, including those related to psychological and cognitive abilities, are regulated by the activity of neurons and their connections. It supports the thesis that human behavior and experience do not necessarily emerge in the brain but emerge from the integrated activity of cellular networks within the body and submits the interplay between neural structure and function as a keystone in the operations of the nervous system (Steinhart et al., 2001). Moreover, the interaction between the endocrine and nervous systems determines the function played by neurons in stress response and homeostasis. The endocrine system is coordinated with the nervous system through neural loops and hormones to facilitate communication in the form of control over body functions. Stressful conditions lead to a cascade of responses under the control of both systems, wherein the brain regulates the body's response by secreting stress hormones. It indicates the need for an integrated approach towards the physiology of human beings since understanding the role of the brain only may be losing critical interactions with other systems. The existing gap between clinicians treating the brain and those treating body responses highlights the need for an integrated model that values the profound impact of stress on disease and health (ILC Workshop Report,2006). From this point of view, the physiology and morphology of neurons reveal their pivotal role in not just neural processes but also in the broader context of organismal response and health (Figure 9.7).

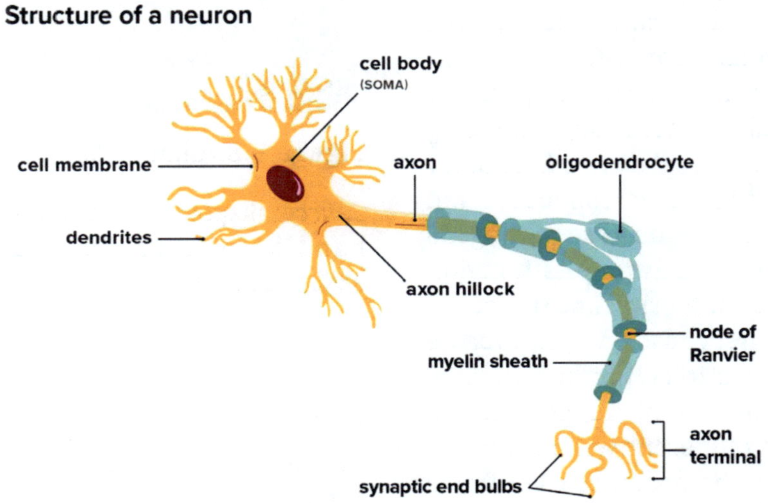

Figure 9.7. Structure of neurons.

Source: https://i0.wp.com/post.healthline.com/wp-content/uploads/2022/01/1932990_An-Easy-Guide-To-Neurons-01.jpg?w=1155&h=2007.

9.2.2. Glial Cells: Support and Protection

Glial cells, also referred to as the unsung heroes of the nervous system, have significant roles in the support and protection of neurons, enabling the intricate actions of the nervous and endocrine systems. Unlike neurons, whose function is primarily the transmission of signals, glial cells reinforce the overall structure of the nervous system by forming a supporting network. They facilitate homeostasis, regulate the chemical environment, and modulate neural activity, allowing neurons to function optimally. Also, some types of glial cells, , such as astrocytes, provide metabolic nourishment and enable communication between neurons and blood vessels, vital for the well-being and activity of neurons within the various body's systems. In addition to their supportive role, glial cells also play a significant role in responding to damage within the nervous system. When neurons are damaged, glial cells rapidly respond to the site of damage, where they participate in protective activities. These include the formation of a glial scar, which is employed to isolate the site of damage from the rest of the tissue and establish a physical barrier against encroaching pathogens. Moreover, glial cell-released neurotrophic factor secretions promote survival and neuronal regeneration, enhancing recovery mechanisms. Microglia are still another glial cell that serves as immune responders and actively search for and phagocytose pathogens and debris, further demonstrating its function in maintaining neural integrity and function, especially for the nervous and endocrine systems. Their interactions further involve biochemical communication, suggesting an advanced relationship inherent within the two systems. For instance, the presence of proteins like chromogranin A (CGA) in glial cells has been ascribed to regulatory mechanisms influencing neuronal function and endocrine signaling. Studies indicate that CGA may be implicated in the signal transduction of the motor system since it is found in CNS neurons and structures derived from adrenal tissue. Deciphering these processes is imperative because they demarcate the intricate roles of glial cells in signaling and neuroprotection, which are essential for the general functionality and responsiveness of the nervous and endocrine systems. The integration of this knowledge could lead to gains in therapeutic strategies for endocrine diseases and neurodegenerative disease and highlight the essence of protective functions of glial cells in human health (Figure 9.8).

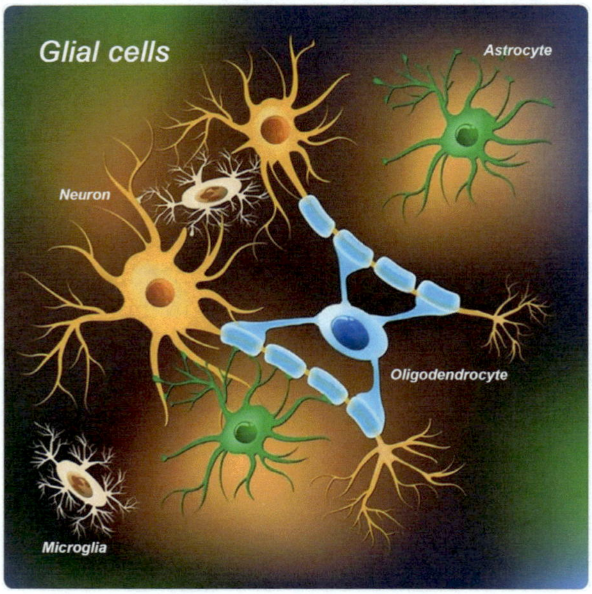

Figure 9.8. Glial cells.

Source: https://www.news-medical.net/image-handler/picture/2019/10/%40shutterstock_241178524.jpg.

9.2.3. Synapses and Neurotransmitters

In the intricate system of the nervous system, synapses are essential connections between neurons for the transfer of signals throughout the body. At synaptic sites, the presynaptic neuron releases neurotransmitters, chemical messengers that bind to postsynaptic receptors on the target neuron. This binding initiates a cascade of events that not only compels signal transfer but also regulates neuronal excitability. Neurotransmitters belong to several classes, including amino acids, monoamines, and neuropeptides, with their specific functions in various physiological processes. Glutamate is the primary excitatory neurotransmitter, while gamma-aminobutyric acid (GABA) is the primary inhibitory neurotransmitter, an indicator of the delicate balance of excitatory and inhibitory inputs for peak brain function. The exploration of individual neurotransmitters has unveiled their pervasive involvement in emotional and cognitive processes. For example, dopamine is an important neurotransmitter in reward and motivation, while serotonin is implicated in mood balance and emotional stability. As such, dysregulation of these neurotransmitters is the underlying cause of a variety of neuropsychiatric conditions ranging from depression to anxiety. Significantly, the synthesis and release of certain neurotransmitters are contingent on extrinsic factors like stress and environmental exposure. The two-way exchange between the nervous system and the external world points to the complexity of these biochemical interactions. Fascinatingly, studies indicate that stress throughout a lifetime can have profound impacts on neurotransmitter systems and overall health and emphasize the interdependence of the endocrine and nervous systems to provide homeostasis and health (ILC Workshop Report,2006). Moreover, some neurotransmitters, i.e., CGA, are a fine example of the convergence of neurobiology and neuroendocrinology, which emphasizes the interaction between neural networks and hormonal secretion. CGA has been shown to be present in several brain regions, one of them being the olfactory bulb, where it is synthesized and then selectively transported into neuroanatomically distinct cellular compartments of the neuron (Baker et al., 1990). This specificity suggests that CGA is not only an essential component of neurotransmitter storage and release but possibly a part of regulating synaptic plasticity and neural communication as well. The understanding of how various neurotransmitters function, interact, and are regulated continues to grow and contributes to our knowledge of synapses as well as to designing therapeutic interventions for psychiatric and neurodegenerative diseases. Ultimately, these findings bring support to the interconnects among synapses, neurotransmitters, and the big picture of the nervous and endocrine system as being crucial components regulating health and behavior (Figure 9.9).

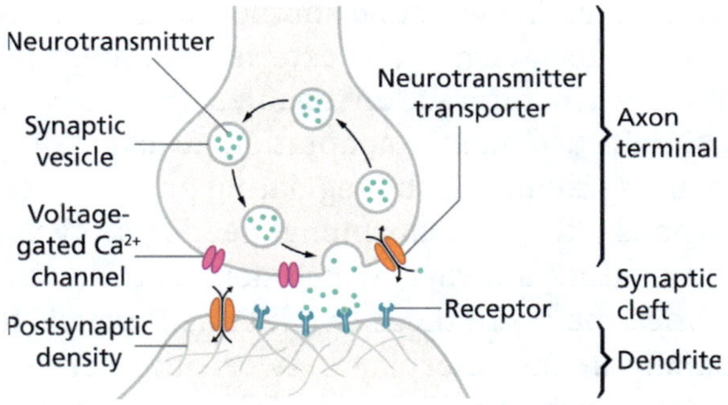

Figure 9.9. Synapses and neurotransmitters.

Source: https://qbi.uq.edu.au/files/7758/Synapse-QBI-brain-neuroscience.jpg.

Test Your Knowledge

1. What are neurons?
2. What is the role of glial cells?

9.3. Overview of the Endocrine System

Learning Outcome

1. Major glands: Pituitary, thyroid, adrenals;
2. Hormones and their functions;
3. Interaction with other body systems.

The endocrine system is an important control network within the human body, coordinating a variety of physiological processes by means of hormone messages. Composed of glands , such as the pituitary, thyroid, adrenal, and pancreas, this system is crucial to obtaining homeostasis. Hormones, chemical messengers secreted directly into the bloodstream, influence growth and metabolism as well as mood, reproduction, and more. The interconnectedness of the endocrine and nervous systems is particularly astounding, with both systems conversing and communicating with one another to collectively respond to internal and external stimuli. This cooperative interaction enables the body to maintain a dynamic adaptation to constantly changing conditions by regulating physiological response through hormonal feedback cycles and neural pathways, which demonstrates the subtlety between these two lifeline systems. In looking at the interplay between the nervous and the endocrine system, it is evident that stress has a significant impact on health results, mediated through hormonal change. The hypothalamus, under the influence of stressors, signals the adrenal glands to release cortisol and adrenaline, hormones that prepare the body for fight or flight. This physiological reaction, though a necessary one for survival, is maladaptive when the stressor is of chronic duration and will lead to long-term health issues , such as cardiovascular disease and immune suppression. Furthermore, the part played by the brain in coordinating these responses makes the holistic model of health provision essential which views emotional health as having a direct effect on endocrine function. It is such understanding that is vital, and it has been emphasized in the reality that throughout the life course, stress is a primary etiologic agent for disease and health (ILC Workshop Report,2006). Exploration of interactions between the endocrine system and immune response also yields strong revelations, for example, concerning hemispheric lateralization and immune functions.

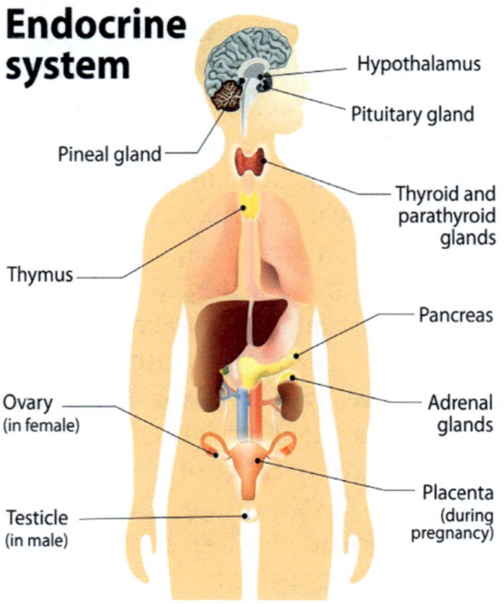

Figure 9.10. Overview of the endocrine system.

Source: https://www.epa.gov/sites/default/files/styles/medium/public/2015-08/endocrine-system.jpg?itok=8pyWl80t.

Earlier studies have established that the functioning of some brain hemispheres could be connected to the effectiveness of immune responses. For example, certain evidence suggests that impaired left hemisphere function might be associated with weaker immunity, while increased right hemisphere activity can stimulate immune response. Such interaction explains the mechanism by which psychological states, as filtered through the brain, influence physical health, an association that remains paramount in clinical and research areas. The examination of these relationships illustrates the complex interplay between psychological process and physiological outcome, underscoring the need for integrative health care and therapy (Gidron et al., 2011) (Figure 9.10).

9.3.1. Major Glands: Pituitary, Thyroid, Adrenals

The endocrine system is crucial in the process of maintaining homeostasis by regulating the hormones released by large glands like the pituitary gland, thyroid gland, and adrenal glands. The pituitary gland, or the master gland, regulates all facets of the hormonal activities through the release of hormones that control other glands in the endocrine system. It is divided between the anterior and posterior lobes, which are both linked to secreting various hormones that influence growth, metabolism, and reproductive processes. Dysfunction of the pituitary may have a significant chain reaction in causing an imbalance in hormones, affecting overall wellbeing and health. Thus, its regulatory functions are crucial in appreciating the intricacies of endocrine relationships, demonstrating the importance of the pituitary gland in the control of disease and health (Brown et al., 1938). The thyroid gland, situated below the Adam's apple, is important in metabolism, energy, and growth control through thyroid hormone secretion, primarily thyroxine (T4) and triiodothyronine (T3). These hormones influence the utilization of energy by the body and influence various physiological processes, including body temperature and heart rate. Imbalances in thyroid hormone can lead to serious health complications, like hyperthyroidism or hypothyroidism, each with its own basket of treatment and management issues. For instance, hyperthyroidism may lead to elevated metabolism and weight loss coupled with anxiety feelings, while hypothyroidism may lead to weight gain and fatigue. Thus, the proper functioning of the thyroid gland is critical to metabolic health, another indication of the interconnectedness of endocrine processes within the larger context of the endocrine and nervous system (Randazzo et al., 2022).

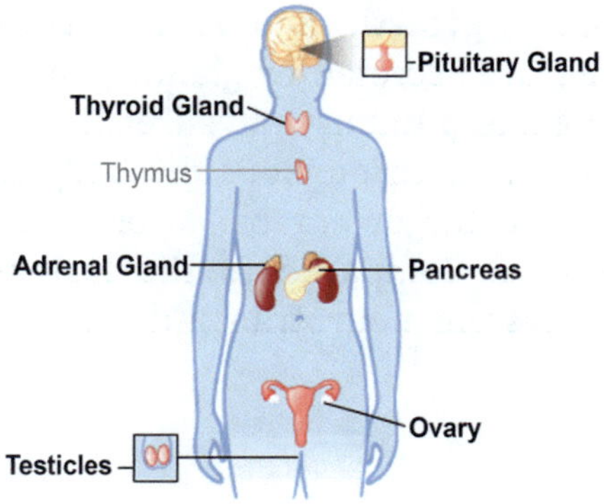

Figure 9.11. Major glands.

Source: https://medlineplus.gov/images/endocrinesystem.png.

The adrenal glands, situated above each kidney, play a vital role in the body's reaction to stress through the production of hormones , such as cortisol, adrenaline, and norepinephrine. These hormones play a crucial role in the body's response to stressors that impact heart rate, blood pressure, and energy. The activation of the adrenal glands results in the fight or flight response, a physical response that prepares the body for danger. Sustained activation of the glands because of ongoing stress can lead to diseases , such as hypertension, anxiety disorders, and metabolic syndrome. With this understanding of the function of the adrenal glands and their interaction with other hormones, it is possible to appreciate the complexity of stress management, highlighting endocrine balance for the prevention of detrimental health consequences that arise from stress (Brown et al., 1938) (Figure 9.11).

9.3.2. Hormones and Their Functions

The intricate relationship between hormones and their function in the body is a fundamental activity of the endocrine system in collaboration with the nervous system. Biochemical messengers known as hormones regulate various physiological activities, including metabolism, growth, and reproduction. The hormones are secreted by glands , such as the hypothalamus, pituitary, and adrenal glands and are transported throughout the body through the blood to target organs, where they exert their effects. Compared to the faster-acting nervous system neurotransmitters, however, hormone effects tend to be of longer duration and more pervasive in effect, highlighting the influence of hormonal balance in maintaining homeostasis. Imbalances of hormones can lead to serious health issues, again pointing up the necessity for a unified system in exploring both hormonal and nervous system regulation of body processes. Furthermore, the interaction between the nervous system and hormones illustrates a complex feedback mechanism that is essential in reacting to stress and overall health. As a reaction to stress, the body activates the HPA axis, leading to cortisol release, an essential stress hormone.

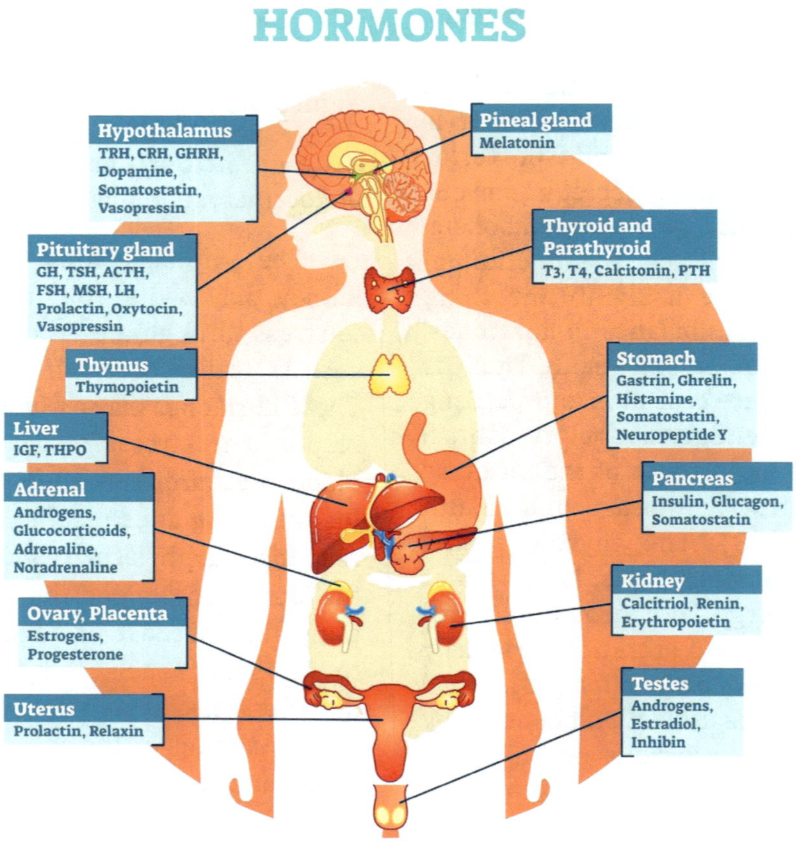

Figure 9.12. Hormones and their functions.

Source: https://www.news-medical.net/images/Article_Images/ImageForArticle_793_17424310885043381.jpg.

This hormonal response not only prepares the body to take instant action but also influences mood and energy, demonstrating the link between psychological and physiological responses to stressors. Since the brain regulates body responses to stress, it is significant for health care practitioners to understand the intricate impacts of hormones like cortisol on body health as well as mental well-being, gaining a better integrated picture of the body as an integrated system (ILC Workshop Report,2006). Moreover, the significance of hormones is that they influence biological rhythms, , such as the circadian rhythm, which indeed belongs to the function of the nervous system. Hormones like melatonin, secreted by the pineal gland, are important factors in transmitting sleep-wake cycles, which then influence cognitive processes, mood, and even metabolic processes. The impairment of these rhythms can result in insomnia and other psychiatric conditions. Additionally, understanding how hormonal fluctuations affect neurobiological mechanisms further our understanding of several health conditions, including those that are concerned with anxiety and depression. The interaction between hormonal activity and nervous system activity is thus paramount in clinical as well as laboratory situations, providing new avenues for innovative treatment and prevention of endocrine disorders (Arnold et al., 2001) (Figure 9.12).

9.3.3. Interaction with Other Body Systems

The interrelation between the nervous system and the other systems of the body demonstrates the nature of human biological complexity. The nervous system communicates with numerous systems, , such as the endocrine system, to regulate physiological processes that are relevant for homeostasis. For instance, signals from the nervous system stimulate the release of hormones from endocrine glands, which in turn modify body functions like metabolism, growth, and stress response. This synergistic action illustrates the way the operation of neurons can influence states within the body well beyond mere motoric response, supporting the integrated nature of human physiology. When people supervene, or exist outside, the processes of computation in nervous networks, this grants a broader range of action and reaction in the body's systems, pointing out that the psychological and biological traits of individuals are not localized in the brain but are copied in the various bodily networks (Steinhart et al., 2001). Further illustrating the interdependence of body systems, the response of the nervous system to external stimuli can have profound effects on endocrine function. For example, in periods of stress, the sympathetic nervous system activates the adrenal glands to release adrenaline and cortisol. These hormones prepare the body to respond to a fight or flight situation, with physiological systems being prepared to respond. The neurotoxicity further demonstrates the way that interference with neural function cascades over to other systems to generate disruptions that mark the tenuous balance sustained by these interactions. Since neurotoxicity has been shown to result from a range of mechanisms in order to target metabolic processes necessary to nervous system function, the common thread here is now set: any interference with neural function resonates throughout the whole organism and affects immune and endocrine responses as well (Brust et al., 2014). Appreciation of the interplay between the nervous, endocrine, and other body systems is key to appreciating human disease and health. The complexity of these interactions plays a significant role in both normal physiology and disease. For instance, long-term stress may lead to endocrine system dysfunction and abnormal hormone levels, which in turn influence mood and cognitive function, which are regulated by the nervous system. Conversely, each failure in the endocrine system would feed back negatively on neurological health as the hormonal imbalance would alter neural pathways for messaging. This interdependency is critical, showing that the human body cannot be understood by the isolated study of its constituent systems. Instead, one needs an integrated strategy in which every system must be viewed in relation to how it interacts with other systems, thus achieving a better knowledge of body disease processes and body health (Steinhart et al., 2001).

Test Your Knowledge

1. What are the major glands?
2. What are the hormones and their functions?

9.4. Mechanisms of Communication

Learning Outcome

1. Electrical signals in the nervous system;
2. Chemical signals in the endocrine system;
3. Comparison of speed and duration of responses.

The details of the communication between the nervous and endocrine systems are complex and multifaceted, which indicates the complexity of human physiological regulation. The nervous system uses a faster mode of communication that involves electrical impulses and neurotransmitters, thus enabling the system to respond to stimuli instantly. The system is especially useful in orchestrating rapid responses, like reflexes, which are important to survival. The endocrine system, in turn, is slower in communication, using the hormone discharge into the blood and its further influence on the target organs over an extended period. These different but complementary strategies indicate that the two systems are required to keep homeostasis and adapt to environmental fluctuations. An effective interaction of these two systems is crucial to coordinate such processes as growth, metabolism, and responses to stress, which guarantees that the body will perform best under different conditions. The neuroendocrine system is one of the examples of the cooperation of the nervous system and the endocrine system and the way they can influence each other in the process of physiological regulation. Hormones secreted by glands must traverse the brain-blood barrier (BBB) to reach their targets, which underscores the challenges posed by this strict communication barrier. To take a few examples, the stomach hormone ghrelin not only affects appetite but also interacts with distinct neuronal populations in the brain. It is crucial to investigate how ghrelin and other hormones targeting the brain do their work to understand their physiological functions. Although the neuroendocrine system is clearly decisive in homeostatic processes, studies indicate that there are knowledge gaps on the way these hormones influence neuronal networks and behavior. Thus, these complex hormonal circuits are an important direction of research in the neuroscience and endocrinology fields, which can shed light on their wider health and pathological significance. Moreover, it is also impossible to overestimate the role of certain hormones in the analysis of the communication processes in the neuroendocrine system. To illustrate, ghrelin, the so-called hunger hormone, has a direct effect on appetite regulation since it affects several brain nuclei that are sensitive to this peptide. But the results have shown that the availability of peripheral ghrelin to the brain is highly restricted, so the question arises how the appetite-stimulating effects of ghrelin appear. These discoveries of peripheral hormone brain targets increase our knowledge of their functions in the regulation of physiological processes other than appetite. Therefore, there is a growing body of research aiming at explaining the complicated neuronal circuits and molecular processes that are part of this communication between the endocrine messages and neural reactions. Such investigations hold potential for advancing therapeutic strategies aimed at addressing appetite disorders and various metabolic conditions, underscoring the importance of understanding the nuances within this communication framework (Cabral et al., 2015, 2020) (Figure 9.13).

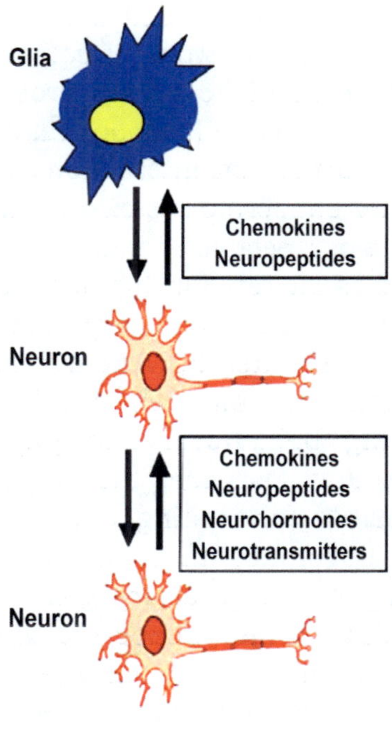

Figure 9.13. Communication pathways in the nervous system.

Source: https://www.researchgate.net/profile/Ellen-Geller/publication/7184458/figure/fig3/AS:277738448539657@1443229414806/Communication-pathways-in-the-nervous-system.png.

9.4.1. Electrical Signals in the Nervous System

The intricate function of the nervous system is largely due to the generation and transmission of electrical impulses, which are the principal mode of communication among neurons. The action potentials or electrical impulses are produced upon exciting a neuron sufficiently, causing a rapid depolarization of the neuronal membrane. This process is crucial since it provides a mechanism for the passing of the electrical signal along the axon to the axon terminals, where the neurotransmitters are released into the synaptic cleft. The speed and effectiveness of this electrical signaling system are high, enabling immediate response to environmental stimuli, thereby ensuring organisms' survival and adaptability. Moreover, the communication among neurons is the origin of higher-order behavior and cognition, and it underscores interdependence between electrical signaling in nervous system physiological processes. The role of electrical signals goes beyond simple communication; they are an essential component of body function overall, influencing the activity of both endocrine and nervous systems. Electrical signals generated in neurons transmit vital information not only to neighboring neurons but also to glandular tissues, thereby coordinating hormonal releases that affect metabolic processes and homeostasis. The crosstalk between electrical signaling in the nervous system and hormonal signals in the endocrine system represents an advanced regulatory system. This junction highlights how information on internal and external environments is conveyed throughout the organism, thereby facilitating adaptive responses. An appreciation of these electrical circuits reveals the complexity of biological processes that result in various psychological and physical effects, suggesting that mind and body are more intimately linked. Recent discoveries concerning the research of electrical signals within the nervous system also emphasize the significance of information in coming to an understanding of consciousness and mental functioning. It has been hypothesized that consciousness occurs as a consequence of functional activity in an informational network of elements distributed throughout the body, encompassing both electrical and non-electrical systems of communication.

It is this view that coheres with the notion that human anatomy is an active information system that never ceases to adapt to changes in the environment via various signaling processes of adaptation. Knowledge of embodiment and disembodiment information processes not only constructs a more comprehensive concept of consciousness but also demonstrates the intricate interdependence of the bodily systems involved in physical as well as mental well-being. In analyzing these integrative processes, we reach an understanding of the significance of electrical signals as primitive components within the overall framework of human biology (Steinhart et al., 2001; Gaiseanu et al., 2019) (Figure 9.14).

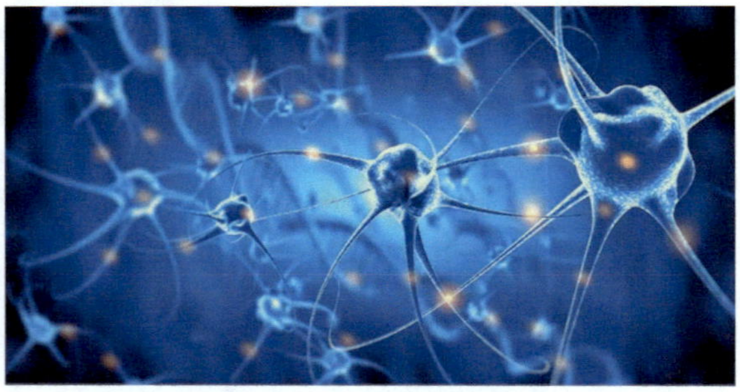

Figure 9.14. Illustration of interconnected neurons transmitting electrical signals, representing neural communication in the nervous system.

Source: https://images.theconversation.com/files/298306/original/file-20191023-119414-19vz68x.jpg?ixlib=rb-4.1.0&q=45&auto=format&w=754&fit=clip.

9.4.2. Chemical Signals in the Endocrine System

The endocrine system basically relies on chemical messages to control an array of processes in the human body. Hormones are chemical messengers that allow communication between several organs and tissues. Hormones are released by other glands, i.e., pituitary gland, thyroid gland, and adrenal glands, and influence diverse functions , such as metabolism, growth, and mood. Unlike the rapid transmission of neural impulses in the nervous system, endocrine signaling over longer durations is required for regulating sustained responses to different stimuli. This characteristic aids in the body's ability to maintain homeostasis, a factor vital for survival. The synergy between the endocrine and nervous systems is a sophisticated network of communication, where networks of hormones crosstalk with neural circuitry to regulate body functions with optimality. Further, the interaction between the nervous and endocrine systems is emphasized through the perspective of neuroendocrinology, which studies the relationship between the two systems. As a multidisciplinary field, this area of study investigates how neurotransmitters and hormones act on each other's functions, determining behavioral and physiological reactions. For instance, the hypothalamus is responsible for linking both systems by releasing hormones that directly modulate the activity of the pituitary gland, which in turn regulates other endocrine glands. The combination of the nervous system's rapid response mechanisms with the longer-lasting effects of endocrine signaling provides the possibility for more delicate regulation of homeostasis. The cooperation is particularly evident in stress response, whereby the nervous system quickly signals the adrenal glands to release cortisol, a hormone that modulates physiological and psychological responses to stressful or threatening situations. In the realm of consciousness and cognition, the function of chemical messengers in the endocrine system assumes further subtlety. Recent research has speculated that consciousness may emerge from the interaction of several bodily informational components, as in the case of the Informational Model of Consciousness. In this model, the participation of neural and hormonal communication pathways is emphasized in the determination of our mental experience. As an example, the varying levels of hormones can affect mood, mental clarity, and overall cognitive function, indicating that chemical messages are an integral aspect of our conscious experience. Recognition of the fact that consciousness is not confined to brain activity. It encompasses the entire biological system is a challenge to the traditional views and highlights the importance of the endocrine system in cognitive process facilitation. Hence, dynamic interaction of chemical messages between systems is the foundation of simple physiological processes as well as complex behaviors, relating the biological and conscious experience (Steinhart et al., 2001; Gaiseanu et al., 2019) (Figure 9.15).

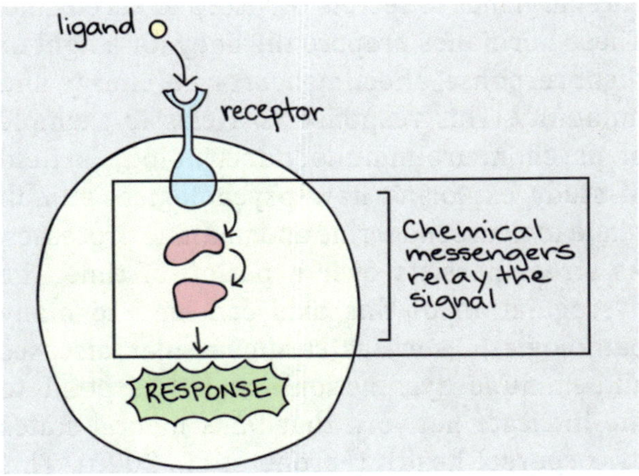

Figure 9.15. Chemical signals in the endocrine system.

Source: https://bpb-us-e1.wpmucdn.com/sites.gatech.edu/dist/6/1810/files/2017/06/675fd80883edf0625857e7e80a6914d4521ac53f-1024x645.png.

9.4.3. Comparison of Speed and Duration of Responses

The comparison of response time and velocity of the endocrine and nervous systems reveals varied physiological processes unique to individual body functions. The nervous system operates with rapid electrical signals transmitted along neurons so that instant reactions can be evoked to stimuli. This rapid communication is essential for reflex action, , such as withdrawing a hand from a hot plate, which occurs almost instantly. On the other hand, the endocrine system coordinates responses by hormonal signaling, efficient as it is, that, since it relies on circulatory delivery of hormones to target tissues, is slower. The time course of endocrine responses is usually long, indicating their role in maintaining long-term processes , such as growth, metabolism, and homeostasis and showing a large difference in operating time scale between these systems. In addition, the relationship between speed and length of response can be illustrated through the application of particular behavioral examples across various organisms. For instance, the research on the fiddler crab, *Uca pugilator*, illustrates a rapid behavioral response to ionizing irradiation that illustrates the rapidity of the nervous system. When exposed to X-rays, the fiddler crab exhibited an immediate halt followed by renewed normal activity, indicating a rapid neural response to environmental stimuli. However, the duration of this alertness can be regulated not only by the nervous system but also by hormonal control that may regulate the overall behavior of the crabs following exposure. This suggests a complex interaction between the short-term neural reaction and the more long-term action mediated by hormonal signaling, illustrating both systems' integrative nature in responding to external threat (Terwilliger et al., 1964). In consideration of the broader implications of speed and duration, one must observe how these differences affect the function and viability as a whole. Rapid response via the nervous system is most important in acute contexts where rapid response is necessary, , such as fleeing from predators or distancing oneself from harmful situations. Conversely, the sustained but slower responses of the endocrine system are critical in periods of growth, reproductive cycles, and stress reactions, enabling adaptive modifications in time. The homeostasis of balance between the two systems is essential in bringing about physiological equilibrium, since nervous overactivity may precipitate exhaustion and underactivity of hormonal signaling can shatter homeostatic equilibrium. Consequently, the comparison between the two systems not only explains their various working processes but also explains how each is dependent on the other for making an organism survive and adjust in a complex world (NASA-S, 1974)

Test Your Knowledge

1. Discuss electrical signals in the nervous system.
2. Explain chemical signals in the endocrine system.

9.5. Interactions Between the Nervous and Endocrine Systems

Learning Outcome

1. The hypothalamus as a link;
2. Stress response: Fight or flight;
3. Regulation of homeostasis.

The rich interaction between the endocrine and nervous systems offers a multifaceted system for controlling a variety of body functions, from behavior to metabolism. This interaction works through multiple pathways and messages in which hormones secreted by endocrine glands influence neural functioning, and neurotransmitters may control hormonal releases. For instance, the hypothalamus is a key interface that bridges the CNS and endocrine system by the release of releasing hormones, which act directly on the pituitary gland. Detection of such interactions is significant to the appreciation of how stress and mental illnesses can lead to metabolic disease or immunologic disorders because regulation of these systems reflects the body's response in sustaining homeostasis under internal and external stress (Cruz et al., 2017). In addition, the mediation of stress responses plays a vital role in the interaction between the endocrine and nervous systems. When the body is fearful, the hypothalamus triggers a cascade of hormonal communications that signals the adrenal glands to secrete cortisol and adrenaline. These hormones prepare the body for a fight or flight response, affecting heart rate, energy, and immunity. This response is a classic example of psychoneuroimmunoendocrinology, a field of study exploring how psychological stimuli relate to neuroendocrine and immune processes. As stress persists over a period of time, the dysregulation of this axis can lead to many pathologies, , such as cardiovascular diseases and immune dysfunctions, and thus point to the intricate network that links mental states to corporeal health (Ferone et al., 2000). This field has seen studies erupt, creating volumes of data along with revealing lacunae in what is known in this regard. Despite advances in the knowledge of the mechanisms by which hormones and neurotransmitters influence each other, there are many bidirectional influences that are poorly understood, often because the pathways involved are multifaceted. Scientists are increasingly researching the molecular and cellular mechanisms underlying these connections with a view to dissecting the roles of individual cytokines and neuropeptides in the modulation of behavior and immune function. These complexities not only add further emphasis to the importance of multidisciplinary models but also highlight the need for future research to resolve conflicting evidence and establish the mechanisms involved. The study of such interactions can ultimately lead to newer treatments for an array of disorders at the intersection of mental and physical health (Ferone et al., 2000) (Figure 9.16).

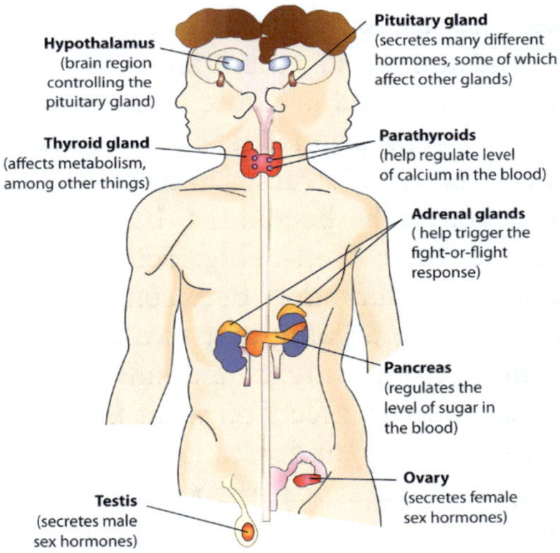

Figure 9.16. Interactions between the nervous and endocrine systems.

Source: https://opentextbc.ca/introductiontopsychology/wp-content/uploads/sites/9/2013/11/68db635fb346f9667c1696de65307587.jpg.

9.5.1. The Hypothalamus as a Link

The hypothalamus is important in connecting the nervous and endocrine systems and it intermediates an enormous variety of physiological activities. This is a small yet powerful part of the brain that keeps the homeostasis by controlling the body temperature, hunger, thirst, and circadian rhythms. It does this by having complex relationships with the pituitary gland, which is the master gland of the endocrine system. The hypothalamus releases releasing hormones in response to certain stimuli, which either stimulate or inhibit the secretion of hormones by the pituitary gland. This complex interaction between the hypothalamus and the pituitary gland is a perfect example of how the nervous system and the endocrine system act together in order to adapt to the internal and external changes of the environment and thus maintain the body functioning at its best. In addition, the hypothalamus acts as a key point of contact with stress reactions, which characterizes its significance in both emotional and physical health. Under stressful conditions, the hypothalamus triggers the release of corticotropin-releasing hormone (CRH), stimulating the pituitary gland to release adrenocorticotropic hormone (ACTH). This subsequently triggers the release of cortisol, which is a hormone that plays a central role in the stress reaction by the adrenal glands. The subtleties of this interaction emphasize the integrative ability, combining neural inputs and hormonal outputs to mediate numerous body processes. Additionally, studies indicate that early life experiences and stress can induce long-term alterations in hypothalamic functioning, affecting individuals' stress responses later in life, ultimately linking this brain region to mental health outcomes (Renard et al., 2014). When considering the dual role of the hypothalamus in controlling the nervous system and the endocrine system, one can begin to see a bigger picture of the effect it has on health. This connection between stress and the hypothalamus means a lot in terms of the susceptibility to diseases, especially those that are related to stress. Chronic stress can dysregulate the HPA axis, leading to conditions , such as anxiety and depression. Observations suggest that alterations in HPA axis functioning due to early life stress can have enduring effects on behavior and endocrine responses (ILC Workshop Report,2006). Consequently, the hypothalamus, not being just a central regulator of vital physiological functions, but also highlighting the unity of mental and physical health, is a graphic example of how successfully it connects the nervous and endocrine systems (Figure 9.17).

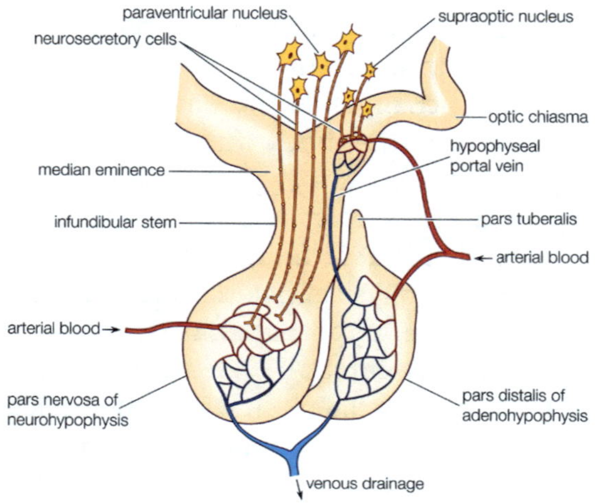

Figure 9.17. The hypothalamus as a link.

Source: https://cdn.britannica.com/26/6526-050-53E03651/anatomy-pituitary-gland-posterior-lobe-anterior-structures.jpg.

9.5.2. Stress Response: Fight or Flight

Fight or flight is not just a vital physiological reaction that is triggered by the endocrine and nervous systems, but also to stressors or threats perceived by a person. After an individual confronts a stressful situation, the hypothalamus triggers this complex process and instructs the adrenal medulla to release catecholamines, with adrenaline and norepinephrine as the main agents. These hormones prepare the body for action by increasing heart rate, dilating airways, and diverting blood flow to critical muscles. This intricate interaction underscores the body's evolutionary reaction to be able to survive immediately threatening dangers, illustrating how highly integrated the nervous and endocrine systems are. This answer should be interpreted, particularly with the acknowledgement of how frequent activation of this system can lead to adverse health effects because stress disrupts homeostasis and makes a negative contribution to mental and physical well-being, the precursor to a variety of disorders as outlined by Estrine et al. (2006). Moreover, frequent activation of the stress system has significant physiological and psychological effects. The discharge of cortisol, a steroid hormone released with prolonged stress, is critical in regulating metabolism, immune response, and inflammation. While a short-term release of cortisol is beneficial in responding to immediate threats, long-term exposure could potentially suppress the immune system and promote disease states of anxiety and depression. This adaptive response indicates that while stress is initially protective, it can eventually become a bad influence that exacerbates well-being. Scientists have studied this interaction between stress and health in depth, arguing that stress is defined differently across contexts and individuals and thus has different effects on what is perceived as well-being. By understanding the nature of complexity in stress responses, scientists aim to create a synthesis among mental and physical health disciplines, as shown by Awasthi et al. (2015). Additionally, the stress response carries implications for individual, social, and environmental contexts. Chronic stress systems may have implications for broader public health consequences, affecting community productivity and quality of life. Disputes on the definition of stress and its implications need to take an integrative approach, merging psychophysiology and epidemiologic perspectives to assess stressors in various settings. As social pressures and environmental factors present new stressors, the reaction to stress must be revised to meet such complexities. Addressing these complex aspects can lead to more effective interventions and health care planning meant to offset the effects of stress negatively. Thus, examining the fight or flight response in a window of opportunity to analyze the key roles played by the nervous and endocrine systems in ensuring health amid increasing stressors provides the foundation for further studies and applications (Figure 9.18).

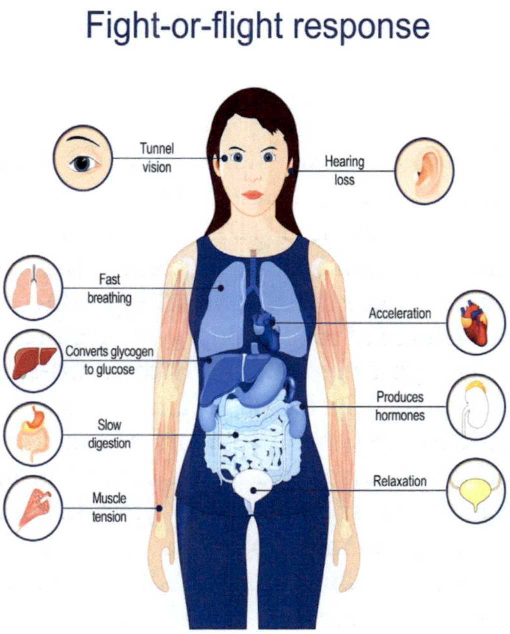

Figure 9.18. Stress response: Fight or flight.

Source: https://images.squarespace-cdn.com/content/v1/62057597e37fee129b460681/184b2d8f-404d-4b2d-999b-a49605ffec47/stress-response-infographic.jpeg.

9.5.3. Regulation of Homeostasis

Homeostatic control is a fundamental aspect of biological regulation to keep the body in homeostasis despite changes in the external environment. The key mechanism to this regulation is the interplay between the nervous and endocrine systems, with the coordinating body response to keep things in balance. The ANS, with its sympathetic and parasympathetic branches, plays a critical role in this process. For instance, increased sympathetic outflow is typically associated with stress responses, while parasympathetic activation supports restorative processes. Such dualism demonstrates how the two systems collaborate to produce quick changes in varied bodily functions , such as heart rate, blood pressure, and digestion and thus achieve a stable internal environment for optimum cellular function and overall health, acting as a feedback process to resist homeostatic disruption (Prins et al., 1996). In addition to the ANS, homeostasis is also supplied by hormonal control. Hormones released by the endocrine system are secreted that modulate processes , such as metabolism, growth, and fluid balance to alterations in the internal environment. Insulin and glucagon, two hormones secreted by the pancreas, are a prime example of this interaction; they maintain glucose homeostasis through regulation of blood sugar levels. Besides, salt and fluid balance is also strictly regulated by hormones like aldosterone, which regulate blood sodium and potassium concentrations. Not only do these hormonal responses ensure short-term correction of homeostatic imbalances, but also induce long-term responses that can affect the metabolic status of the organisms as a whole, thereby linking endocrine function with the preservation of homeostasis directly (Neal et al., 2016). Furthermore, research has identified the complex feedback mechanisms involved in homeostatic processes, highlighting the importance of both neural and hormonal communication. For instance, overfeeding has been shown to result in altered autonomic function, demonstrating how derangements in nutrition influence hormonal balance and sympathetic tone, , such as increased parasympathetic activity and baseline rates of norepinephrine in animal models (Prins et al., 1996). This association points to the pivotal function of the feedback loops in the homeostatic process, where a change in one system is selectively influences the other. These interactions, in turn, prompt questions regarding how other body needs, like sodium loading, crosstalk with the regulation of blood pressure and fluid homeostasis in general. Knowledge of these complex interrelations enhances our respect for the fine balance struck by the endocrine and nervous systems in the control of homeostasis, which reflects their sheer importance to disease and health (Neal et al., 2016).

Test Your Knowledge

1. What is the stress response: Fight or flight?
2. What are the regulations of homeostasis?

9.6. Disorders of the Nervous and Endocrine Systems

Learning Outcome

1. Common disorders of the nervous system;
2. Common disorders of the endocrine system;
3. Impact of disorders on overall health.

A complete understanding of nervous and endocrine disorders requires a recognition of the interconnectedness inherent in these systems. Dysregulation of the hormonal systems can seriously affect neurological functioning because the systems rely on a highly integrated biochemistry of response and signal. For instance, the endocrine system secretes hormones which may influence the activity of neurons, mood, cognition, and overall behavior. Conversely, neurological disease can also impact endocrine function, showing a two-way relationship. Cushing's syndrome's excess production of cortisol can lead to impairments in cognition as well as psychological disturbances, showing the interconnectedness of these physiological systems. This interaction necessitates an integrated approach to assess and treat individuals with co-occurring nervous and endocrine system diseases, emphasizing the necessity of delivering whole person care individually tailored to each patient. Mitochondrial disorders (MTDs) are a stark reminder of how disruptions in energy generation can influence the nervous and endocrine systems. MTDs result from defects in the mitochondrial or nuclear DNA, which adversely affect the ETC in ATP production, a crucial source of cellular energy. Therefore, MTD patients typically present with multi-system effects, primarily on the brain and endocrine organs. Notably, MTD patients are inclined to present with disorders in the control of blood glucose levels,, such as hypoglycemia. In a survey, 18.97% of sampled patients were reported to have had a minimum of one hypoglycemic attack, with a high risk in comparison to the general population. Hypoglycemia's occurrence can result in neurological symptoms , such as seizures and developmental delay and thus necessitates greater glucose monitoring among such patients, especially during the neonatal period (Moats et al., 2019). The interaction of nervous and endocrine systems is also extended to the functioning of neuroendocrine factors in psychological health. Psychological processes and traits are not only confined to the brain but are influenced by biochemical functioning in the whole body, highlighting the point that psychological and physiological aspects are interconnected (Steinhart et al., 2001). Disease in these systems can manifest unevenly, quite often complicating diagnosis and treatment. For instance, dysregulation of hormonal pathways brought on by stress can worsen mood disorders and anxiety, obscuring differentiation between psychological and physiological presentation. Additionally, treatments for endocrine imbalance most often improve neurological symptoms, thus highlighting the necessity of treatment through a multidimensional approach. By considering the integrated nature of these systems, clinicians can better manage the complexities that are emerging in patients with disorders of both nervous and endocrine systems and, in turn, improve patient outcomes and quality of life.

9.6.1. Common Disorders of the Nervous System

The complex interplay of the endocrine and nervous systems is required for maintaining homeostasis and responding to the environment. The majority of nervous system disorders may profoundly disrupt such functions, and this leads to severe illnesses. Some of these include multiple sclerosis, Parkinson's disease, and Alzheimer's disease, which illustrate the grave consequences of neurological imbalances. These disorders typically result from a combination of environmental, lifestyle, and genetic factors that cause varying degrees of neurodegeneration and functional impairment. For instance, multiple sclerosis is a disorder in which the immune system mistakenly attacks myelin, leading to communication issues between the brain and the body. Thus, understanding the processes that govern such disorders is important in designing adequate treatments and coping strategies for sufferers as well as countering the more general effects on public health. Chronic stress is also a key contributory cause of prevalent nervous system disorders.

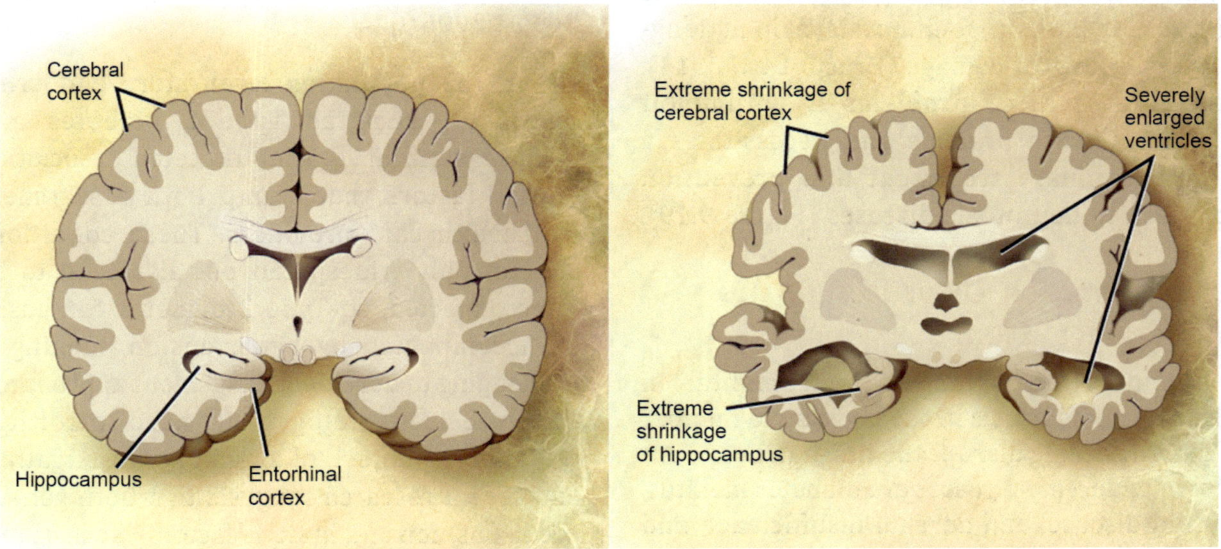

Figure 9.19. Disorders of the nervous system.

Source: https://opentextbc.ca/biology/wp-content/uploads/sites/96/2015/03/Figure_35_05_01.jpg.

The brain coordinates stress response and closely integrates the nervous and endocrine systems. When under stress, the body releases hormones , such as cortisol, which not only have short-term but also long-term effects on physical health. Long-term exposure to stress can lead to neuroinflammation and a decline in intellectual capacity, much like symptoms of Alzheimer's disease. The relationship between stress and neurological disease highlights the need for clinicians to adopt a generalist approach to the care of patients that involves psychological as much as physiological considerations. As has been debated in literature, "Throughout the life course, stress plays a major role in health and disease," meaning that the impact of stress on the nervous system cannot be ignored in clinical practice (ILC Workshop Report, 2006). Neurotoxicity becomes an equally important area of inquiry in the study of prevalent nervous system disorders. A wide variety of chemicals can disrupt neural function, leading to peripheral neuropathy or even neurodegenerative illness. The neurotoxicity can be rapidly or slowly progressive, causing a wide range of symptoms that can significantly decrease an individual's quality of life. To add insult to injury, the body's recuperative capacity is inconsistent as some of the neurotoxicity will cease once exposure is abated, but some will lead to permanent harm. Interestingly, the research indicates that "neurotoxicity is usually self-limiting once exposure has ceased," which means that detection and prevention of toxic exposure is imperative in not developing long-term neurological damage (Brust et al., 2014). Detection of these problems is paramount in public health practice and necessitates a multidisciplinary treatment and prevention approach to neurotoxic diseases (Figure 9.19).

9.6.2. Common Disorders of the Endocrine System

The endocrine system is instrumental in ensuring homeostasis in the body, but a number of disorders may disrupt the fine balance. Well-known endocrine diseases are diabetes mellitus, thyroid disease, and adrenal insufficiency, and each of them has its own etiology and impact on the general health state. Diabetes, especially Type 1 and Type 2, has been an ever-increasing phenomenon in the world, with genetics, lifestyle and obesity being some of the factors that have led to its increase. Thyroid disease, whether it is hyperthyroidism or hypothyroidism, influences metabolism, energy, and even mood, which means that many physiological processes are affected. These disorders help to underline the importance of hormones in the regulation of body functions since their imbalance may cause serious health problems, which need to be addressed and managed. Moreover, stress also significantly influences the endocrine system, as it activates the HPA axis, leading to the release of cortisol and other stress hormones. Stress can be a trigger or worsen such conditions as adrenal insufficiency or type 2 diabetes, which shows the unity of the endocrine and nervous systems. Notably, despite being commonly viewed as a mental health problem, stress has a significant impact on physical health, which occurs via hormonal mechanisms. Studies have shown that persistent stress may lead to insulin insensitivity, thus predisposing one to metabolic syndrome , such as obesity and hypertension. This relationship underscores the need for a holistic view of health that integrates both mental and physical health in understanding and treating endocrine disorders (ILC Workshop Report,2006).

At the population level, more endocrine diseases , such as Type 2 diabetes and hypertension, have a multifaceted association of risk factors that stump both researchers and health care providers. These conditions are not only widespread but also tend to be interconnected, so Type 2 diabetes can easily be accompanied by hypertension, making it challenging to develop management approaches. The precise etiology of these disorders remains a subject of ongoing investigation, and recent research suggests that both genetic predisposition and lifestyle factors, , such as diet and physical inactivity, play significant roles in their development (Zheng et al., 2010). This means that the treatment of these disorders needs to be multifaceted, involving modification of lifestyle, teaching of the patient and probably drug therapy as well, which means that the nervous system and the endocrine system must be viewed together in order to promote better health (Figure 9.20).

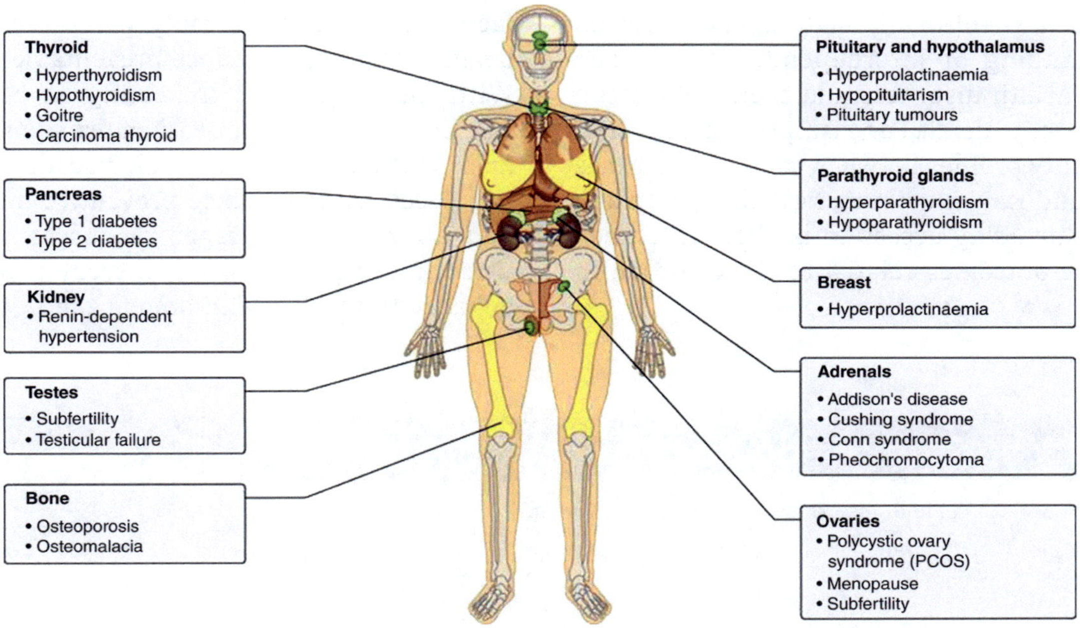

Figure 9.20. Endocrine disorders.

Source: https://i0.wp.com/pocketdentistry.com/wp-content/uploads/2021/11/c22f002.gif?w=960.

9.6.3. Impact of Disorders on Overall Health

The interaction of endocrine and nervous system disease can have a significant impact on general well-being, influencing mortality and overall quality of life in populations. For instance, diseases , such as mental and behavioral disorders are anticipated to grow in prevalence, especially among older individuals, as evidenced by recent projections. The increase in deaths from these conditions captures the complex interplay between mental and physical well-being, necessitating an urgent social imperative to reinforce health services. These illnesses are generally chronic, contributing to the burden on the health system and necessitating the use of more resources to ensure appropriate care to those who have been impacted. It should be noted that as the mortality pattern changes, so must health management, particularly in high-risk groups, lest it exacerbate existing health inequalities. Apart from mental illness, such chronic diseases as diabetes and hypertension and their interplay in the endocrine system, can adversely affect neurological health. For instance, modulation of glucocorticoids and their impact on metabolic well-being is particularly important in lowering systemic inflammation that often accompanies most neurodegenerative diseases. Assessment of morbidity rates indicates that patients with endocrine disorders are likely to have cognitive as well as emotional impairments, thereby worsening their disease. This linkage highlights the importance of an integrative care model addressing endocrine and neurological well-being issues. Addressing these intertwined disorders through individually tailored treatment, healthcare providers can significantly enhance patient outcomes and optimize overall health, with significant gains for the aging population with complex health profiles. The scenario of death has also drastically transformed due to evolving health trends, so researchers and health professionals have had to adapt their methods according to evolving trends. A multiple cause-of-death analysis allows for an understanding of interdependency between various health disorders and their unforeseen associations in death. For example, the difference in mortality profiles between countries , such as France and Italy illustrates how the systematic factors, including access to medical care, may influence outcomes related to the disorder. In addition, determining the common latent risk factors that cause mortality will direct public health programs to reduce preventable deaths. With the projections indicating an alarming rise in some causes of death, most significantly from mental and behavioral disorders, there is a necessity to enhance preventive measures and appropriately target resources to curb these future health issues, thereby promoting enhanced overall health among the populace.

Test Your Knowledge

1. What are the common disorders of the nervous system?
2. What are the common disorders of the endocrine system?

SUMMARY

A journey into the intricate interaction of the nervous and endocrine systems, it is evident that the two systems work in concert to regulate numerous physiological processes essential to the process of homeostasis. The interaction between the neuroendocrine enables significant responses to environmental changes, which are facilitated by hormonal messages and neural impulses. The modulatory action of such mechanisms, as in the case of balneotherapy, points to the relative significance of such mechanisms in regulating not only stress reactions but also immunity. In fact, significant correlations were found between changes in both electroencephalogram (EEG) patterns and hormone levels, pointing to the very potent effect of treatment intervention on neuroendocrine function and overall health (Popovych et al., 2021). Moreover, the consequences of mitochondrial diseases for the endocrine system illustrate how failure in one system spreads to others with consequent complex clinical presentations. Research has indicated that patients with mitochondrial diseases are at higher risk of endocrine dysfunctions, , such as hypoglycemia. This fact reminds clinicians of the interconnectedness of the systems. Because neural circuits can infringe on hormonal homeostasis and vice versa, such interaction implies that pathology along one dimension of human physiology has broad implications for several systems, particularly in the case of mitochondrial malfunction (Moats et al., 2019). Future research should be directed by the clarification of how certain pathways and mechanisms of interaction between the endocrine and nervous systems operate. More insight into these interactions could introduce new therapeutic possibilities and better clinical management schemes in neuroendocrine-immune dysfunction patients or resulting diseases. It is thus apparent from findings linking treatment practice with extensive physiological changes within neurohormonal and immune function that the application of integrated treatment approaches is central. An integrative model that accounts for the dynamic interplay between the nervous and endocrine systems will not only enhance our comprehension of human health but also direct prevention and treatment strategies that prioritize systemic well-being (Popovych et al., 2021).

The cooperation between the endocrine and nervous systems is responsible for homeostasis and responding to environmental changes. The nervous system, composed of the brain, spinal cord, and peripheral nerves, enables rapid communication through electrical impulses. The endocrine system employs hormonal messages communicated through the bloodstream, governing slower but longer regulatory responses. Both of these processes enable the body to react quickly to immediate threats through the nervous system while controlling long-term change through hormonal processes. How efficient each can be is realized in how both utilize numerous mechanisms of physiology, like stress reaction, metabolism, and reproduction, because each relies on the other to enable optimal body functioning. Comprehension of the interaction between these systems is of particular relevance to the comprehension of stress management and health impact. Stress activates the HPA axis, giving insight into how outside stimuli can be capable of inducing a cascade of neuroendocrine reactions. This mechanism highlights the importance of solving health issues in an integrated fashion, transcending purely intellectual understanding of treatment to incorporate the role of hormonal signaling in overall health. This integration allows for a more expansive approach to health care in an encompassing environment, with mental health and physical health as interdependent areas of consideration (ILC Workshop Report, 2006). This paradigm is core in generating efficient stress-related disorder management strategies as well as the general promotion of well-being. The consequences of such systems extend to research support and the design of novel therapeutic technologies. Investment in medical research not only enhances understanding of disease but also propels the advancement of new therapies. As shown by the high-return donations from diverse funding sources including charity groups to government bodies, interdisciplinary efforts need to be undertaken in order to promote a healthy research environment for medicine. The interdependence of public, private, and philanthropic sources of finance emphasizes the complementarity of donations towards putting emphasis on the importance of continued investment in research studies (Crane et al., 2014). With the passage of time, success in the nervous and endocrine systems will necessitate coordinated efforts and dedication to new medical science frontiers.

KEY TERMS FROM THIS CHAPTER

Adrenal glands
Brain
Endocrine glands
Endocrine regulation
Feedback mechanism
Hormones
Motor neurons
Nervous regulation
Neuron
Neurotransmitter
Pancreas
Pituitary gland
Reflex arc
Sensory neurons
Spinal cord
Synapse
Target cells
Thyroid gland

CRITICAL THINKING QUESTIONS

1. **What is the basic functional unit of the nervous system?**
   a. Neuron
   b. Hormone
   c. Gland
   d. Muscle fiber
2. **Which part of the nervous system controls voluntary movements?**
   a. Autonomic nervous system
   b. Somatic nervous system
   c. Sympathetic nervous system
   d. Parasympathetic nervous system
3. **Hormones are produced by:**
   a. Nerves
   b. Glands
   c. Muscles
   d. Bones
4. **The endocrine system regulates body processes by releasing:**
   a. Enzymes
   b. Electrical signals
   c. Hormones
   d. Antibodies
5. **Which gland is often called the "master gland" because it controls other endocrine glands?**
   a. Thyroid gland
   b. Adrenal gland
   c. Pituitary gland
   d. Pancreas

Short Answer Questions

1. What are the main functions of the nervous system?
2. What is the basic structural and functional unit of the nervous system?
3. Describe the main parts of a neuron and their roles.
4. How do neurons transmit signals? Explain the role of action potentials.
5. What is the difference between the central nervous system (CNS) and the peripheral nervous system (PNS)?

Answers to Critical Thinking Questions

1. (a); 2. (b); 3. (b); 4. (c); 5. (c)

REFERENCES

1. (NASA-SP-7011(129)) AEROSPACE MEDICINE N74-29442 'AND BIOLOGY: A CONTINUING BIBLIOGRAPHY WITH INDEXES, SUPPLEMENT 129, JUNE 1974Aerospace Medicine and Biology: A Continuing Bibliography with Indexes (supplement 297). https://core.ac.uk/download/pdf/42837264.pdf (accessed on 13th June 2025).
2. Agostinelli, E., Altissimi, G., Artico, M., et al. (2017). Trophic and neurotrophic factors in human pituitary adenomas (Review). https://core.ac.uk/download/127587365.pdf (accessed on 13th June 2025).
3. Alingh Prins, A., Scheurink, A., Steffens, A. B., Steffens, A. B., Steffens, A. B., Steffens, A. B., Steffens, A. B., et al. (1996). Overfeeding, autonomic regulation and metabolic consequences. https://core.ac.uk/download/pdf/12942087.pdf (accessed on 13th June 2025).
4. Arias Cruz, A., Elizondo Villarreal, B., González Díaz, S. N., et al. (2017). Psychoneuroimmunoendocrinology: Clinical implications. https://core.ac.uk/download/322372786.pdf (accessed on 13th June 2025).
5. Arnold, B. S., Gross, T. S., McDonald, K., et al. (2001). Relations of environmental contaminants, algal toxins, and diet with the reproductive success of American alligators on Florida lakes. https://core.ac.uk/download/11016994.pdf (accessed on 13th June 2025).
6. Awasthi, H. H., & Dwivedi, M. K. (2015). Physiology and anatomy of stress. https://core.ac.uk/download/354380655.pdf (accessed on 13th June 2025).
7. Bergmann, M., Gratzl, M., Lahr, G., et al. (1992). Chromogranin A in neurons of the rat cerebellum and spinal cord: Quantification and sites of expression. https://core.ac.uk/download/12168028.pdf (accessed on 13th June 2025).
8. Brown, M. M. (1938). The inheritance factor in endocrine dysfunction. https://open.bu.edu/bitstream/2144/9711/1/inheritancefacto00brow.pdf (accessed on 13th June 2025).
9. Brust, V., Creeley, C., Feldman, R., Grandjean, P., Herken, H., Panzica, G., A., et al. (2014). Neurotoxicity. https://core.ac.uk/download/286729125.pdf (accessed on 13th June 2025).
10. Cabral, A., De Francesco, P., & Perelló, M. (2015). Brain circuits mediating the orexigenic action of peripheral ghrelin: Narrow gates for a vast kingdom. https://core.ac.uk/download/153565281.pdf (accessed on 13th June 2025).
11. Cabral, A. S., De Francesco, P. N., & Perelló, M. (2020). Brain circuits mediating the orexigenic action of peripheral ghrelin: Narrow gates for a vast kingdom. https://core.ac.uk/download/325983853.pdf (accessed on 13th June 2025).
12. Crane, P., Farrell, C., Garau, M., et al. (2014). Exploring the interdependencies of research funders in the UK. https://core.ac.uk/download/30608686.pdf (accessed on 13th June 2025).
13. Desesquelles, A., Meslé, F., Frova, L., Pappagallo, M., Salvatore, M. A., Pace, M., & Egidi, V. (2025). Revisiting the mortality of France and Italy with the multiple-cause-of-death approach. https://core.ac.uk/download/pdf/6544416.pdf (accessed on 13th June 2025).
14. Ferone, D. (2000). Physiological and pathophysiological role of somatostatin receptors in the human thymus. https://core.ac.uk/download/18515504.pdf (accessed on 13th June 2025).
15. Fica, S., Malinici, E., Popa, M., et al. (2021). Linking the brain and bone through fat. https://core.ac.uk/download/397483679.pdf (accessed on 13th June 2025).

16. Gaiseanu, F. (2019). The informational model of consciousness: Mechanisms of embodiment/disembodiment of information. https://core.ac.uk/download/287611919.pdf (accessed on 13th June 2025).

17. Gidron, Y., Kishore, U., Nowicky, A. V., et al. (2011). Hemispheric lateralization and immune function: A systematic review of human research. https://core.ac.uk/download/20496329.pdf (accessed on 13th June 2025).

18. Hennigar, S. R. (2010). Exposure to polybrominated diphenyl ethers (PBDEs) suppresses the release of pro-inflammatory products by alveolar macrophages in vitro. https://core.ac.uk/download/215515896.pdf (accessed on 13th June *2025).*

19. *H*idalgo, V., Puig-Pérez, S., Pulópulos Tripiana, M., et al. (2018). Psychophysiological response to social stressors: Relevance of sex and age. https://core.ac.uk/download/158345410.pdf (accessed on 13th June 2025).

20. Hirz, J., Schmock, U., & Shevchenko, P. V. (2015). Forecasting leading death causes in Australia using extended CreditRisk+. http://arxiv.org/abs/1507.07162 (accessed on 13th June 2025).

21. Iida, M., & Takemoto, K. (2018). A network biology-based approach to evaluating the effect of environmental contaminants on human interactome and diseases. Ecotoxicology and environmental safety, 160, 316–327. https://doi.org/10.10*16/j.ecoenv.2018.05.065(accessed on 13*th *Ju*ne 2025).

22. ILC Workshop Report(2006). Stress: Putting the brain back into medicine. An Interdisciplinary Workshop of the INTERNATIONAL LONGEVITY CENTER–USA Sponsored by International Longevity Center-USA Institute for the Study o*f Aging Can*yon Ranch Canyon Ranch Institute Alliance for Health & the Future. https://core.ac.uk/download/71356692.pdf (accessed on 13th June 2025).

23. Johnson, B. A., LaJeunesse, D. R., Zapotoczny, G., Presnell, J., & Catignas, K. K. (2009). Regulated peristalsis into the acidic region of the Drosophila larval midgut is co*ntrolled by a no*vel component of the autonomic nervous system. https://core.ac.uk/download/pdf/288861.pdf (accessed on 13th June 2025).

24. Lahr, G., Heiss, C., Mayerhofer, A., Schilling, K., Parmer, R. J., O'Connor, D. T., & Gratzl, M. (1990). Chro*mogranin* A in the olfactory system of the rat. Neuroscience, 39(3), 605–611. https://doi.org/10.1016/0306-4522(90)90245-y (accessed on 13th June 2025).

25. Li, A., Goldberg, A. M., Bal-Price, A., Coecke, S., Cooper, H. M., Owen, D., Silbergeld, E., et al. (2007). Workgroup report: Incorporating in vitro alternative methods for developmental neurotoxicity into international hazard and risk assessment strategies. https://core.ac.uk/download/pdf/7623193.p*df (accessed on 13th June 2025).*

26. *Moats, A. (2019). Hypoglycemia in mitochondrial disord*ers. https://core.ac.uk/download/228022820.pdf (accessed on 13th June 2025).

27. Neal, C. M. (2016). Evaluating techniques in tissue clarification using CLARITY imaging and investigating where sodium is sensed in the body. https://core.ac.uk/download/142066226.pdf (accessed on 13th June 2025).

28. Popovych, A., Popovych, D., Popovych, I., Ruzhylo, et al. (2021). Relationships between changes in the state of neuroendocrine regulation and immunity in patients of Truskavets' spa. https://core.ac.uk/download/587245576.pdf (accessed on 13th June 2025).

29. Prins, A. (2011). The brain-gut interaction: The conversation and the implications. https://core.ac.uk/download/478383279.pdf (accessed on 13th June 2025).
30. Randazzo, A. (2022). Stressed out and stressed in: Regulating the stress response at an endocrine level through dance/movement therapy. https://core.ac.uk/download/519916653.pdf (accessed on 13th June 2025).
31. Renard, G. M., & Rivarola, M. A. (2014). Qué sabemos sobre las consecuencias a largo plazo de la separación materna temprana y la respuesta neuroendocrina al estrés. https://core.ac.uk/download/159291663.pdf (accessed on 13th June 2025).
32. Steinhart, E. (2001). Persons versus brains: Biological intelligence in human organisms. https://core.ac.uk/download/287611886.pdf (accessed on 13th June 2025).
33. Terwilliger, R. C. (1964). The behavioral responses of the fiddler crab, *Uca pugilator*, to ionizing irradiation. https://open.bu.edu/bitstream/2144/32857/1/Terwilliger_Robert_1962_web.pdf (accessed on 13th June 2025).
34. Thanigaivelan, N. K. (2023). Adaptation of the human nervous system for self-aware secure mobile and IoT systems. https://core.ac.uk/download/556979741.pdf (accessed on 13th June 2025).
35. Vasconcelos, M., Stein, D. J., Gallas-Lopes, M., Landau, L., & de Almeida, R. M. M. (2020). Corticotropin-releasing factor receptor signaling and modulation: Implications for stress response and resilience. *Trends in Psychiatry and Psychotherapy, 42*, 195–206. https://doi.org/10.1590/2237-6089-2018-0027 (accessed on 13th June 2025).
36. Volpe, M., Gallo, G., & Rubattu, S. (2022). Endocrine functions of the heart: From bench to bedside. *European Heart Journal, 44*, 643–655. https://doi.org/10.1093/eurheartj/ehac759 (accessed on 13th June 2025).
37. Zheng, X., & Zhou, X. (2010). Neuron impairment or loss in brain may be responsible for type 2 diabetes and essential hypertension. https://core.ac.uk/download/pdf/288998.pdf (accessed on 13th June 2025).

REPRODUCTION AND DEVELOPMENT

10

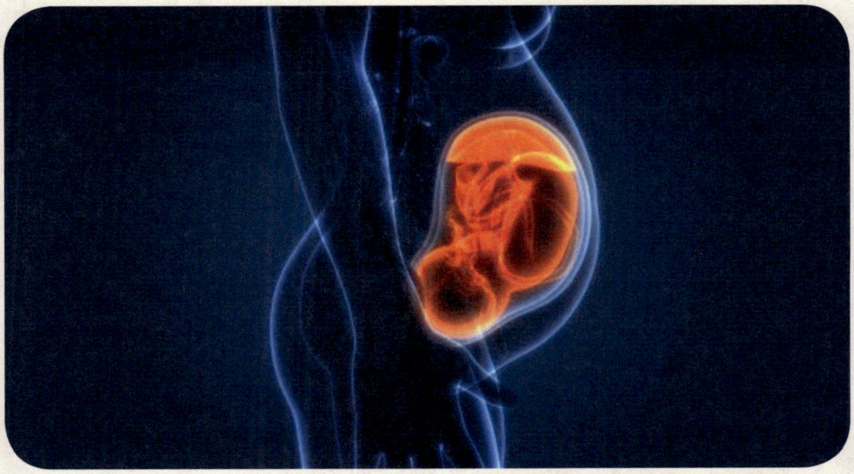

LEARNING OBJECTIVES

After studying this chapter, you will be able to:

1. Explain the types of reproduction;
2. Define gametogenesis;
3. Explain fertilization;
4. Explore embryonic development;
5. Discuss developmental biology;
6. Focus on the evolutionary perspectives on reproduction.

Introduction

Reproduction and development are fundamental biological processes responsible for the continuity of life in all organisms. Reproduction refers to the various mechanisms used by organisms to produce offspring, resulting in genetic variation and species survival. Development is the series of processes through which organisms grow from fertilization to maturity, characterized by growth and differentiation. The interaction between these two processes is critical, as reproductive strategies can directly influence developmental processes. As species develop in environments, the intersection of reproductive success and successful development is a rich one with ecological, genetic, and evolutionary elements. Examining these interactions not only sheds light on biological complexities within a species but also informs larger-scale ecological processes that influence biodiversity and ecosystem function. There are numerous reproductive methods that take place throughout the plant and animal kingdoms, each delicately adapted to alleviate specific environmental issues.

Sexual reproduction, for instance, typically involves genetic recombination, gracefully illustrated by the process of meiosis, which peaks genetic variation among offspring. Asexual reproduction, like apomixis in plants, including St. John's wort (Hypericum perforatum), allows for genetically similar offspring production through seeds, sidestepping the complexities of sexual reproductive mechanisms. This reproductive adaptability allows the organism to develop quickly in the presence of environmental stressors while maintaining its status in the ecosystem. All forms of reproduction include evolutionary advantages that control survival and adaptability as much as the evolution of developmental processes. This variation is necessary, as it shows interaction between reproductive success and development strategies important to the long-term survival of wide-ranging species. The study of reproduction and development is still a significant division of biological research, with applicability continued into agricultural practices, conservation, and understanding of ecological relationships. Research into the molecular and genetic processes that regulate such phenomena, e.g., the research on transcripts and expressions of apomixis from H. perforatum flower transcriptome, provides great depth into the mechanisms. As emphasized in contemporary studies, the establishment of genes regulating reproductive strategies can be a point towards upcoming innovations in agriculture aimed at crop resilience and yield. By demystifying reproduction and development, scientists offer fundamental information that informs biodiversity conservation and ecosystem management sustainability strategies. Therefore, it is critical to possess solid knowledge of such biological events to address the ecological consequences of global alterations and species invasions, an area of major interest in contemporary biology (Galla et al., 2015; Prospero & Cleary, 2017) (Figure 10.1).

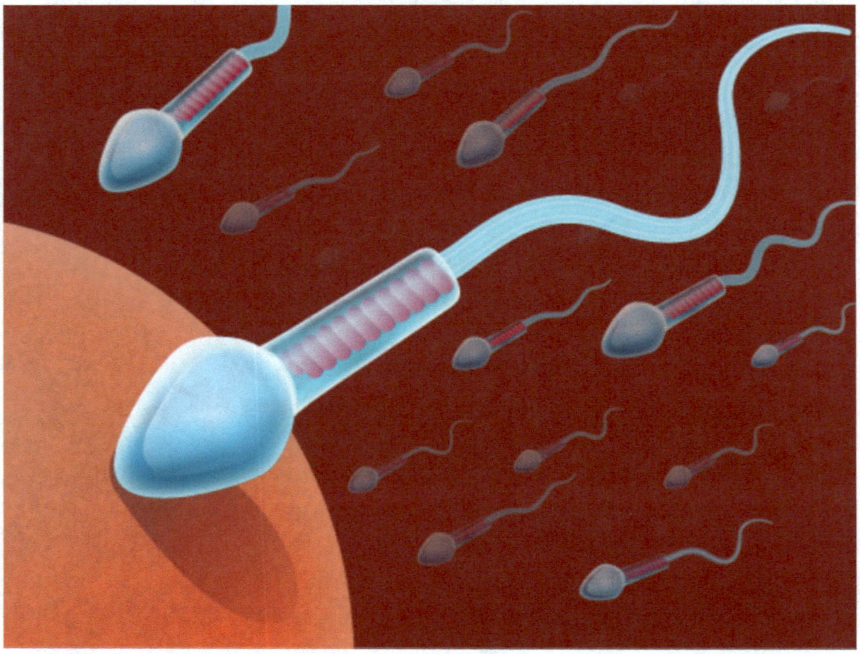

Figure 10.1. Reproduction and human development.

Source: https://dr282zn36sxxg.cloudfront.net/datastreams/f-d%3A1fad473e86b40ae4a13dab525075942775fb82260f447801a1734c4d%2BIMAGE%2BIMAGE.1.

Definition of Reproduction

Reproduction is a fundamental biological process, continuing species from generation to generation. Basically, it involves the mechanism by which organisms produce offspring, hence genetic continuity. Reproduction can be broadly classified into two categories: sexual and asexual reproduction. Sexual reproduction includes the combination of gametes of two parental organisms and leads to genetic diversity among the offspring. However, asexual reproduction, as seen in bacteria and certain plants, allows for offspring to be generated from one parent without the exchange of genetic information. This difference points out how mechanisms of reproduction can shape the pattern of population life and the direction of evolutionary trends, significantly influencing the biodiversity of ecosystems. Reproduction is not only a concept in biology, as social and cultural components are involved, particularly when dealing with human societies. Within most socio-cultural environments, reproduction is accompanied by issues of social continuity, heritage, and identity. For instance, the mechanisms for forming family units have a constitutive impact on social membership as well as justice within communities. Just like in the context of Cuban society, reproduction is not only a biological process but also a part of broader social processes reflecting justice and active participation of families in building their community (del Secade et al., 1994). Expectations and roles that are associated with reproduction can substantially vary, influencing individuals' personal choices and societal expectations, hence forming a complicated relationship between reproductive practices and social structures. Further, the study of reproduction is essential in understanding developmental processes that can be applied to humans and other organisms. In recent discourse on reproduction, particularly in fields , such as management development, the need for localizing concepts is widely recognized. For instance, while Western management is typically the origin of developmental models, localization, or consideration of cultural and traditional ways of learning is unavoidable for use and usefulness (Oppong, 2013). This resonates through the overall concept of reproduction, emphasizing that good development practice requires more than knowing definitions, a diversified approach that embraces both biological imperatives and socio-cultural environment in which organisms exist. The definition of reproduction is, therefore, a rich field of research that brings together biological processes with social forces and developmental theory (Figure 10.2).

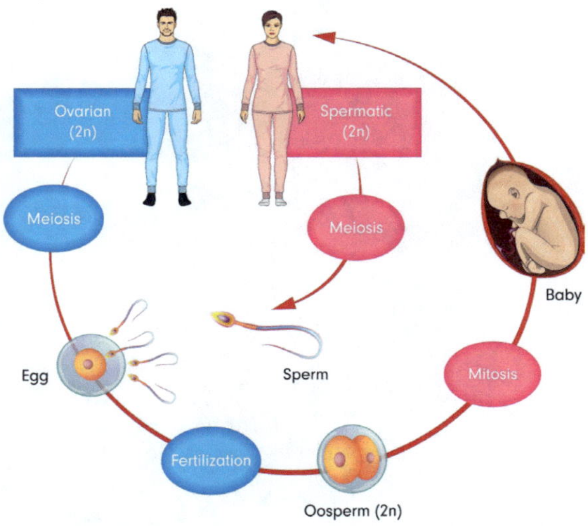

Figure 10.2. Definition of reproduction.

Source: https://cdn.entrance360.com/media/uploads/2019/07/16/shutterstock_1312942451.jpg.

Importance of Reproduction in Biological Systems

Reproduction is a fundamental process that underlies biological processes, allowing species to perpetuate from one generation to the next. In addition to survival, reproduction triggers genetic variability by means of sexual reproduction, which increases fitness in adaptive environments. The variation is essential for evolution and the integrity of ecosystems over long periods. For example, the evolution of new characteristics can render populations resistant to disease or climate change. Besides, improving knowledge in reproduction enables scientists to regulate such processes for agriculture and environmental applications. Such regulation is through the production of GMOs that can carry traits to enhance growth or disease resistance. As an organizing axis of the biological processes, reproduction's importance has a profound impact on other operations, , such as development, where the early parts of an organism's life are decided by reproductive strategies and genetic inheritance of the genes, highlighting the interdependence of life. The priority of reproduction is realized through model organisms, which have been pivotal in increasing the progress of the biological processes. Widely utilized in genetics research, inbred research animals like the laboratory mouse provide significant information regarding human health and disease processes.

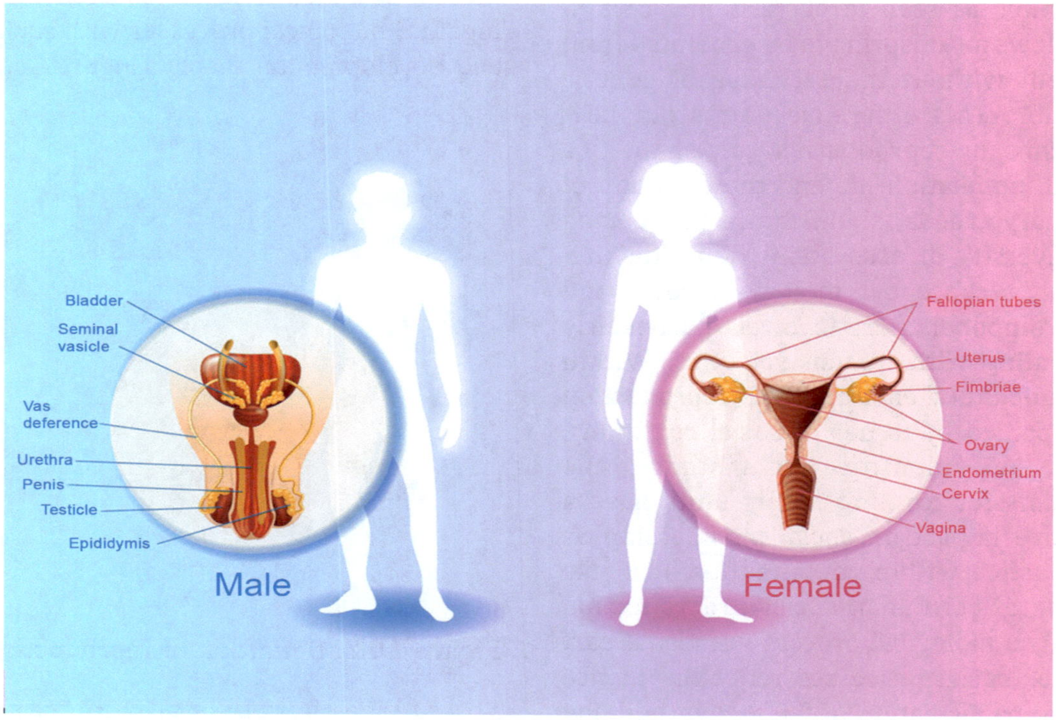

Figure 10.3. Reproduction in biological systems.

Source: https://blogcdn.aakash.ac.in/wordpress_media/2022/03/Blog-Image-11.jpg.

In this case, the role of standardization is key, as researchers seek to create homogeneous populations for intense experimentation. However, these techniques often clash with the intrinsic variability of the biological organism. Findings regarding surprise outcomes in experiment populations highlight the irony between controlled reproduction and the chance nature of biological existence. The findings illustrate that while standardization is fundamental to systematic research, the acknowledgment of the luck and one-of-a-kind nature of reproduction also applies, e.g., to the other tales of inbred mouse strains where the fortune of surprise litters can disrupt attained effects (Burt et al., 2013). In insects , such as the medfly, the sophistication of reproductive biology emphasizes its role in both pest control and broader ecological processes. Scientific study of the medfly's genetics demonstrates how reproduction is coordinated with behavioral traits, , such as mating behavior and sensory response. For instance, research has demonstrated that sex determination genes are crucial and enable reproductive success and population dynamics directly. An understanding of such mechanisms allows for the development of specific biological control strategies, offering environmentally safe substitutes for pesticide chemicals. The use of genetic research not only offers means of crop pest control but also teaches us about evolutionary processes that underlie species adaptation. Thus, reproductive research extends beyond theoretical use, as its functional importance moves both to ecological wholeness and agricultural production, illustrating the variable but critical role reproduction plays in biological systems (Bonaldo et al., 2008) (Figure 10.3).

Overview of Development Processes

Development processes in living organisms are complex, comprising a series of processes by which one fertilized ovum forms a whole organism. Cleavage is the process by which the zygote is rapidly divided and forms a multicellular body called the blastula. Cells begin to differentiate during this process, setting the basis for specialized tissues and organs to be formed later. Most importantly, this stage is important as it is the stage where the homogeneously structured cells branch out into a more differentiated cellular structure. As the embryonic development goes on, during gastrulation, the blastula is transformed into a three-layered gastrula, which gives rise to the three primary germ layers: the ectoderm, the mesoderm, and the endoderm. These layers later develop into all-over organism organ systems, demonstrating the intrinsic relationship between the early stages of development and the final phenotype of the adult organism. As the organism continues to develop through additional growth, organogenesis becomes prominent, where the various germ layers begin the elaborate process of organ and system development. This phase is marked by cells within the germ layers proliferating, migrating, and differentiating in very stereotypic ways to make functional tissues. Hormonal signals, patterns of gene expression, and cell-cell interactions play central roles in controlling these processes. Of special interest is coordinating these signals to have organs develop in appropriate positions and at certain times. Most importantly, these developmental stages are not only of biological significance but are also influenced by extrinsic forces , such as environmental conditions and maternal health. The effects of these forces can be extensive in that they can disrupt the normal process of development and lead to congenital defects. Thus, understanding organogenesis is significant for basic biology and even future medical applications. Developmental processes occur in the backdrop of not only individual organisms but also influence evolutionary processes. Developmental principles have shed light on how genetic diversity leads to phenotypic variation within populations and have provided a foundation for evolutionary biology. The interplay between developmental biology and evolutionary theory substantiates the principle that species change not only because of genetic changes but also due to changes in developmental processes themselves. For instance, developmental changes in gene regulation in the embryo can result in dramatic morphological changes. The process highlights the interdependence between evolution and development, or as it is more popularly described, evo-devo. Convergences of this kind bring to the fore fundamental questions as to how cascades in development might be rewired generation by generation so that species can respond to different forms of environmental pressures and ultimately to the enormous extent of biodiversity today. Therefore, learning about development processes is important not just to understand reproduction and lifecycle but also to unravel the intricate tapestry of evolution and adaptation.

10.1. Types of Reproduction

Learning Outcome

1. Asexual reproduction;
- Sexual reproduction;
- Comparison of reproductive strategies.

The biological process of reproduction in which organisms give rise to offspring can be divided into two main types: asexual and sexual reproduction. Asexual reproduction is a process that involves a single parent organism that produces offspring that are identical to the parent in terms of genetic makeup. This is an efficient method that enables the population to grow fast, especially in a stable environment where the parents' characteristics are favorable for survival. Binary fission, budding and vegetative propagation are some common types of asexual reproduction that allow species like bacteria and some plants to prosper without fertilization. Nevertheless, although asexual reproduction may be favorable under certain conditions, it reduces genetic variations, which makes the populations more susceptible to environmental fluctuations and diseases, which affects their long-term flexibility and ability to survive. Conversely, sexual reproduction entails recombination of the genetic material of two parents, making their offspring to have genetic variation. This type of reproduction occurs through mechanisms , such as meiosis, where gametes (sperm and egg cells) are produced, followed by fertilization. Sexual reproduction also enhances genetic variation that is essential to the evolution and adaptation of species in the changing environment. The different genetic characteristics may benefit the general resilience of the population, where individuals have different abilities to respond to environmental stresses. A notable example of sexual reproduction is observed in flowering plants, which produce intricate reproductive structures that facilitate mating between different individuals, ensuring a greater pool of genetic combinations that may lead to improved survival rates as the environment fluctuates over generations (Marc et al., 2015). Within the wider scope of reproduction and development, a knowledge of asexual as well as sexual reproduction is central to unravelling the mysteries of evolutionary biology. These differences in costs and benefits of the different reproductive strategies can determine the course of evolution of organisms, their morphology, behavior, and ecological niche. To give a few examples, organisms using asexual reproduction can have characteristics enabling them to colonize new habitats rapidly, whilst those using sexual reproduction have frequently evolved specialized structures to produce and protect gametes, since in diverse environments, higher reproductive success is required. The integration of these reproductive types into developmental biology illustrates how the mechanisms of reproduction can inform our understanding of larger biological processes, including organismal development and ecological interactions, and underscores the importance of studying these systems within a framework that emphasizes their underlying physiological and evolutionary principles (Zhou et al., 2019) (Figure 10.4).

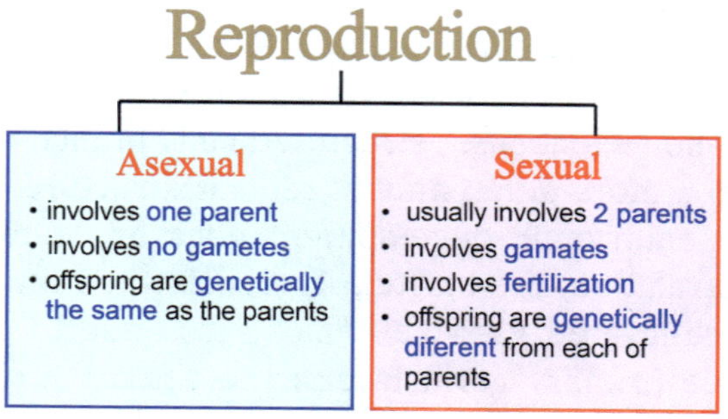

Figure 10.4. Types of reproduction.

Source: https://www.careerpower.in/blog/wp-content/uploads/sites/2/2023/08/29120311/types-of-reproduction.png.

10.1.1. Asexual Reproduction

Asexual reproduction is one of the major methods of propagating most organisms so that they can inhabit different habitats. Although sexual reproduction has the complication of merging genetic material of two parents, asexual reproduction has the process of growing offspring from one individual to reproduce fast without the complication of mate selection and blending of genes. This mode of reproduction is especially useful in permanent environments where adapting to change is far less critical. Most unicellular organisms, for instance, bacteria and protists, thus predominantly employ methods like binary fission, budding, or fragmentation towards reproduction. These are illustrations that show a good resource investment because they allow organisms to maximize reproductive effort when circumstances are favorable and save energy typically expended on searching for and attracting mates. For evolutionary biology, there are fascinating implications of asexual reproduction for genetic diversity and adaptability. While asexual organisms produce clones of the parent, the process can still lead to enormous evolutionary adaptations through mechanisms , such as HGT, mutation, and environmental pressures. For instance, certain species consist of fission followed by regeneration, demonstrating their ability to recover from injury while simultaneously reproducing. The studies on *Convolutriloba longifissura* uncover genetic mechanisms behind asexual reproduction and regeneration processes. Here, the combination of signaling pathways like BMP (Bone Morphogenic Protein) signals uncovers the way regeneration in asexual reproduction processes can reorganize developmental axes to fuel the organisms' adaptability when it comes to responding to environmental changes (Aziz et al., 2023). Besides, asexual reproduction cannot be limited to unicellular life forms but also takes place in more complex multicellular organisms, , such as certain plants and animals. The evolution from simple asexual processes to more complex reproduction processes illustrate how organisms refine their mode of reproduction to suit their environment. For example, the evolutionary trend towards alternation of generations has demonstrated the flexibility that can arise from varied modes of reproduction, emphasizing the role of environmental interaction in shaping such processes. Asexual reproduction in plants can be explained in terms of their mode of dispersal, where propagative reproduction results in new plants that are genetically the same as the parent. They enable long-term genetic stability, allowing these organisms to outcompete niche habitats effectively and ensure survival and resilience to environmental changes (Willemse et al., 2009) (Figure 10.5).

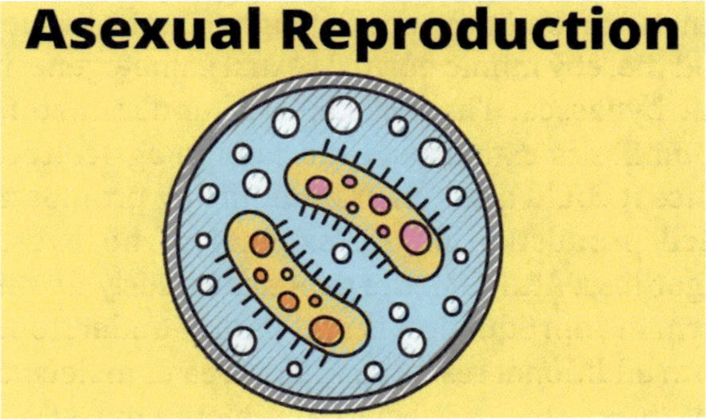

Figure 10.5. Asexual reproduction.

Source: https://i.ytimg.com/vi/F89XgjHyt0Q/maxresdefault.jpg.

10.1.2. Sexual Reproduction

Sexual reproduction is an elementary mechanism of evolution of varied forms of life, offering a fundamental pathway for gene diversity. The process is based on meiotic division, by which the gametes containing half as many chromosomes as in the parent cell are formed, to permit the combination of the male and female gametes to form a zygote. With the organism becoming accustomed to varying environments, sexual reproduction is added, which heightens genetic variability to enable populations to endure and adjust under varying conditions. Notably, the role of organisms in gamete union and with environmental factors plays a major part in reproductive success. These processes uncover sexual reproduction not just as a process of reproduction but also as a fundamental driving force of evolutionary change, and they highlight the intricate interaction between reproduction, development, and environmental adaptation (Willemse et al., 2009). The emergence of apomixis is a notable feature of sexual reproduction, particularly in some plant species , such as St. John's wort (Hypericum perforatum). , such asexual reproduction allows for the production of viable seeds without fertilization, which can be employed for cloning the parent plant type. Aposporic gametogenesis allows for the somatic cells in the ovule to produce functional embryo sacs through mitotic division and thereby mimic normal sexual gametogenesis, but bypasses. The relevance of understanding apomixis is extensive, especially in agriculture, since it could alter plant breeding by promoting seed production at the expense of no hybrid vigor loss. Although the genetics underlying this form of reproduction is still poorly understood, with additional research in the area of molecular pathways that segregate apomictic and sexual modes being required (Barcaccia et al., 2015). Besides enhancing genetic diversity and adaptation planning, sexual reproduction is also crucial in the maintenance of ecological interactions. One of the best examples is the complex interaction between plants and animals for pollination, where the majority of flowering plants rely on sexual reproduction via animal services for successful gene exchange. These biotic interactions are a fine example of the interdependence of organisms whose reproductive success can have a significant impact on other organisms in the ecosystem. Moreover, sexual reproduction can sculpt populations for the promotion of evolutionary development for traits increasing reproductive fitness and gamete recognition. The foundations formed by these unions emphasize the core position of sexual reproduction not just as a biological process but as a basis of ecological networks that sustain various forms of life (Willemse et al., 2009) (Figure 10.6).

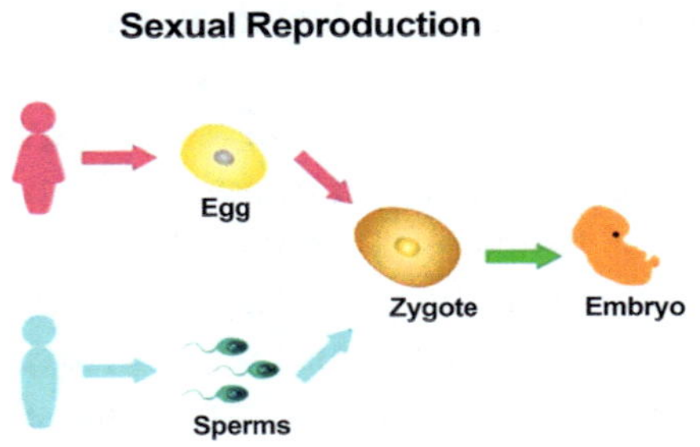

Figure 10.6. Sexual reproduction.

Source: https://www.careerpower.in/blog/wp-content/uploads/sites/2/2023/08/31151258/Sexual-Reproduction-2.png.

10.1.3. Comparison of Reproductive Strategies

Reproductive strategies are important in explaining the different mechanisms through which species reproduce their lineage, particularly under conditions of environmental stress. Sexual and asexual reproduction have sharp contrasts, each with some advantages and disadvantages. Sexual reproduction favors genetic variation through gene mixing, which has an advantage for adaptability in ever-changing environments. On the other hand, asexual reproduction enables quick population to increase quickly since it does not require finding a partner, although it usually results in lower genetic variety. The contrast gets more complicated when alternatives , such as parthenogenesis and fission are taken into account, the latter enabled by organisms , such as *Dugesiidae planarians,* possessing favorable regenerative power. By looking at the repercussions of such reproductive strategies, it can be observed how evolutionary pressures shape reproductive mechanisms and influence the genetic fitness of species. The theory of investment in reproduction complicates the comparative study of reproductive strategies even further. Studies on crustaceans like Gammarus illustrate how female body size influences reproductive output through the impact on egg size and egg numbers. In aquatic ecosystems, the egg size versus number of eggs produced trade-off is evident; generally, larger eggs will produce fewer but more surviving offspring. This strategic allocation of resources is a reflection of an evolutionary compromise between fecundity and survivorship of offspring. Investigation suggests that freshwater species are exhibiting fecundity patterns according to their environments, but comprehensive knowledge regarding reproductive effort is lacking within their clades. Research on such patterns emphasizes the significance of the environment in shaping reproductive strategies and the general layout of reproductive biology. Of particular interest is the evolutionary significance of genetic mechanisms like the Meselson effect and Muller's Ratchet, particularly in asexually reproducing organisms. Genetic variation in fissiparous types, for instance, certain planarians, arises from the incomplete inheritance of somatic mutations to result in characteristic genetic mosaicism outcomes. This characteristic of asexual reproduction is bivalent; while it tends to lead to the accumulation of deleterious mutations in the long run, it also allows for acclimatization in established systems. Thus, studies of the genetic effects of fissiparous reproduction, especially when contrasted with sexual reproduction, introduce new issues for ecological and evolutionary study. The genomic frameworks behind such reproduction strategies not only influence the fitness of the organism but also have broader contexts in terms of biodiversity and species survival in a shifting environment.

Test Your Knowledge

1. What is asexual reproduction?
2. What is sexual reproduction?

10.2. Gametogenesis

Learning Outcome

1. Process of sperm production;
2. Process of egg production;
3. Role of hormones in gametogenesis.

The process of gametogenesis, which plays a key role in sexual reproduction, entails the formation of gametes through a series of complex biological processes. The male and female gamete-forming pathways in the animals are spermatogenesis and oogenesis, respectively. Spermatogenesis in the testes entails diploid spermatogonia going through mitotic divisions and then meiosis to produce haploid sperm cells. This is a process of enormous differentiation and maturation, which results in elongated cells specialized for motility. On the other hand, oogenesis, taking place in the ovaries, involves the formation of oocytes, and just one haploid egg forms from each meiotic cycle, and the others (polar bodies) are absorbed. Understanding the controlling mechanisms of gametogenesis is crucial since they reveal the evolutionary compromises and reproductive strategies that ensure species continuation. In plants, gametogenesis occurs in a typical manner that is usually within dedicated structures referred to as gametophytes. Microgametogenesis and megagametogenesis are the two principal mechanisms of gamete development in plants, yielding sperm and egg cells, respectively. Microgametogenesis involves the production of pollen grains through a series of mitotic cell divisions and meiotic activity, ultimately leading to the formation of haploid microgametes. In contrast, megagametogenesis occurs in ovules and involves one megasporocyte that undergoes meiosis and produces four megaspores that normally develop into one viable egg in an embryo sac.

The plant gametogenesis process is frequently affected by environmental conditions as well as genetic control, which can be inclusive of mechanisms common to sporogenesis. Therefore, knowledge of plant gametogenesis requires an in-depth examination of both ecological and genetic contexts. Our knowledge of gametogenesis has been significantly broadened through recent insights into apomixis, a type of asexual reproduction. In species , such as Hypericum perforatum, the aposporic gametogenesis process illustrates how somatic cells can bypass the normal meiosis route to produce embryo sacs that reflect the outcomes of sexual reproduction without genetic recombination.

This alternative reproductive mechanism is particularly significant in agriculture for enhancement purposes since it facilitates the reproduction of desirable genes, all without the resulting intricacies linked to sexual reproduction. The genetic control of the apomictic processes is poorly explained; however, mis-expression of the genes controlling the sexual reproductive pathway has been shown to trigger apomixis. A study of the gametogenesis processes elucidated by Galla et al. (2017) and Barcaccia et al. (2015) provides a foundation for future research on potential applications in crop breeding and ecological conservation (Figure 10.7).

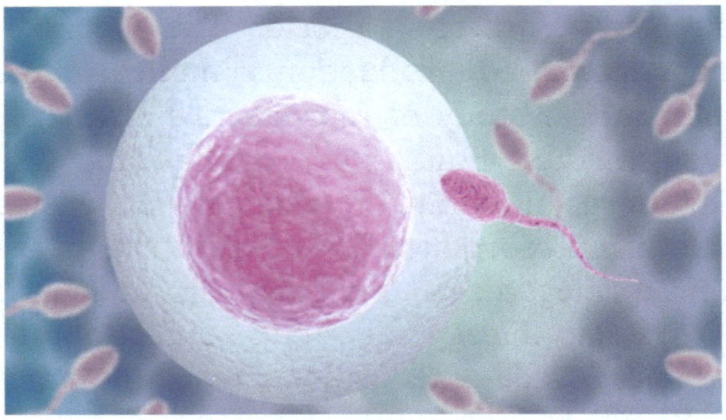

Figure 10.7. Gametogenesis.

Source: https://media.springernature.com/w580h326/nature-cms/uploads/collections/GMTG_Hero_image-6242e1d7797232fc9f85b250fd63b51c.jpg.

10.2.1. Process of Sperm Production

Sperm production or spermatogenesis can be defined as a complicated and strictly controlled sequence of events that takes place in the seminiferous tubules of the testes. This complex process starts with spermatogonia, the undifferentiated germ cells which undergo multiple divisions and differentiations to finally produce mature spermatozoa. An important part of spermatogenesis is the process of meiosis, during which spermatogonia cells produce haploid spermatids after two rounds of cell division. These spermatids then undergo significant morphological transformation, whereby they obtain flagella and acquire an acrosome that plays a crucial role in fertilization. The whole process of spermatogenesis requires about 64–74 days, and the effectiveness of this process greatly influences the male fertility and reproductive well-being. After spermatogenesis, sperm cells need to undergo a process of vital maturation called capacitation to become competent to fertilize an ovum. This stage mostly takes place in the female reproductive system, where the sperm is exposed to certain biochemical factors that make them capable of fertilizing an ovum. Notably, bicarbonate ions (HCO_3^-) play a crucial role in this maturation process, as they are involved in activating the adenylate cyclase enzyme, leading to increased cyclic AMP (cAMP) levels within the sperm. This elevation in cAMP promotes essential physiological changes, , such as membrane hyperpolarization and alkalinization of the cytoplasm (Austin et al., 2017). These changes are essential to the motility and hyperactivated movement of the sperm, which primes them to succeed in interacting with the oocyte in the process of fertilization. Moreover, the functioning of mitochondria is incorporated in the energetic requirements of sperm motility and the success of fertilization in general. Mitochondria, the powerhouses of the cell, generate ATP, which is vital for flagellar movement and the capacitation process. Studies have indicated that a sufficient number of mitochondria and their proper distribution are imperative for optimal sperm function and subsequent fertilization outcomes (Abeydeera et al., 2006). In this regard, the wellbeing of mitochondria is strictly interconnected with the success of sperm maturation. An intact mitochondrial network serves to emphasize that sperm have the energy required to navigate the female reproductive tract and eventually reach a successful fertilization, thus demonstrating the unity between sperm production, maturation, and general reproductive potential (Figure 10.8).

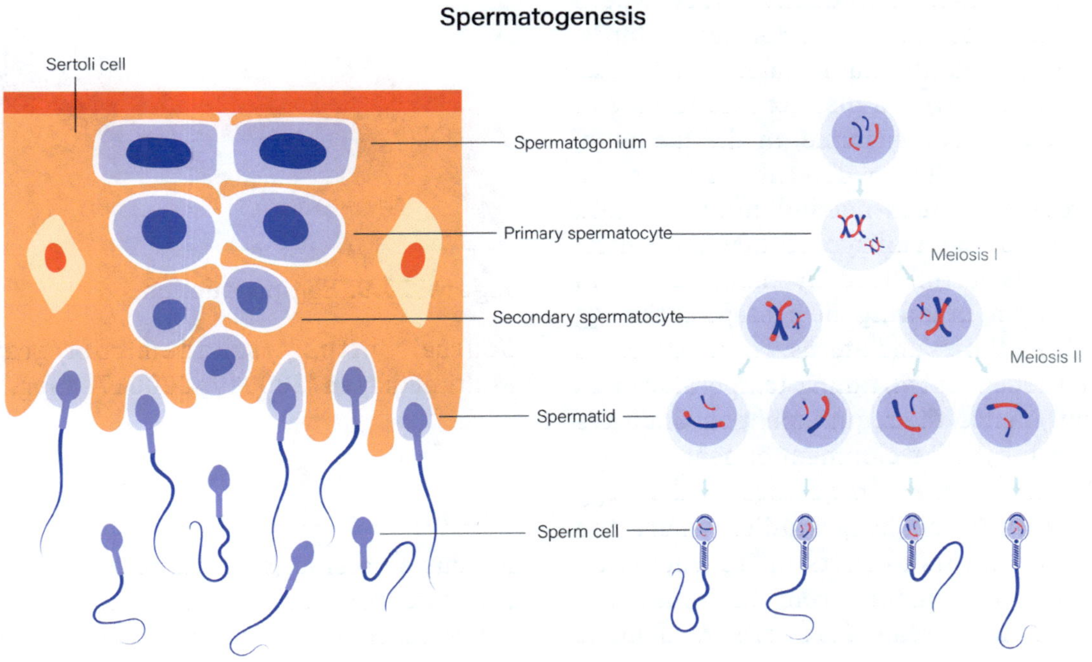

Figure 10.8. The sperm production cycle.

Source: https://cdn.prod.website-files.com/66ee2fbb803ca97841036c9f/6724440b82de17a60bde8bed_6724416a3a2179d6e806579c_spermatogenisis-iStock-1264913191%2520(1).jpeg.

10.2.2. Process of Egg Production

Egg production, particularly in oviparous animals , such as birds and reptiles, is a complex biological process characterized by typical reproduction mechanisms. It starts in female organisms with the production of oocytes in the ovaries, regulated by a complex hormonal control. Follicle-stimulating hormone (FSH) induces the growth of ovarian follicles, while luteinizing hormone (LH) induces ovulation. As these follicles mature, they are reshaped by mechanisms that lead to the eventual release of eggs. This hormonal interaction not only integrates maturation but also ensures that eggs are adequately prepared to accept fertilization and subsequent growth. The subtle aspects of this hormonal effect on egg development are important to determine because they reveal a window to reproductive health as well as fertility issues in many animal species. Having been released from the ovaries, the eggs travel through the female reproductive tract, where several other processes are performed. In birds, this is via the infundibulum, magnum, isthmus, and uterus, each with a distinct role to play in egg production. For example, in the magnum, albumen or egg white is deposited around the yolk, providing much-needed nutrients and protection to the embryonic developmental process. This is followed by membrane and shell development during the passage of the egg through the isthmus and uterus. These processes are highly dependent on timing, temperature, and conditions; hence, the requirement of conducive conditions for the development of healthy eggs. Interruptions in these processes lead to egg defects as observed from studies where the existence of adverse effects on reproduction systems and egg quality from environmental pollutants in bird organisms avows the delicate balance required for successful reproduction (Hermansson et al., 2007). Completion of egg formation culminates in egg laying, a critical time that is determinant in that various factors determine reproductive success. For example, the presence of environmental stressors , such as pollutants, can inhibit physiological processes that are crucial to egg formation, leading to suboptimal outcomes. It has been discovered in studies that being exposed to contaminants in the environment , such as EE2 and o,p′-DDT, can lead to thinning of eggshells and low fertility, demonstrating the relationship between environmental health and reproductive fitness. Moreover, metabolic investment in egg formation and maternal well-being is proportionate to the quality and quantity of eggs laid. Therefore, advancing our understanding of mechanisms of egg formation adds not just to the knowledge of reproductive biology but also conservation and biodiversity upkeep, since egg health is paramount to surviving in species (Edward et al., 2007) (Figure 10.9).

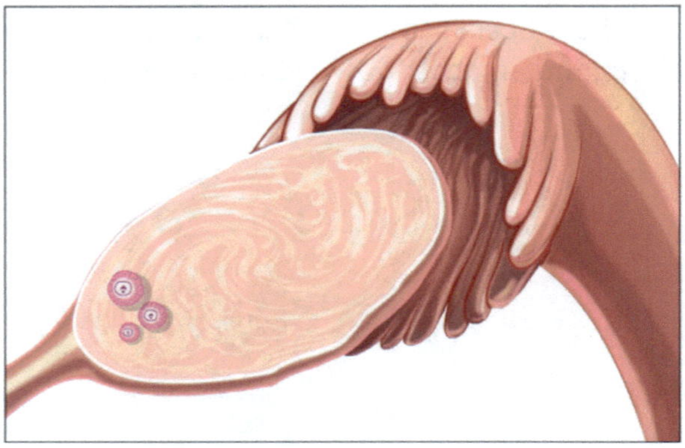

Figure 10.9. Egg production.

Source: https://ssl.adam.com/graphics/Multimedia/en/200017/200017ed.jpg.

10.2.3. Role of Hormones in Gametogenesis

A crucial element of gametogenesis is the intricate process of hormones that control reproductive processes. Central to this regulatory mechanism are gonadotropins, i.e., FSH and LH, which play a critical role in the maturation and discharge of gametes in females and males. These are secreted by the anterior pituitary gland and exert their effect after receiving gonadotropin-releasing hormone (GnRH) from the hypothalamus. FSH primarily triggers spermatogenesis in the gonads in men by activating Sertoli cells to promote sperm development, while in women, it supports follicular growth and maturation of the oocyte.

The efficacy of these processes relies heavily on coordination and balance among hormones, and hence, the primary role of the gonadotropic-releasing activity of the hypothalamic-pituitary-gonadal (HPG) axis in governing reproductive well-being and gametogenic activity becomes imperative. Proceeding further, the participation of transcription factor cFOS mirrors the complexity of hormonal regulation during gametogenesis. This component plays a vital role in the control of the gonadotropin subunit genes to allow the secretion of FSH and LH in the pituitary gland's gonadotropes. It has been demonstrated that cFOS deficiency disrupts normal gametogenesis, even with sex-steroid levels, hence explaining its selective control on the reproductive axis. The decrease in cFOS-null females in the number of Kiss1 neurons and in abundance of Kiss1 mRNA suggests that cFOS has a unique role in kisspeptin regulation, a component that is absolutely necessary for GnRH neuron activation. This tiered interaction is important in illustrating how cascades of hormonal signaling and transcriptional regulation play a pivotal role in the maintenance of reproductive function, thereby justifying the significance of examining these hormonal dynamics within the broader context of reproduction and development (Coss et al., 2015). Moreover, estrogens' impact on gametogenesis also illustrates the function of hormones beyond the gonadotropic axis. Estrogens, through their respective receptors, have complicated roles in regulating reproductive processes in all vertebrates. Exploration of the expression pattern of G protein-coupled estrogen receptors (GPERs) in teleost fish , such as the European sea bass, identifies their distinct sex-specific and seasonal regulatory functions within the pituitary and hypothalamus. Notably, Gpera changes with the reproductive cycle, with increases during vitellogenesis, while Gperb maintains stable levels, reflecting their complementary functions. These findings underscore that hormonal regulation, e.g., by estrogens, subtly regulates gametogenesis by neuroendocrine pathways, reflecting the fine-tuning process occurring throughout reproduction and development (Alvarado et al., 2018).

Test Your Knowledge

1. What is the process of sperm production?
2. What is the role of hormones in gametogenesis?

10.3. Fertilization

Learning Outcome

1. Mechanisms of fertilization;
2. External vs. internal fertilization;
3. Factors influencing fertilization success.

Fertilization is the most important process in sexual reproduction where the union of the male and female gametes leads to a zygote. As part of this intricate biological process, the sperm must undertake a journey through the female reproductive tract to the ovum, where it unites. The process, as mentioned, is the commencement of development because it initiates a sequence of cell division and differentiation that ultimately leads to a viable embryo. Successful factors involve sperm viability and motility, environmental conditions, as well as health of the female reproductive tract. Integrity of the sperm DNA is also an important factor since damaged genetic content may result in low rates of fertilization and poor-quality embryos, further illustrating the fine balance necessary for effective reproduction (García-Ferreyra et al., 2015). In the context of anatomy involved in fertilization, the oviduct (Fallopian tube) is an important site where fertilization and early embryo development take place. Historically neglected in assisted reproductive technologies, the oviduct has been recognized for facilitating fertilization and further embryo development. It has been established that cells from the oviduct offer vital support to embryos, allowing them to be better positioned to develop successfully. While *in vitro* fertilization (IVF) methods often bypass the natural anatomy through the direct uterine implantation of embryos, understanding the function of the oviduct is still necessary for improving assisted reproduction outcomes. Coculture systems *in vitro* have been employed in experiments to reproduce the nurturing environment of the oviduct so that researchers can examine the embryotrophic factors necessary for development (Lee et al., 2002). Further, the effects of fertilization are not confined to mere genetic union but also affect subsequent stages of development and reproductive outcome.

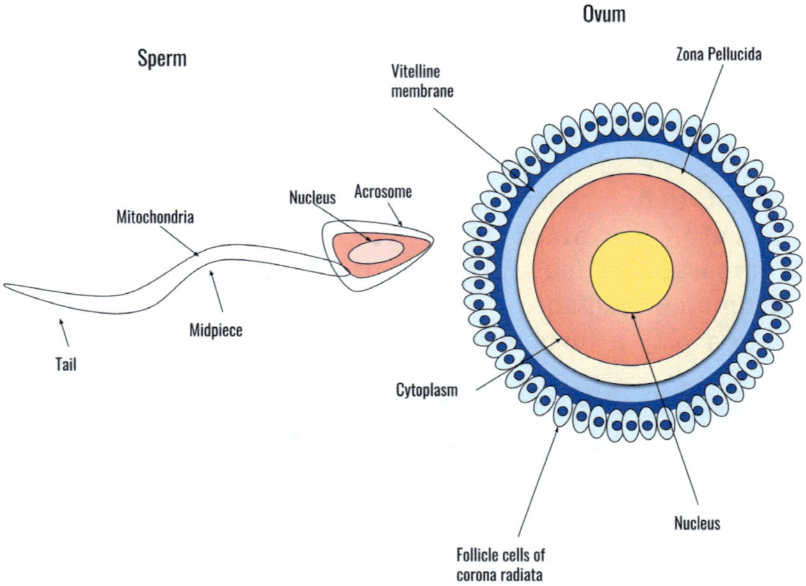

Figure 10.10. Fertilization.

Source: https://upload.wikimedia.org/wikipedia/commons/thumb/5/5f/The_sperm_and_ovum_during_fertilization.svg/1200px-The_sperm_and_ovum_during_fertilization.svg.png.

Both parents' genetic contributions are vital, while sperm quality is decisive, with DNA integrity as an indicator of outcome in assisted reproductive procedures. Extreme sperm DNA fragmentation has been linked with compromised embryo quality and reduced rates of fertilization, affecting the success of pregnancy and live birth chances (García-Ferreyra et al., 2015). Therefore, fertility tests would now include, on a routine basis, measurements of sperm DNA integrity, disclosing the interconnectedness of fertilization mechanics and reproductive well-being. With ongoing research and new reproductive technologies, more understanding of fertilization can enrich strategies for the treatment of infertility and maximizing outcomes with assisted reproduction, thereby constraining the significance of this biological process into the broader context of development and reproduction (Figure 10.10).

10.3.1. Mechanisms of Fertilization

The mechanisms of fertilization are a point of intersection between reproductive biology and developmental processes. Essentially, fertilization is simply the joining of male and female gametes, producing a zygote. This initiation of embryogenesis is followed by a cascade of developmental processes , such as cleavage, differentiation, and morphogenesis. Fertilization mechanisms can vary greatly between species and are an evolutionary rich tapestry of strategies. For instance, external fertilization that is typically observed in aquatic animals appears to contrast starkly with terrestrial animals' means of internal fertilization. These elements emphasize the adaptability of reproductive mechanisms to prevailing environmental conditions, ultimately determining the directionality of developmental paths. Thus, understanding the complexity of fertilization mechanisms may enlighten general principles of development and reproduction. The complex fertilization process can be further divided into numerous important steps, namely gamete recognition, fusion, and later activation of the zygote.

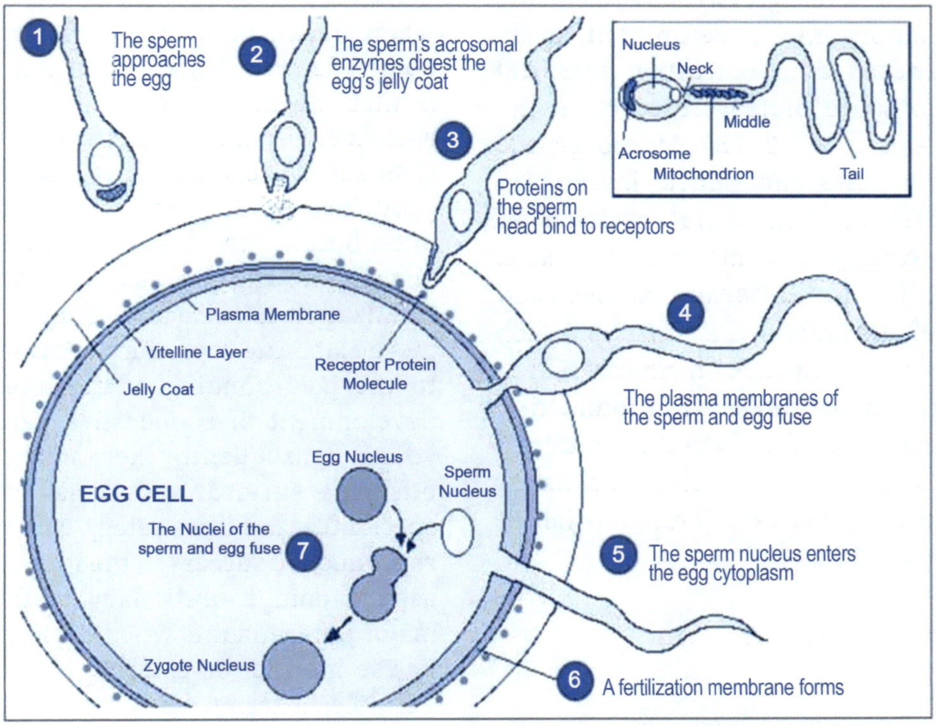

Figure 10.11. Mechanisms of fertilization.

Source: https://zoolibs.com/wp-content/uploads/2023/10/Fertilization-process-in-animal.jpg.

First, sperm have to penetrate through protective coats enveloping the egg, an achievement met by both mechanical and biochemical forces. Sperm adhesion to certain receptors on the egg's surface initiates a cascade of cellular activities, allowing for membrane fusion. Following fusion, the zygote is converted into a newly activated phase of active cell division and metabolic activation. This turning point not only marks the fulfillment of fertilization but also the initiation of intricate developmental processes. These processes are kept strictly in check by various pathways of signaling that play an important role in completing embryogenesis successfully, further reinforcing the significance of fertilization as a determinant of reproductive success. The adaptive role of fertilization mechanisms has consequences far more than mere gametic union; it defines the entire life history of organisms. Variations in fertilization strategies allow for different adaptations to environmental stress, which define reproductive success in particular ecological niches. As an illustration, developmental biology research has investigated how microgravity environments may affect embryonic development early in embryogenesis, demonstrating potential significance to space biology research in the future (Halstead et al., 2025). Moreover, the sexual selection roles and gamete investment costs underpinning the differentiation of reproductive strategies in males and females define the dynamic interaction between developmental outcomes and fertilization. Such findings highlight the need for further research on the fertilization mechanisms as a foundation for further developing the general narrative of reproduction and organism development across different ecological contexts (Franklin-Hall et al., 2020) (Figure 10.11).

10.3.2. External vs. Internal Fertilization

The modes of fertilization have great impacts on the reproductive tactics and developmental results in several organisms. External fertilization is typical of aquatic species and entails the gametes being released into the environment and the fertilization taking place outside of the female's body. By doing this, a high number of eggs can be fertilized at the same time, and this gives high chances of survival of offspring. Nonetheless, external fertilization is greatly subject to environmental factors, including water temperature, salinity, which may influence the viability of gametes. In addition, fertilized eggs and larvae may have reduced survival due to the danger of predation and other environmental risks. Even though species employing external fertilization, like most fish and amphibians, can take advantage of genetic diversity because of the sheer numbers of gametes produced during breeding seasons, increasing their adaptability to fluctuating environments. Conversely, internal fertilization offers a more regulated condition for the offspring's development, as is the case with the terrestrial vertebrates, including mammals and reptiles. This is a procedure whereby sperm is transferred to the female reproductive tract, where fertilization can take place inside the body. The main benefit of internal fertilization is that the developing embryos are shielded against environmental stressors, thereby raising their survival chances. Also, the process usually produces fewer, but more advanced young ones, because the female is able to put a lot of energy into the process of gestation. Internal fertilization can also lead to the development of more elaborate reproductive strategies, leading to the pair bonding of the animals and the development of cooperative parenting of the young. Consequently, there might be increased offspring survival and social interactions of species that use this strategy, which leads to their reproductive success in the long run. Ecological aspects and evolutionary modifications are major determinants when deciding on whether to use internal or external fertilization. As an example, external fertilization can be favored by aquatic habitats, where gametes are more diluted in water and large-scale dispersals are more feasible. On the other hand, on land where environmental stresses play an important role, internal fertilization is selected as it provides protection. The specific adaptations in embryonic development, mating behaviors and parental investment are the results of the evolution of certain reproductive strategies. Studies on organisms like the Portuguese oyster, which undergoes external fertilization, highlight the complexity involved in their reproductive cycles and the ecological pressures they face, including habitat restoration (Melo et al., 2015). Conversely, advances in cryopreservation techniques in species with internal fertilization underscore the intricate balance between evolutionary strategies and their practical applications in conservation efforts (Ackerman et al., 2018). The two approaches explain the active interaction between reproduction and the development requirements of organisms in different ecological niches.

10.3.3. Factors Influencing Fertilization Success

Successful fertilization is, in a great deal of ways, contingent upon some behavioral as well as environmental elements that are important to reproductive processes. Among these, reproductive migration timing and related behaviors of said populations can be deciding factors when it comes to fertilization trial results. Amphibians, for instance, will have distinctive courtship behaviors along with enhanced development of secondary sexual characters at breeding time. These phrases are crucial as they not only bring the future mates but also make insemination and fertilization successful. The duration people remain in the breeding habitat has a direct relationship with the success of these reproductive processes. Fundamentally, the interactions between environmental conditions, , such as temperature and quality of water, and the duration of migrations play a major influence on fertilization success and, by extension, the reproductive success of the concerned species, as noted in the case of temperate amphibians (Geltsch et al., 2007). Sperm quality is also an important factor influencing the success of fertilization across many species. There has been recent research to demonstrate that sperm phenotype can have significant effects on male fertility success and the quality of offspring. Sperm can be altered by environmental stressors in terms of quality, affecting traits like motility and viability. In experiments with the internally fertilizing fish Poecilia reticulata, aging of the sperm was found to negatively affect fertility capability, and old sperm were characterized by lower velocity and competitive disadvantage when competing to fertilize eggs. The implications of the research are not restricted to individual mating success; sperm quality can have long-lasting effects on offspring, influencing not just short-term reproductive success but also the subsequent fitness of the offspring. Developing an understanding of how environmental conditions or reproductive behavior impact sperm quality is therefore essential for the unravelling of the complexity of fertilization success in various species (Adriaenssens et al., 2017). The relationship between sperm quality and maternal environment is still a major area of concern in reproductive biology. Even with research emphasizing the significance of sperm phenotype, there are challenges in unraveling the effect of maternal factors on fertilization success. The importance of how different environments can affect sperm and egg viability and possibly cause epigenetic modifications influencing the development of offspring cannot be overlooked. Moreover, an understanding of how maternal environmental circumstances can make contributions elucidates evolutionary adjustments in populations. Integrating these components into a unified framework for fertilization success provides researchers with a sense of the intricate relationship between male reproductive strategies, environmental settings, and resulting offspring fitness. Analysis of these dynamics is of broader significance for conservation and management of biodiversity and underscores the value of multidimensionality for comprehending reproduction and development (Figure 10.12).

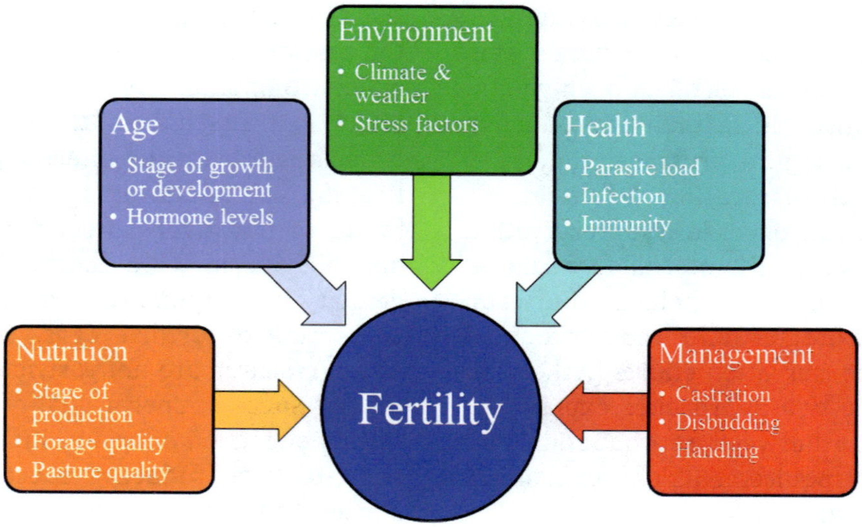

Figure 10.12. Factors influencing fertility.

Source: https://www.researchgate.net/profile/Holly-Evans-7/publication/361678431/figure/fig1/AS:11431281198892680@1697470053895/Factors-influencing-fertility-A-multitude-of-factors-can-affect-the-fertility-of-an.ppm.

Test Your Knowledge

1. Illustrate the mechanisms of fertilization.
2. What are the factors influencing fertilization success?

10.4. Embryonic Development

Learning Outcome

1. Stages of embryonic development;
2. Role of genetic and environmental factors;
3. Importance of cell differentiation.

Embryogenesis is an essential process of reproduction that occurs after fertilization and leads to the formation of a fully mature organism. At its center is the intricate interplay of genetic and environmental factors governing the process from one fertilized egg cell to a multicellular organism. The initial stages of cleavage, gastrulation, and organogenesis are made up of precise cellular divisions and various differentiations. During cleavage, the one-cell zygote is mitotically dividing at a high frequency, resulting in a blastula that will be the precursor to the remainder of the morphogenetic movements. Gastrulation is particularly important, as it establishes the three germ layers, ectoderm, mesoderm, and endoderm, each of which will further specialize into specific tissues and organs. Such an orderly series of steps is indicative of the highly regulated nature of embryonic development and how critical it is to reproductive success. In mammals, the roles of specific genes and pathways in embryonic development have been of considerable interest. Secreted frizzled-related proteins (Sfrps), for instance, have been described as key mediators of signaling pathways that are crucial in development. Sfrp1 and Sfrp2 have been reported to have critical roles in embryonic sexual development and regulate the morphology and gonad positioning within the developing embryo. Conditional gene knockout experiments confirmed that embryos lacking these proteins exhibit apparent defects like misplaced testes and reduced gonadal volume (Bogani et al., 2009).

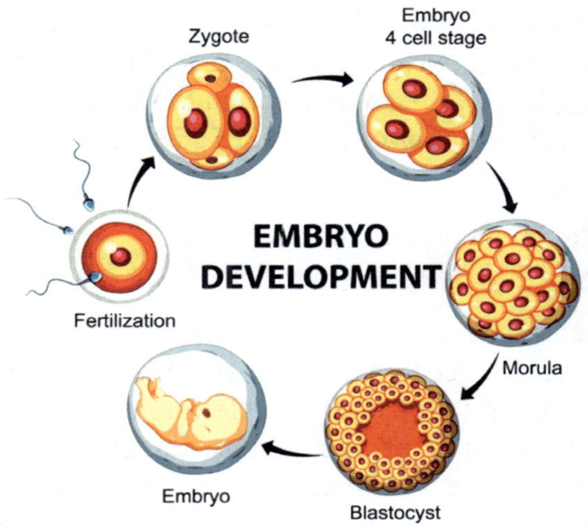

Figure 10.13. Embryonic development.

Source: https://www.ferty9.com/wp-content/uploads/2025/04/Monsoon-Skin-Hair-Care-Tips-for-Kids-2023-03-28T192341.4121.png.

The findings go to highlight the importance of non-canonical Wnt signaling in the embryogenesis process, observing that its disruption at this stage would lead to drastic reproductive alterations and potential sterility among mammals. Outside of mammals, embryonic development also happens in a unique manner in many plants, demonstrating the diversity of reproductive processes in the living world. Perhaps the most dramatic case is St. John's wort (*Hypericum perforatum*), with its fascinating form of asexual reproduction known as apomixis. In the course of this process, somatic cells within the ovule are developmentally plastic and form embryo sacs that evade the usual process of fertilization and thereby facilitate cloning by seeds (Barcaccia et al., 2015). This not only goes against conventional reproductive biology but is also of significant interest in agricultural production. By describing the molecular basis of apomixis and genetic control, researchers envision using this information to enhance crop breeding techniques, noting how the diverse mechanisms of embryogenic development shape reproductive strategy in plants and animals alike (Figure 10.13).

10.4.1. Stages of Embryonic Development

The stages of embryonic development are critical to the grand scheme of reproduction and development since they signal the transition from one fertilized cell to a multicellular organism. Fertilization is the starting point of the process, where the sperm and egg unite to form a zygote. The zygote goes through a series of rapid mitotic divisions known as cleavage, leading to the formation of a blastula, a hollow ball of cells. This is succeeded by gastrulation, a developmental stage at which the cells of the blastula reorganize to form three primary germ layers: ectoderm, mesoderm, and endoderm. These further differentiate to form tissues and organs, laying the groundwork for building the elaborate forms that are the hallmark of developed organisms. This choreography of cells is crucial, as any deviations could lead to developmental malformations that could compromise organismal viability and functionality. With development in embryonic growth, organogenesis phases are in action. In this phase, the three germ layers further specialize to form the organs and systems that characterize the organism. The development of structures, including the skin and nervous system, results from the development of the ectoderm, whereas muscles, the circulatory system, and skeletal system develop from the mesoderm. In contrast, the endoderm is largely responsible for the development of the gastrointestinal tract and associated organs, , such as the liver and pancreas. Accuracy in timing and communication among these layers is essential because it ensures that organ systems develop cooperatively and functionally. Moreover, molecular signaling pathways and genetic control are underlying these processes, which orchestrate the intricate dance of cellular differentiation and positioning.

A clearer understanding of these developmental stages is required, as defects within these processes can cause congenital defects that occur in millions of individuals on this planet. Recent studies have shown the evolutionary significance of embryonic developmental stages, particularly in vertebrate evolution. Experiments have confirmed that DNA methylation reprogramming during embryogenesis is a force driving the differentiation between invertebrates and vertebrates, with patterns specific to various taxa. In vertebrates, for instance, the dynamic nature of DNA methylation across the various stages of embryonic development is highly correlated with gene expression patterns essential for the correct morphogenesis and function of organs. Research in model species , such as the cat shark, has revealed conserved axon scaffold structures to be necessary for nervous system development and indicates evolutionary continuity across vertebrates. Clarifying these molecular events not only sheds light on processual mechanisms of normal development but also reveals the potential of therapeutic intervention in treating developmental disorders in terms of the embryonic development steps (Aluru et al., 2019; Schubert et al., 2014) (Figure 10.14).

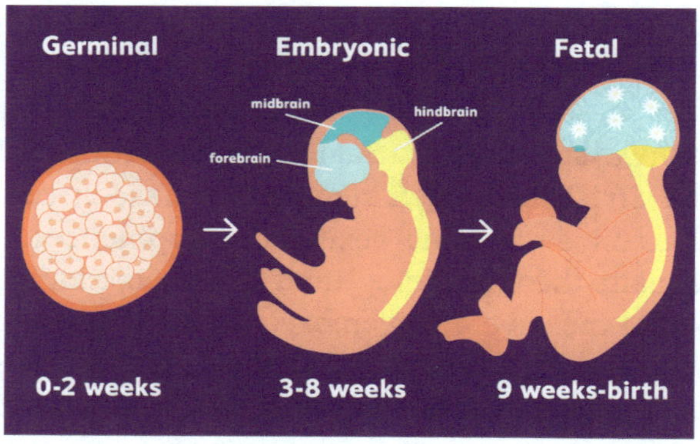

Figure 10.14. Stages of embryonic development.

Source: https://www.verywellmind.com/thmb/bOGQenRSHLRpMRkpJ0e8PlM0-u4=/1500x0/filters:no_upscale():max_bytes(150000):strip_icc()/2795073-stages-of-prenatal-development-01-5a3040f6eb4d5200362d5553.png.

10.4.2. Role of Genetic and Environmental Factors

The interaction between environmental and genetic forces is key to the understanding of reproduction and development. Heritable determinants at the genetic level influence phenotypic characters that underpin reproductive success. For example, genetic variation can influence the timing of reproduction, fertility levels, and offspring viability. A recent body of evidence has revealed the significance of telomere length (TL), a heritable character with significant individual variability. Research points out that maternal TL had a major contribution to offspring TL, identifying the maternal lineages fundamental role in reproductive outcome. The genetic contribution to our understanding of the complex interactions between inherited characteristics and the underlying mechanism for reproductive development is further elucidated in how genetics predisposes individuals to specific reproductive strategies and outcomes. Environmental factors cannot be dismissed as they also play a major role in affecting reproductive processes. The environment can influence the availability of resources, presence of stressors, and overall health, all of which can influence reproductive efficiency. Environmental toxins, presence of nutrients, and social structure, for example, can lead to variations in reproductive strategies between species. Furthermore, maternal conception age has also been directly related to TL offspring, showing that prenatal environmental conditions, , such as maternal health, diet, and stress levels, also play a large role in development. This kind of interaction explains the complex interplay of environmental influences on genetic predisposition, ultimately shaping reproductive success and developmental outcome in offspring, lending support towards a holistic approach to the study of reproduction and development. Research into the genetic and environmental causes of reproduction continues, highlighting the value of interdisciplinary collaboration involving genetics, ecology, and developmental biology. Understanding the heritability of TLs and the implications for reproductive health is one among many illustrations of the need to consider genetics and environment simultaneously. For instance, although genetic variations with the maintenance of telomeres have been discovered, their expression is likely to be influenced by external factors , such as stress and diet that can modulate their effect over time. This convergence of genetic and environmental factors can lead to a larger picture of reproductive processes, which shows how both factors are accountable for the success of species under varying environments. Further investigation into these variables will increase our knowledge of evolutionary dynamics, population genetics, and ultimately, the complex phenomena of reproduction and development in general.

10.4.3. Importance of Cell Differentiation

Cell differentiation is an essential feature of the reproduction and development processes that guarantee the production of specialized cells out of the undifferentiated stem cells. This specialization plays a central role in the evolution of multicellular organisms, in which the various types of cells perform various functions that are essential to survival. In the embryonic development process, undifferentiated cells are stimulated by a number of signals, , such as chemical signals and environmental signals, to assume a particular role. Improper differentiation may result in developmental abnormalities or pathology, which highlights the significance of the strict regulation of differentiation processes. For instance, disruptions in epigenetic mechanisms, , such as those induced by environmental toxins like bisphenol A (BPA), can adversely impact reproductive health by altering gene expression related to oocyte maturation, thereby intimately linking cell differentiation to overall reproductive and developmental success (Carnevali et al., 2016). Cell differentiation is also a key factor in the development of complex organs and systems, especially when the nervous system is taken into consideration. As organisms get advanced, complex cellular coordination and communication is necessary. GFs, , such as insulin-like growth factor-I (IGF-I), are instrumental in guiding the differentiation of various cell types, including those within the auditory system. Cells have to receive the right signaling pathways to survive, differentiate and mature in the right way. In studies involving murine models, deficits in IGF-I have been shown to result in dramatic underdevelopment of auditory neurons, revealing that specific cellular differentiation processes are fundamental for the formation of sensory organs and for preventing conditions like sensorineural deafness (Sanchez-Calderon et al., 2010). Therefore, the consequences of cell differentiation can be felt not only on the organization of the cells but on the functioning of the whole system. Moreover, the genetic-epigenetic interactions that occur during cell differentiation are crucial to the flexibility and robustness of organisms to fluctuating environments. Studies have underscored the disruption of normal differentiation processes by exogenous agents that cause long-term consequences on reproductive health and development. For example, BPAs interference with histone modifications can critically alter oocyte maturation signals, raising concerns over environmental impacts on reproductive capabilities (Carnevali et al., 2016). Furthermore, elucidation of the signaling networks that control cell differentiation may be applied to the design of therapeutic interventions in several diseases, as well as infertility and developmental disorders. The analysis of cell differentiation, therefore, not only explains some basic biological mechanisms, but it also outlines possible ways of intervention and enhancement of reproductive and developmental health.

Test Your Knowledge

1. What are the stages of embryonic development?
2. What are the roles of genetic and environmental factors?

10.5. Developmental Biology

Learning Outcome

1. Key concepts in developmental biology;
2. Influence of stem cells on development;
3. Applications of developmental biology in medicine.

Developmental biology plays a key role in understanding the intricate processes involved in reproduction and the development of organisms. The cornerstones of this science are developmental control genes responsible for governing significant morphological changes during the lifecycle of an organism. Of special interest are the TFs , such as LEAFY and MADS-domain proteins that regulate crucial processes in plant reproduction as a prime example of the role that genetic regulation plays in development. These proteins regulate the switch to the floral phase in angiosperms and the reproductive organ specification in gymnosperms and angiosperms. The knowledge derived from the genome-wide profiling and structural biology also sheds more light on the complex dynamics of these regulatory factors to enable understanding of how development processes have shaped plant reproductive structures (Brennich et al., 2016). Beyond botany, developmental biology is heavily influenced by environmental factors that affect embryonic development, a factor well highlighted by the work of early biologist Ernest Everett Just. His scientific investigation of marine biology and cytology discovered the significance of ectoplasm and cellular cytoplasm in determining developmental outcomes. Just's method puts the need for *in vivo* conditions in perspective, offering a novel perspective on the role of external conditions in determining the development of organisms. His research led to the establishment of ecological developmental biology, which considers the interaction between developmental events and environmental stimulation. Accounts like Basic Methods for Experiments on Eggs of Marine Mammals and The Biology of the Cell Surface remain core books in cell biology, looking back at the necessity of understanding internal and external controls of embryonic development (Connor et al., 2013). Advances in developmental biology provide timely perspectives into the reproduction mechanics of different forms of life, from flowering plants to sea creatures. By studying the genetic and environmental contexts that shape developmental trajectories, researchers enlighten us on how organisms adapt and thrive in an ever-changing world. The interaction between development and reproduction is not just a matter of genetic determination but is also influenced by ecological drivers, which emphasizes a broader integration of disciplines in biology. Such integration warrants cross-disciplinary work, enabling enhanced understanding of basic biological processes. Ultimately, the synergy between genetics, epigenetics, and environmental influences in developmental biology not only informs basic scientific knowledge but also holds potential implications for fields , such as conservation biology and biotechnology, emphasizing the interconnectedness of all life forms (Connor et al., 2013; Brennich et al., 2016) (Figure 10.15).

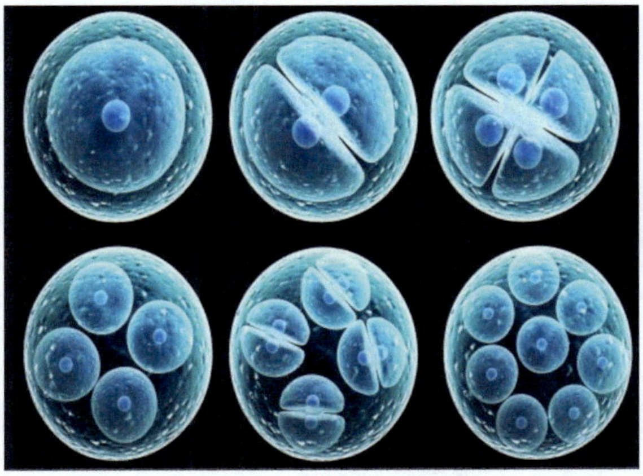

Figure 10.15. Developmental biology.

Source: https://www.creative-diagnostics.com/upload/image/developmental-biology.png.

10.5.1. Key Concepts in Developmental Biology

Developmental biology is the observation of several phenomena that serve to account for the sophistication of reproduction and development. A key aspect is embryonic development, which illustrates how a single fertilized egg cell can give rise to a complex organism. This is done through a cascade of finely regulated events, including cell division, differentiation, and morphogenesis. During cell division, they are instructed on what to do and become, so that they fit perfectly into several tissues and organs. Such processes are most crucial in developmental biology since they form a basis for understanding congenital illnesses and potential therapeutic interventions. This is significant because it illuminates how genes and the environment interact in developmental processes, showing the dynamic nature of biological processes and setting the stage for a deeper exploration of epigenetics in the determination of developmental trajectories. A second salient idea in developmental biology is that gene and environmental interactions have a significant role to play in determining development. The field has made a lot of progress in realizing that development is not only a consequence of inheritance by genes but also a consequence of complex interactions with the environment. For instance, when development is happening at key times of development, extrinsic factors , such as nutrition, exposure to toxins, and interactions with others can significantly influence gene expression and resulting phenotypic effects. This dialogue also emphasizes the importance of the epigenetic landscape, which determines gene expression in response to environmental stimuli. By studying these interactions, researchers can better understand how developmental processes can become disrupted and result in ensuing health effects. With that being said, the convergence of genetics and environmental influences contributes to a larger understanding of the laws governing reproduction and development. Aside from genetic and environmental influences, the processes of evolution also play an important role in shaping developmental processes throughout life in all species. Evolution theory based on natural and sexual selection provides a platform for explaining how developmental routes were maintained or altered through time. Main theories like heritable variation and selective forces explain how characters evolve within species, ultimately guiding reproductive strategy and developmental timing. This evolutionary perspective ties developmental biology to the general biological sciences, revealing that our understanding of development and reproduction must be modified to account for historical circumstances. Furthermore, because biology and culture are bound together, their connection allows us to study development and behavior in a finer-grained way. As highlighted in current discussion, the incorporation of evolutionary observations into sociocultural environments adds depth to the causes of development in favor of an even broader understanding of influences, firmly establishing the validity of these concepts within developmental biology (Chasiotis et al., 2011; Hanisch & Eirdosh, 2020).

10.5.2. Influence of Stem Cells on Development

The intricate interrelation between developmental processes and stem cells is responsible for the expertise of reproduction and development. Stem cells possess a unique ability to differentiate into various cell types, a necessity for the development and functioning of specialized tissues during embryogenesis. Specifically in the ovarian scenario, the somatic cell differentiation , such as granulosa and theca cells, into luteal cells plays a pivotal role in the production of progesterone, an essential hormone for maintaining pregnancy.

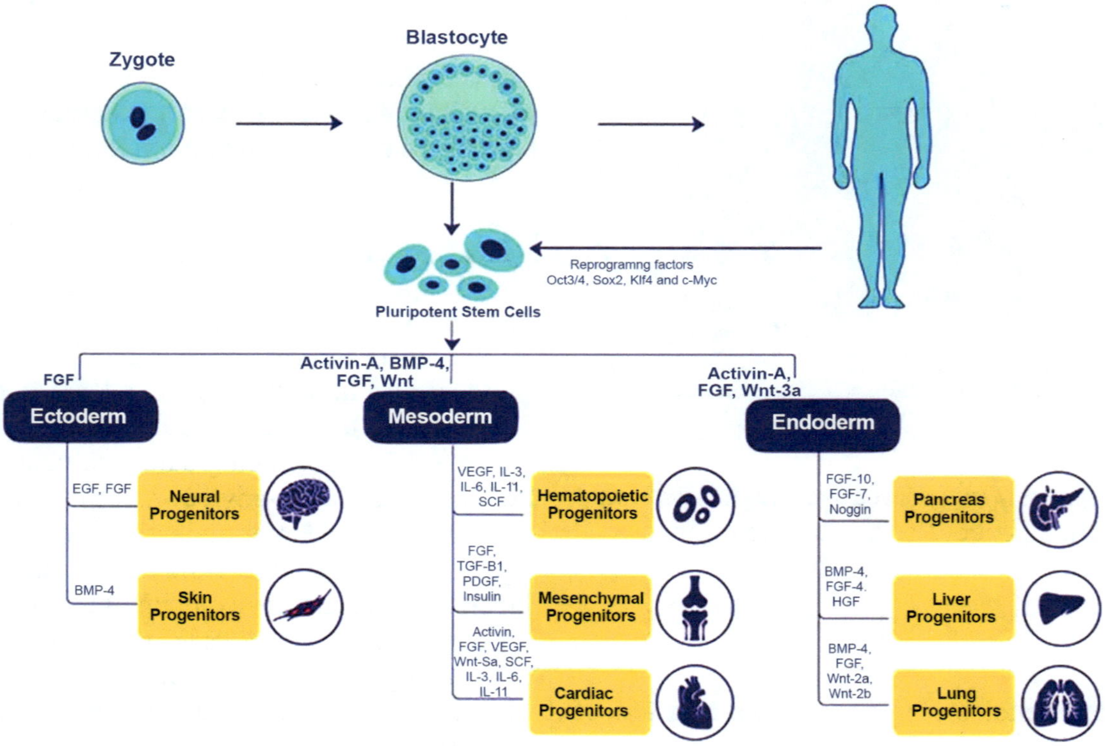

Figure 10.16. Influence of stem cells on development.

Source: https://www.sigmaaldrich.com/deepweb/assets/sigmaaldrich/marketing/global/images/technical-documents/articles/cell-culture-and-analysis/stem-cell-culture/growth-factors-in-stem-cell/growth-factors-in-stem-cell.jpg.

When there is no appropriate differentiation and corpus luteum development, fertility is jeopardized, and infertility results. Anovulation, which refers to the failure to release an oocyte, highlights the importance of regulation of stem cells in reproductive biology and corpus luteum function, and consequently on the ultimate outcome of fertility in mammalian females (Abedel-Majed et al., 2019). Besides, chaotic dynamics of cellular interactions and differentiation processes demonstrate the complexity of stem cell systems. The "chaos hypothesis" is the idea that stem cell differentiation is a random process, controlled by chaotic intra-cellular dynamics and cell-cell interactions. The randomness in this means that stem cells do not necessarily evolve linearly to differentiation; their destinies are fixed by environmental cues and random events. Such a system assures robustness and versatility, enabling an organism to cope with fluctuating environments effectively. This stochastic differentiation also emphasizes the irreversibly lost multipotency in stem cell conversion into more specialized cell types, resulting in a loss of cellular chemical diversity. The development of a functional reproductive system thus hinges on the precarious balance between these stochastic processes and stem cell dynamics (Furusawa et al., 2000). Last, the applicability of stem cell research to reproductive medicine and developmental biology is tremendous. By understanding the molecular and genetic forces that induce stem cell differentiation, researchers can create goal-directed therapy for reproductive dysfunctions, including those resulting from granulosa or theca cell dysfunction. Emerging findings reveal gene expression patterns that differentiate among the somatic cells of one another during the luteal transition, specifying possible pathways for therapeutic intervention in infertility. Furthermore, unraveling stem cell dynamics messiness not only increases our knowledge of cell differentiation but also suggests new directions for amplifying fertility and maximizing reproductive success. Stem cell biology is thus an epicenter of seeking augmented reproductive health and elucidating the complex mechanisms of development (Abedel-Majed et al., 2019; Furusawa et al., 2000) (Figure 10.16).

10.5.3. Applications of Developmental Biology in Medicine

The investigation into developmental biology has been the cause of enormous breakthroughs in medicine, particularly regenerative treatments. The application of MSCs, which are said to possess outstanding regenerative capacities, is among the biggest applications. These cells are plastic and may differentiate into any type of cell, and this is a key factor in tissue regeneration and repair. For instance, stem cells from the oral cavity in the guise of human periapical cyst mesenchymal stem cells (hPCy-MSCs) have been hailed as a promising source for clinical application due to their accessibility and versatile nature. Obtaining hPCy-MSCs from surgically removed periapical cysts is an environmentally friendly procedure that not only recycles biological waste but also spares damage to neighboring healthy tissues, hence rendering them more favorable for regenerative medicine (Codispoti et al., 2017). Besides, the scope of developmental biology is also reflected in the management and comprehension of congenital diseases, which are caused by aberrant development. By inspecting the molecular and cellular mechanisms of development, researchers can identify significant pathways that become pathologic in various conditions. Using this information, they can establish targeted therapies that may normalize or circumvent these developmental anomalies. For instance, CRISPR gene editing technologies have become effective agents for repairing genetic defects that cause congenital malformations. By refining such techniques, scientists can correct developmental errors in the embryonic stage, significantly improving the fate of afflicted individuals (Arnold et al.,

2001). The incorporation of developmental biology in medical practice not only encourages novel therapeutic strategies but also underlines the significance of personalized medicine. As scientists get to unravel the intricacies of human development, the potential of approaching treatments on the basis of individual genetic and developmental parameters becomes more real. Precision medicine strategies have the capacity to transform the treatment offered to patients, more so for reproductive health and developmental interventions. By applying knowledge derived from the study of developmental pathways, clinicians can deliver customized interventions that attend to the unique needs of each patient. As a whole, the value of developmental biology in medicine offers a revolutionary connection between biological inquiry and clinical practice, opening opportunities for improved diagnostic and therapeutic intervention that elevates the quality of health outcomes in heterogeneous populations (Codispoti et al., 2017).

Test Your Knowledge

1. What are the key concepts in developmental biology?
2. How influence of stem cells on development?

10.6. Evolutionary Perspectives on Reproduction

Learning Outcome

1. Evolution of reproductive strategies;
2. Impact of sexual selection;
3. Coevolution of reproductive traits.

An evolutionary approach to the study of reproductive strategies provides great explanations on the diversity of life forms. All species develop species-specific reproductive adaptations that increase their survival and reproductive success under different environmental conditions. As an example, the variations between sexual and asexual reproduction illustrate that species incorporate reproductive tactics depending on ecological conditions and resources. Although sexual reproduction is energetically and temporally expensive, it assists in genetic variation, which is vital in surviving in dynamic environments. Conversely, asexual reproduction enables a population to grow rapidly but has limited genetic variation. Understanding these dichotomies is crucial for comprehending broader evolutionary patterns, including speciation and the evolution of complex life forms, ultimately raising essential questions about what constitutes a biological individual and how these individuals interact within their ecosystems (Barker et al., 2019). The diversity in reproductive strategies between the sexes in the animal kingdom is highly diverse and this makes it an exciting field to explore in evolutionary view of reproduction. The adaptive differences between the sexes, , such as sexual dimorphism where one sex is visibly or sizeably different to the other, can frequently indicate the adaptive strategies evolved to ensure reproductive success. As an illustration, in animals, male rivalry in the quest to find a mate can lead to the development of ornate features that increase their attractiveness or aggressiveness, whereas females can evolve to be choosy in regard to their mates depending on these ornaments. This interactive play poses critical questions as to what biological sex is. Although the conventional categories of male and female may seem basic, they tend to conceal more advanced evolutionary puzzles. As such, these classifications may offer limited explanatory power, underscoring the need for a nuanced understanding of sex as a "historical explanatory kind" that can significantly influence reproductive outcomes and population dynamics (Franklin-Hall et al., 2020). The discussion of evolutionary view on reproduction also requires addressing the issue of wider implication to the interaction of species and ecological system. Mating tactics have an enormous influence on population dynamics, predator-prey associations, and community organization. As an example, the timing and synchrony of reproductive phenomena may affect food webs and resource distribution in ecosystems. Moreover, the combination of reproductive isolation and gene flow may result in the speciation process that adds value to the biodiversity. These aspects demonstrate that reproduction is a key force behind evolution, and it shapes both microevolutionary tendencies, including genetic drift and natural selection, and macroevolutionary tendencies, including the development of new species. Hence, research into evolutionary perspectives on reproduction transcends individual species, emphasizing the interconnectedness of life and the critical role that reproductive strategies play in shaping evolutionary trajectories (Barker et al., 2019) (Figure 10.17).

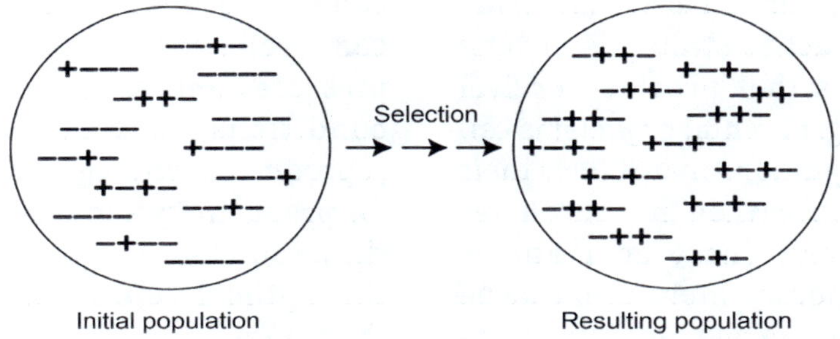

Figure 10.17. Evolutionary perspectives on reproduction.

Source: https://www.nature.com/scitable/content/ne0000/ne0000/ne0000/ne0000/56091/otto_evolution_sex-f2_LARGE_2.jpg.

10.6.1. Evolution of Reproductive Strategies

Reproductive strategy evolution is an essential theme of the investigation into organismal development and adaptation. Organisms have developed diverse reproductive modes through evolutionary history, responding to various pressures from the environment. Reproductive strategies can be broadly classified into sexual and asexual reproduction, with their respective advantages and disadvantages. Genetic recombination, sexual reproduction, promotes greater genetic diversity so that populations are capable of evolving in response to the environment. However, reproduction in asexual organisms is capable of generating rapid population growth, and this is a strategic advantage where environmental conditions are stable and available resources are abundant. The interaction between these strategies is important since it directly influences the long-term survival and evolutionary history of a species, thereby emphasizing the importance of reproductive strategy in the grand canvas of reproductive success and fitness. Environmental factors have a critical role to play in the regulation of reproductive strategies, influencing the timing and also the modality of reproduction across species. For instance, seasonal climate changes may initiate extreme changes in the reproductive habits, leading to types of adaptations that optimize the offspring's chances of survival. In *Aiptasia diaphana*, research indicates that temperature and photoperiod have a significant influence on sexual and asexual reproductive cycles. The sea anemone shows elevated asexual reproduction levels during warmer seasons, while its sexual reproductive cycle coincides with the first onset of spring heat. Seasonal reproductive plasticity thus provides Aiptasia with an adaptive advantage by allowing them to harness environmental cues in the advancement of reproductive success under conditions of changing environments. Such findings demonstrate the eloquence of reproductive strategies to ecological interactions and pressures as exhibitions of their evolutionary significance. Furthermore, the mechanisms of reproductive strategies include epigenetic adaptations that provide survival advantages to offspring. Parental organisms can invest in heritable adaptations that render their offspring more competitive when they encounter environmental stressors. These adaptations, as posited in contemporary literature, can be epigenetic in nature and include modifications that do not alter genetic composition but influence developmental processes and fitness traits. This phenomenon not only points to the flexibility of reproductive tactics but also underlines how parental investment molds generations. As species are subjected to increasingly dynamic environments, familiarity with these epigenetic reactions and their consequences is becoming ever more essential. By examining these adaptive processes, researchers are better positioned to comprehend the evolution of reproductive and developmental dynamics and ultimately contribute insights into conservation and management practices for biodiversity conservation in the context of environmental change (Burt et al., 2010; Gulyas et al., 2019).

10.6.2. Impact of Sexual Selection

The effect of sexual selection is great and impacts reproductive success and developmental pathways across a broad spectrum of species. Sexual selection operates through mechanisms that favor the enhancement of particular traits or behaviors that are attractive for mating. For instance, in the majority of species, males with superior body characteristics or complex courtship behaviors have higher probabilities of winning mates from females, which results in increased reproductive success. This process is a demonstration of how sexual selection pressures can stimulate the evolution of fitness-enhancing qualities, thus shaping gene pools generation after generation. Through this, the interaction between sexual selection and evolution emphasizes the importance of being aware of reproduction strategy in the development context, to observe how traits and behavior can evolve under particular environmental pressures and social contexts. Under farming conditions, sexual selection influences reproduction strategies with direct implications for productivity and sustainability. For example, the profitability of dairy farming depends on the optimal reproductive performance, and thus, effective sexual selection mechanisms are required in ranching for breeding. Because conventional methods may not suffice due to the complexity of animal genetics, integration with functional genomics is one crucial strategy. This approach provides evidence of physiological mechanisms underlying reproductive traits. With the application of genomic selection techniques, farmers can improve the reproductive abilities of animals, bypassing subfertility problems. These modifications are critical to economic viability as well as animal welfare and sustainability, showing that sexual selection influences agricultural advancement. Sexual selection also gives evidence about the adolescent phase of human life, demonstrating the influence of the environment on how reproductive strategies are shaped. Change from juvenility to adult life represents plastic adaptation, wherein energy availability and social pressure could influence maturation processes. Adolescence is the age when characteristics predisposing to match changing social demands are usually established, establishing reproductive advantages. For example, psychosocial stress can drive a faster life history trend, with early maturity being the better option than extended longevity. These processes highlight the importance of sexual selection within an evolutionary-developmental framework, with a focus on how reproductive strategy not only reflects genetic transmission but also responds to complex social and environmental cues. This inter-disciplinary approach offers new directions for human reproduction and developmental biology studies, inviting scholars to admire the delicate balance between the biological necessities and adaptive strategies responding to environmental stress (Figure 10.18).

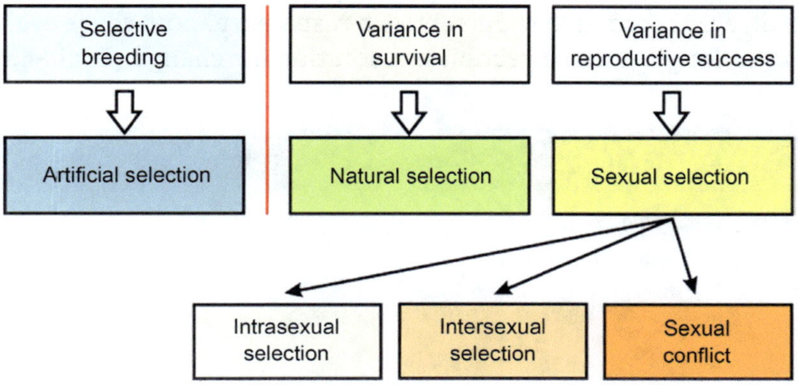

Figure 10.18. Impact of sexual selection.

Source: https://media.springernature.com/lw685/springer-static/image/chp%3A10.1007%2F978-3-030-82879-0_8/MediaObjects/509426_1_En_8_Fig2_HTML.png.

10.6.3. Coevolution of Reproductive Traits

Coevolution of reproductive traits is one of the driving forces in the definition of species interaction dynamics and the overall evolutionary landscape. This process centers around life cycle coordination among species, initiated by the necessity for optimal reproduction timing and resource availability. This synchrony can be achieved through pairwise direct contact, two species evolving in synchrony, or indirectly via ecosystem engineers, which make altered environments. This type of interaction illustrates that traits co-evolve not just in isolation but as part of a greater ecological community that is composed of varied environmental conditions and interactions with many other different species. Understanding such complexity is crucial, as it determines reproductive strategy potential for responses to changing ecological environments, thus impacting overall species survival and reproduction (Diniz-Filho et al., 2017). Postcopulatory sexual selection dynamics also emphasize how male-male competition will create evolutionary developmental changes in reproduction. This process occurs in the female reproductive tract, as sperm competition escalates stakes for male fertility characters and leads to a variety of morphological and physiological adaptations. These adjustments can be through modifications in sperm morphology, copulatory behaviors, as well as mechanisms of female choice. This intrasexual conflict typically creates an evolutionary arms race and leads to increased diversity of adaptations. However, to explain the full implications of these changes, researchers ought to investigate the genetic origins of male and female characters involved in sperm selection. Identification of the specific genes or genomic regions governing such characters is important to understand coevolutionary trajectories between sexes fully (Civetta et al., 2002). When coevolutionary processes are at their culmination, they present an intriguing dynamic balance between selection pressures and reproduction strategies. For instance, the labels *semelparity* and *iteroparity* reflect varied approaches to evolution that are employed by organisms in a bid to optimize reproductive success. Environmental stability and the availability of resources decide the strategies, whereby semelparity or single-lifetime reproductive events can occur in unstable environments, but iteroparity provides room for repeated reproductive events over a life history. Such approaches express an intricate link between behavior, physiology, and morphology during reproduction. This richness underpins the notion that evolutionary adaptations are not so much the result of optimizing traits but an emergent product of an interrelated group of selective pressures leading to diverse reproductive outcomes in response to both the prevailing environmental challenges and also the longer-term direction of evolutionary change (Diniz-Filho et al., 2017).

Test Your Knowledge

1. What is the impact of sexual selection?
2. What is the coevolution of reproductive traits?

SUMMARY

In reconciling the complex interplay between reproduction and development, it is evident that scientific rigor takes precedence in being able to move forward in our comprehension of these topics. The use of Minimum Information guidelines, specifically MIASE, shows how important it is to give clear details regarding experimental simulations to enable reproducibility in biological studies. As explained, the MIASE framework emphasizes the need to describe the protocols that are applied in simulation experiments, thereby encouraging the openness of science efforts. Following guidelines , such as these, scientists can reproduce results easily, hence strengthening the foundation on which theories of reproduction and development in biology are built. This codified framework not only allows for accurate representation of primary work but also allows scientists to collaborate, enabling the co-emergence of knowledge that can reverse arduous biological phenomena (Hucka et al., 2011). In looking at the consequences of development and reproduction, the similarities between biological processes and computational methodology seem to be significant thinking mechanisms. Computational theories explaining reproduction enable scientists to study biological requirements with more specificity, utilizing algorithmic theory to explore the intricacies of developmental stages. This convergence is encapsulated by the Cosmological Natural Selection theory, where the biological processes elucidate the intricate relationships of living organisms and their environments. As has been explained in the physical constants and beginnings analysis, these analogies of life are fresh perspectives that form the core of how the fundamental principles of evolutionary adaptations are understood. Therefore, applying these understandings gives a deeper perspective of reproductive strategies and development regimes, which further shape the trajectory of life (Carr et al., 2010). In short, the combination of stringent scientific methodology and innovative analogical thinking marks the fields of reproduction and development. By ensuring that studies conform to published MIASE guidelines, scientists enhance reproducibility and thus build reliable and useful knowledge. In addition, using computational and biological analogies tells us more, offering novel explanations for complex processes that control life. This dualism not only makes scientific inquiry stronger but also contributes to gaining a better holistic vision towards how development and reproduction occur in the broader canvas of biological science. With these considerations in place, the relationship between empirical observations and theoretical frameworks is even more important to ensure progress that can still unpeel the mysteries of life (Carr et al., 2010; Hucka et al., 2011).

In terms of development and reproduction, the interaction among various biological and environmental factors plays a significant role in determining outcomes in all species. One crucial factor is the gestation and early development periods, setting the foundation for continued development and health. During these time frames, maternal nutrition factors like food intake, levels of stress, and toxin exposure can have highly pertinent influences on offspring outcomes. The relevance of controlled trials, like those investigating interventions to improve balance and mobility in specific populations, is representative of the imperative to understand these interactions. Such research offers valuable information about how extrinsic factors can modify developmental trajectories, with note taken of the importance of strategic intervention at key moments in development (Andrade et al., 2019). Of equal neglect should also be the influence of caregiver participation, particularly in developmental settings. Carer-directed intervention, , such as cognitive stimulation therapy for individuals with dementia, has been both promising and limited in effectiveness to enhance cognitive functioning and quality of life. The interventions point to the potential for reaching out to caregivers in more meaningful ways, thereby recognizing their efforts toward facilitating healthy habits of development. While cognitive stimulation can lead to improved relational patterns between individuals with dementia and caregivers, outcomes in terms of direct enhancement of cognitive functioning appear to be limited. This evokes a pertinent consideration of the subtleties of developmental interventions and more universal considerations regarding their ramifications for holistically facilitating growth (Burns et al., 2015). Further, the concept of developmental feasibility trials emerges as something worthy of serious consideration in expanding our understanding of reproduction and development. Trials of new programs, , such as the BRiMS exercise intervention for individuals with MS, are valuable models for testing the efficacy of expert interventions in specific populations. The knowledge derived from such trials not only sheds light on the interpenetration of developmental health, but also on the importance of tailored strategies specifically designed to intervene on the specific needs of diverse groups. With researchers continuing to advance their approach and incorporating qualitative data, we have a better grasp of the influences that provide the best reproductive and developmental outcomes across the life course (Andrade et al., 2019; Burns et al., 2015).

KEY TERMS FROM THIS CHAPTER

Asexual reproduction	Blastula	Embryo
Fertilization	Gametes	Gastrulation
Gestation	Hormonal regulation	Larva
Meiosis	Metamorphosis	Mitosis
Oogenesis	Organogenesis	Ovary
Placenta	Sexual reproduction	Spermatogenesis

CRITICAL THINKING QUESTIONS

1. **Which type of reproduction results in offspring genetically identical to the parent?**
   a. Sexual reproduction
   b. Asexual reproduction
   c. Fertilization
   d. Meiosis
2. **In sexual reproduction, the fusion of sperm and egg is called:**
   a. Mitosis
   b. Fertilization
   c. Budding
   d. Pollination
3. **Which structure provides nutrients to the developing embryo in mammals?**
   a. Placenta
   b. Amniotic sac
   c. Yolk sac
   d. Uterus
4. **During which stage does an embryo develop three primary germ layers?**
   a. Fertilization
   b. Gastrulation
   c. Cleavage
   d. Organogenesis
5. **What is the main advantage of sexual reproduction over asexual reproduction?**
   a. Faster reproduction
   b. Produces genetically diverse offspring
   c. Requires less energy
   d. Only one parent is needed

Short Answer Questions

1. What is the difference between sexual and asexual reproduction?
2. What are the main stages of the human menstrual cycle?
3. Describe the process of fertilization in humans.
4. What is the function of the placenta during pregnancy?
5. Name and briefly describe the three stages of prenatal development.

Answers to Critical Thinking Questions

1. (b); 2. (b); 3. (a); 4. (b); 5. (b)

REFERENCES

1. Abdellaoui, A., Amin, N., Arakelyan, A., et al. (2019). WGS-based telomere length analysis in Dutch family trios implicates stronger maternal inheritance and a role for RRM1 gene. https://core.ac.uk/download/275656279.pdf (accessed on 13th June 2025).
2. Abedel-Majed, M. A., Cupp, A. S., Davis, J. S., et al. (2019). Perturbations in lineage specification of granulosa and theca cells may alter corpus luteum formation and function. https://core.ac.uk/download/286364499.pdf (accessed on 13th June 2025).
3. Abeydeera, H., Alves, D., Anderson, J., Barritt, G., et al. (2006). Mitochondria directly influence fertilization outcome in the pig. https://core.ac.uk/download/29820079.pdf (accessed on 13th June 2025).
4. Ackerman, J., Ailstock, M., Alexandre, G., et al. (2018). Light is more important than nutrient ratios of fertilization for *Cymodocea nodosa* seedling development. https://core.ac.uk/download/186632092.pdf (accessed on 13th June 2025).
5. Adriaenssens, A., Aitken, R. J., Almbro, M., Bonduriansky, R., Boschetto, C., Crean, A. J., et al. (2017). Sperm storage by males causes changes in sperm phenotype and influences the reproductive fitness of males and their sons. https://core.ac.uk/download/168406257.pdf (accessed on 13th June 2025).
6. Galla, G., Zenoni, S., Avesani, L., Altschmied, L., Rizzo, P., Sharbel, T. F., & Barcaccia, G. (2017). Pistil Transcriptome Analysis to Disclose Genes and Gene Products Related to Aposporous Apomixis in *Hypericum perforatum* L. *Frontiers in plant science*, *8*, 79. https://doi.org/10.3389/fpls.2017.00079 (accessed on 13th June 2025).
7. Aluru, N., Chen, X., Ci, W., et al. (2019). Evolutionary transition between invertebrates and vertebrates via methylation reprogramming in embryogenesis. https://core.ac.uk/download/288430933.pdf (accessed on 13th June 2025).
8. Alvarado, M. V., Andrade, A., Estêvão, M. D., et al. (2018). Duplicated membrane estrogen receptors in the European sea bass (*Dicentrarchus labrax*): Phylogeny, expression and regulation throughout the reproductive cycle. https://core.ac.uk/download/288958632.pdf (accessed on 13th June 2025).
9. Andrade, J., Barton, A., Berrow, M., Creanor, et al. (2019). A self-management programmed to reduce falls and improve safe mobility in people with secondary progressive MS: The BRiMS feasibility RCT. https://core.ac.uk/download/293867254.pdf (accessed on 13th June 2025).
10. Arnold, B. S., Gross, T. S., McDonald, K., et al. (2001). Relations of environmental contaminants, algal toxins, and diet with the reproductive success of American alligators on Florida lakes. https://core.ac.uk/download/11016994.pdf (accessed on 13th June 2025).
11. Austin, C., Berger, B., Bobadilla, M., Buffone, G., et al. (2017). Essential role of CFTR in PKA-dependent phosphorylation, alkalinization, and hyperpolarization during human sperm capacitation. https://core.ac.uk/download/159287934.pdf (accessed on 13th June 2025).
12. Aziz, F. A. (2023). Investigating the role of TGF-beta signaling in acoel regeneration & fission. https://core.ac.uk/download/604719555.pdf (accessed on 13th June 2025).
13. Barcaccia, G., Sharbel, T. F., Galla, G., et al. (2015). *De novo* sequencing of the *Hypericum perforatum* L. flower transcriptome to identify potential genes that are related to plant reproduction *sensu lato*. https://core.ac.uk/download/53524393.pdf (accessed on 13th June 2025).

14. Barker, M. J., & Wilson, R. A. (2019). Biological individuals. https://core.ac.uk/download/286027681.pdf (accessed on 13th June 2025).

15. Beerda, B., Pas, M. F. W., Veerkamp, R. F., et al. (2008). Expression profiles of genes regulating dairy cow fertility: Recent findings, ongoing activities and future possibilities. https://core.ac.uk/download/pdf/29259921.pdf (accessed on 13th June 2025).

16. Bogani, D., Brixey, R., Dean, C. H., et al. (2009). *Sfrp1* and *Sfrp2* are required for normal male sexual development in mice. https://core.ac.uk/download/pdf/82514591.pdf (accessed on 13th June 2025).

17. Bonaldo, M. F., Dimopoulos, G., Gasperi, G., et al. (2008). Gene discovery in an invasive tephritic model pest species, the Mediterranean fruit fly, *Ceratitis capitata*. https://core.ac.uk/download/pdf/8090803.pdf (accessed on 13th June 2025).

18. Brennich, M., Hugouvieux, V., Jourdain, A., et al. (2016). Evolution of the plant reproduction master regulators LFY and the MADS transcription factors: The role of protein structure in the evolutionary development of the flower. https://core.ac.uk/download/52673410.pdf (accessed on 13th June 2025).

19. Burns, A., de Waal, H., Henderson, C., et al. (2015). Individual cognitive stimulation therapy for dementia: A clinical effectiveness and cost-effectiveness pragmatic, multicenter, randomized controlled trial. https://core.ac.uk/download/151161307.pdf (accessed on 13th June 2025).

20. Burt, A., Schlesinger, A., Weismann, A., Heyward, A. J., Schlesinger, A., Rinkevich, B., et al. (2010). Sexual plasticity and self-fertilization in the sea anemone *Aiptasia diaphana*. https://core.ac.uk/download/pdf/8436247.pdf (accessed on 13th June 2025).

21. Burt, J., Canguilhem, G., Cooper, M., Deleuze, G., Derrida, J., Grosz, E., Haraway, D., et al. (2013). Mobilizing experimental life: Spaces of becoming with mutant mice. https://core.ac.uk/download/18460454.pdf (accessed on 13th June 2025).

22. Carnevali, O., Cobellis, G., Dalla Valle, L., et al. (2016). BPA-induced deregulation of epigenetic patterns: Effects on female zebrafish reproduction. https://core.ac.uk/download/95163508.pdf (accessed on 13th June 2025).

23. Carr, B., Duve De, C., Vidal, C., Leary, D. E., Hofstadter, D. R., Chaisson, E. J., et al. (2010). Computational and biological analogies for understanding fine-tuned parameters in physics. http://arxiv.org/abs/1002.3905 (accessed on 13th June 2025).

24. Chasiotis, A. (2011). Evolution and culture. https://core.ac.uk/download/10687091.pdf (accessed on 13th June 2025).

25. Civetta, A., Singson, A., Carvalho, A. B., Bateman, A. J., Chippindale, A. K., et al. (2002). Postcopulatory sexual selection. https://core.ac.uk/download/59809.pdf (accessed on 13th June 2025).

26. Codispoti, B., Marrelli, M., Pacifici, A., et al. (2017). Potential use of human periapical cyst-mesenchymal stem cells (hPCy-MSCs) as a novel stem cell source for regenerative medicine applications. https://core.ac.uk/download/153370357.pdf (accessed on 13th June 2025).

27. Coss, D., Jonak, C. R., Kauffman, A. S., et al. (2015). Gonadotropin and kisspeptin gene expression, but not GnRH, are impaired in cFOS deficient mice. https://core.ac.uk/download/323311126.pdf (accessed on 13th June 2025).

28. del Carmen Cano Secade, M. (1994). Reflections on the processes of justice and social participation from the perspective of family. https://core.ac.uk/download/13636097.pdf (accessed on 13th June 2025).

29. Diniz-Filho, J. A. F., Martins, R. P., & Tidon, R. (2017). The evolutionary ecology of interactive synchronism: The illusion of the optimal phenotype. https://core.ac.uk/download/286139287.pdf (accessed on 13th June 2025).

30. Hanisch, S., & Eirdosh, D. (2020). Conceptual clarification of evolution as an interdisciplinary science. EdArXiv. https://doi.org/10.35542/osf.io/vr4t5Franklin-Hall, L. (2020). The animal sexes as historical explanatory kinds. https://core.ac.uk/download/199234750.pdf (accessed on 13th June 2025).

31. Furusawa, C., & Kaneko, K. (2000). Theory of robustness of irreversible differentiation in a stem cell system: Chaos hypothesis. http://arxiv.org/abs/nlin/0009007 (accessed on 13th June 2025).

32. Galla, G., Vogel, H., Sharbel, T. F., & Barcaccia, G. (2015). *De novo* sequencing of the *Hypericum perforatum* L. flower transcriptome to identify potential genes that are related to plant reproduction *sensu lato. BMC Genomics, 16*. https://doi.org/10.1186/s12864-015-1439-y (accessed on 13th June 2025).

33. García-Ferreyra, J. (2015). Sperm DNA fragmentation and its relation with fertility. https://core.ac.uk/download/322424996.pdf (accessed on 13th June 2025).

34. Geltsch, I. C., Hartel, T., Pernetta, A., et al. (2007). The reproductive dynamics of temperate amphibians: A review. https://core.ac.uk/download/63345.pdf (accessed on 13th June 2025).

35. Gulyas, L., & Powell, J. R. (2019). Predicting the future: Parental progeny investment in response to environmental stress cues. https://core.ac.uk/download/232537188.pdf (accessed on 13th June 2025).

36. Halstead, T. W., & Souza, K. A. (2025). NASA Developmental Biology Workshop: A summary. https://core.ac.uk/download/pdf/42843557.pdf (accessed on 13th June 2025).

37. Hermansson, A. (2007). Effects on the reproductive system in domestic fowl (*Gallus domesticus*) after embryonic exposure to estrogenic substances. https://core.ac.uk/download/11695140.pdf (accessed on 13th June 2025).

38. Hucka, M., & Waltemath, D. (2011). Minimum information about a simulation experiment (MIASE). https://core.ac.uk/download/4888899.pdf (accessed on 13th June 2025).

39. Lee, C. K. F., Xu, J. S., & Yeung, W. S. B. (2002). The oviduct and development of the preimplantation embryo. https://core.ac.uk/download/37882883.pdf (accessed on 13th June 2025).

40. Leria, L., Riutort, M., Solà, E., et al. (2019). Outstanding intraindividual genetic diversity in fissiparous planarians (*Dugesia*, Platyhelminthes) with facultative sex.

41. Melo, F. F. A. P. (2015). Development and standardization of a protocol for sperm cryopreservation of two important commercial oyster species. https://core.ac.uk/download/61526649.pdf (accessed on 13th June 2025).

42. Moshi Amsi, M. (2015). The politics of teachers' work in the context of curriculum resources marketisation policy reforms in three secondary schools in Tanzania: A thesis presented in partial fulfilment of the requirements for the degree of Doctor of Philosophy (PhD) at

Massey University, Manawatu Campus, Palmerston North, New Zealand. https://core.ac.uk/download/148643742.pdf (accessed on 13th June 2025).

43. O'Connor, W. G., Williams, K. M., Willis, M. S., et al. (2013). Ernest Everett Just, PhD: Pioneer in ecological developmental (eco-devo) biology. https://core.ac.uk/download/217660041.pdf (accessed on 13th June 2025).

44. Oppong, N. Y. (2013). National managerial talent development in Ghanaian gold mining industry: Comparing Western and Indigenous approaches. https://core.ac.uk/download/234681251.pdf (accessed on 13th June 2025).

45. Pearce, E. J., Jung, E., Maizels, R. M., & Freitas, T. C. (2007). TGF-β signaling controls embryo development in the parasitic flatworm *Schistosoma mansoni*. https://core.ac.uk/download/pdf/7585134.pdf (accessed on 13th June 2025).

46. Prospero, S., & Cleary, M. (2017). Effects of host variability on the spread of invasive forest diseases. *Forests, 8*, 80–80. https://doi.org/10.3390/f8030080 (accessed on 13th June 2025).

47. Sanchez-Calderon, H., Varela-Nieto, I., Pichel, J. G., Rodriguez-de la Rosa, L., Milo, M., Holley, M., & Linden, R. (2010). RNA microarray analysis in prenatal mouse cochlea reveals novel IGF-I target genes: Implication of MEF2 and FOXM1 transcription factors. https://core.ac.uk/download/50033.pdf (accessed on 13th June 2025).

48. Schmid, M. W., Schmidt, A., & Grossniklaus, U. (2015). The female gametophyte: An emerging model for cell type-specific systems biology in plant development. *Frontiers in Plant Science, 6*. https://doi.org/10.3389/fpls.2015.00907 (accessed on 13th June 2025).

49. Schubert, F., Ware, M., & Waring, C. (2014). Development of the early axon scaffold in the rostral brain of the small spotted cat shark (*Scyliorhinus canicula*) embryo. https://core.ac.uk/download/29588230.pdf (accessed on 13th June 2025).

50. Sutcliffe, D. W. (1993). Reproduction in *Gammarus* (Crustacea, Amphipoda): Female strategies. https://core.ac.uk/download/11020068.pdf (accessed on 13th June 2025).

51. Templeton, A., Arquitt, A. B., Migliano, A. B., Pesonen, A. K., Parent, A. S., Bogin, B., Campbell, B., et al. (2013). Evo-devo of human adolescence: Beyond disease models of early puberty. https://core.ac.uk/download/18528095.pdf (accessed on 13th June 2025).

52. Willemse, M. T. M. (2009). Evolution of plant reproduction: From fusion and dispersal to interaction and communication. https://core.ac.uk/download/pdf/29248993.pdf (accessed on 13th June 2025).

53. Zhou, Y., Zhou, B., Pache, L., Chang, M. W., Hadj Khodabakhshi, A., Tanaseichuk, O., Benner, C., et al. (2019). Metascape provides a biologist-oriented resource for the analysis of systems-level datasets. *Nature Communications, 10*. https://doi.org/10.1038/s41467-019-09234-6 (accessed on 13th June 2025).

POPULATION AND COMMUNITY ECOLOGY

11

Introduction

Within the intricate matrix of environmental science, population and community ecology are dominant paradigms for understanding the processes of life on Earth. These disciplines examine how species function in their worlds, with emphasis on the interactions among individuals within the same species and among different species within communities. As human endeavor is constantly transforming natural landscapes, the importance of investigating these ecological processes becomes increasingly central to understanding the world around us. Through the examination of population processes, birth, death, and migration, ecologists can see how populations grow, decline, or stabilize over time and thus make observations about larger community structures. Not only does this type of analysis shed light on the behavior of the single species, but also on how these processes collectively are responsible for ecosystem health and function, and the interdependence of all living things. With such interactions, the integration of community and population processes into ecological modeling

LEARNING OBJECTIVES

After studying this chapter, you will be able to:

1. Explore the key concepts in population ecology;
2. List the factors affecting population growth;
3. Discuss population interactions;
4. Describe the key concepts in community ecology;
5. Explain community interactions;
6. Illustrate human influence on population and community ecology.

is a critical component to understanding carbon dynamics and ecosystem responses to varied change. Such complexities are illuminated by studies that illustrate how colonization and mortality rates affect biomass accumulation in successional stages, introducing novel feedback processes into play that regulate carbon storage and release in ecosystems (Carlos et al., 2008).

Because species differ in their characteristics, , such as rates of growth and death, their interactions create patterns of carbon processes that cannot be modeled realistically by averaging over simple models. Thus, their omission leads to huge uncertainties that influence our predictions about future environmental conditions. Thus, a deep understanding of population and community ecology is crucial in making the science of ecosystem management and in guiding conservation efforts. Besides the understanding gained from exploring relationships between species, theoretical models are crucial in furthering our understanding of community structure and dynamics. Various models, from static to more advanced dynamic and species assembly methods, explain the intricacies of food webs and species interactions. Static models provide a simple explanation by linking with respect to only rudimentary rules; however, they fall short in describing the richness of ecological stability (Drossel et al., 2002). Dynamic models incorporate population dynamics between multiple interacting species, with the capacity to explore community stability and resilience richly. By using evolutionary techniques, researchers can simulate the convergence of disparate species throughout periods of time and give relevant perspectives on how complicated ecological networks develop. Ultimately, these models contribute extensively to our understanding of ecosystems, informing conservation measures and policy management in a dynamically changing world (Figure 11.1).

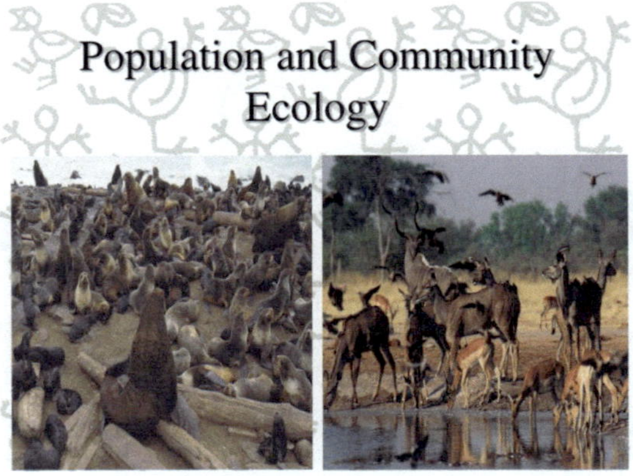

Figure 11.1. Population and community ecology.

Source: https://image1.slideserve.com/2077136/population-and-community-ecology-l.jpg.

Definition of Population Ecology

Population ecology is a significant branch of ecology that deals primarily with species population dynamics and their relationship with the environment. It examines how these processes , such as birth, death, immigration, and emigration affect the population size and distribution through time. It explains the factors causing population change, providing information on species viability as well as ecosystem conditions. Conceptually, population ecology integrates ecological theory and mathematical models to predict population dynamics under varying environmental conditions. These theoretical models play an important role in conservation biology since they allow one to identify key thresholds and possible risks of extinction for endangered species. Knowledge about these processes is significant, especially in the context of the Anthropocene, where human activities strongly alter natural environments and break classical population models (Martins et al., 2017). Another critical component of population ecology is the interactions between organisms in a specific population. The species will compete with other species for limited resources, and these interactions can have significant effects on population growth and viability. Predation and mutualism, where two species live together in harmony, are also crucial factors in defining population dynamics. These interactions are central to understanding the broader ecological system since they determine not only the destiny of specific species but also the condition of ecosystems in general. But since complexity within these relations has a tendency to get boiled down to simplicities, our comprehension is often veiled by overly simplistic categorizations. The finer interplay of mutualism, often dwarfed by competition and predation, reveals that between-species cooperation can also have beneficial ecological impacts, demonstrating the importance of adopting a multiple-variable method for population studies (Boucher et al., 1982). Lastly, population ecology has an important role in guiding conservation policy and management methods that can minimize loss of biodiversity.

Figure 11.2. Population ecology.

Source: https://interfaithsustain.com/wp-content/uploads/2021/11/population-ecology.jpg.

The mounting loss of biodiversity under anthropogenic pressures demands a comprehensive understanding of population structure and ecological niched they are found in. Population ecology principles may be utilized by scientists to model species responses to environmental change and human disturbance. Such predictability is required in the planning of effective management interventions to preserve species and recover habitats. In addition, facilitation of communication between scientists and policymakers, particularly with respect to the technicalities of ecological concepts, is important in enabling well-informed decisions. Efforts made to clarify definitions and utilization that pertain to ecological concepts , such as niche serve to close the divide between theoretical ideas and conservation action in the long term, towards better ecological outcomes as we navigate the Anthropocene challenges (Martins et al., 2017) (Figure 11.2).

Definition of Community Ecology

Community ecology is a fundamental subdiscipline of the general field of ecology, with its intricate discussion of interactions among various species within a given geographic area. As a science, it attempts to account for the patterns and processes of coexistence and species diversity. By examining how different organisms interact either by competing, being preyed on, or coexisting as symbionts, community ecology can provide insight into the stability and resilience of ecological communities. One major problem the discipline has is the theories that underpin it, which have been traditionally focused on unitary organisms and have made it difficult to apply these ideas to modular organisms like filamentous fungi. In a bid to close this gap, researchers have begun integrating modern coexistence theory and movement ecology into the community ecology approach, with the emphasis being placed on the importance of spatial dynamics and movement pathways in promoting species persistence and community assembly (Bielčik et al., 2022). The complexities of community ecology are also amplified by the intrinsic diversity of ecological niches that species inhabit in communities. Ecological niche theory is helpful in understanding how species dynamics regulate community structure, but it is of a loose nature, and thus it is difficult to apply theoretical models in practice. It is made even more difficult in the Anthropocene, where there has been unprecedented biodiversity loss and habitat alteration, and ecology must be applied in conservation. The challenge of creating valid ecological models that reflect the intricate, multi-causal processes involved is further complicated by increasing examination of niche definitions and structures. It is assumed by researchers that resolving these philosophical matters in ecology can facilitate better communication between theoretical ecologists, conservationists, and policymakers and lead to more effective biodiversity conservation management (Martins et al., 2017). With these complexities in mind, community ecology is not a straightforward theoretical endeavor; it has significant practical uses. Learning about community dynamics can inform conservation biology, ecological restoration, and land management strategies. With climate change and human activities continuing to pummel biodiversity, information gleaned from community ecology becomes increasingly important. The application of sophisticated ecological theories, most notably those that can accommodate the unique characteristics of unitary and modular organisms, enables ecologists to make use of approaches that can alleviate the decline in biodiversity. Thus, by reconciling theoretical concepts with practical application, community ecology aims to build a better appreciation for ecological interactions, ultimately contributing to the preservation of the earth's biological diversity (Martins et al., 2017) (Figure 11.3).

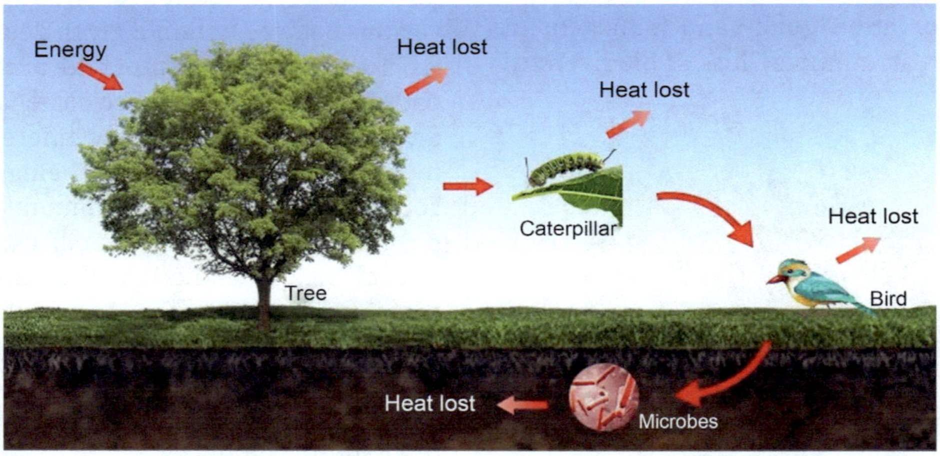

Figure 11.3. Community ecology.

Source: https://media.geeksforgeeks.org/wp-content/uploads/20240206132210/Community-Ecology.webp.

Importance of Studying These Fields in Understanding Ecosystems

A defining factor influencing the integrity of ecosystems is the interactive nature of the relationship between community and population ecology. Both areas of study in combination reveal how species interactions structure community organization and subsequently determine ecosystems' health. Information regarding population dynamics yields essential insights into species growth rate, reproductive strategies, and mortality and therefore how those factors scale up to larger-scale community interactions. For instance, the function of keystone species can be explained through the investigation of population ecology, thus gaining a greater understanding of their impact on composition and stability of the community. Further, identification of population thresholds is useful in evaluating ecosystem resilience to disturbances, guiding conservation practices for the purpose of preserving biodiversity within changing ecosystems. One of the prominent research areas in population and community ecology is seed dispersal, an important process in the maintenance of plant community dynamics and distribution. As indicated by recent evidence, knowledge of seed dispersal is key to predicting the persistence of plant populations and global change response (Aslan et al., 2020). Extensive research in this field can unravel the mechanisms governing seed movement and their influence on plant fitness, as well as community assembly. By establishing networks , such as the CoDisperse Network, researchers are drawing boundaries around the intricate relationships between seed dispersal, population processes, and ecosystem processes. As a result, encouraging mechanistic understanding of these interactions is crucial in informing effective conservation policies that address the issues precipitated by environmental changes. In addition, the comparison of climate change and community and population ecology underscores the significance of these fields in accounting for ecosystem resilience. Historical alterations during the Quaternary epoch illustrate how climatic-induced changes have impacted community assemblages, evoking evolutionary reactions, extinctions, and shifts in species distributions (Stewart, 2010). Investigating how current populations interact with climatic fluctuations underscores the ongoing applicability of these ecological fields. By examining how climate conditions are associated with population dynamics, ecologists can forecast future community composition and resilience measures. The core position of population and community ecology is thus emphasized, insofar as it provides essential conceptual frameworks for revealing changes in the environment, informing conservation strategies, and informing policy in an era of rapid environmental transformation.

11.1. Key Concepts in Population Ecology

Learning Outcome

1. Population dynamics;
2. Carrying capacity;
3. Reproductive strategies.

The dynamics of species populations and what determines their size, distribution, and survival through time are the basic issues of population ecology. Important notions are population density, the number of people per square unit, and population distribution, which measures the way people are distributed over geographic areas. These concepts are useful to ecologists in forecasting population growth dynamics and the carrying capacity of environments, i.e., the maximum size of a population that an ecosystem can support indefinitely. This is especially relevant in the interpretation of the tenets of interspecific interactions that drive the dynamics of communities. For instance, the relationships between competing species can be significantly influenced by shifts within their microbial communities, revealing a complex interplay that ecologists must decipher to understand broader ecological systems (Bernstein et al., 2019). The other important feature of the population ecology is the reproduction strategies of various species which may differ significantly among the organisms. These strategies are roughly divided into r-selected and K-selected species which determine the rates of population growth and reaction to fluctuating environmental conditions. Many insects are R-selected species, and tend to have large numbers of offspring, but invest little in them, enabling the population to grow rapidly in suitable habitats. Conversely, K-selected species, , such as elephants, have a high investment in fewer offspring, which promote high survivability in steady conditions.

A thorough grasp of these reproductive tactics allows ecologists to analyze community resilience and adaptability, especially how microbial communities in systems like anaerobic digesters can respond to environmental stresses and fluctuations in nutrient availability (Bocher et al., 2016). Moreover, ecological niche is a major concept that determines the dynamics of populations and their interaction within communities. Every species is said to be in a niche, defined by habitat, resource utilization and its place in the ecosystem. Knowledge of these niche dimensions guides ecologists to evaluate how species compete with one another, which affects biodiversity and the stability of ecosystems. With ecosystems under the dual stresses of climate change and habitat loss, niche adaptation of species is becoming one of the key factors in species survival and population maintenance. The insights gained from studying these dynamics, especially in relation to microbial interactions and community structure, can augment our comprehension of population behaviors and support efforts in conservation and sustainable management of natural resources (Bernstein et al., 2019) (Figure 11.4).

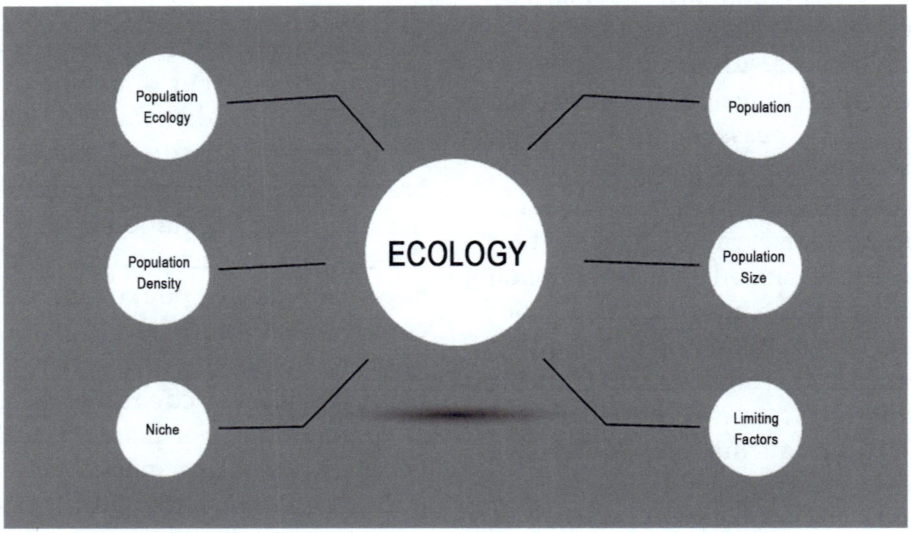

Figure 11.4. Population ecology concept.

Source: https://0701.static.prezi.com/preview/v2/oxnzrnq7hzfxq5w2sagvs6qbdh6jc3sachvcdoaizecfr3dnitcq_3_0.png.

11.1.1. Population Dynamics

Population dynamics examines population size and structure changes over time as a result of various biological and environmental drivers. Ecologists should learn about these dynamics when examining the interactions among species in communities and how species respond to habitat and climate change. Birth rates, death rates, immigration, and emigration significantly contribute to population increase and decrease. As a case in point, an unanticipated influx of nutrients within an ecosystem will produce population explosions in certain species, whereas predation or disease will cause population crashes. Changes do not just impact individual species but cascade through the whole community, influencing interactions among species, resource availability, and ecosystem functioning. The complexity of population dynamics is further enhanced by the interdependence of species and environments. With the evolution of ecosystems, the delicate balance that is attained between various species might be disrupted by extrinsic conditions, , such as climate change. With observations made in the past, climatic oscillations during the Quaternary period have reshaped ecological communities profoundly, influencing species adaptation, evolution, and even extinction levels (Stewart, 2010). Such alterations show that populations do not function in isolation; their population is inextricably linked to wider ecological processes. Moreover, the changes in one population of a species can have a cascade effect on predator-prey, competition, and resource availability, highlighting the significance of understanding these processes to ensure effective conservation and management interventions. The consequences of population dynamics extend beyond theoretical views and comprise practical uses as well as policymaking. For example, the long-term survival of human societies depends significantly on healthy ecosystems, which are organized by the interaction of population dynamics and soil diversity. Healthy populations contribute to ecosystem services , such as nutrient cycling, water purification, and habitat provision (Van der Putten et al., 2010). Soil diversity care is particularly vital because it subsidizes the ecosystem functions that allow life. By examining population dynamics in the context of community ecology, therefore, we can gain a deeper understanding of the complex relationships that ensure ecological balance. Such an understanding is crucial to devising mitigation strategies for the effects of environmental change and for promoting the long-term sustainability of ecosystems and human societies (Figure 11.5).

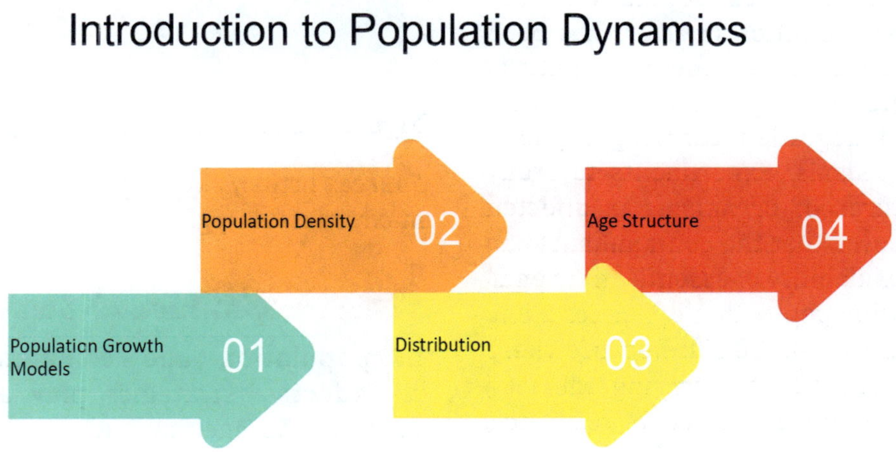

Figure 11.5. Understanding population dynamics.

Source: https://fastercapital.co/i/Population--Understanding-Population-Dynamics-through-Systematic-Sampling--Introduction-to-Population-Dynamics.webp.

11.1.2. Carrying Capacity

The concept of carrying capacity is central to explaining population dynamics in the field of population and community ecology. It refers to the maximum number of individuals within a species that can be supported by an environment in a sustainable way, depending on the availability of accessible resources like food, water, and habitat. When populations exceed their carrying capacity, there are ecological implications that include depletion of resources, increased competition, and increased death rate, which ultimately result in the population's reduction. In addition, the effects of carrying capacity are not linear due to the diverse response of various species according to their adaptability and mode of reproduction. For instance, species with nonlinear responses to the availability of resources have competitive relationships affecting their coexistence. As noted in recent studies, this relative interaction can give rise to nonlinearity of competition, allowing stable coexistence of species regardless of resource scarcity (Dieckmann et al., 2014). Extraneous to intrinsic ecological drivers, there are extrinsic determinants, and among them, human actions play a dire effect on carrying capacity. Anthropogenic determinants in the form of pollution as well as resource depletion possess the capability to alter environments at the cost of Indigenous populations. For example, nutrient pollution in rivers has been discovered to destabilize food web natural balance and biodiversity, resulting in local extinctions and degraded carrying capacity for aquatic organisms. Experiments to study isotopic signatures of organisms in affected ecosystems uncover that the contaminants can transform the availability of essential nutrients, thereby influencing population dynamics along the river (Calizza et al., 2017). Because these external pressures modify carrying capacity, ecological communities are tested for resilience, and therefore researchers have pointed to the necessity for effective monitoring and management to ensure the sustainability of such ecosystems. Ultimately, being aware of carrying capacity extends way beyond census-taking; it is an interactive pattern of relationships among ecological communities. Availability of resources, competition, and extrinsic stressors combine in complex ways to create a subtle matrix that underlies ecosystem health. By decomposition of the carrying capacity in terms of environmental stress and species interactions, ecologists can predict with greater accuracy trends in biodiversity and ecosystem process. This coordinated approach is particularly critical in conservation efforts and ecosystem management, in which the integrity of habitats has to be maintained not only to protect single species but also to preserve the fine interconnections that constitute the basis of their ecological communities. Strong strategies can only be developed to cope with the impact of human pressure and assure ecological resilience under the omnipresent forces of environmental change with sophisticated carrying capacity understanding (Figure 11.6).

Figure 11.6. Ecological carrying capacity.

Source: https://i.ytimg.com/vi/XV2-2Ym3IIc/maxresdefault.jpg.

11.1.3. Reproductive Strategies

In population and community ecology, reproductive strategies play a key role in defining the success and survival of species under various habitats. They can be collectively classified as r-selected and K-selected traits, each with unique attributes. R-selected species, , such as the majority of insects, produce many offspring with minimal parental care, thus enabling populations to grow faster when environments fluctuate. On the other hand, K-selected species like elephants invest much in few offspring, which ensures high survival rates in settled environments. The dichotomy influences population structure and interactions within communities because r-selected species will exploit resource availability following disturbance, while K-selected species compete in settled communities. The information regarding the reproductive strategy gives a clue about how species maintain themselves and interact with one another within a complex ecological network. Reproductive strategy studies discover that there are certain physical characteristics that are responsible for promoting reproductive abilities among different species. For instance, cladocerans have indeterminate growth, as they simply continue to grow when they reach reproductive maturity. The growth leads to increasing reproductive output, providing an evolutionary trade-off between intrinsic rates of increase maximization and total reproductive output (Gabriel et al., 1985). Lab experiments have demonstrated that variations in resource availability can drastically impact these patterns of growth and methods of reproduction. Depending on variation in environment, advantages of size-dependent reproduction are achieved. Larger organisms will be more reproductive, illustrating the importance of resource allocation in reproduction and how such strategies react to varying ecological environments. In terrestrial ecosystems, variation in seed mass is of prime significance in species fitness and reproductive strategy. Light and soil competition and allocation of resources directly act on the reproductive success of plant species, creating dynamic feedback among communities. Seeds are predisposed to grow and compete if they acquire greater amounts of light per unit biomass under some resource conditions. Small seeds, on the other hand, are efficient under resource-poor conditions due to their faster germination and potential for colonization (DeMalach & Kadmon, 2018). This variation in reproductive strategy illustrates how communities are capable of altering functional diversity based on the availability of external resources, revealing the significant role that environmental gradients play in reproductive characteristics. Overall, this information about the reproductive strategies in both plants and animals highlights the intricate dynamics behind population processes and community organizations (Figure 11.7).

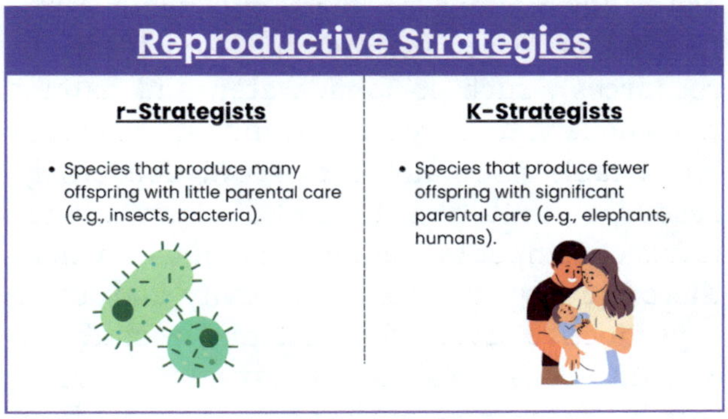

Figure 11.7. Reproductive strategies.

Source: https://images.examples.com/wp-content/uploads/2024/07/Reproductive-Strategies.png.

Test Your Knowledge

1. What is population dynamics?
2. What is carrying capacity?

11.2. Factors Affecting Population Growth

Learning Outcome

1. Biotic factors;
2. Abiotic factors;
3. Human Impact on population growth.

Understanding what controls population growth is important to understanding community ecology and ecosystem dynamics. One of the prime determinants of population growth rates is the availability of resources, which has a direct effect on birth and death rates. In productive ecosystems, there are plenty of resources , such as food, water, and habitat that can lead to increased reproductive success and enable populations to grow. Conversely, resource limitation tends to cause higher levels of competition among individuals, where there are higher rates of mortality and lower reproductive levels. For instance, a study on Sanriku Coast seagrass ecosystems reveals that biological and environmental factors both interact with one another in intricate manners to modulate spatial distribution and productivity of indigenous species, which impacts overall growth in these vital ecosystems (Keiko et al., 2003). This delicate balance suggests the importance of resource availability for individual species as well as for the structure of entire communities. In addition to resource availability, genetic and ecological factors have important roles to play in population dynamics. Genetic population differentiation can influence population prospects of adapting to pressures imposed by the environment, and these, in turn, affect growth rates. For example, genomic research on *Daphnia pulex* x *pulicaria* populations emphasizes how gene-influenced traits modulate niche invasion and extinctions among aquatic ecosystems. Success in invading established zooplankton communities by various populations differs and illustrates that reproductive and survival success in a specific ecological context may be modulated by genetic diversity (Juenger et al., 2011). This reveals that assembly of communities is genetically sensitive, meaning that evolutionary processes ought to be included in our perception of population dynamics. Hence, both ecological interactions and genetic variation are key factors in deciding whether populations increase, stabilize, or decline over time. Environmental change also plays a significant role in determining population growth and can greatly alter resource availability and quality of the habitat.

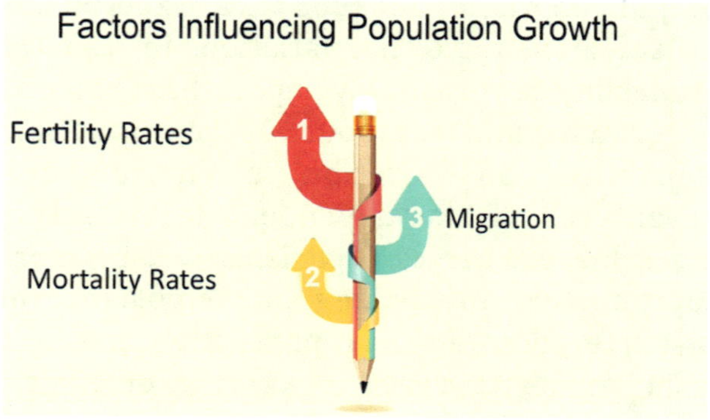

Figure 11.8. Factors influencing population growth.

Source: https://fastercapital.co/i/Population-growth--Demographic-Analysis--Understanding-the-Dynamics-of-Population-Growth--Factors-Influencing-Population-Growth.webp.

The impacts of pollution, habitat destruction, and climate change are drastic because they have the potential to modify well-documented ecological relations and processes. For instance, shifts in temperature and precipitation patterns can favor some organisms at the cost of others, leading to a change in community structure and a general reorganization of populations. The connection between the biotic and abiotic factors constitutes a delicate network that encompasses population growth and persistence. Moreover, understanding how populations respond to environmental stressors will be useful for conservation efforts aimed at maintaining biodiversity and ecological integrity. With a determination of these intricate factors, ecologists can develop more effective solutions to ecosystem management and population stability with respect to anthropogenic threats (Figure 11.8).

11.2.1. Biotic Factors

Biotic factors are very critical in population and community ecology, and they influence the dynamics of ecological systems. Such factors include the interactions between living organisms, , such as competition, predation, mutualism, and parasitism which affect population size, distribution, and evolution. As an example, competition over the scarce resources including food and habitat promotes natural selection, which prompts the adaptations that allow species to coexist. In addition, the predator and prey pairings are examples of the balances that exist in the ecosystem where the rise and fall of one population have a direct result on the other. These interactions highlight the complexity of life whereby biotic factors do not only determine the survivability of the species but also enforce the evolutionary processes that are essential to the ecological integrity. This knowledge is critical in the establishment of effective conservation measures which take into account the interdependence of biological organisms in an ecosystem. The importance of biotic factor is not restricted to the interaction between individual species; they also affect the entire ecological community. For example, the impacts of climate change over the Quaternary period serve as a compelling illustration of how biotic factors can shift community composition and biodiversity (Stewart, 2010). Throughout this period, differences in climate conditions placed pressures that led to the adaptation and evolution of species and in other instances extinction. The shift in species distribution emphasizes the adaptability of biological communities to environmental pressure, and it indicates that biotic entities are not stationary but robust and flexible. With the change of environments, the association, and the dependence of the species on each other may change drastically and hence a thorough knowledge of these factors will be needed to predict the possible changes that may happen in the ecological balance and the outcome of biodiversity following global change. Finally, the idea of biotic factors embraces the crucial connection between biodiversity and ecosystem services that are the key to the human well-being.

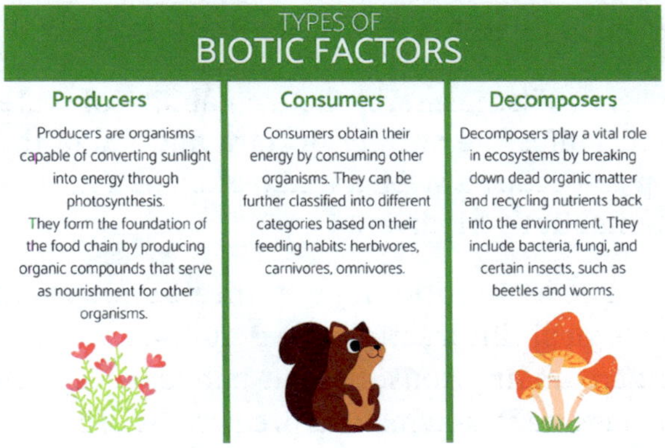

Figure 11.9. Biotic factors.

Source: https://cdn0.thedailyeco.com/en/posts/0/3/2/what_are_biotic_factors_230_orig.jpg.

Healthy ecosystems offer a multiplicity of services, including food production, water purification, and climate regulation, based in the complex webs of biological interactions that support them. Soil biodiversity exemplifies this connection; it serves as the foundation for many ecosystem services and is critical for maintaining soil health, fertility, and resilience (Van der Putten et al., 2010). Societies may protect the ecosystem services on which they rely by conserving the complex diversities of life in the soil, and by stabilizing these biotic interactions. Such interdependence highlights the fact that biodiversity conservation is not only an environmental issue, but also a human prosperity issue, which explains the necessity of integrated conservation strategies based on an understanding of the biotic factors (Figure 11.9).

11.2.2. Abiotic Factors

Abiotic factors are building blocks that define the structure and dynamics of populations and communities in various ecosystems. Non-living components, , such as temperature, water availability, light, and soil quality, determine the physiological and behavioral acclimations of organisms. For instance, temperature directly affects metabolic rates, influencing growth and reproductive success in various species. Furthermore, the availability of water determines the variety of flora and fauna that can co-occupy a given habitat, often determining the habitat's biodiversity.

These abiotic factor interactions and ecological interactions are fundamental to understanding conservation habits, ecosystem management, and making predictions of climate change effects on biodiversity, as these factors control not only the distribution of species but also species interaction in ecological communities. The interactions among abiotic and biotic components can potentially strongly alter community structure and outcomes of biodiversity. In grassland ecosystems, evidence indicates that large herbivores significantly affect arthropod communities through diverse mechanisms, e.g., direct predation and modification of vegetation structure and soil properties. For example, grazing intensification has been shown to have varied effects on plant diversity but generally a detrimental effect on arthropod diversity (van Klink et al., 2015). This is a sign of the necessity to maintain equilibrium in herbivore densities to maximize heterogeneity in both types of life in these ecosystems. With variation in herbivore density, it would have cascading effects on the food web that ultimately decides the diversity and stability of the entire community. Therefore, such complex interactions must be thoroughly understood while developing feasible management and conservation plans. Urban ecosystems, with their drastic abiotic changes, also exemplify how physical conditions decide the structuring of ecological communities. Urbanization introduces artificial influences , such as light pollution, which significantly changes the temporal patterns of bird populations. Studies have shown that this anomaly stabilizes abiotic conditions, reducing diurnal and seasonal variability among bird assemblages in urban settings compared to rural areas (Leveau et al., 2018).

The inhibition of such natural rhythms leads to the persistence of specific bird species that may not be as strong in changing landscapes. The implications of such stabilizations underscore the need for ongoing research into the effects of urbanization on community structure and the related issues impacting wildlife in increasingly urbanized habitats. By examining the way abiotic elements regulate community structure, ecologists can provide for improved predictions and buffering of the anticipated effects of rapid urbanization on biodiversity (Figure 11.10).

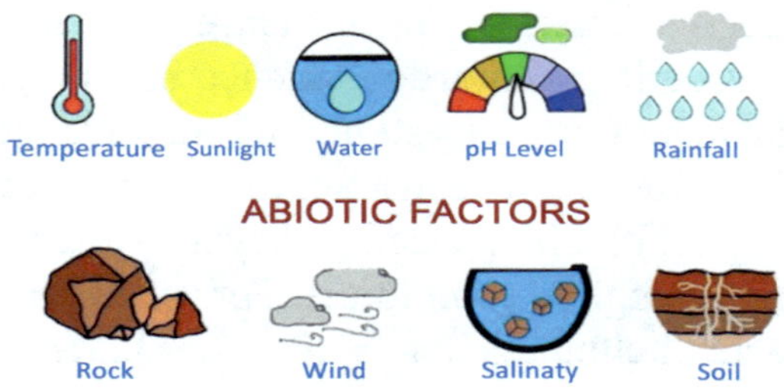

Figure 11.10. Abiotic factors.

Source: https://b2716464.smushcdn.com/2716464/wp-content/uploads/2023/02/Abiotic-factors-678x381.jpg?lossy=1&strip=1&webp=0.

11.2.3. Human Impact on Population Growth

Human actions have greatly influenced patterns of population growth in a range of species, fundamentally transforming the structure and well-being of ecosystems in the process. Overexploitation, loss of habitat, and pollution pressure have led to most populations declining significantly, with a knock-on effect through ecological communities. For instance, diadromous fishes that are species of great importance for trophic dynamics and nutrient cycling have suffered unprecedented declines in numbers because of human activities like river alteration and pollution. These changes compromise their breeding grounds, affecting their reproductive success directly as well as their population stability in general. This loss of such a crucial species not only poses a threat to biodiversity but also to the economic and ecological roles they play, thus forming a vicious cycle where poor population management brings about more environment degradation and species loss, thus indicating that improved conservation techniques coupled with sustainable measures are the way to go (Costa-Dias et al., 2009).

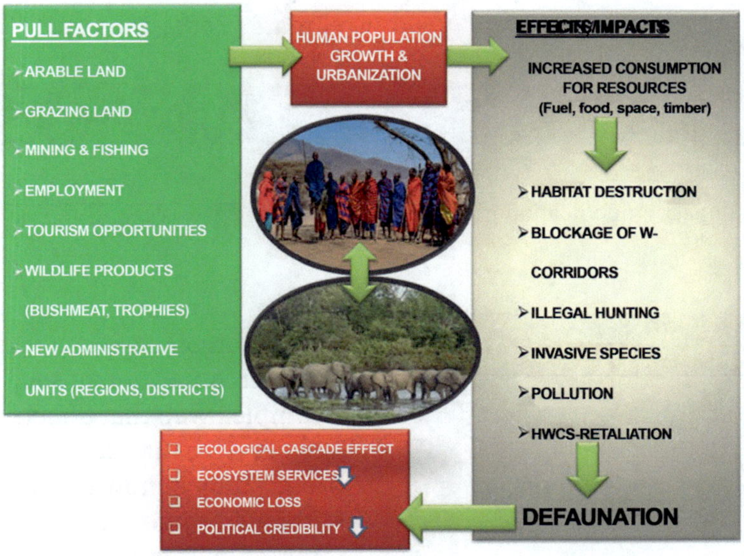

Figure 11.11. Human impact on population growth.

Source: https://www.researchgate.net/profile/Jafari-Kideghesho/publication/341544965/figure/fig25/AS:918674848358419@1596040560809/The-pull-factors-and-effects-of-human-population-growth-and-urbanization.ppm.

Moreover, climate change contributes to the already existing issues brought about by human action since temperature and precipitation changes further destabilize the ecosystem. The changes introduced by climatic conditions might result in the adaptation or extinction of species, which is a dynamic interaction among ecological communities and external stressors. As explained in Quaternary period research, the interplay between climate change and species evolution has caused community composition to undergo dramatic shifts. Contemporary examples exhibit the same process; species that fail to adapt to changing circumstances experience population decline. This indicates the critical need for an interdisciplinarity in ecology that considers human impacts along with the natural evolutionary processes affecting population dynamics (Stewart, 2010). Thus, avoidance of these newly emerging problems requires a holistic understanding of the interconnection between human activities, ecological health, and climatic adaptability. To effectively address the adverse effects of human-induced population changes, there is a necessity to integrate multi-dimensional management systems that provide focus to ecosystem integrity. Implementation of sustainable harvesting strategies, rehabilitation of habitats, and pollution controls can facilitate the recovery of impacted populations. For instance, habitat enhancement for inland fisheries will have a direct benefit to fish species , such as the salmonids and the European eel, whose declines have been linked to human encroachment and water pollution. Not only do these efforts contribute to recovery of economically valuable fish populations, but they also eventually contribute to the conservation of biodiversity within these systems. By creating ecosystems that can sustain diverse species interactions, it is conceivable to create a more sustainable balance which honors both ecological processes and human economic requirements. This integrated approach is not only important for maintaining biodiversity but also for the fact that our ecological communities must be able to adapt and flourish in an increasingly changing climate (Costa-Dias et al., 2009) (Figure 11.11).

Test Your Knowledge

1. What are the biotic factors?
2. What is the human impact on population growth?

11.3. Population Interactions

Learning Outcome

1. Competition;
2. Predation;
3. Symbiosis.

Population interaction dynamics are the basis of ecological community structure and function. Population interaction dynamics are defined in terms of different types of interaction, e.g., predation, competition, mutualism, and commensalism, each having a distinct impact on population size as well as community structure. For instance, predation has the potential to regulate prey populations, producing oscillatory dynamics that influence not only the directly interacting species but also their immediate ecological community. Consequently, local biodiversity can increase or decrease, depending on the outcome of such interactions. Ultimately, the interconnection of such relationships creates a complex web that affects ecosystem processes and services, underscoring the importance of a broad study of how species coexist and compete within the same areas. The impact of population interactions extends beyond single species to have significant effects on ecosystem services on which human societies depend. For instance, herbivore/plant balance can regulate vegetation structure, which in turn regulates soil health and nutrient cycling. Healthy soil ecosystems are vital to food, clean water, and other resources necessary for sustainable human existence. As highlighted in the European Commission report, the health of soil biodiversity has a pivotal role to play in these processes, underlining the interdependence between ecological processes and human well-being (Van der Putten et al., 2010). When species relationships are disrupted, either through anthropogenic pressures or natural phenomena , such as climate change, these are experienced across entire ecosystems and lead to losses in biodiversity and ecosystem functioning. Furthermore, learning from previous climate pressures caused by climate change can also contribute to understanding population interactions in contemporary ecological environments. The Quaternary period is an excellent example of how ecosystem community dynamics were shaped by climatic changes, which formed adaptive and extinction evolutionary processes. These changes illustrate interdependence among species and their environments and highlight the persistence or vulnerability of populations depending on their interaction with one another (Stewart, 2010). Through the observation of such historical interactions in combination with current dynamics, ecologists can draw lessons from patterns and predict future trends, ultimately enhancing conservation practice on maintaining biodiversity in an ever-changing world. As we strive to maintain ecosystems in their wholesomeness, monitoring the intricacies of population interactions is an inherent quest for ecological research as well as practical conservation practice (Figure 11.12).

Population Interactions

Species A	Species B	Name of Interaction
+	+	*Mutualism*
–	–	*Competition*
+	–	*Predation*
+	–	*Parasitism*
+	*0*	*Commensalism*
–	*0*	*Amensalism*

Figure 11.12. Population interactions.

Source: https://blogger.googleusercontent.com/img/b/R29vZ2xl/AVvXsEg0KEEUftG_r-7J69m8CYl_phOUgO4DwWvv1MG-fJ0mxbegQ5V49mzN9xgPE1Vug5mVc1dRRJCrqmx2U-oU_S77z2ekNWm3K7XfD2HwGgFLNS1q6NIs6UU_Mz4yQw9fryEpgwNdNdcjGNTvNyeIYvg7JpwDPl1Cex8RJOiN1MMW6SCpsmFzWmEe_J4/s1679/Population%20Interactions.jpg.

11.3.1. Competition

In population and community ecology, competition is a key interaction that influences several ecological processes. Competition occurs when organisms or species are competing for limited resources, which may be food, space, light, or nutrients. The conflict could result in various outcomes, ranging from resource partitioning to local extinctions, that structurally construct community and influence species distributions. Competition, for instance, is a dominant driver of both community structure and biogeochemical cycle operation in microbial communities. Understanding these intricate dynamics is not just crucial for theory development, but also for the creation of effective conservation strategies that are sensitive to competitive dynamics of organisms with their ecological niches. By elucidating competition mechanisms in diverse ecological environments, researchers can make inferences about potential outcomes of species interactions under shifting environmental regimes, enhancing the predictive power of ecological models. Competition can also be more comprehensively understood by its differential impacts upon various taxa within a community. For example, in deep-sea benthic ecosystems, competition is experienced in a different way by organisms based on the life history characteristics of the organisms involved. Research carried out through spatial point process analyses (SPPA) has presented evidence that fixed organisms , such as glass sponges and octocorals experience competition differently based on their spatial arrangement and densities. Although octocorals exhibited higher intra-specific competition at high-density levels, glass sponges proved to be immune to density-dependent mortality due to their specialized features. This serves as a demonstration of how competing forces can lead to divergent responses among co-occurring species and the intricacy in community relationships that scientists have to consider when addressing population dynamics (Harris et al., 2020). In addition, competitive consequences range beyond resource allocation within ecosystem processes to influence more pervasive ecological functions , such as nutrient cycling. In microbial communities, competitive processes can govern primary biogeochemical cycles, particularly in soil ecosystems. The interplay of density dependence and biotic influences , such as competition and predation has been identified as primary regulators of microbial function. As uncovered through simulations, these interactions not only shape the structure of microbial communities but also their functionality in soil carbon formation and nutrient acquisition. By incorporating a population and community ecology framework, researchers can develop more advanced theoretical models that take into account competition and its manifold ecological implications and hence provide insights into ecosystem dynamics and the processes that underpin biodiversity (Bradford et al., 2017).

11.3.2. Predation

The role of predation in structuring populations and communities in ecological systems is important and has a strong influence on the interaction between species and community structure. Predation, being a process whereby one organism feeds on another, may cause both evolutionary changes and population control. Predators can be agents of natural selection and drive features in prey species that increase their survival, including camouflage, speed, or behavioral shifts. Such an interactive relationship does not only influence individual species but also plays a role in structuring and functioning of whole ecosystems. Predator-prey relations are crucial to the preservation of biodiversity and the well-being of the ecosystem, making the research on predation as a part of population and community ecology significant. Furthermore, the complexity of habitats significantly influences predation dynamics. Other researchers have shown how habitat structure may provide refuge to prey species and change predator accessibility and foraging success. An important feature of this interaction is the size structure of predator populations which may cause the variation in the rate of predation. Bigger predators can overpower smaller ones resulting in a situation known as intraguild predation where predators prey on the same type of resources. These interactions may play an important role in structuring the predator guild in a community in terms of composition and density. Prey density is often greater in extremely complex habitats because such habitats offer necessary refuges that can cause prey to survive, which influences population dynamics and community stability. Understanding these habitat-mediated effects is vital for effective conservation and management strategies in ecosystems (Carter et al., 2018). Also, the inclusion of the size structure of predators and complexity of habitat in ecological research improves the interpretation of the community dynamics. Studies have shown that size structure in predators may directly and indirectly affect rates of predation and determine community interactions. As an example, the hunting behavior of bigger predators might differ with that of the smaller ones, affecting the total predation pressure and the availability of prey. Also, predator-prey interactions, , such as cannibalism and interspecific competition, may reduce the rate of predation, thus changing the structure of the community in the long run. By acknowledging these intricate relationships and the role of environmental heterogeneity, ecologists can develop more comprehensive models that reflect the realities of food webs and contribute to the preservation of biodiversity (Carter et al., 2018) (Figure 11.13).

Figure 11.13. Crucial role of predators in ecosystems.

Source: https://media.licdn.com/dms/image/v2/D4D12AQELqh6lUSwInA/article-cover_image-shrink_600_2000/article-cover_image-shrink_600_2000/0/1697170573129?e=2147483647&v=beta&t=-rvgqtNitOP2V-iD05BJV8jP_7sd8t4cxAUohckkKIY.

11.3.3. Symbiosis

Symbiosis, a key process of ecological interaction, has a significant role to play towards population and community dynamics. The term itself derives from the Greek meaning living together and can be used for a range of interactions between species , such as mutualism, commensalism, and parasitism. While competition and predation have been central to ecological concepts for many years, symbiosis is less prominent in spite of its key role in ecosystem functioning. Mutualism, a symbiotic relationship in which both individuals benefit, has been the focus of interest due to its many varied forms in nature. Ecological resilience and stability are enhanced by the diversity of symbiotic associations, with intricate webs of association facilitating biodiversity. Since symbiosis fosters interdependence among species, it is necessary to grasp this term as part of the overall comprehension of population and community ecology as symbiosis reveals how species live and affect one another's survival and reproduction in various environments. Symbiotic interactions among fungi, microorganisms, and plants in soil biota demonstrate the significant influence of symbiosis on ecological processes. AMF form important associations with higher terrestrial plants that benefit nutrient acquisition, particularly phosphorus. This symbiosis occurred even in the earliest times and emphasizes the importance of mutualistic associations in plant nutrition and growth and soil fertility and structure improvement. Recent studies have proven the gene and functional diversity of the fungi, which opposes previous assumptions of their species poor nature. These findings illustrate the complexity of symbiotic relationships and their function in controlling ecosystem processes, , such as nutrient cycling. In addition, the participation of AMF in biogeochemical cycles reflects the interdependence of organisms in soil ecosystems and the means by which these connections can influence our understanding of ecosystem processes, ultimately affecting higher-level population and community organization (Barrett et al., 2005). The consequences of symbiosis extend beyond the straightforward swapping of nutrients; they also have influence on ecological resilience and community structure. For instance, certain symbiotic relationships can confer competitive advantage to certain species, influencing their establishment and success in diverse habitats. This is seen in the case of invasive species, where specific symbiotic relationships may facilitate or hinder the introduction of non-native taxa. Appreciation for these interaction mechanisms allows ecologists to be able to predict community response to environmental change. Functional redundancy, the concept in which multiple species can provide similar functions within a system, is another concept that enhances symbiosis. It also allows our appreciation of symbiotic relationship complexity to augment our understanding of population dynamics and community structure, integrating the study of individual species with patterns at larger scales (Boucher et al., 1982) (Figure 11.14).

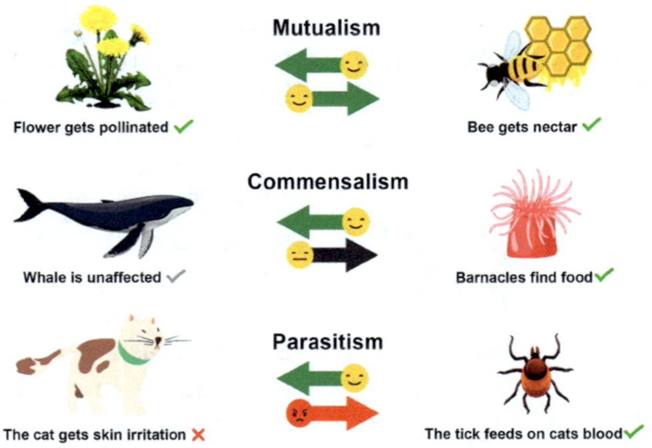

Figure 11.14. Symbiosis of ecological interactions.

Source: https://media.geeksforgeeks.org/wp-content/uploads/20231030113751/Symbiosis.png.

Test Your Knowledge

1. Explain about predation.
2. What is symbiosis?

11.4. Key Concepts in Community Ecology

Learning Outcome

1. Community structure;
2. Species diversity;
3 Ecological niche.

Community ecology embraces some of the most significant ideas necessary to explain species interactions in a given habitat. Some of these ideas are species richness, diversity, and community structure, all of which play central roles in deciding ecosystem dynamics. Species richness merely concerns itself with the number of various species present in a community, while diversity encompasses not only species richness but also the evenness of species abundance. Such high diversity is also generally accompanied by greater resilience in ecosystems, which allows communities to resist environmental changes. In addition, hierarchical complexity of ecological interactions demonstrates the role played by abiotic and biotic factors in community structures. Through the recognition of such patterns, ecologists will be better able to comprehend ecological function and the processes governing such function that underlie community dynamics, providing a foundation for conservation planning. Temporal dynamics is a crucial aspect of community ecology research, since communities are continuously influenced by variables that vary over time, , such as climate, levels of resources, and species interaction. Understanding these temporal dimensions allows ecologists to determine how community compositions change and evolve, revealing the interactive dynamic between drivers and responses. Hierarchical complexity of temporal structure is a notion that serves as a canvas for how multiple interactions manifest themselves at multiple scales of time, from which predictive modeling and experimental design can be enhanced in ecological science. Under this premise, community ecology not only accounts for the present organization of communities but also sheds light on predicting future ecological change. This integrative strategy focuses on the imperative of understanding both spatial and temporal dynamics in community contexts (Aguilar-Trigueros et al., 2019). Phenotypic plasticity is increasingly recognized as a central community ecological process explaining how organisms can adjust their phenotypes to variability in the environment. This adaptability has the potential to influence community dynamics through modification of species interaction and resource utilization, thereby altering community structure and composition. For instance, plasticity of flower shape can have immediate impacts on pollinator visitation rates, which can propagate to influence plant community composition. Through facilitating the integration of such ideas as natural selection and niche construction, phenotypic plasticity enhances our understanding of species interactions and ecosystem functioning. With rising environmental pressure on ecosystems, it is important to include accommodation of phenotypic plasticity to make management and conservation efforts effective. Hence, recognizing and incorporating plasticity in the theoretical models of community ecology provides valuable insight into maintaining biodiversity and ecological resilience (Martins et al., 2018) (Figure 11.15).

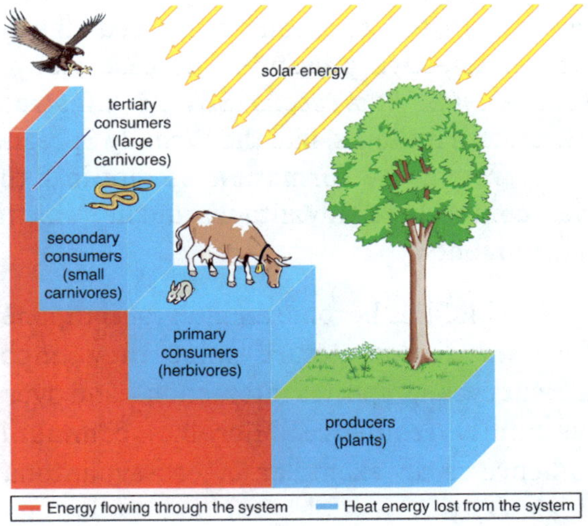

Figure 11.15. Community ecology.

Source: https://cdn.britannica.com/47/6547-050-1C80C83D/energy-Transfer-ecosystem-level-proportion.jpg.

11.4.1. Community Structure

Community structure theory is also core in population and community ecology as it addresses the organization and inter-action of species in a habitat. Community structure has a composition, abundance, and distribution of species that is influenced by various biotic and abiotic factors. The examination of community structure allows ecologists to identify ecological relationships , such as predation, competition, and symbiosis, which influence the dynamic stability of ecosystems. For example, some species will have features that enable them to be dominant in certain niches, thus changing community structure in the long term. The development of particular traits in populations tends to propel adaptive evolution, which often increases biodiversity, supporting ecosystem resilience. Lastly, the study of community structure illustrates the intricate web of life interactions, which reveal how the associations keep ecological equilibrium in check despite the flux in the environment. Community structure in microbial worlds can be highly dynamic, as observed in the creation of isolated lineages through selective pressure. The case study of *Pseudomonas fluorescens* SBW25 illustrates how certain mutants, like the Wrinkly Spreader, exist through the formation of biofilms that offer competitive advantages during shifting environments.

The biofilms become an integral component of community structure since they impact nutrient cycling, species interactions, and overall community resilience. Microbial community resilience is an exemplar of eco-evolutionary dynamics, where ecological and evolutionary processes become inextricably linked to facilitate rapid scale change in community structure. As populations grow and become more diverse, community structures can become increasingly complex, and a deeper appreciation of the ecological processes underlying them is required. This emphasizes the significance of community structure in microbial ecology but also shows its broader applicability in furthering our understanding of ecosystem function. Effective conservation and management of biodiversity are generally based on accurate assessment of community structure, which could be difficult to obtain because of sparse data. Many species show population declines that go unnoticed until late, so it is challenging to conserve them. As a counter to such an obstacle, scientists can employ known information about species physiology, life history, and ecological interdependence, integrating such information into advanced population models. By applying multispecies approaches that take advantage of the congruence of evolutionary traits, ecologists can enhance test strength and better forecast community change. The process allows for conservation decision-making grounded in the appreciation of the interconnectedness of organisms in ecosystems and the realization that changes in a single species will propagate across the community. Finally, broad insights into community structure are necessary for building effective strategies that pick-up issues of biodiversity decline and promote ecological sustainability.

11.4.2. Species Diversity

Species diversity is central to ecosystem stability and functionality, thus the necessity for holistic understanding in population and community ecology. The fluctuation in species richness and abundance determines ecosystem resilience, allowing communities to recover from disturbances and respond to change. This principle of ecology is especially important when considering the relationships between various trophic levels and their impact on species dynamics. For instance, the presence of large grazing herbivores can have far-reaching effects for species diversity in grasslands. Research has indicated that arthropod diversity is affected by grazing intensity differently compared to the effect of grazing intensity on plant diversity, where increasing grazing leads to lowered arthropod diversity despite maintaining plant diversity (van Klink et al., 2015). This suggests the complexity of species relationships and the need for cautious conservation practices. It is important to learn about mechanisms of species diversity in an effort to manage ecosystems effectively. Seed mass range among plant species and its implications for community structure as well as competition for resources is one area of research in this field. Studies indicate that seed mass variation relates to a variety of ecological strategies, which are generally dichotomized into two groups: ones examining intra-community as well as inter-community seed mass variations. A model predicting the dynamics of annual community competition for resources suggests that factors , such as asymmetry of light and growth allometry are of crucial importance in determining such patterns. Specifically, large seedlings have an advantage with increased light capture, while small-seeded species do poorly under reproductive success-favorable conditions. Notably, the model explains greatest variation in community seed mass under intermediate amounts of soil resources, highlighting the intricate relationship between resource supply and functional diversity (DeMalach & Kadmon, 2018). The implications of these processes extend beyond theoretical understanding and into practical applications in conservation and management efforts aimed at promoting biodiversity. For example, maximizing arthropod diversity as well as plant diversity requires the use of grazing management practices that are species-specific.

Additionally, herbivore population control, maintaining heterogeneous management practices, and rotational grazing assist in ecological stability. These practices not only promote the coexistence of diverse species but also enhance the ecosystem services that are critical to human societies. Therefore, the study of species diversity, particularly from population and community ecology contexts, fosters increased knowledge of biodiversity conservation and sustainable resource utilization, thus promoting healthier and stronger ecosystems (Figure 11.16).

Figure 11.16. Species diversity.

Source: https://www.ekolojik.com.tr/images/tur-cesitliligi-ne-demek.jpg.

11.4.3. Ecological Niche

The concept of ecological niche is core in the discussion of population and community ecology. An ecological niche is the description of the role that a species has in its ecosystem, including its habitat preferences, resource consumption, interactions with other species, and its place in the food web. Niche concept does not remain the same, it is dynamic and may change with variation in environmental conditions or by the arrival of invasive species. As an illustration, introduction of exotic predators may greatly change community dynamics, by changing species interaction and trophic relations. This shows how the ecological niche is important in explaining species coexistence, competition and biodiversity within a community as the niches are taken to be the multidimensional space of ecological roles of species that determine survival and reproduction of the species. Historical contingencies, or priority effects, often cause the dynamics of species interactions within an ecological niche. Such effects may influence the structure of communities based on the order of species arrival as well as their individual effect on the environment. How species exploit the shared resources is also important; species that have high resource overlap can have greater priority effects and can be competitively excluded or can niche differentiate. As an example, considering the nectar-inhabiting microorganisms, one can see that the way various species influence their habitat is connected with the ecological niche. Such interactions can elevate priority effects, adding complexity to community dynamics and highlighting the intricate relationships among species that are crucial for preserving ecological balance, as discussed in theories of community assembly (Fukami et al., 2013). The impacts of invasive species on community ecology can be explained by taking into consideration the ecological niches. When a non-native species is successfully established, it may alter current ecological niches, which is what happened with the introduction of such species as Palaemon elegans and Rhithropanopeus harrisii in the Baltic Sea. Such invasive species do not only inhabit similar niches as native species, but they are also capable of restructuring local food webs and ecosystem processes. Studies have also suggested that invasive species may tend to exploit resources in a different way to their native counterparts which may cause more ecological effects and alterations of species interactions. Thus, examining how these invasive species model their ecological niches provides a clearer understanding of how ecosystems may transform in response to species introductions and why the preservation of native biodiversity is critical in maintaining ecological stability (Kuprijanov, 2017) (Figure 11.17).

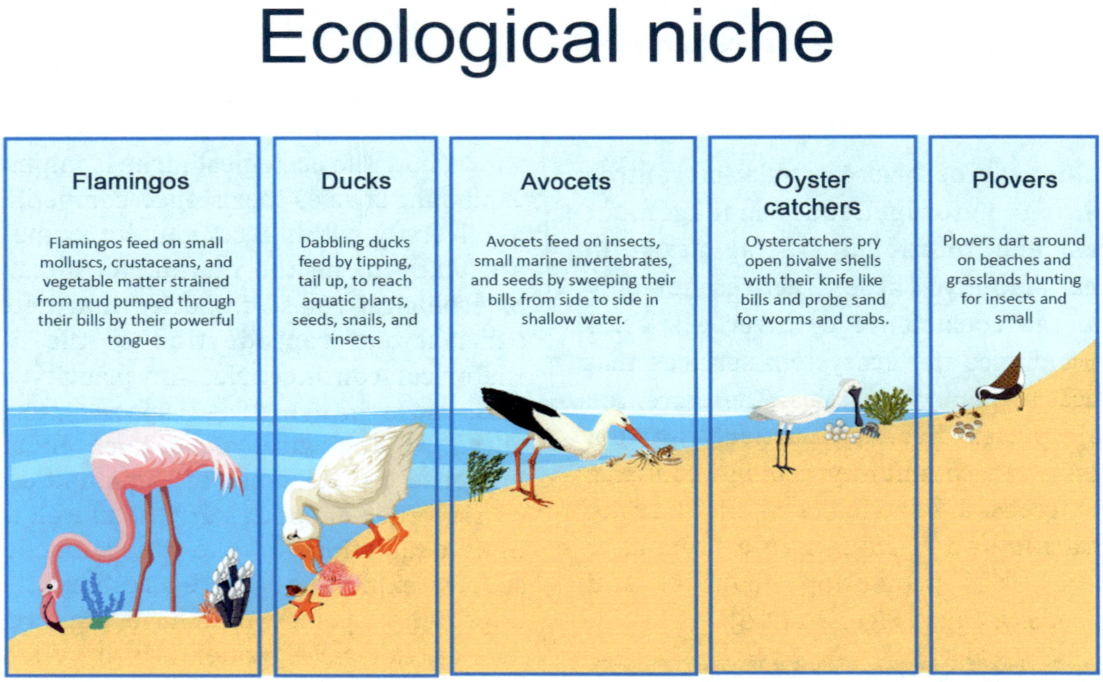

Figure 11.17. Ecological niche.

Source: https://media.geeksforgeeks.org/wp-content/uploads/20240529145704/Ecological-niche-(1).png.

Test Your Knowledge

1. What do you understand by species diversity?
2. What is ecological niche?

11.5. Community Interactions

Learning Outcome

1. Trophic levels;
2. Keystone species;
3. Ecological succession.

In Population and Community Ecology, community interactions are a fundamental determinant of ecosystem function. Interactions range from competition between species for resources to mutually symbiotic relationships whereby both partners' benefit. This diversity is evidenced by ecological relationships that define coexistence among species and promote diversity. Community interactions not only influence the survival of specific species but also what the structure of the community is. For instance, the fine balance between the numbers of prey and predators tends to regulate species numbers and influence the energy flow within the ecosystem. A complete understanding of such interactions is vital, particularly in the context of ongoing environmental changes that can disrupt prevailing dynamics and lead to composition changes in communities and species fitness, bringing into focus the intricate web of life that characterizes ecological communities (Stewart, 2010). A further examination of community interactions suggests that these interactions are strongly regulated by climatic processes and human activities. When ecosystems endure the impacts of climate change, species distribution and behavior changes are revealed, altering the previously established community patterns. For example, during the Quaternary epoch as great climate changes happened, ecological communities changed abruptly that affected species evolution as well as extinction directly. These changes provide evidence for mutual impact between climatic conditions and community interactions that show how the environment shapes biodiversity. In addition, human-induced changes, , such as land use alteration and biodiversity loss, complicate these interactions, having a tendency to lead to homogenization of the ecosystems. As the population adapts to living with these alterations, an appreciation of community interactions is imperative in predicting ecological future projections and designing conservation practices that reverse loss of biodiversity (Stewart, 2010). Community interaction dynamics also comprise the basic functions of various species in an ecosystem, particularly herbivores and arthropods. As per research, large herbivores with grazing intensity considerably influence arthropod diversity, demonstrating the dependency of various species on each other in a community. Research demonstrates that although rising grazing intensity will maintain plant diversity, it typically has adverse impacts on arthropod populations. This is due to shifts in vegetation structure, plant community composition, and overall availability of resources. Effective management practices that enhance habitat heterogeneity can retard the adverse impacts on arthropod diversity, with the utmost requirement for site-specific conservation techniques. Through these multifaceted interactions, ecologists are better equipped to advise management strategies to promote ecosystem well-being and stability in the presence of both intrinsic and extrinsic stressors, while highlighting the importance of incorporating arthropod monitoring into conservation programs for biodiversity.

11.5.1. Trophic Levels

Trophic levels are fundamental in understanding the intricate interactions that encapsulate population and community ecology. The levels, which demarcate the flow of energy and nutrients within an ecosystem, typically consist of primary producers, primary consumers, secondary consumers, and tertiary consumers. Primary producers, , such as plants and phytoplankton, obtain solar energy, via the process of photosynthesis, serving as the foundation for all other trophic levels. The energy the producers take in is transferred to primary consumers—herbivorous plants that eat the producers. This tiered system defines the line of energy and creates a framework for analyzing biodiversity and species relationships in ecosystems. Furthermore, as illustrated by Lindeman's trophic dynamics, understanding these layers says a great deal regarding the stability and health of ecological systems, thus necessitating further research into how various organisms rely on one another to survive (Hamm, 1972). At the level of primary consumers, the occurring interactions are very significant in determining the population size dynamics and the structure of communities. Herbivores play a crucial role in controlling plant populations and, as a result, influence the diversity and abundance of producers. The process is influenced by various ecological forces, including predation, competition, and resource availability. For instance, density-dependent processes, as described by Smith and others, can stabilize primary consumer populations by regulating their size in relation to food availability, avoiding overgrazing, and maintaining ecosystem health (Kolding et al., 2025). In the second instance, these interactions generate complex feedback mechanisms where population alteration in one species has the potential to trigger cascading effects throughout the food web, underscoring community interdependence. These intricacies illustrate how trophic levels are not simply static classifications but are the foundation of ecological resilience. As one moves up the trophic pyramid, energy transfer becomes increasingly more complex, particularly if one considers secondary and tertiary consumers. Occupants of these roles often have advanced hunting strategies and socialization driven by necessity to acquire sustenance in light of competitive forces.

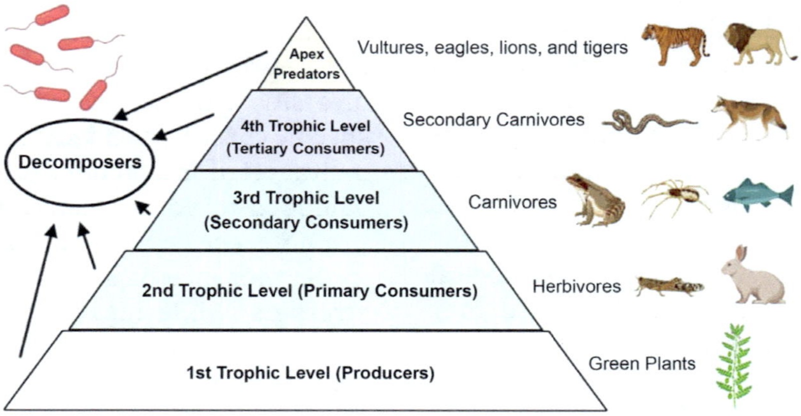

Figure 11.18. Trophic levels.

Source: https://microbenotes.com/wp-content/uploads/2021/02/Trophic-Level-pyramid.jpeg.

They not only regulate populations of primary consumers but also add diversity and evolutionary pressure to ecosystems. Apex predators, for example, will regulate communities with even community composition by eliminating weaker individuals and enhancing the health of the prey. Moreover, other theorists like Birch have aimed at the role played by genetic determinants in the formation of characters influencing consumer success in food procurement and successful competition (Hamm, 1972). Thus, trophic level, therefore, is not merely a topic of energy flow; it represents a more general understanding of ecological interactions, prompting scientists to investigate how these interactions affect species persistence and evolution in diverse settings (Figure 11.18).

11.5.2. Keystone Species

The keystone species hypothesis forms the core in the interpretation of ecological processes and community structure from the vantage of population and community ecology. Keystone species, while typically found in small numbers, have a disproportionate role to play in contributing to the integrity and diversity of their ecosystems. For instance, in the aquatic ecosystem, certain herbivore species of sea urchins can significantly determine the makeup of underwater vegetation. The sea urchin *Paracentrotus lividus* demonstrates how keystone herbivores may overgraze in seagrass beds and macroalgal rocky communities, thereby affecting the numbers of a large number of other species within the environment. This exchange brings into focus the intricate relationships in ecosystems, where the occurrence or absence of one species can cascade through multiple trophic levels, impacting not merely the surrounding community but altering the habitats overall form and function as well (Alcoverro et al., 2012).

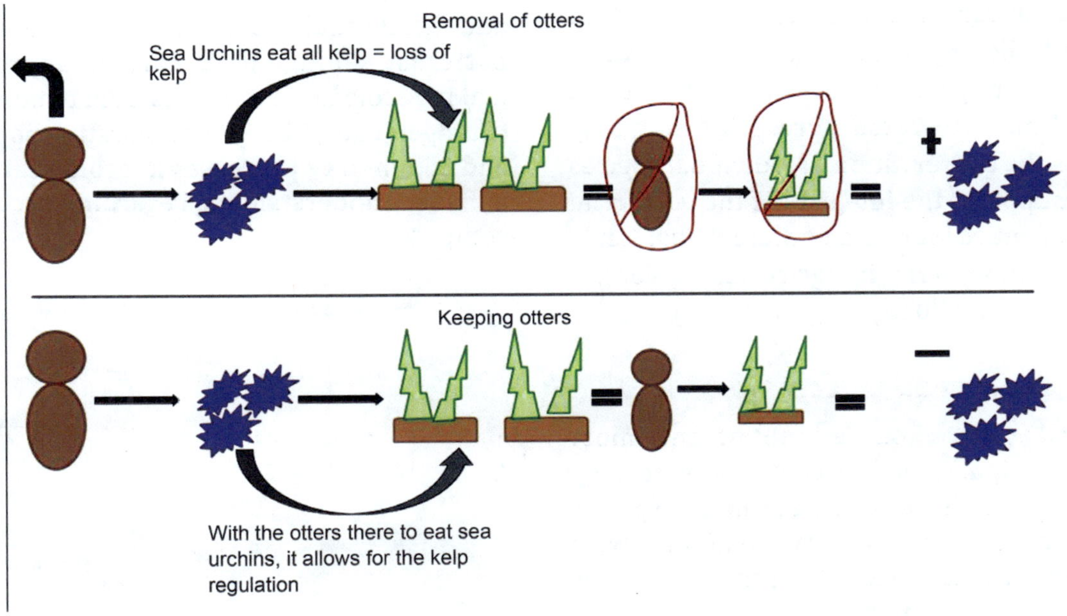

Figure 11.19. Keystone species.

Source: By R3nee321 - Own work, CC BY 4.0, https://commons.wikimedia.org/w/index.php?curid=165814000

Analyzing such dynamics typically demands inspection of food webs and the effects of species extinctions upon them. Keystone species extinction has the potential to bring about drastic changes in community processes because their unique niches tend to ensure the stability and resilience of many ecological associations. Food web model experiments show that while loss of non-keystone species might not directly affect the stability of ecosystems, the extinction of keystone species might lead to drastic changes. The adaptability of predator-prey interactions and the reliance of species on various trophic levels illustrate the responsiveness of these networks. Empirical studies show that in the removal of keystone species, cascades can be initiated with loss of biodiversity and transformation of community structure, even if according to the model there is minimal overall instability without their presence (Abrams et al., 2004). Recognition of keystone species has added applied conservation practice to ensure ecological balance. Effective habitat management must consider these cornerstone organisms and their roles in upholding community structure. Conservation of keystone species, particularly with marked demographic bottlenecks, requires an interdisciplinary approach. For instance, understanding the habitat requirements and spatial dynamics of organisms like *P. lividus* guides more comprehensive ecological action to address threats like overfishing or habitat destruction. Effective ecosystem management should therefore incorporate practices that cover both the organisms and the complicated, interdependent processes over which they exert control. This integrated perception not only promotes the conservation of individual species but also supports the integrity of the ecological makeup of entire biomes and thereby maintains long-term ecosystem integrity and diversity (Alcoverro et al., 2012) (Figure 11.19).

11.5.3. Ecological Succession

Ecological succession is one of the most important population and community ecology processes, which is the gradual change of species assemblage and community structure with time. It often takes place in a series of stages, beginning with the pioneer species, occupying the new or disturbed habitats. They modify the environment such that it becomes more suitable for other species to inhabit. For example, in primary succession, vegetation , such as lichens and mosses pave the way for the rest of the vegetation since they facilitate the development of soil and the accumulation of nutrients. Knowledge of these processes is necessary because it shows how ecosystems develop and transform in response to both biotic and abiotic factors and finally end up forming a climax community in equilibrium. Various methods are employed in the research of ecological succession, and chronosequences are particularly important. Chronosequences allow ecologists to study plant assemblages and soil development across different time scales and thereby obtain a conceptualization of temporal dynamics. Care must be taken, however, in applying this technique due to potential misinterpretation. For example, certain characteristics like species richness can be investigated reliably via chronology, whereas others like species composition show very high variability that renders interpretations a challenging task. As emphasized in recent research, the applicability of using chronosequences can be limited by factors like biodiversity, disturbance rate, and individual trajectories of successional patterns, calling for a careful approach to making ecological inferences from temporal data (Bardgett et al., 2010). Incorporating population and community processes in ecological models enhances understanding regarding successions complexities.

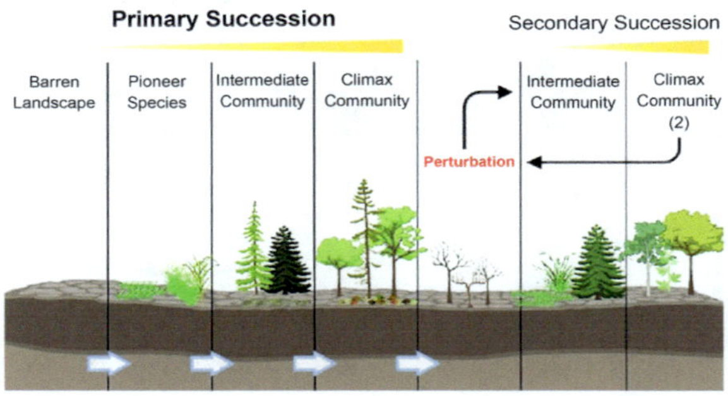

Figure 11.20. Ecological succession.

Source: https://static.wixstatic.com/media/8d410a_98bc532f9988463c93b513501d079a4d~mv2.jpg/v1/fill/w_980,h_551,al_c,q_85,usm_0.66_1.00_0.01,enc_avif,quality_auto/8d410a_98bc532f9988463c93b513501d079a4d~mv2.jpg.

Studies indicate that at the population scale, colonization and death rates significantly regulate total biomass achievable in succession sequences. For instance, the rate of colonization determines the rate at which carbon accumulation begins, whereas death rates determine yearly fluctuations of biomass. Additionally, long-term changes in community structure introduce complexity to the carbon dynamics, demonstrating that species with distinct traits can generate patterns not representative when averaged over species. This underscores the need to include population and community variables in ecological modeling because excluding these variables introduces an enormous uncertainty in the prediction of ecosystem behavior and future carbon dynamics (Carlos et al., 2008) (Figure 11.20).

Test Your Knowledge

1. What are the keystone species?
2. What is ecological succession?

11.6. Human Influence on Population and Community Ecology

Learning Outcome

1. Urbanization;
2. Climate change;
3. Conservation efforts.

The human impact on population and community ecology is a major cause of ecological results because of a number of processes, especially habitat changes and climate change. Habitat fragmentation caused by urbanization, agriculture and industrialization disturbs the natural occurrence of species distribution and negatively influences the dynamics of ecosystems. The resulting isolated populations due to fragmented habitats may limit gene flow and decrease genetic diversity, which may further increase extinction rates. Also, when human action alters the landscape, it tends to bring invasive species which outcompete the native plants and animals causing further imbalance in the community structure. These interactions are important to understand since they exemplify how anthropogenic factors result in changes in population dynamics and community assemblage, which leads to an effect on biodiversity. With the altering environmental conditions, the consequences are most significant to the conservation management approaches and thus a sound knowledge of the rearrangement of the ecological systems by the human factor is required. In addition to the habitat change, the significance of behavioral ecology implies the need to comprehend how single species react to the anthropogenic changes in the environment. Ecological effects of such alterations, it has been recently pointed out, are frequently mediated by means of behavior, and this behavior may differ greatly among individuals of a single species. As an example, changed behavioral patterns because of human intrusion can affect reproduction, survival and interspecific interactions which shows how behavior is an important connection between environmental change and population dynamics. The rejection of strictly adaptationist views on behavior ecology permits scientists to pay attention to the intricacies of behavioral phenomena in anthropogenically-induced landscapes. By analyzing individual behavioral responses, ecologists can develop more accurate models predicting population-level impacts and community-level disruptions, thus ensuring that conservation efforts are tailored and effective in mitigating human impacts on ecosystems (Bro-Jørgensen et al., 2019). In addition, the active climate change and ecological community structure interaction underlines the importance of the long-term evolution processes comprehension. The Quaternary period shows how severe climatic oscillations influenced the development of species and the formation of communities. The adaptation and extinction of species that occurred in the period proves that climate factors are major elements that cause biodiversity and ecological interactions.

Figure 11.21. Conceptual illustration of global population distribution, representing human population growth and its ecological implications.

Source: https://simply.science/images/content/biology/evolutionary_biology/community_and_population_ecology/conceptmap/image7.gif.

The current climate change is also causing changes in the same way that community dynamics can be changed because changes in temperature can cause mismatch in species interactions including predator-prey pairs and plant-pollinator interactions. With the increasing rate of human-induced climate change, it is crucial to determine the extent to which the changes will impact further on the ecological network and population resilience. Grounding conservation biology in comprehensive behavioral ecology insights equips researchers and policymakers to develop strategies that address these pressing challenges effectively (Stewart, 2010) (Figure 11.21).

11.6.1. Urbanization

Urbanization is a good predictor of population dynamics and community structures in ecological landscapes. Human population expansion into urban areas typically results in loss of habitat, habitat fragmentation, and alteration of natural habitats. These alterations are likely to bring about significant biotic and abiotic changes, therefore affecting plant communities and their ecological interactions. For instance, urbanization has been shown to affect the genomics of plant populations and reproductive success, , such as in research involving species like *Brassica incana*. In such instances, urban stresses have been proven to influence the genetic diversity and adaptive characteristics of plants, demonstrating the complex interactions of ecological forces taking place in urban environments. Urbanization not only constrains habitat quantity but also affects the environment under which these species co-occur, thus affecting the evolutionary trajectory of plant communities in urban environments (Laccetti et al., 2025). The effects of urbanization, in addition, are not just upon plant communities but spill over to other animals, particularly insects. Urban environments are likely to be abundant in a host of resources, which in turn may support unique insect communities. But with such benefits are challenges, as habitat alteration tends to result in changes in species richness and community composition. In a Rome study, it was found that the richness of predator and parasitoid flower-visited insects showed significant variability according to urbanization gradients. While generalist species , such as some plants benefited from urbanization, evenness and abundance of specialist predators were adversely affected by the structural constraints of the urban scenes. This indicates that urbanization directly and indirectly affects insect populations and calls for multi-scale evaluation of these ecological transitions to value their significance in terms of biodiversity and ecosystem functioning (Caruso et al., 2019). The intricate relationship between urbanization and ecological processes points towards the importance of knowledge-based conservation. As cities continue to grow, an appreciation of how species react to such modified environments is critical to ensure biodiversity. Conservation must consider the biocomplexity of urban landscapes, as understanding that urban green spaces may be an essential source of resources for numerous species, both plant and insect. Connectivity-promoting planning and heterogeneous plantings must be employed to enhance insect populations by offering necessary resources throughout all phases of their life cycle. Secondarily, preservation of species richness and functional diversity must be accomplished for fostering robust urban systems. Last but not least, addressing the environmental consequences of urbanization requires an interdisciplinary approach, coordinating conservation and city planning initiatives in pursuit of promoting sustainability in rapidly growing urban centers (Caruso et al., 2019; Laccetti et al., 2025).

11.6.2. Climate Change

The climate change process has become synonymous with a threat-representing challenge to population processes and community biology, remodeling the relation between organisms and their ecosystems. Global warming and erratic precipitation patterns induce stressors directly affecting biodiversity in communities. In freshwater communities, for instance, increases in water body temperatures in freshwater ecosystems have impacts on fish habits , such as habitat use and spawning that ultimately change population processes by altering metabolic and growth rates. These changes at the individual level cascade through populations since these changes cascade, altering community structures, which lead to changes in species abundance and distribution. Thus, ecological effects extend beyond the direct species interaction to influencing energy pathways and nutrient cycling among communities and highlighting complex interconnectivities between climate change and ecological response in community contexts (Alofs et al., 2017). Moreover, climate change is entangled with the dynamics of species invasions, which becomes an issue and an opportunity for ecologists to make sense of ecological effects. Increasing temperatures and altered hydrological regimes add to the susceptibility of native ecosystems to invasion by exotic species that further aggravate loss of biodiversity. In addition, the changing climate adds new opportunities for invaders to establish themselves in what previously were unsuitable habitats and cause further competition among resident species for resources. Interactions between invasions of species and climate change need a multidisciplinary approach since it is crucial to investigate how environmental change is likely to alter ecosystems' susceptibility to invasive species, as well as the implications those invaders are likely to bring within their new ecological context. This highlights the value of studying both invasive species and drivers behind their dissemination to prevent their effects on native populations (Alofs et al., 2017). In addition, the effects of climate change permeate into the health arena, specifically through zoonotic diseases that can impact population dynamics in wildlife as well as human populations. Climate conditions significantly affect the distribution and movement of various animal species, thereby having a potential impact on the transmission of parasites and pathogens. For example, as the environment changes or becomes fragmented due to climate change, interactions between wildlife, vectors, and human individuals may become more intensified, thereby increasing disease transmission , such as Ebola. Where environmental changes in food systems brought about by climate erode customary food sources, human populations may turn more to bushmeat as a substitute, making the risk for zoonotic spillover even higher. For this reason, addressing health implications of climate change in the general population and community ecology context is then significant, which requires concerted effort in enhancing health systems that can adapt to ecosystem changes.

11.6.3. Conservation Efforts

Of concern in the conservation field have been freshwater species, particularly mollusks, due to their ecological significance and a critical downward slide. There existed a concerted strategy to address the conservation of freshwater mussels originally laid out in a 1998 strategy paper, which identified the core issues that are responsible for their survival decline. Subsequently, significant progress has been made with the finalized strategy of the Freshwater Mollusk Conservation Society in 2011. The updated work highlights the criticality of the knowledge of distribution, taxonomy, and ecological dynamics of the mollusks. It highlights the significance of conservation actions in restabilizing productive populations that can sustain themselves, which shows the changeover from mere identification of issues to embracing executable plans towards rehabilitation and sustainability. These priorities must be addressed to facilitate the resilience of freshwater ecosystems in response to ongoing environmental challenges (Bradley et al., 2016). Human-carnivore conflict (HCC) is another key issue in the conservation debate, where human activity crosses with the habitat of wildlife to reflect the intricacies that result from such an overlap. As urbanization and agricultural production advance into areas that are historically occupied by carnivores, instances of conflict, e.g., livestock depredation and possible attacks on humans, have become more common. This is not only dangerous for wildlife populations but also imperils the livelihoods of concerned communities. Conservation efforts must hence adopt a holistic stance that reconciles the interests of human populations with maintaining biodiversity. By embracing holistic approaches that address both ecological and sociocultural dimensions of HCC, it is possible to reduce conflict while promoting harmonious co-existence between human beings and carnivores and, in doing so, offering ecological integrity (Russell, 2020). The integration of scientific research in conservation is vital to being able to successfully address the multi-dimensional issues facing various species and their habitats.

Figure 11.22. Community members engaging in tree planting activities as part of a local reforestation effort.

Source: https://blog.grow-trees.com/wp-content/uploads/2019/07/imgonline-com-ua-CompressToSize-kGvKfXSnkOC.jpg.

For instance, emerging stressors like climate change and habitat destruction require adaptive management responses through ongoing research and observation. Conservationists must tackle not only the currently imminent hazards but also forecast impending challenges even before these are actually felt. This is done through teamwork among stakeholders like policymakers, researchers, and local communities. Education and training of future scientists and resource managers is also a key function that contributes to building a conservation culture that values diversity. With the dynamics of ecosystems evolving, sustained investment in learning and strategy updating will serve to rejuvenate conservation action, enabling ecosystems to persist under anthropogenic stress (Bradley et al., 2016) (Figure 11.22).

Test Your Knowledge

1. How climate change influences the population dynamics and community ecology?
2. What is the role of conservation efforts in population and community ecology?

SUMMARY

The intricate dynamics at the population and community ecological level accentuate the central role that biodiversity plays in maintaining ecosystem services and processes. As observed, for instance, soil biodiversity underpins the governance of ecosystem services essential to human life like clean water and air, climate regulation, and food provision. The significance of these ecosystem services becomes essentially critical in the context of the mounting danger to ecosystems around the world induced by human behavior, climate change, and deforestation. Soil biodiversity review highlights the need for strategies to preserve biological diversity to render these services sustainable, in accordance with calls for greater awareness and conservation, and notably in the context of the International Year of Biodiversity in 2010 (Van der Putten et al., 2010). Further, herbivore-arthropod diversity interaction is indicative of the complexity nature of ecological interactions. The arthropod community experiments measuring the effect of large grazing herbivores reveal that while plant diversity can withstand a certain level of grazing, arthropod diversity is often severely impacted. This is caused by direct predation, altered vegetation structure, and altered resource abundance. Such reciprocal dependence between varying taxa puts special emphasis on the necessity for overall management procedures considering the needs of a variety of species, particularly arthropods, whose roles in ecosystems are often overlooked. By drawing attention to the importance of monitoring and controlling such interactions, conservation strategies might better be suited to preserve biodiversity (van Klink et al., 2015). Lastly, the general conclusions drawn from population and community ecology favor the development of conservation strategies that acknowledge the subtle balance in ecosystems. Since terrestrial and aquatic ecosystems continue to be exposed to unprecedented threats, the significance of understanding the patterns of species interactions—herbivores and the rest of the community, for example—grows. The use of management strategies that enhance heterogeneity of habitats, along with managed levels of herbivores, holds the promise to substantially enhance biodiversity values. This type of research not only enhances our scientific understanding but also contributes to pragmatic action toward ecological resilience and sustainability, ensuring future generations leave behind a healthier world filled with biodiversity (Van der Putten et al., 2010; van Klink et al., 2015).

In community and population ecology, key considerations are likely to emerge that elucidate the intricate relationships between organisms and their environment. There must be a holistic understanding that can underpin several ecological theories and models to make sense of such dynamics. Transition theory, for instance, posits that communities are constantly changing as a result of numerous perturbations and adaptive processes. This strategy allows the revelation of how socio-ecological systems, , such as those observed within Alaskan society, react to abrupt changes, for example, fluctuations in fauna and economic alterations. As revealed through the research from Anaktuvuk Pass, households demonstrate resilience through the adaptation of their resource management in accordance with external stresses, thereby ensuring cultural consistency while adapting to new challenges (Ye). Another such critical theme in population and community ecology is the regulation of ecological interactions, particularly the herbivore-natural enemy relationship. This intricate interaction is crucial to maintaining ecosystem well-being, particularly in agroecosystems like coffee plantations. The top-down and bottom-up control paradigm recognizes how every herbivore, , such as the green coffee scale and coffee berry borer, is regulated by its predators and pathogens (Martin et al., 2014). Such sophistication supports the need to incorporate pest management interventions alongside the interconnected nature of these ecological factors. Ecologists can then develop expert interventions based on these interactions to optimize agricultural production while sustaining ecological harmony. Lastly, the study of adaptive trajectories of societies reveals the importance of resilience in the face of environmental changes. Families in the changing socio-ecological settings are likely to describe innovative strategies that integrate traditional knowledge and new opportunities, , such as is observed in the Alaskan example (Martin et al., 2014). Such flexibility is crucial to the development of social resilience since communities can adapt to the uncertainties of climate change, resource constraints, and economic hardship. Similarly, in agroecosystems, interconnectedness of organisms is expected to predict that positive feedback processes can lead to complex system behavior , such as hysteresis or critical transitions (Armbrecht et al., 2019). Therefore, population and community ecology must incorporate consideration of resilience, interconnectedness, and adaptive capacity to craft resilient ecosystems in the long term.

KEY TERMS FROM THIS CHAPTER

Biodiversity	Birth rate	Carrying capacity
Commensalism	Community	Competition
Death rate	Ecosystem	Emigration
Exponential growth	Habitat	Immigration
Logistic growth	Mutualism	Niche
Parasitism	Population	Population density

CRITICAL THINKING QUESTIONS

1. **What is a population in ecology?**
   a. All the living organisms in an area
   b. A group of individuals of the same species living in the same area
   c. All the ecosystems on Earth
   d. The non-living components of an environment
2. **What term describes the role an organism plays in its environment?**
   a. Habitat
   b. Niche
   c. Community
   d. Ecosystem
3. **Which of the following interactions benefits one species and harms the other?**
   a. Mutualism
   b. Commensalism
   c. Parasitism
   d. Neutralism
4. **What is carrying capacity?**
   a. The maximum population size that an environment can sustain
   b. The number of offspring produced by an organism
   c. The range of habitats a species can live in
   d. The total biomass in an ecosystem
5. **Which of the following is an example of a density-dependent factor affecting population size?**
   a. Natural disasters
   b. Disease
   c. Weather changes
   d. Habitat destruction

Short Answer Questions

1. What is a population in ecological terms?
2. What factors affect population size and growth?
3. Explain the difference between exponential and logistic population growth.
4. What is carrying capacity, and what happens when a population exceeds it?
5. Define the following terms: birth rate, death rate, immigration, and emigration.
6. What are limiting factors? Give examples of density-dependent and density-independent limiting factors.
7. What is a community in ecology?

Answers to Critical Thinking Questions

1. (b); 2. (b); 3. (c); 4. (a); 5. (b)

REFERENCES

1. Abrams, P., Arditi, R., Borrvall, C., Caldarelli, G., Carlton, J., Drake, J., et al. (2004). Deleting species from model food webs. http://arxiv.org/abs/q-bio/0401037 (accessed on 13th June 2025).
2. Aguilar-Trigueros, C. A., Muller, L. A. H., Pinek, L., et al. (2019). Basic principles of temporal dynamics. https://core.ac.uk/download/225124652.pdf (accessed on 13th June 2025).
3. Aioi, K., Kouchi, N., Nakaoka, M., Omori, Y., Tanaka, N., & Tatsukawa, K. (2003). Distribution, productivity, life history and biodiversity of seagrass community along Sanriku Coast: A review. https://core.ac.uk/download/197087282.pdf (accessed on 13th June 2025).
4. Alcoverro, T., Ceccherelli, G., Farina, S., et al. (2012). Habitat and scale shape the demographic fate of the keystone sea urchin *Paracentrotus lividus* in Mediterranean macrophyte communities. https://core.ac.uk/download/11693358.pdf (accessed on 13th June 2025).
5. Alofs, A., Angilletta, M., Baerum, B., Bauer, C., Belovsky, G., et al. (2017). Conceptualizing the interactive effects of climate change and biological invasions on subarctic freshwater fish. https://core.ac.uk/download/84365642.pdf (accessed on 13th June 2025).
6. Armbrecht, I., De la Mora, A., Ennis, K. K., et al. (2019). The community ecology of herbivore regulation in an agroecosystem: Lessons from complex systems. https://core.ac.uk/download/286728897.pdf (accessed on 13th June 2025).
7. Aslan, C. E., Beckman, N. G., & Rogers, H. S. (2020). Introduction to the special issue: The role of seed dispersal in plant populations: Perspectives and advances in a changing world. https://core.ac.uk/download/478906517.pdf (accessed on 13th June 2025).
8. Bardgett, R. D., Clarkson, B. D., Walker, L. R., et al. (2010). The use of chrono sequences in studies of ecological succession and soil development. https://core.ac.uk/download/29197851.pdf (accessed on 13th June 2025).
9. Barrett, D. J., Beijerinck, M. W., Brookes, P. C., Dunger, W., Finlay, B. J., Fitter, A. H., et al. (2005). Darkness visible: Reflections on underground ecology. https://core.ac.uk/download/59119.pdf (accessed on 13th June 2025).
10. Bernstein, H. C., Fredrickson, J. K., Haruta, S., et al. (2019). Minimal interspecies interaction adjustment (MIIA): Inference of neighbor-dependent interactions in microbial communities. https://core.ac.uk/download/286729903.pdf (accessed on 13th June 2025).
11. Bielčik, M. (2022). The role of space, dispersal and active movement in fungal community assembly. https://core.ac.uk/download/541748666.pdf (accessed on 13th June 2025).
12. Bocher, B. T. W., Maki, J., Venkiteshwaran, K., et al. (2016). Relating anaerobic digestion microbial community and process function. https://core.ac.uk/download/213083395.pdf (accessed on 13th June 2025).
13. Boucher, D. H., James, S., & Keeler, K. H. (1982). The ecology of mutualism. https://core.ac.uk/download/323061299.pdf (accessed on 13th June 2025).
14. Bourgarel, M., Kassie, D., & Roger, F. (2015). Climate change and Ebola outbreaks: Are they connected? http://agritrop.cirad.fr/577071/7/ID577071p.pdf (accessed on 13th June 2025).
15. Bradford, M. A., Buchkowski, R. W., Grandy, A. S., et al. (2017). Applying population and community ecology theory to advance understanding of belowground biogeochemistry.

https://core.ac.uk/download/603976412.pdf (accessed on 13th June 2025).

16. Bradley, M., Butler, R. S., Dunn, H. L., et al. (2016). A national strategy for the conservation of native freshwater mollusks. https://core.ac.uk/download/539653868.pdf (accessed on 13th June 2025).
17. Bro-Jørgensen, J., Franks, D. W., & Meise, K. (2019). Linking behavior to dynamics of populations and communities: Application of novel approaches in behavioral ecology to conservation. https://core.ac.uk/download/323059897.pdf (accessed on 13th June 2025).
18. Calizza, E., Careddu, G., Costantini, M. L., et al. (2017). Effect of habitat degradation on competition, carrying capacity, and species assemblage stability. https://core.ac.uk/download/98347468.pdf (accessed on 13th June 2025).
19. Carter, S., Crumrine, P., & Vodopich, D. (2018). Heterogeneity in body size and habitat complexity influence community structure. https://core.ac.uk/download/270056548.pdf (accessed on 13th June 2025).
20. Caruso, V., Cerretti, P., Corcos, D., et al. (2019). Impact of urbanization on predator and parasitoid insects at multiple spatial scales. https://core.ac.uk/download/196295259.pdf (accessed on 13th June 2025).
21. Costa-Dias, S., Laffaille, P., Lobón-Cerviá, J., et al. (2009). The decline of diadromous fish in Western Europe inland waters: Main causes and consequence. https://core.ac.uk/download/12043566.pdf (accessed on 13th June 2025).
22. DeMalach, N., & Kadmon, R. (2018). Seed mass diversity along resource gradients: The role of allometric growth rate and size-asymmetric competition. http://arxiv.org/abs/1709.09391 (accessed on 13th June 2025).
23. Dieckmann, U., Hartig, F., Johst, K., et al. (2014). On the sympatric evolution and evolutionary stability of coexistence by relative nonlinearity of competition. https://core.ac.uk/download/44737119.pdf (accessed on 13th June 2025).
24. Drossel, B., & McKane, A. J. (2002). Modelling food webs. http://arxiv.org/abs/nlin/0202034 (accessed on 13th June 2025).
25. Dulvy, N. K., Horswill, C., Juan-Jordá, M. J., et al. (2018). Overcoming the data crisis in biodiversity conservation. https://core.ac.uk/download/161514092.pdf (accessed on 13th June 2025).
26. Fukami, T., Vannette, R. L., & Wootton, T. (2013). Historical contingency in species interactions: Towards niche-based predictions. https://core.ac.uk/download/323069283.pdf (accessed on 13th June 2025).
27. Gabriel, W., Hrbáček, J., & Lampert, W. (1985). Reproductive strategies of two similar Daphnia species. https://core.ac.uk/download/12165873.pdf (accessed on 13th June 2025).
28. Hamm, Pauline Estelle, "Taxonomic Composition, Biomass, and Trophic Structure of Grassland Invertebrates at Cottonwood, South Dakota" (1972). Electronic Theses and Dissertations. 4700. https://openprairie.sdstate.edu/etd/4700(accessed on 13th June 2025).
29. Harris, S., & Mitchell, E. G. (2020). Mortality, population and community dynamics of the glass sponge dominated community "The Forest of the Weird" from the Ridge Seamount, Johnston Atoll, Pacific Ocean. https://core.ac.uk/download/341386909.pdf (accessed on 13th June 2025).

30. Jerdan, R., Lungin, O., Moshynets, *O. V.*, et al. (2020). Extending an eco-evolutionary understanding of biofilm formation at the air-liquid interface to community biofilms. https://core.ac.uk/download/287584064.pdf (accessed on 13th June 2025).

31. Juenger, T. E., Leibold, M. A., & Pantel, J. H. (2011). Population differentiation in Daphnia alters community assembly in experimental ponds. https://core.ac.uk/download/211330877.pdf (accessed on 13th June 2025).

32. Kolding, J., & van Zwieten, P. (2025). Improving productivity in tropical lakes and reservoirs. https://core.ac.uk/download/pdf/7084044.pdf (accessed on 13th June 2025).

33. Kuprijanov, I. (2017). Kohalike ja invasiivsete röövtoiduliste suurselgrootute elupaigakasutus ja toitumissuhted Läänemere põhjaosas. https://core.ac.uk/download/143642946.pdf (accessed on 13th June 2025).

34. Laccetti, Lucrezia; Frachon, Léa; Arrigo, Luca; Scopece, Giovanni (2025). Urbanization affects population connectivity, reproductive success and p*henotyp*ic traits in the Mediterranean cliff species Brassica incana (Brassicaceae). Urban Forestry Urban Greening, 104:128627. DOI: https://doi.org/10.1016/j.ufug.2024.128627(accessed on 13th June 2025).

35. Leveau, L. M. (2018). Urbanization, environmental stabilization and temporal persistence of bird species: A view from Latin America. https://core.ac.uk/download/287883393.pdf (accessed on 13th June 2025).

36. Martin, S. (2014). Indigenous social and economic adaptations in northern Alaska as measures of resilience. https://core.ac.uk/download/286999879.pdf (accessed on 13th June 2025).

37. Martins, R. P. (2017). To what degree are philosophy and the ecological niche concept necessary in the ecological theory and conservation? https://core.ac.uk/download/286139274.pdf (accessed on 13th June 2025).

38. Martins, R. P. (2018). The conceptual structure of evolutionary biology: A framework from phenotypic plasticity. https://core.ac.uk/download/286139307.pdf (accessed on 13th June 2025).

39. Olff, H., van der Plas, F., van Klink, R., et al. (2015). Effects of large herbivores on grassland arthropod diversity. https://core.ac.uk/download/55846534.pdf (accessed on 13th June 2025).

40. Sierra, C. A., & Harmon, M. E. (2008). Effects of processes at the population and community level on carbon dynamics of an ecosystem model. https://core.ac.uk/download/pdf/287840.pdf (accessed on 13th June 2025).

41. Stewart, J. R., 2010. Understanding evolutionary processes during past Quaternary climatic cycles: Can it be applied to the future? In: Grant, F., Mergeay, J., Santamaria, L., Young, J. and Watt, A.D., eds. Evolution and Biodiversity: The evolutionary basis of biodiversity and its potential for adaptation to global change. Report of an e. Mallorca, Spain: Centre for Ecology & Hydrology, INBO, IMEDA, CSIC-UIB.(accessed on 13th June 2025).

42. van der Putten, W. H., Mudgal, S., Turbé, A., de Toni, A., Lavelle, P., Benito, P., & Ruiz, N. (2010). Soil biodiversity: functions, threats and tools for policy makers. Bio Intelligence Service. http://ec.europa.eu/environment/archives/soil/pdf/biodiversity_report.pdf(accessed on 13th June 2025).

ECOSYSTEMS AND CONSERVATION BIOLOGY

12

LEARNING OBJECTIVES

After studying this chapter, you will be able to:

1. Explain the types of ecosystems;
2. Discuss biodiversity and its importance;
3. List the threats to ecosystems;
4. Define conservation biology: An overview;
5. Describe the conservation strategies;
6. Define the role of technology in conservation..

Introduction

Ecosystems are complex relationships between organisms and their physical environment, made up of biotic and abiotic components. Fundamentally, ecosystems are dynamic systems where energy transfer and nutrient cycling are all-important processes. Producers , such as plants and phytoplankton capture solar energy and convert it into chemical energy through the process of photosynthesis to form the basis upon which food webs are supported. These consumers are reliant on by predators and herbivores, and decomposers , such as fungi and bacteria are key in breaking down organic matter, returning vital nutrients to the soil. Such interactions are critical to be comprehended for conservation since disturbing a part of the system has the possibility to cascade across the system, creating imbalances within ecology destabilizing ecosystem services and biodiversity. Ecosystems' value extends beyond their intrinsic ecological value; they provide valuable services that maintain human life and economic endeavors. Ecosystems contribute to food production, clean water supply, climate regulation, and recreation and hence are critical for sustainable development. However, as revealed by recent research, the extinction rates of species are alarmingly high due to habitat loss, climate change, and other human-induced factors. For example, New Zealand biodiversity is under strong threat despite continued efforts aimed at conservation to protect it. The challenge signifies the necessity of integrating conservation biology into ecosystem management plans for enhancing the sustainability and resilience of these natural systems to ensure they can provide much-needed services to humankind (Jay et al., 1998). The training of future biology teachers plays a key role to instill conservation awareness and sustainable practices required to preserve ecosystems.

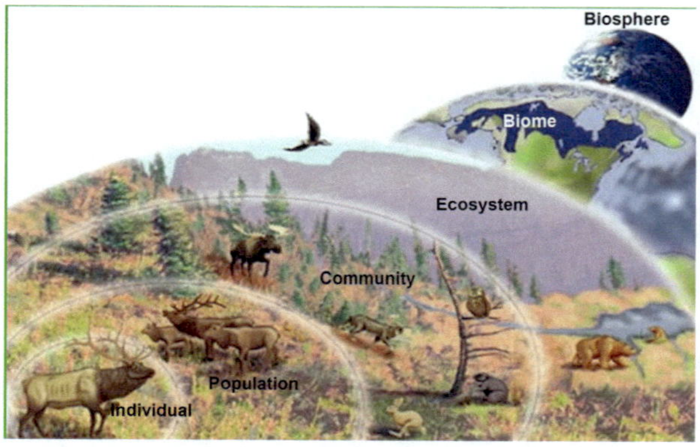

Figure 12.1. Ecosystems.

Source: https://o.quizlet.com/rs9ulWrG0AOBZqWUKgiE-A_b.png.

The acquisition of professional and terminological skills by future biology teachers helps them gain appropriate skills to transmit knowledge about biodiversity and its conservation to students. Courses and programs that emphasize the acquisition of educators' capacities in terms of ecological vocabulary and sustainable practices are fundamental in orienting the next generations understanding towards ecosystems. The integration of creative teaching methods and courses specialized in environmental issues can increase students' participation and devotion to sustainability. If future biology teachers graduate with a strong foundation in the principles of ecology, then they will be better positioned to advance the cause of conservation efforts at home and engage in the discourse about biodiversity and sustainability in the context of global environmental concern (Britchenko et al., 2020) (Figure 12.1).

Definition of Ecosystems

Ecosystems are an intricate interaction of associations among living organisms and physical environments that form a unifying theme in conservation biology. Generally speaking, ecosystems are the constantly changing associations of plants, animals, microorganisms, soil, water, and air that occur in a specific habitat. This association indicates how energy flows through trophic levels and nutrients flow through biogeochemical cycles. Significantly, ecosystems vary in size and complexity, ranging from small ponds to vast forests or entire biomes, and can be classified by overlying vegetation types or geography. Conservation requires knowledge of these systems because the integrity and stability of ecosystems influence biodiversity and, in turn, human health. The emphasis on preserving ecosystems grows more imperative as human endeavors place unprecedented demands on ecosystems, perilously pushing them towards degradation and even collapse. In response to ongoing environmental change, ecosystem evaluation has come to the forefront of conservation biology. The International Union for Conservation of Nature (IUCN) is constructing a red list of ecosystems (RLE), which attempts to rank ecosystems in a way consistent with the already existing Red List of Species. This program puts a lot of stress not just on the importance of proper ecosystem classification but also on identifying some of the major drivers of change , such as climate change and habitat destruction. However, the proposed framework for the RLE has been criticized based on its scientific merit and operational efficacy. It has been faulted for only focusing on broad vegetation types and not accounting for the details of ecosystem processes and their resistance to global change drivers. This raises strong questions about the use of the RLE to inform conservation policy and management practice to sustain ecosystems, underlining the need for proper assessment before extensive application (Boitani et al., 2014). Moreover, appropriately planned RLE would balance generality, accuracy, realism, and simplicity in their approach towards evaluating risk and managing ecosystems. The IUCNs criteria attempt to implement an international standard in quantifying ecosystem vulnerability, yet the challenge is finding efficient methods that involve the key determinants of ecosystem health, dynamics, and resilience. Lessons learned from recent national red lists of threatened ecosystems demonstrate the potential value added in applying a systematic assessment framework to policy-making and environmental management. An effective RLE could enable anticipatory conservation action, guiding action towards maintaining ecosystem integrity and service delivery. Later revisions of the RLE can take advantage of the iterative development that has been experienced by the Red List of Threatened Species so that it remains dynamic and responsive to emerging challenges in conservation. Ultimately, a successful RLE can provide valuable lessons in the management of landscapes and seascapes and in promoting the appreciation of how ecosystems are connected to the services they provide (Adler et al., 2015).

Components of Ecosystems: Biotic and Abiotic Factors

The intricate web of association that defines ecosystems is composed of both biotic and abiotic components, both of which play essential roles in the overall functioning and maintenance of these systems. Biotic factors, living things , such as plants, animals, fungi, and microorganisms, are interconnected with themselves and their surroundings in intricate ways. These organisms, for instance, participate in nutrient cycling, energy flow, and species diversity, all of which are fundamental to ecosystem health. Conversely, abiotic factors like climate, soil, water, and sunlight comprise the environment necessary for these biotic interactions to occur. Understanding the interaction between these two categories is fundamental to conservation biology because disruption of either has significant effects on ecological balance and biodiversity, and therefore on the delivery of essential services by ecosystems to humanity and the world at large. One of the most important facets of their importance is their contribution to soil health, which is an integral element of ecosystem functioning and service provision. Soil is a habitat of central significance that harbors a diverse array of invertebrates and microorganisms and is actively involved in nutrient supply and the decomposition of organic matter. Healthy soil biodiversity is crucial in the management of various ecosystem services, as addressed in the report findings presented to the European Commission and emphasizing the role of soil biodiversity in sustaining human activity sustainably. By preserving soil health, ecosystems can better support various forms of life and their interactions and ultimately end up with healthier and functioning environments. Thus, the protection of these abiotic elements must become part of conservation action to bring about ecosystem stability and conservation of biodiversity (Van der Putten et al., 2010). Moreover, biotic interactions, particularly between various taxa, reveal the complexity of ecosystem dynamics as well as the finer balance that is required to sustain them. The relationship between arthropod diversity in grasslands and large grazing herbivores is a compelling case study; these herbivores generally maintain vegetation structure and prevent overgrowth, but can destroy arthropod diversity conversely by modifying habitat and competing. Maxing out grazing intensity has been shown to enhance habitat diversity, according to a scientific study, but this balance is fragile, with sensitive management needed to sustain arthropod populations. Conservation practices aimed at the management of herbivore populations and the production of heterogeneity in vegetation and abiotic habitats are key to mitigating the negative effects on arthropod communities. Integrated approaches like these not only aid species richness but also enhance the stability and resilience of ecosystems against anthropogenic influence (Olff et al., 2015) (Figure 12.2).

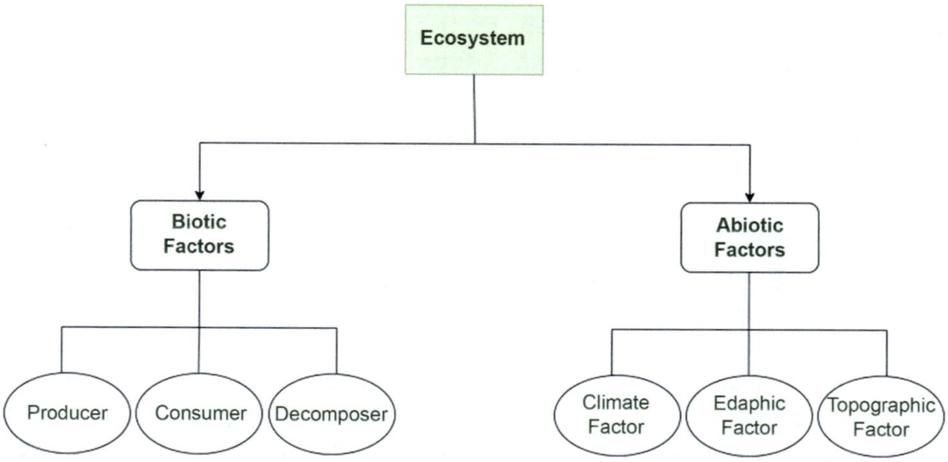

Figure 12.2. Components of ecosystems.

Source: https://media.geeksforgeeks.org/wp-content/uploads/20221220143326/Ecosystem.jpg.

Importance of Ecosystems in Maintaining Biodiversity

Ecosystems help to conserve biodiversity, acting as complex networks that support a rich variety of life. Well-being in an ecosystem is inherently linked with species diversity, resulting in resistance to ecological change and stress. This diversity enhances ecosystem productivity and stability, allowing various species to have unique ecological roles. Since there are numerous species with diverse functions, from carbon sequestration to pollination, biodiverse ecosystems are better placed to recover from perturbations, and hence ensure necessary services are sustainable. As mentioned, species diversity is particularly important in ecosystem integrity, and a prerequisite for long-term sustainable development (Jay et al., 1998). Such a link focuses attention on the pivotal role of conserving diverse ecosystems because their loss fosters increased species extinction and ecosystem degradation, leading to human survival risks. Loss of biodiversity is directly linked with habitat loss and environmental degradation, which are rapidly changing ecosystems around the world. Urbanization, agriculture, and industry place monstrous pressure on natural habitats, reducing species populations. In regions , such as New Zealand, where evolution has proceeded independently for plants and animals, this degradation poses a very serious threat. Domestic indigenous biodiversity is particularly vulnerable to invading species, since they will compete with native species for space (Armitage et al., 2004). With each step of habitat deterioration, the intricate processes holding ecosystems together disintegrate, resulting in degraded biodiversity functions and services, e.g., water filtration, nutrient cycling, and habitat provision for diverse species. The snowball effect necessitates an immediate conservation effort to ensure the remaining diversity for the well-being of our world and its future generations. Also, conservation of the ecosystem promotes not only biological diversity but also socio-economic well-being within communities that depend on them. Maximum health in ecosystems offers products , such as food, clean water, and raw materials, which are vital to regional and global economies. Ecosystem services offered by biodiverse ecosystems enable communities to thrive and contribute even to cultural and recreation values. Human and ecosystem interdependence is clear: when ecosystems are harmed, impacts flow through social, economic, and environmental systems. Consequently, all approaches to conserving ecosystems must prioritize sustainable practices that balance human interests with ecosystem health. As indicated in most studies, preserving biodiversity through effective conservation is crucial in fostering resilience of ecosystems while ensuring that human well-being is safeguarded for future generations (Jay et al., 1998).

12.1. Types of Ecosystems

Learning Outcome

1. Terrestrial ecosystems: Forests, grasslands, deserts;
2. Aquatic ecosystems: Freshwater and marine;
3. Urban ecosystems: Characteristics and significance.

The major types of ecosystems are terrestrial and aquatic, although each of these divisions consists of many subdivisions that perform distinct ecological roles. Terrestrial ecosystems incorporate biomes that entail forests, grasslands, deserts, and tundra, and all these have considerable dissimilarity in climatic conditions, plants, and animals. Examples include the fact that forest ecosystems have a dense canopy that provides high biodiversity in terms of numerous species of plants and animals. On the other hand, grasslands, which are open country with predominance of grasses, are necessary to sustain large herbivores and predators. These habitats are examples of the complex inter-relationships between species where changes to one level can have impacts right through the ecosystem. Understanding the specific characteristics and dynamics of these ecosystems is vital for effective conservation strategies that seek to maintain or restore ecological balance in the face of anthropogenic pressures and climate change, where the interaction between species can lead to evolutionary changes over time (Olff et al., 2015). Freshwater and marine aquatic ecosystems are also important in the general wellbeing of the earth. Rivers, lakes, and wetlands are key freshwater ecosystems that are important to drink water, agriculture, and biodiversity, supporting many species that depend on the habitats to survive. Of special note is the capacity of wetlands to filter pollutants as well as serving as important breeding areas to a number of aquatic species. Marine ecosystems on the other hand are diverse with intertidal to deep ocean environments and contain complex food webs. Water quality and temperature are key factors that determine the health of these ecosystems, and they are usually affected by human activities , such as pollution and climatic change. Research indicates that changes in marine biodiversity have far-reaching consequences beyond individual species, affecting ecosystem functioning and services (Olff et al., 2015). Therefore, there is an urgent need to protect these ecosystems to ensure ecological integrity and livelihoods of people. The terrestrial and aquatic ecosystems are interconnected, thus the need to embrace the holistic approach to conservation biology. Anthropogenic activities tend to cause confusion between these ecosystems, and some of the challenges encountered include habitat degradation, extinction of species, and a reduction in the ecosystem services. The problem of climate change is a super-threat that adds to these problems, which influences the community structure and resilience in the ecosystems. Studies have shown that alterations in climate can influence species distribution and interactions, leading to shifts in biodiversity patterns (Stewart, 2010). Conservation measures therefore should not only aim at protecting individual ecosystems but also aim at promoting resilience to climate variability. The best management practices must consider including adaptive management which takes into consideration the dynamics of the ecosystems, and therefore there is need to ensure that ecological processes and functions are maintained in the face of the current changes which are being experienced in the environment. This extensive conservation is what is needed to make both the ecosystem and the human society sustainable (Figure 12.3).

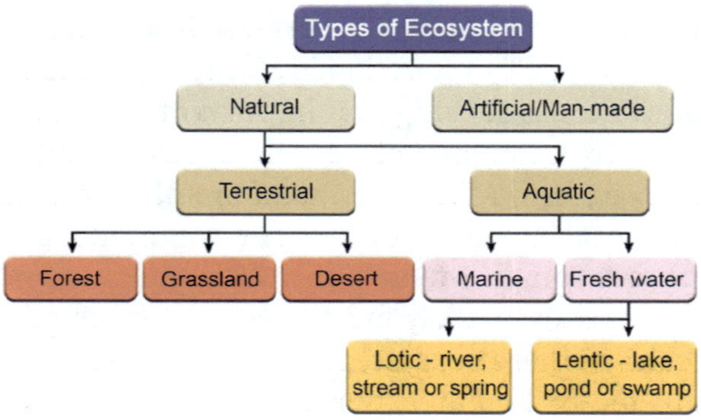

Figure 12.3. Types of ecosystems.

Source: https://www.nextias.com/blog/wp-content/uploads/2024/08/types-of-ecosystem-2.png.

12.1.1. Terrestrial Ecosystems: Forests, Grasslands, Deserts

Forests are a critical component of terrestrial ecosystems, playing an important function in the maintenance of global biodiversity as well as climate control. Characterized by closed tree canopies, forests harbor an unmatched diversity of vegetation and fauna, supporting ecological resilience as well as the provision of ecosystem services. Forest ecosystems, e.g., tropical rainforests, temperate forests, as well as boreal forests, are under threat from various drivers , such as deforestation as well as climate change. Notably, nitrogen deposition from the atmosphere poses a critical threat to plant diversity in these systems. Excess nitrogen may alter species composition by augmenting competitive interactions and diminishing the abundance of sensitive species, specifically where nitrogen is not typically limiting. Therefore, it is essential to understand the implications of nitrogen accumulation in forest societies to formulate conservation biology interventions to safeguard global biodiversity and regulate the detrimental effects of human-created environmental change (Alkemade et al., 2009). Grasslands blanket vast areas across continents and are essential ecosystems characterized by expansive open grasslands, rather than trees. These ecosystems are critical in supporting varied wildlife, including big herbivores and various birds, and also serve essential functions in carbon sequestration and soil conservation. The grasslands, though, increasingly face threats from agricultural intensification and urban expansion that degrade their ecological health. Conservation must focus on recovering and sustainably using such ecosystems to support biodiversity. Moreover, the contribution of atmospheric nitrogen to grasslands cannot be ignored as it can lead to altered composition of species and less diversity in plants. Incorporation of the critical load concept can help direct management practices in such a way that grassland ecosystems are preserved in their natural setting and the ecological services they provide are sustained (Alkemade et al., 2009). Deserts, when considered as those barren wastelands, blanket varied environments highly sensitive to environments of dryness and extremes of temperature. In contrast to grasslands and forests, deserts contain specialized vegetation and animals with remarkable water-saving adaptations and survival mechanisms in harsh environments. These environments are, however, becoming increasingly vulnerable to the impact of human activities, including urbanization and exploitation of resources. Protected areas have greater meaning when used in desert conservation because they serve as invaluable havens for endemic flora and fauna and maintain integrity within an ecosystem. The world has, in the recent past, recognized that protected areas must be created and extended to preserve varied ecosystems, including deserts. By setting measurable objectives for protected areas, particularly in hotspots of biodiversity, we can make conservation more effective and help build resilience to climate change and habitat destruction (Figure 12.4).

Figure 12.4. Terrestrial ecosystems.

Source: https://cdn0.thedailyeco.com/en/posts/7/5/0/what_is_a_terrestrial_ecosystem_1057_600.jpg.

12.1.2. Aquatic Ecosystems: Freshwater and Marine

Freshwater and marine aquatic environments are essential components of the Earths biosphere, and they sustain a wide variety of life forms. Though their marine environments occupy most of the attention, freshwater ecosystems like lakes, rivers, and wetlands do their fair share of contribution towards ecological stability and diversity. Freshwater biodiversity richness examples include the diadromous fishlike salmon and eels. However, these species are increasingly subjected to human impact through habitat alteration, pollution, and overfishing. The decline in their populations not only threatens ecological balance but also disrupts economic activities based on these fish stocks, focusing on the inevitable interdependence between human prosperity and biodiversity, as per the current condition of Western European inland fisheries (Costa-Dias et al., 2009). In contrast, marine ecosystems, covering over 70% of the Earth's surface, contain an astonishing diversity of plants and animals, contributing immensely to global biodiversity. The intricate processes of marine ecosystems regulate nutrient cycling and energy flows, which are essential for the well-being of the planet. However, these systems are seriously jeopardized by climate change, pollution, and overexploitation, which have led to potential losses in biodiversity that could impact trophic processes and biogeochemical cycles. Having sustainable practices and conservation measures in place under the various human impacts imposed upon these systems is vital. Such demand is underscored by the implications of losing key species regulating ecosystem processes, pointing to the imperative for comprehensive marine conservation practices prioritizing ecological integrity along with balancing human demands. The freshwater networks in boreal forests also give important contributions toward mitigating climate change impacts, further underscoring their multifaceted significance. These ecosystems are vital carbon sinks for storage and, as such, natural buffers against climate change and a source of essential resources for rural communities. These aquatic ecosystems are dependent upon for food, cultural, and livelihood purposes by indigenous and rural communities, an indicator of human livelihoods and ecosystem health interdependence. Measures to stem harmful industrial activity in these ecosystems are essential to uphold their integrity and their functions. As noted, conservation of such ecosystems not only ensures biodiversity but also fortifies the foundations that enable resilience to climate change and for community well-being, illustrating the profound interdependence between conservation biology and sustainable management of freshwater and marine ecosystems (Wells et al, 2011) (Figure 12.5).

Figure 12.5. Vibrant coral reef ecosystem showcasing diverse marine life and coral species, highlighting the biodiversity of aquatic ecosystems.

Source: https://www.sciencing.com/sciencing/description-four-types-aquatic-ecosystems-8145/24eef1f06b5d4f6c920f979dd4f20a3b.jpg.

12.1.3. Urban Ecosystems: Characteristics and Significance

Urban ecosystems are a multifaceted interaction of anthropogenic and natural factors, which set them apart as distinct environments in high contribution to total biodiversity. With a mix of habitats , such as parks, gardens, and green roofs, these cities host a wide range of plant and animal species. By providing important habitats between concrete structures, cities have become an important conservation role due to the capacity for supporting species that would otherwise be extinct because of habitat destruction in rural areas. The presence of greenspaces in urban areas not only enhances biodiversity but also ecological services like carbon sequestration and pollution filtration, which emphasizes the dual benefits of these ecosystems in urban areas. Further, some urban features , such as patch area and connectivity are shown to have profound effects on biodiversity and ecosystem health, and this points towards the use of sustainable urban design and development that integrates natural components organically into urban areas (Beninde et al., 2015). The value of urban ecosystems goes beyond mere biodiversity; they also enhance the health of urban residents. Green spaces provide recreational activities and improve the physical and mental health of inhabitants through provision of sites for recreation, relaxation, and socialization. It has been documented that urban dwellers with access to nature are less stressed and have improved general health outcomes, which further illustrates the intrinsic connection between human health and environmental quality. Urban ecosystems also facilitate educational activities, as they can be utilized as living laboratories for ecological studies and conservation practices. These features underscore the necessity of developing urban environments that prioritize and integrate ecological principles, with urban ecosystems being focal points in broader conservation biology practice (Alison et al., 2014).

In the face of rapid urbanization, understanding the characteristics and significance of urban ecosystems is essential for effective conservation planning and biodiversity management. Urban areas are expanding at a rate previously unseen, and this has led to fragmentation of habitats and species loss. To tackle these issues, there is a requirement for continued study of urban biodiversity and management techniques so that plans can be developed to ensure the quality and connectivity of habitats are optimized. Studies indicate that habitat management, , such as green corridor development and vegetation structure improvement, is key to the high level of biodiversity found in cities. Such interventions, in addition to countering the adverse impacts of urbanization, also promote community activism and sensitization towards ecological sustainability. With urban ecosystems advancing further, integration of ecological science into urban planning will become crucial in crafting resilient cities that coexist with nature, hence contributing to overall conservation biology goals (Beninde et al., 2015) (Figure 12.6).

Figure 12.6. Urban ecosystems.

Source: https://images.theconversation.com/files/94269/original/image-20150909-18672-539wsg.jpg?ixlib=rb-4.1.0&q=20&auto=format&w=320&fit=clip&dpr=2&usm=12&cs=strip.

Test Your Knowledge

1. What are the terrestrial ecosystems?
2. What do you understand by aquatic ecosystems?

12.2. Biodiversity and Its Importance

Learning Outcome

1. Definition of biodiversity;
2. Role of biodiversity in ecosystem stability;
3. Economic and cultural benefits of biodiversity.

Biodiversity encompasses the richness in variety in life on this planet and includes diversity within species, ecosystems, and landscapes. Biodiversity transcends its beauty; biodiversity is important in providing ecosystem stability and resilience. Healthy biological communities can better resist environmental stressors , such as climate change, natural disasters, and human activities. For instance, ecosystems with high biodiversity are likely to be more productive, which provides essential services , such as clean air, water filtration, and nutrient cycling. These ecosystem services are important for the existence of humankind and the interdependence between human health and biological diversity. While scientists examine urban ecosystems, evidence shows that conserving what is left of natural habitats within highly modified landscapes can preserve high levels of plant diversity, with implications of the importance of urban biodiversity conservation efforts (Kowarik et al., 2020).

Ultimately, at the end of it all, there is a need for sound knowledge about biodiversity function to implement effective conservation efforts to protect ecosystems and human livelihoods. Apart from intrinsic ecological motivations, ecosystem functioning holds deep implications for biodiversity. The biodiversity-ecosystem functioning concept implies that ecosystem process is largely governed by species richness and other elements of biodiversity. However, most research has centered on single ecosystems while neglecting the basic interconnectivity present between environments. It is crucial to comprehend such functional linkages to be able to more fully appreciate how biodiversity enriches ecosystem productivity and resilience, especially in coupled ecosystems. For example, terrestrial-aquatic organism exchanges of energy and matter offer channels by which biodiversity dictates ecosystem function, suggesting the importance of highly interconnected landscapes in enabling the overall well-being of ecosystems (Beisner et al., 2022).

These are important findings for conservation and restoration efforts that aim to enhance biodiversity in various ecosystems so that interventions are effective in promoting ecological integrity and balance. The value of preserving biodiversity is also appreciated by its role in ensuring food security and economic prosperity. Ecological intricate systems have a tendency to provide the fundamental resources supporting agriculture, fisheries, and forestry.

Ecosystems with high biodiversity are more likely to ensure stable populations of wild and domesticated species, hence a reliable supply of food under different environmental conditions. Loss of biodiversity will lead to increased food system exposure, since monocultures and low genetic diversity leave crops open to disease and pests. Furthermore, the economic benefits that accrue as healthy, biodiverse ecosystems are not limited to food crop production but also have utility in sectors , such as ecotourism and pharmaceutical research. These sectors rely on ecological health and species diversity to a significant degree, and therefore the economic link with biodiversity. Consequently, conservation of biodiversity is not only a matter of environmental imperative but also a key economic strategy to sustainable development (Figure 12.7).

Figure 12.7. Biodiversity and its importance.

Source: By I, Luc Viatour, CC BY-SA 3.0, https://commons.wikimedia.org/w/index.php?curid=998412

12.2.1. Definition of Biodiversity

Biodiversity is the concept that embraces all the life forms on the Earth, genetic, species and ecosystem level. These dimensions form part of a web of relationship without which there will be no ecological balance and resilience. Genetic diversity is the difference within the species, and this enables the populations to move with the changes in the environmental conditions. Species diversity entails the abundance of dissimilar species which inhabit diverse environments which facilitates stability and functionality of the ecosystem. Ecosystem diversity echoes the intricate mosaic of habitat and ecosystem that sustain life on earth. Their interdependence highlighted by the fact that these layers have on each other, shows that biodiversity is not just important to the ecological health but also to the human sustenance and health. A critical part of biodiversity comprehension is its implication regarding ecosystem services, or the benefits people get out of nature. The international year of biodiversity 2010 raised awareness on the connection between biodiversity and human well-being, and the important services given to us by the ecosystems, including food provision and provision of clean water. Of special importance is soil biodiversity because many ecosystem processes rely on it, , such as nutrient cycling or soil formation because these processes are driven by the multitude of organisms that live in the soil. By recognizing the integral role of biodiversity in providing these essential services, we can appreciate the urgent need for conservation initiatives that sustain and protect the myriad forms of life that contribute to human existence and ecological integrity (Van der Putten et al., 2010). Furthermore, biodiversity has links with cultural diversity that results in an in-depth perception of the conservation issues and possibilities. The distinctions that existed in the past between nature and culture have in many cases acted as barriers to the acknowledgement of the connections that existed between biological and cultural diversity. This makes conservation harder, because local knowledge and cultural practice usually guide sustainable relationships with the environment. New knowledge indicates that biological diversity loss is intimately connected with disappearance of cultural practices and languages through which ecosystems are informed and guided in their management. Consequently, next-generation conservation policy should be based on an integrated approach in which biological and cultural diversity are equally appreciated as the key to promoting resilient ecosystems. By recognizing these intricate connections, we can develop strategies that nurture both the natural world and the cultural narratives that sustain human societies (Adams et al., 2009).

12.2.2. Role of Biodiversity in Ecosystem Stability

The intricate relationship between ecosystem stability and biodiversity lies at the heart of ecological integrity maintenance. Diversity of species and genetic variation within an ecosystem is critical to building resilience to environmental perturbations. For instance, ecosystems with a greater number of species are generally more capable of withstanding changes , such as climate fluctuations or foreign species invasions. Such systems also exhibit compensatory dynamics where the removal of a species can be replaced by others playing a comparable ecological function and hence be capable of maintaining overall stability. This dependence in ecology illustrates how diverse ecosystems are more capable of handling perturbations compared to their counterparts that are impoverished. It is significant to realize this relevance not only for conservation but also for planning how to reverse the impact of human activity and climate change on biodiversity, emphasizing the need to preserve varied habitats to preserve long-term ecological balance. Additionally, biodiversity contributes to ecosystem services that are vital to the existence of humans, including clean air, purification of water, and climate regulation. It is the complicated interactions among various species that make such services feasible, making them more than byproducts of ecological interactions; they are essential to sustaining life and economies. For example, salt marsh systems, as emphasized in the case of long-term research at Mont-Saint-Michel Bay, were until recently unappreciated. Around 20 years of research unveiled their indispensable role in keeping estuarine systems together ecologically, socially, and economically. The marshes are not only habitats for different species; they also protect coastlines, remove pollutants, and act as carbon sinks. The intricate role of biodiversity in such a scenario point to its importance in the development of ecosystem resistance and the urgency for anticipatory conservation measures to protect such precious assets from ongoing environmental changes (Bouchard et al., 2003). Further, the evolutionary dynamics combined with biodiversity is important in determining how ecosystems function. Climate change, for instance, shapes ecological community organization both as an agent and byproduct of evolutionary processes , such as adaptation and species extinction. When ecosystems are subjected to altered climatic regimes, species fortunes are determined by their ability to adapt to new situations. This has significant implications for ecosystem stability because such adaptations, whether through natural selection or genetic variation, have the potential to stabilize or destabilize community resistance. The character of the interactions necessitates multi-faceted conservation strategies that address not only current biodiversity loss but also ecological problems of the future. The recognition of the role of biodiversity as a driver of stability in ecosystems allows conservation biology to better direct processes for ensuring ecological resilience in the context of altered environmental stress (Stewart, 2010) (Figure 12.8).

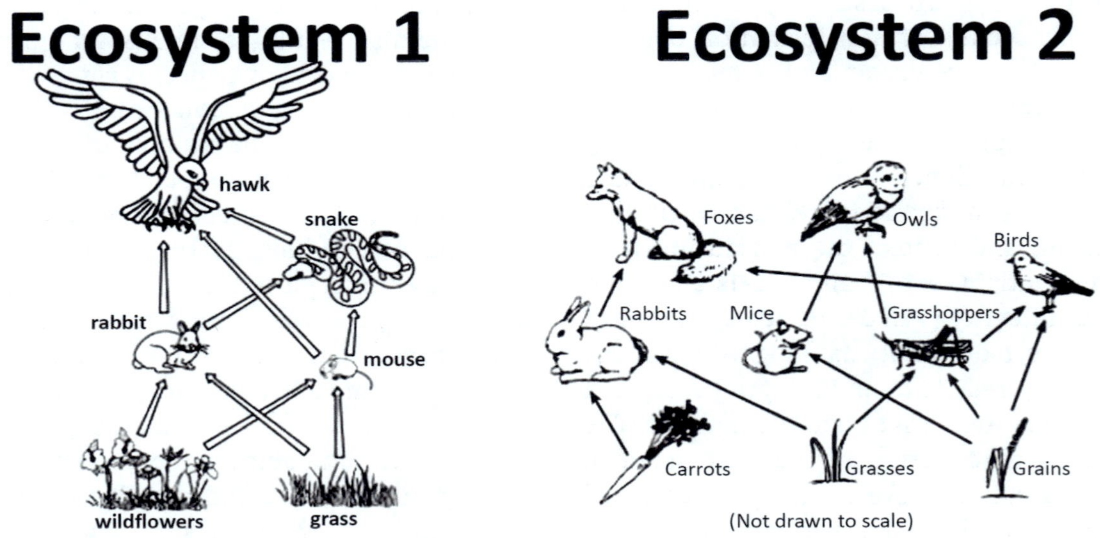

Figure 12.8. Biodiversity and stability.

Source: https://media.baamboozle.com/uploads/images/680326/1647436066_44344.jpeg.

12.2.3. Economic and Cultural Benefits of Biodiversity

The financial worth of biodiversity is various and plays a very important function in sustenance of human welfare. Biodiversity sustains various industries, including agriculture, medicine, and tourism, by providing the genetic resources required for producing disease-resistant agriculture and discovering new medicine. Agriculture relies particularly on a vast array of species to ensure sustenance security and disease and climate change resilience. Also, biodiversity is credited with ecosystem services that augment agricultural productivity, , such as pollination, pest control, and soil fertility. Decline of these ecological services due to decline in biodiversity has the potential to significantly impact economic outputs, citing the urgency of integrative conservation approaches that embrace the inherent economic worth that biodiversity enjoys in economic systems (Beyene et al., 2016). In addition to economic returns, biodiversity holds profound cultural value to many groups of people across the world. Cultures are often inextricably linked with local ecosystems, and diverse practice, tradition, and identity with natural systems and species they host. For Indigenous people, biodiversity is not biological diversity; it is part of their cultural tradition, practice, and belief, demonstrating the ethical imperative to safeguard these natural resources. Loss of biodiversity threatens such cultural constructs and can lead to the decay of customary knowledge systems. This highlights the need for applying conservation interventions that engage local communities and respect cultural values so that biodiversity conservation enhances ecological integrity as well as cultural continuity (Beyene et al., 2016). Lastly, integrating biodiversity into urban design and rural planning is essential in a bid to create sustainable communities that are both economically and culturally prosperous. Green infrastructure approaches that reconcile rural and urban areas foster synergies between the two, enabling better balancing of human demands with awareness of nature. Urban societies will also benefit from additional green spaces, which enhance living standards through recreational activities, higher air quality, and protection of biodiversity. At the same time, countryside regions can use their natural resources to develop eco-tourism and sustainable agriculture. Such comprehensive methods not only enhance economic growth but also enhance the cultural importance of biodiversity by promoting community stewardship and appreciation of native ecosystems.

Test Your Knowledge

1. Define biodiversity.
2. What is the role of biodiversity in ecosystem stability?

12.3. Threats to Ecosystems

Learning Outcome

1. Habitat destruction and fragmentation;
2. Climate change and its impact on ecosystems;
3. Pollution and its effects on biodiversity.

Ecosystems, whose role is essential in the preservation of biodiversity and provision of vital services, are vulnerable to various factors that undermine their integrity and processes. Habitat destruction, pollution, climate change, and invasive alien species are some of the key drivers of ecosystem degradation. Habitat destruction due to predominantly urbanization, agriculture, and deforestation leads to massive loss of diversity and impairment of ecological processes. Such fragmentation not only causes the rest of the species to struggle more to survive, but it also decreases the ability of ecosystems to provide crucial services , such as clean air and water, carbon sequestration, and soil fertility. The interdependence between various elements in an ecosystem makes the necessity to minimize such detrimental activities even greater. Mitigation of these threats is crucial to conservation biology, which seeks to understand and apply mechanisms for the maintenance of biodiversity and ecosystem resilience in relation to external stress. Pollution is yet another profound threat to ecosystems that occurs in many forms , such as chemical runoff, plastic litter, and air pollutants. These pollutants deteriorate the state of soil, water, and air, adversely affecting plant and animal health, which can further result in reduced biodiversity. The relationship between soil biodiversity and ecosystem services highlights the importance of maintaining soil health as an integral aspect of ecological balance. Soil organisms play a key role in recycling nutrients and regulating environmental conditions necessary for plant growth and survival. To this, climate change often adds its effects, further increasing the burden on ecosystems to be more susceptible to collapse. The recognition of these intertwined threats is important in the development of effective conservation practices that are in balance between ecosystem conservation and restoration programs. The development of tools , such as the IUCN RLE is a proactive measure to assess and combat such threats. This approach aims to provide a structured framework for quantifying the risk of ecosystem decline to guide policy and management decisions. Integrating a comprehensive analysis of ecosystem dynamics, degradation, and potential collapse, the Red List enhances understanding of threats to ecosystems and underscores the significance of individual conservation interventions. Employment of such tests not only assists in identifying vulnerable ecosystems but also creates public awareness about the need for immediate action. With the right policies and good management practice prioritizing ecosystem preservation, we can contain the existing threats to ecosystems and enable the natural services upon which human society remains strongly reliant, as profiled in biodiversity and soil ecosystem reports (Van der Putten et al., 2010) and in the IUCNs endorsement of risk assessment processes (Adler et al., 2015) (Figure 12.9).

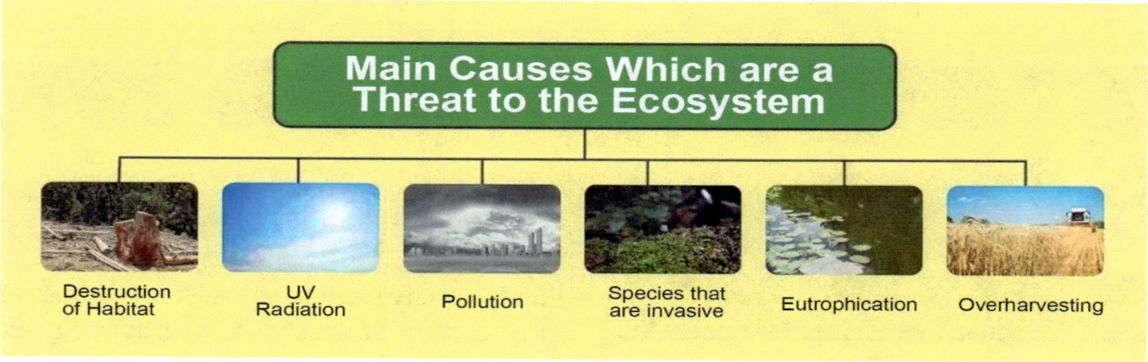

Figure 12.9. State the main causes which are a threat to the ecosystem.

Source: https://media.geeksforgeeks.org/wp-content/uploads/20230615175120/main-causes-which-are-a-threat-to-the-ecosystem-(2).webp.

12.3.1. Habitat Destruction and Fragmentation

Habitat degradation and fragmentation are major threats to biodiversity and ecosystem stability and are among the biggest concerns in conservation biology science. The natural habitats of species tend to degenerate due to activities , such as agriculture, urbanization, and deforestation, disrupting the fragile balance that ecosystems maintain. Such habitat collapse translates to the reduction of a species' population, particularly those species dependent on large areas or certain conditions for survival. For instance, migratory fish species like shads and eels have been observed to undergo extreme declines due to habitat degradation and water quality issues arising due to both overfishing and invasive introductions (Costa-Dias et al., 2009). Decline of such keystone species significantly affects ecological processes like nutrient cycling and trophic dynamics, reflecting an alarming shift towards ecosystem health. The fragmentation of the habitat in the majority of instances results in fragmented ecosystems, which are significantly limited in the movement of species between habitats, and thus genetic diversity is lost, while extinction risk increases. This was starkly evident in aquatic ecosystems where human settlement fragmented natural channels to give water bodies that limited the migratory nature of the majority of aquatic species (Costa-Dias et al., 2009). The impact of affected species goes beyond their individual populations, as their loss can destabilize an entire ecosystem. Additionally, the effects of habitat fragmentation are magnified by climate change, providing additional stressors that organisms have to contend with. Given these challenges, conservation needs to concentrate on creating wildlife corridors and safeguarding key habitats to ensure species connectivity and resilience. In the context of ecosystems in the world, including those found in boreal forests, there is no need higher than the maintenance of intact habitats. Such ecosystems are not only essential to biodiversity conservation but also to the provision of vital services , such as climate regulation and rural community provision of resources. Such degradation can therefore have calamitous economic and cultural impacts, especially for individuals who sustain their livelihoods based on natural resources. Effective conservation efforts should incorporate sustainable land-use policies that mitigate the impact of industrial activity while advancing both human and ecosystem stability. Through a multi-phased approach of incorporating habitat conservation, restoration, and community participation, one may curb the multidimensionality of habitat loss and fragmentation and construct a sustainable future for human and biodiversity populations (Figure 12.10).

Figure 12.10. Roosevelt elk in a deforested area, highlighting the impact of logging on wildlife habitats and forest ecosystems.

Source: https://www.treehugger.com/thmb/erXhOrTZwpMOJC0ofPEyBBomg3E=/1500x0/filters:no_upscale():max_bytes(150000):strip_icc()/roosevelt-elk-cervus-elaphus-roosevelti-in-a-cut-forest-east-coast-near-telegraph-cove-vancouver-island-british-columbia-canada-july-590532428-5897a5553df78caebc299a6b.jpg.

12.3.2. Climate Change and Its Impact on Ecosystems

Climate change has many dimensions of influence, which create serious problems in the ecosystem of the world in terms of biodiversity and the complexity of life. With the increase in temperature and alteration in the weather patterns, the species are triggered to migrate or adjust, thus causing a transformation in the community structure. Because ecosystems are dependent upon a variety of biological interactions, they are particularly vulnerable since such changes may cause food web disruption and habitat availability changes. Also, loss of biodiversity enhances adverse effects of climate change because when there is no genetic variation, a species capacity to adjust to new surroundings is minimal. Consequently, the necessity to conceptualize climate change as a driver of ecological disturbance is critical to the field of conservation biology, hence the importance of implementing proactive measures that would serve as interventionist approaches to both climate mitigatory practices and biodiversity protection. The consequences of the climate change are not just the migration of the species; it triggers a wave of the evolutionary processes that may endanger the integrity of the ecosystem. The past precedent, e.g., the changes in the Quaternary, shows the impact of the climate oscillations on the evolution, adaptation, and extinction of the species spuriously. With the changing of the ecosystems due to the climatic pressures, some species will not be able to adjust, and they will either decline in numbers or become extinct, reducing the biodiversity. Such loss may be cascading, creating vulnerability in the ecosystem, and diminishing the capacity of surviving species to prosper. Here, scientists and conservationists are important to highlight the threatened species and apply the protection strategies that may help to build an adaptive resilience in the ecosystems and ensure their sustainability in the period of climatic uncertainty. Education on strategic community empowerment can be identified as one of the essential strategies to address the constant danger of climate change on ecosystems. By increasing biological literacy through programs, people can get valuable knowledge about how humans and nature are connected, and this helps people to better appreciate biodiversity. Environmental awareness can result in behavior change and the mobilization of communities to take part in conservation efforts, which would further support the resilience of the ecosystem. As an example, local initiatives, which promote afforestation, waste disposal, and the preservation of endemic species, can play a significant role in the reduction of climate effects on the local level. By harnessing the collective efforts of communities, it becomes possible to not only conserve existing biodiversity but also adaptively manage ecosystems, aligning ecological conservation objectives with societal needs as climate challenges continue to intensify (Stewart, 2010; Attruk et al., 2024) (Figure 12.11).

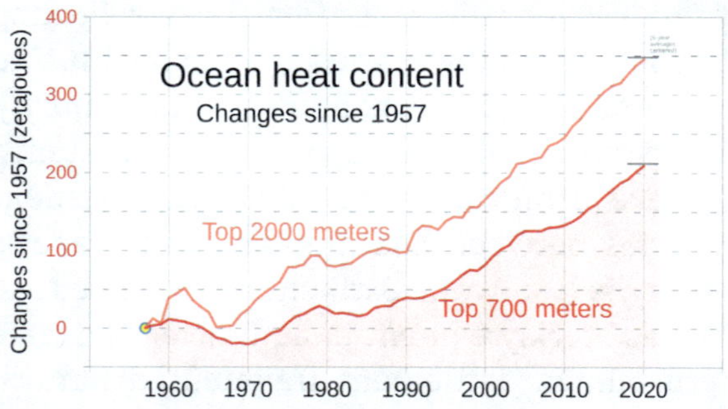

Figure 12.11. Climate change and its impact on ecosystems.

Source: By RCraig09 - Own work, CC BY-SA 4.0, https://commons.wikimedia.org/w/index.php?curid=129163862

12.3.3. Pollution and Its Effects on Biodiversity

Pollution becomes a harsh threat to biodiversity, impacting ecosystems in a sophisticated variety of ways. Soil pollution is often the most insidious of all types of pollution because it disrupts the sensitive balance of life that it supports. Soil biodiversity underpins ecosystem processes that support human livelihoods, , such as agriculture and climate regulation. Healthy soil, as underscored in European Commission (Environment) report, is essential to sustainable human existence and the survival of varied kinds of life that render ecosystems resilient. Soil quality change resulting from pollutants not only poses danger to the organisms inhabiting it but also diminishes the lands capacity to provide key ecological services, hence having broader implications for society and nature as a whole (Van der Putten et al., 2010). Hence, understanding and mitigating pollutions impact on soil biodiversity is critical to establishing resilient ecosystems. Atmospheric pollution by nitrogen deposition is another aspect of complexity to the existing biodiversity crisis. Terrestrial ecosystem nitrogen accumulation has been recognized as a significant threat that could potentially restructure species composition in ecosystems across the globe. Existing estimates indicate that the detrimental effects of nitrogen deposition are witnessed differently across ecosystems, from arctic tundras to rainforests. Nitrogen excess could lead to competitive inequalities towards the advantage of some few privileged species on the cost of other species, thus lowering plant diversity. Just as was observed in the results of terrestrial plant responses, the phenomenon can lead to drastic changes in ecosystem processes and between-species interactions, which are essential for ensuring biodiversity (Alkemade et al., 2009). Fighting nitrogen pollution through conservation policies that are well informed becomes necessary in the safeguarding of vulnerable ecosystems and the preservation of their long-term health. The double dynamic between biodiversity and pollution serves to emphasize the requirement of inclusive conservation action that fights multiple ecological stresses. With ecosystems exposed to the double impact of soil pollution and atmospheric deposition, conservation biology is compelled to go whole systems in its strategy that recognizes interdependence of species and their niches. For instance, the protection of sites with varied plants and soils not only protects particular species but also enhances ecosystems as a whole in terms of their resilience against the adverse effects of pollution. Policy and educational measures to reduce exposure to pollution and encourage sustainable means are necessary for this. By prioritizing in the context of biodiversity's well-being, we create a platform on which ecosystems can thrive and continue to provide services that are essential and thus provide a sustainable future for both natural communities and human societies (Alkemade et al., 2009; Van der Putten et al., 2010) (Figure 12.12).

Figure 12.12. Large accumulation of plastic and solid waste at the edge of a forest illustrates the severe impact of pollution on natural ecosystems and biodiversity.

Source: https://rolleatusa.com/wp-content/uploads/2024/02/pollution-and-biodiversity-1200px-x-800px.jpg.

Test Your Knowledge

1. What is climate change and its impact on ecosystems?
2. What is pollution and its effects on biodiversity?

12.4. Conservation Biology: An Overview

Learning Outcome

1. Definition and goals of conservation biology;
2. Historical context and development of the field;
3. Key principles guiding conservation efforts.

Conservation biology is a critical field of research that addresses the mounting challenges facing ecosystems around the world. With ongoing human activity placing a burden on biodiversity, it becomes crucial to understand the mechanisms determining ecosystem resilience. Freshwater ecosystems provide a good case in point for the necessity of conservation since they account for less than 1% of the Earth's surface but support more than 10% of all animal species that are known to science. The quick decline in overall biodiversity as well as freshwater species has attracted increasing attention from scientists as well as policymakers, leading to the launch of various programs aimed at conservation and the enhancement of ecosystem services. These monitoring activities and research projects are generating substantial amounts of data, and intuitive platforms like the Freshwater Information Platform to make it easier to access information of this nature have become a necessary requirement (Bremerich et al., 2019). In measuring conservation work, resilience appears as an abstract and multidimensional goal at times. Resilience comprises not just the capacity of ecosystems to recover from perturbations but also drivers , such as biodiversity, network structures, and adaptive capacities. Identification of these for effective management of ecosystems is crucial. Researchers have proposed the resilience trinity framework to avoid dealing with diverse contexts and time scales in ecosystem management. This conceptualization defines three central contexts of decision-making: reactive, adjustive, and provident. Through the identification of these contexts, conservation professionals can prioritize that maintains ecosystem services amidst impending near-term threats while waiting out long-term uncertainties (Auge et al., 2020). Familiarity with the diversity of interpretations of resilience is essential for the formulation of functional conservation strategies. Last, ecosystem conservation dynamics establish the importance of informative and goal-oriented interventions.

Figure 12.13. Conservation biology.

Source: https://www.researchgate.net/profile/Oluwakayode-Coker/publication/336124583/figure/fig2/AS:808310399832065@1569727625807/Structure-of-conservation-biology-and-the-position-of-genetics-in-it-Source-Frankham-et.ppm.

Effective conservation relies on an integrated understanding of ecological processes and the drivers of ecosystem functioning. To address these complexities is critical for preserving biodiversity as well as for ensuring the sustainability of ecosystem services upon which human societies rely. Continuing collaborative efforts between scientists, policymakers, and the public are necessary to ensure that conservation measures become more effective.

As our understanding of ecological resilience evolves, so too must our strategies, allowing us to adapt to both imminent threats and longer-term challenges. Such an integrative approach will serve as the foundation for fostering healthy ecosystems and ensuring their vitality for generations to come, reinforcing the imperative of conservation biology in today's environmental landscape (Figure 12.13).

12.4.1. Definition and Goals of Conservation Biology

Conservation biology entails the scientific study of conservation and management of biodiversity whose main objective is protecting ecosystems, species, and genetic diversity from threats to extinction and degradation. Conservation biology is an interdisciplinary approach that integrates insights from ecology, genetics, as well as social sciences to develop plans towards making natural environments sustainable. In its core, conservation biology aims to prevent human-inflicted loss of biodiversity in the form of habitat destruction, pollution, invasive species, and climate change. Conservation biologists strive to find effective management practices by understanding intricate relationships within ecosystems, thus allowing ecosystems to cope with continuous environmental changes while maintaining their ecological integrity. Conservation biology goals extend beyond conserving species alone; it also entails restoring existing habitats as well as promoting sustainable utilization that involves local populations. At the heart of planning for conservation is the development of ecosystems and species management plans that are often informed by frameworks , such as protected area designations, as in the case of the development of special protection areas (SPA) and sites of community importance (SCI) within the European Union's Natura 2000 network (Apeldoorn et al., 2009). Such activities attempt to follow management strategies that not only seek to conserve species but also address the overall health of the habitats in which such species reside. Collaboration among government agencies, NGOs, and local communities is essential to promote a sense of shared stewardship and promote biodiversity conservation goals across various landscapes. In addition, conservation biology emphasizes the significance of integrating scientific knowledge with practical applications in addressing complex problems of ecosystems. For instance, knowledge about how invasive species work can be beneficial for repairing native ecosystems, enabling the coexistence of non-native and native species (Esler et al., 2012). Conservation practices incorporating outcomes from both invasion and restoration ecology can lead to more integrative approaches, minimizing the detrimental impacts of alien species while enhancing ecosystem recovery. By promoting more interdisciplinary collaboration, conservation biology can provide a more advanced perspective on ecosystem management, ensuring that different ecological needs are met while progressing toward larger ends of conservation. Finally, the fusion of practice and research is the foundation of successful decision-making in the quest to safeguard the planet's rich biological heritage.

12.4.2. Historical Context and Development of the Field

The history of conservation biology and ecosystems is marked by a growing understanding of the complex interdependence between organisms and their surroundings. The mid-20th century was the turning point when ecological science coalesced around the concept of ecosystems as dynamic systems rather than as collections of species. Early studies were centered on species interdependence, guided by pioneering work from scientists , such as Arthur Tansley, who defined the term ecosystem in 1935. This model served as the basis for the subsequent study of the conservation practice that later gained prominence with industrialization and urbanization threatening diverse habitats. In contemporary conservation practice, though, the science has expanded to include systemic levels that consider both ecological integrity and human socio-economic processes, effectively integrating natural systems and anthropogenic processes. The creation of global networks and collaborative scientific efforts has tremendously developed the science of conservation biology. Specifically, the Assembling the Tree of Life in Europe (AToLE) initiative emphasized international coordination for creating solutions to maintaining biodiversity through heightened ecological awareness and collaborative research (Hassanin et al., 2009). The initiative indicates recognition of global interconnectivity of ecosystems, fostering more integration in conserving ecologies. Moreover, as the field matured, scientists and policymakers began to adopt holistic measures towards ecosystem management, addressing complex issues , such as climate change and loss of habitats. The crisis caused by declining biodiversity has indicated a shift towards a paradigm that integrates ecological, economic, and social aspects of conservation, which are reflective of an adaptive approach that fulfills contemporary environmental needs. The IUCN's development of new Red List criteria is an institutionalization of ecosystem analysis standards, pointing to the advancement in conservation biology (Adler et al., 2015). The criteria were designed to maximize the reliability and applicability of ecosystem-level risk analyses, aiming to strike a balance between complexity and utility in terms of policy application. This shift in methodology is a reflection of the field's continuing evolution amidst growing demands for effective conservation measures. Through experience, namely repetitive refinement of the Red List of Threatened Species, conservation biology can now promote more holistic policies aimed at sustainable outcomes for ecosystems. As the discipline continues to evolve, sustained action will ensure conservation policy remains responsive to the ecosystem's needs and promotes resilience to anthropogenic pressures.

12.4.3. Key Principles Guiding Conservation Efforts

The pillar of any sound conservation measure can be summed up in the understanding of the inherent value of biodiversity and its importance in the stability of the ecosystem. Conservation efforts should focus on preserving the ecological relationships that enable species to flourish so that they become resilient to environmental fluctuations. This needs a multi-pronged solution that takes into account the intricate interdependencies between flora, fauna, and their habitats. As an example, it is important to preserve the native aquatic life and regulate the introduction of better fish strains through fish farming to ensure the protection of the biodiversity in the vulnerable ecosystems especially in areas , such as Africa that have great freshwater potential. By focusing on the conservation of native species and the ecosystems they inhabit, the risks of invasive species undermining local biodiversity can be mitigated, ensuring a holistic approach to natural resource management and conservation efforts (Gupta et al., 2004). Another vital principle guiding conservation efforts is the integration of ecological restoration techniques, specifically through strategies , such as forest landscape restoration (FLR). This is a strategy aimed at correcting the destruction of forest landscape and at the same time improving the socioeconomic welfare of the locals. The adaptive application of FLR is the key to its success and it may range between the passive restoration of the natural processes and the active reforestation programs. Studies have supported the significance of developing effective community participation and creating local capabilities to ensure that the restoration efforts are sustainable and successful. Moreover, the proper institutional support and regulation frameworks can also play a vital role in transcending the difficulties related to the involvement of local stakeholders. Ultimately, embracing FLR and its principles can lead to improved ecosystem services, making it a fundamental component of contemporary conservation biology (del Castillo et al., 2012). The future progress of conservation initiatives requires a collaborative and inclusive approach, especially when trying to resolve various problems that ecosystems across the world have to deal with. The involvement of a broad range of stakeholders (governmental agencies, non-profit organizations, and local communities) will help to make sure that conservation strategies are informed and practically useful. Indigenous knowledge combined with scientific research can improve adaptive management practices to create resilience in shifting environmental landscapes. Besides, building fair partnerships will enable the local stakeholders to actively participate in decision-making mechanisms, which will increase the sense of responsibility and action. As a result, conservation efforts can be designed in such a way that they address the local needs and at the same time facilitate the greater ecological stability. By emphasizing collaboration and inclusivity, the conservation community can more effectively navigate the complexities of ecological management and bolster the resilience of the world's ecosystems (Gupta et al., 2004; del Castillo et al., 2012) (Figure 12.14).

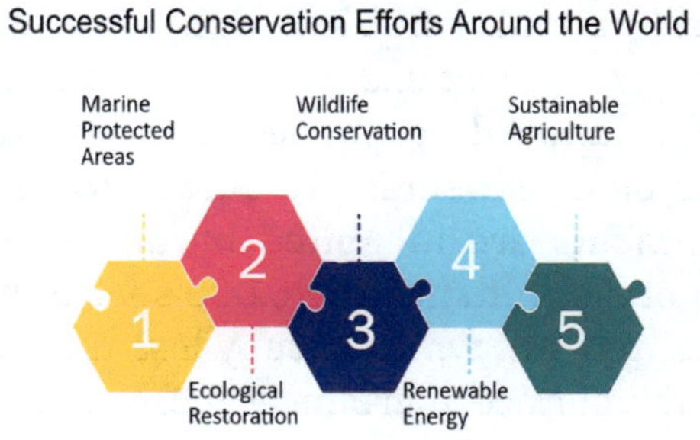

Figure 12.14. Key principles guiding conservation efforts.

Source: https://fastercapital.co/i/Conservation--Preserving-Arer-s-Beauty--The-Importance-of-Conservation--Successful-Conservation-Efforts-Around-the-World.webp.

Test Your Knowledge

1. What are the goals of conservation biology?
2. What are the key principles guiding conservation efforts?

12.5. Conservation Strategies

Learning Outcome

1. Protected areas and wildlife reserves;
2. Restoration ecology and habitat rehabilitation;
3. Sustainable practices in agriculture and fisheries.

The need for proper conservation practices has been emphasized by increasing anthropogenic impacts on ecosystems. The present paradigms have to be enhanced to fit the rapidly changing environments faced by many species. Traditional practices of ecosystem conservation in their native states are unsustainable. Conservation biology is shifting to an adaptive management paradigm with a focus on ecosystem resilience and functional competence instead. This transition requires an encompassing insight into long-term ecological processes that regulate species and habitat relations. Further, resolving the cumulative risks of increasing human population, overconsumption, pollution, and climate change is crucial for integrated conservation measures. By merging emergent findings from numerous scientific fields, next-generation conservation measures can become firmer and more effective in maintaining ecosystem persistence and resilience in the context of evolving environmental processes (Ackerly et al., 2017). Conservation of freshwater ecosystems is an example of the need for tailored responses to acknowledge specific environmental pressures. Freshwater environments are inhabited by an incredible variety of life, with nearly 36,000 species of fish documented, but which occupy less than 1% of the Earth's surface. Alarmingly, these ecosystems are also being subjected to unprecedented stress through habitat loss, contamination, and effects of climate change. The rapid loss of wetlands—three times as rapid as forest loss—emphasizes the importance of innovative conservation measures. Effective conservation schemes will have to be based on a thorough knowledge of the ecological needs of freshwater organisms, their vulnerabilities, and determinants of resilience. Interdisciplinary cooperation among specialists will be crucial to synthesize ecological data with socio-economic parameters in developing holistic management plans. Without adequate investment in the research in these freshwater environments, conservation efforts may fall behind schedule, imperiling the biodiversity that sustains these important systems (Miranda, 2022). Successful action of conservation interventions will require an interdisciplinary framework that conciliates ecological needs with people's interests. Conservation organizations and governments need to come together to craft sustainable goals that recognize complexity that is involved in conserving biodiversity. The process needs to establish frameworks that place a premium on effective resource allocation and local people's participation, with the active inclusion of local stakeholders as participatory partners in conservation. Closing knowledge gaps in the region also needs to be done; decision-making strategies must be informed by ongoing research and up-to-date scientific understanding of the operation of nature. Encouraging collaboration among scientists, decision-makers, and Indigenous peoples will ensure socially more inclusive and ecologically more suitable conservation strategies. Lastly, fostering public awareness and commitment towards conservation of biodiversity will be important to create a sustainable future where societies learn to coexist in harmony with nature (Ackerly et al., 2017; Miranda , 2022) (Figure 12.15).

Figure 12.15. Conservation strategies.

Source: https://fastercapital.co/i/Land-legacy--How-to-Leave-a-Legacy-with-Your-Land-and-Impact-Future-Generations--Conservation-Strategies-and-Practices.webp.

12.5.1. Protected Areas and Wildlife Reserves

Protected areas and wildlife reserves are key elements within the greater context of ecosystem and conservation biology. These designated areas function to preserve habitats and the respective wildlife, reversing the effects of habitat loss, climate disruption, and exploitative practices. By acting as refuges for wildlife, these locations foster biodiversity and help ensure the ecological well-being of the world. In addition to sheltering endemic species, they ensure ecosystem resilience so that natural processes can continue unbroken. Reserving areas is necessary for species like the western gorilla, whose survival relies on varying food supplies, which are normally under threat due to human interference. Understanding such species' diets and seasonal change can inform conservation, stressing the importance of protected areas in avoidance of risks associated with dwindling wildlife populations (Bashir et al., 2015). The success of marine protected areas (MPAs) shows how legal frameworks and community engagement are capable of creating effective conservation activities. The goal of MPAs is to preserve marine biodiversity by limiting the activities of human beings that harm marine ecosystems. Countries , such as Canada have used this approach via legislation , such as the Oceans Act, which enables local communities to take a frontline role in marine conservation. Activities studied in these MPAs, including those at Gilbert Bay and Eastport, show that tight interdependent relationships between government agencies, local inhabitants, and the scientific community can deliver considerable ecological benefits. MPA establishment also holds special opportunities to study marine wildlife and the interaction of wildlife with the ecosystem, further underscoring proof that science led by or for indigenous locals can expand knowledge about environmental complexity (Jones et al., 2007). But future initiatives for protected areas and wildlife reserves are contingent upon adaptive management strategies that respond to changing ecological and social conditions. With growing human pressures, balancing conservation objectives with surrounding socioeconomic conditions becomes more imperative for surrounding communities. Strategies must promote not just the conservation interests of biodiversity but also in a way that there are real benefits accruing to local people from the existence of protected areas. New strategies, like sustainable tourism and community-based conservation projects, can indeed empower local stakeholders and also build a sense of ownership of natural resources. These efforts can lead to improved conservation of protected areas, solidifying their status as keystones to ecosystem conservation and the associated biodiversity (Bashir et al., 2015) (Figure 12.16).

Figure 12.16. A Bengal tiger crossing a forest road in a protected wildlife reserve, illustrating the intersection of conservation and human infrastructure.

Source: https://static.toiimg.com/img/98858208/Master.jpg.

12.5.2. Restoration Ecology and Habitat Rehabilitation

The discipline of restoration ecology is concerned with restoring degraded and destroyed ecosystems that result from anthropogenic activities to reinstate their natural function and biodiversity. The discipline works on the belief that ecosystems can heal through an integration of passive and active restoration methods, which can involve habitat reconstruction, reintroduction of species, and restoration of natural processes. One of the most important elements of restoration is knowing the ecological past of the site, as well as the particular species and relationships that formerly inhabited the ecosystem. By focusing on these factors, ecologists will be able to better create interventions that are not merely effective in restoring the habitat but also long-term sustainable. But as observed in numerous projects worldwide, ecosystem complexity tends to surprise, and it is for this reason that adaptive management methods must be undertaken at all levels of restoration. An effective method utilized in restoration ecology is the application of Ecohydrology, where physical factors in an ecosystem are modified to enable ecological recovery. Ecoengineering seeks to create proper conditions for biotic colonization by restoring hydrological and sedimentological conditions in the ecosystem, ultimately leading to the reactivation of ecological processes. For instance, in estuaries, successful Ecoengineering has led to the rehabilitation of various aquatic fauna, demonstrating the positive impacts of such measures in the recovery of habitats. However, not all restoration efforts are positive. While there will be some gains in the immediate term, these may also contribute to creating habitats that are not entirely suitable for the hosting of native species interactions, thus calling into question the sustainability of the measure (Elliott et al., 2016). In addition, these measures highlight a key trade-off between ecological success in the short and long term. Terrestrial fisheries decline in critical species also underlines the imperative for target restoration in conservation biology. The majority of diadromous fishes, , such as salmonids and European eel, have undergone extreme declines in population due to habitat changes altered by humans, degrading water quality, and overfishing. The species are important components of nutrient cycling and trophic function, and their extinction can severely impair ecosystem functioning. While restoration ecology attempts to turn these declines around, it does so with an eye towards making sure that its methods prioritize balancing biodiversity upkeep with financial considerations. Proposals for habitat restoration management for the purposes of recovering environments for these crucial fish species place on the table the delicate balance conservationists must navigate between ecological health and human impact (Costa-Dias et al., 2009). In searching for such restoration services, one should be aware of the particular demand of different species so that not only are they restored but also aquatic habitats' strength (Figure 12.17).

Figure 12.17. Ecological restoration.

Source: https://cdn.britannica.com/88/224788-050-3E0DF400/Forester-at-the-pine-tree-plant-for-reforestation-in-the-forest.jpg.

12.5.3. Sustainable Practices in Agriculture and Fisheries

Sustainable fisheries and agricultural practices are the key strategies to maintaining ecological balance while meeting human needs. Agronomic practices , such as crop rotation, polyculture, and agroforestry in agriculture have aims of enhancing soil health, biodiversity, and reducing dependence upon chemical pesticides and fertilizers. These practices build a lasting ecosystem that is able to withstand pests and disease and still add organic matter to the soil. Also, the inclusion of sustainable livestock management assists in reducing greenhouse gas emissions as well as enhancing animal welfare. Likewise, for fisheries, sustainable techniques focus on fish stock management via ecosystem approaches whose aims are to put an end to overfishing and habitat destruction. Thus is the emphasis on achieving the interconnectedness of ecosystems and realizing that the well-being of the life of the ocean influences not only fisheries but coastal communities whose living depends on these resources. EBM or ecosystem-based management of fisheries is a revolutionary approach to handling the complexity of marine ecosystems. EBM relies on a thorough understanding of the way that fisheries are linked to other ecological processes, considering the effects of other human activities aside from fishing itself. Adaptive management practices flexible enough to respond to changing environmental and socio-economic conditions are one of the key principles of EBM. The FAO Code of Conduct for Responsible Fisheries serves as an essential blueprint in such a context, establishing the ground rules under which the use of EBM can be carried out effectively. By linking management to ecological realities, we can open up opportunities to maintain both fish stocks and the people whose livelihoods they represent. The challenges of implementing EBM underscore the need for stakeholder cooperation, for instance, among national fishers, government organizations, and environmental organizations, to undertake overall management objectives (Zerbi et al., 2003). Research activities, , such as conducted by the Rockefeller Foundation, indicate the need for cross-disciplinary answers to resolve the pressing concerns of fisheries and farm systems. By examining the effects of deteriorating ocean health as a result of overfishing, climate change, and pollution, these efforts aim to help guide future interventions. Knowing how vulnerable communities rely on marine ecosystems for food and livelihood provides a more targeted intervention to ensure sustainability. The intersection of policies, local management practices, and innovative financing gives rise to an integrated framework that can serve both marine conservation and community resilience (Goldberger et al., 2013). The integrated perspective calls for reconciling human well-being and environmental health and demonstrates the sustainable agriculture and fishery practices are not only preferable but essential to the future food security at the global stage and environmental sustainability.

Test Your Knowledge

1. Differentiate restoration ecology and habitat rehabilitation.
2. What are the sustainable practices in agriculture and fisheries?

12.6. Role of Technology in Conservation

Learning Outcome

1. Use of remote sensing and GIS in ecosystem monitoring;
2. Genetic tools for species conservation;
3. Citizen science and public engagement in conservation efforts.

The involvement of technology in the conservation has transformed the way of looking at the management of the ecosystem and the preservation of the biodiversity. Emerging tools , such as remote sensing, geographic information systems (GIS), and drones have enhanced the capacity to monitor environmental changes and assess the health of ecosystems. As an example, precision habitat mapping with the use of these technologies can result in conservationists knowing the key areas that need protection and restoration. Moreover, virtualization of data implies that conservation efforts are informed by data, making it easy to make informed decisions regarding resource distribution and policy making. Despite these advances, there is an ongoing debate regarding the commodification of biological resources through technological means, as highlighted by ecological critiques of value representation in conservation practices, which prompts a need for broader ecological discourse in policy implementation (Aslaksen et al., 2012). Besides, technology has allowed a more participatory kind of conservation, where the local people are involved in sustainable practices, which combine traditional ecological knowledge with contemporary practices. Awareness and education through the digital medium enable the Indigenous population to use their ancestral knowledge to take care of the environment. An example that can be given here is the situation of Dayak people in Kalimantan who practice shifting cultivation and agro forestry systems, which highlights the need to combine technological changes with local knowledge. Their strategy shows how technology can assist the conventional ways of conserving biodiversity as well as dealing with the economic needs of the local people. Such strategies not only promote ecological balance but also challenge prevailing narratives that often vilify Indigenous practices, emphasizing instead their potential in sustainable resource management (Setyawan et al., 2010). Finally, the ethics of using technology in conservation are also worth paying attention to. Technology holds tremendous promise as a means to increase the effectiveness of conservation, but also evokes accessibility, equity and the loss of complex ecological interactions that are simplified to data points. The move to measure biodiversity and ecosystem services threatens to swamp the more intangible values of nature that are not readily amenable to measurement. This view is in line with the demands of a more integrative approach, where biophysical indicators are coupled with social and ethical concerns. The issue is to find the balance between technological innovations and philosophical approaches that would take into account the multiple values of ecosystems and keep conservation measures holistic and comprehensive. As the discourse surrounding technology in conservation evolves, it is crucial that practitioners remain vigilant to the potential pitfalls of reductionism (Aslaksen et al., 2012) (Figure 12.18).

Figure 12.18. The role of technology in conservation.

Source: https://fastercapital.co/i/Preservation--Preserving-the-Splendor--The-Conservation-of-Crown-Jewels--The-Role-of-Technology-in-Conservation.webp.

12.6.1. Use of Remote Sensing and GIS in Ecosystem Monitoring

The integration of remote sensing and GIS has revolutionized the monitoring and management of ecosystems, providing scientists and conservationists with valuable tools for assessing environmental change over time. Remote sensing allows information collection from satellites and aerial platforms, enabling observation of large and often inaccessible areas. High-resolution images can reveal landscape patterns, vegetation health, and land cover change that are fundamental to knowing ecosystem dynamics. GIS provides such spatial information to be mapped and analyzed, making it possible for the identification of trends, determining relationships between ecological variables, and facilitating decision-making on conservation programs.

These are technologies that one cannot do away with when it comes to the threat that faces us from human activities and global warming, which is the premise of the need for effective measures in ecosystem monitoring in conservation biology. Ecosystem monitoring using remote sensing and GIS has played a definite role in the evaluation of target ecosystems, , such as the mangroves, which act as cradles for biodiversity and coastal protection. For instance, the application of satellite time-series information, e.g., previous land cover maps, enables researchers to track mangrove area alterations and assess their health over time. One study conducted in the Bhitarkanika Wildlife Sanctuary presented remote sensing techniques that were capable of measuring biophysical parameters , such as leaf chlorophyll, leaf area index, and gross primary productivity, providing insights into mangrove ecosystem resilience against climate change stress. The deduction drawn between meteorological observations and biophysical characteristics further demonstrates how such technologies can be used to project future conditions and facilitate anticipatory management action to preserve key habitats. Employing such techniques, conservation practice can be more appropriately designed to address specific ecosystem concerns, improving biodiversity and world ecological health outcomes. In addition, remote sensing and GIS are not only employed in the monitoring of static habitats; they are utilized to observe the movement patterns and habitat utilization of species. Information concerning the behavior of endangered species, , such as the western gorilla, is crucial for the development of efficient conservation actions. Satellite data have the potential to demarcate the spatial pattern of gorilla density relative to resource density, allowing scientists to identify high-priority areas for conservation.

The ability to analyze variability of resources within the diet of these frugivorous animals, as determined from various remote sensing parameters, has provided novel opportunities in ecosystem monitoring. The linkage between changes in food availability and data collected by satellites allows scientists to better understand nutritional dynamics driving these animals and ultimately translates into conservation applications towards maintaining ecosystem integrity. With the compulsion of persisting environmental fluctuation, such technologies constitute a necessary framework for subsequent research and management strategies in conservation biology.

12.6.2. Genetic Tools for Species Conservation

The application of genetic tools in species conservation is currently a vital strategy in the control and preservation of biodiversity. With improved molecular genetics, conservation biologists can utilize techniques , such as DNA barcoding, genomic sequencing, and population genetics analysis to assess the genetic diversity of endangered species and describe unique populations. These techniques enable scientists to monitor genetic health, flexibility, and resilience to environmental change, all of which are key attributes to survival in species that cannot interbreed. The genetic composition of populations can be understood to conservation biologists, enabling them to more effectively plan management, holding resources aside for the genetically most healthy groups. This information not only aids in conservation but also enhances the management of ecosystems reliant on these organisms for ecological function and balance, and thus general ecosystem health. Genetic tools are also important in providing information on the influence of climate change on biodiversity, which is significant in the current conservation situation. When climatic change reshapes habitats and species ranges, the genetic adaptability of species becomes increasingly vital.

There is evidence that ecological community composition in the Quaternary was significantly influenced by climate change, driving evolutionary processes, adaptation, and species and population extinction dynamics (Stewart, 2010). With the introduction of genetic markers, scientists can assess the ability of species to adapt in modified habitats such that habitat restoration and population management decisions are informed. The use of genetics in conservation biology not only contributes to the conservation of individual species but also enhances ecological resilience to environmental changes at a global level. Moreover, genetic technologies intersect to soil biodiversity, which increasingly is viewed as the foundation of ecosystem services and human well-being. Healthy soil ecosystems are driven by subtle relationships between various organisms, ranging from microbial assemblages to larger animals, all of which together deliver key services including nutrient cycling and water purification (Van der Putten et al., 2010). Genetic methods allow comprehensive assessment of soil biodiversity, which can supply vital information to guide conservation management for the sustenance of soil health. An awareness of the genetic diversity in the populations can facilitate the planning of optimal land use and farming practices that enhance soil resilience without compromising but rather allowing the conservation of many useful species. Thus, the integration of genetic tools and ecosystem management is a sophisticated model of conservation addressing the health of both species and the habitat, ultimately establishing sustainable environments (Figure 12.19).

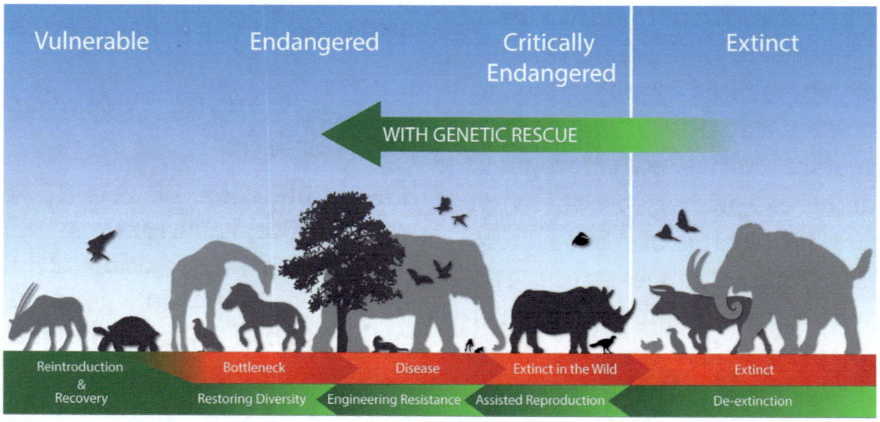

Figure 12.19. Cool genetic tools that could save biodiversity.

Source: https://reviverestore.org/wp-content/uploads/2017/05/genetic-rescue-continuum.jpg.

12.6.3. Citizen Science and Public Engagement in Conservation Efforts

The integration of citizen science in conservation is a paradigm shift in the gathering and application of information in ecosystem management. Through involvement of the general public in systematic observation and data gathering, citizen science fosters an awareness and sense of responsibility on the part of community members towards their own immediate surroundings. Platforms , such as Anecdata.org are at the forefront of such innovation by enabling citizens to measure variables , such as water quality and marine life, facilitating policymakers and conservationists to make critical decisions. Democratization of the scientific process is not only beneficial in bringing parity for the citizen but also expands the volume and diversity of data, which can be an effective tool in solving key environmental challenges. Through these participatory practices, citizens become accountable actors working towards conservation, guaranteeing an enlightened citizenry engaged in the governance of their own ecosystems, resulting in a more resilient nature. Citizen science also makes educational pathways possible, which equip individuals with the capability to understand the complexities of ecological systems and climate change effects. Projects exhibited on platforms like Anecdata.org involve the public through enhanced awareness of environmental patterns and concerns within areas like Maine's coastal ecosystems. Such an educational component is critical as it bridges the gap between the realm of scientific inquiry and society at large, leading to informed discussion in conservation. Through direct participation in data gathering and analysis, citizens are exposed directly to ecological processes, fostering a culture of environmental literacy. This increased consciousness can, in return, drive advocacy movements, as individuals who understand their ecosystem's issues are more likely to act on agendas at the local scale, enriching public engagement in the entire field of conservation biology. Citizen science's impacts do not extend to data collection only; they influence policymaking since they provide an evidence-based ground where environmental issues can be addressed by stakeholders. Citizen data can supplement coastal management plans and decision-making where conventional information sources are lacking. Trend analysis derived from citizen-collected information, for example, can provide significant insights into the state of marine ecosystems that may otherwise remain unnoticed. Agencies can formulate plans responsive to the needs of society and the environment by incorporating such information into the policy-making process. In practice, such citizen-scientist collaboration not only contributes to the science-based robustness of conservation policies but also underscores the importance of people's involvement at the grassroots levels towards achieving effective change in conservation policies and practices on the basis of a shared responsibility for creating and protecting our environments (Charles Stewart Mott Foundation, 2011; Bailey et al., 2023).

Test Your Knowledge

1. What is the use of remote sensing and GIS in ecosystem monitoring?
2. What are the genetic tools for species conservation?

SUMMARY

This conservation biology and ecosystem research, it is critical to factor in the intricate relationships that serve as the foundation for biodiversity and ecological system integrity. Decline in species diversity driven by human-induced factors , such as habitat destruction and climate change threatens conservation in a big way. As noted, diversity among species is fundamental to the wellbeing of ecosystems; without it, resilience in natural systems can significantly be compromised, initiating previously unpredicted effects on the environment and human civilization at large. This acknowledgment of biodiversity as the basis for sustainable development emphasizes the necessity of an effective conservation response in combating the increasingly looming threat of extinction facing numerous species under relentless anthropogenic stress (Jay et al., 1998). In addition, interactions among various species are complex and often tenuous, requiring a conservational approach that acknowledges such interdependencies. The role of climate change, as with prior ecological community shifts within the Quaternary, illustrates how species evolution, adaptation, and extinction are all related processes that result from changing environments. It is a poignant reminder that conservation biology must be not only concerned about the maintenance of currently existing species but also worried about maximizing ecological resilience and responsiveness. Engagement with the realities of climate dynamics will enhance the effectiveness of conservation efforts to maintain biodiversity, ultimately allowing ecosystems to adapt and persevere amidst constant environmental change (Stewart, 2010). Finally, the end of this discussion comes down to the importance of an interdisciplinary response that combines scientific study, policymaking, and community participation in the conservation of the natural world. The stakeholders need to get together and make these approaches more effective by promoting sustainable practices that focus on biodiversity at local and global scales. Public education and awareness of the critical functions' ecosystems perform and the risks they are exposed to can incite people to support conservation, both for us and future generations to benefit from healthy, rich ecosystems. As time has passed, it is evident that sustainable ecosystem management will need to remain an issue because it is the key to long-term ecological balance and conservation of the precious biodiversity of our planet.

In understanding the intricate dynamics of ecosystems, several key points come to the fore that underscore the significance of conservation biology. The impact of climate change on ecological community structure is arguably the most significant, particularly in the Quaternary. The evidence suggests that climate-forced changes had an effect on evolutionary processes, with adaptation and extinction of numerous species and populations resulting. Understanding how ecological composition arises in response to climatic change not only teaches us about historical trends in biodiversity but also tells us why it is so important to conserve today. As these historical data are utilized to inform contemporary conservation efforts, they underscore the interconnectedness of climatic change, species evolution, and ecosystem resilience, offering a foundation for future conservation biology research (Stewart, 2010). Furthermore, the Anthropocene era brings unprecedented challenges to wildlife species, resulting in accelerated population declines worldwide. Wildlife ecologists have come to increasingly appreciate the value of synthesizing technological innovation and large datasets to comprehend these declines. Following the challenges experienced by wildlife in human-dominated environments, scientists have devised novel methods that shed light on animal behavior and movement patterns. However, there is still a need to keep incorporating human factors into wildlife ecology studies because knowing people's beliefs, practices, and the impacts of infrastructure on them is unavoidable if conservation is to be dealt with in a proper manner. The research in this field emphasizes the need for considering a holistic system where one is very careful to include both ecological and human factors in conservation (McInturff et al., 2019). Finally, the importance of anthropogenic effects on ecosystems is pertinent to other areas of study in conservation biology. These include comprehending interactions between human structures and wildlife, with an example being the ecological effects of fencing, a ubiquitous feature found in modern landscapes. Experiments indicate that fences significantly affect the behavior of animals, habitat access, and consequently, species population trends. By the recognition and solving of such complexities, conservation biology can move forward in enhancing biodiversity management in an evolving world. Prioritizing the application of socio-ecological models enhances not only our understanding of human-wildlife interactions but also maximizes the effectiveness of conservation, ultimately contributing to more sustainable ecosystems. Climate-human action-ecosystem interaction therefore, remains a core position in conservation method development (McInturff et al., 2019).

KEY TERMS FROM THIS CHAPTER

Abiotic factors	Biodiversity	Biomass
Biotic factors	Climate change	Conservation
Decomposers	Ecological footprint	Ecological succession
Ecosystem	Endangered species	Energy flow
Food web	Habitat fragmentation	Invasive species
Primary producers	Restoration ecology	Sustainability

CRITICAL THINKING QUESTIONS

1. **What is an ecosystem?**
   a. A group of the same species living together
   b. All the living organisms and their physical environment interacting as a system
   c. A single organism's habitat
   d. Only the non-living components of an environment
2. **Which process describes the movement of energy through an ecosystem?**
   a. Photosynthesis
   b. Energy flow
   c. Decomposition
   d. Nitrogen fixation
3. **Which of the following is a major cause of biodiversity loss?**
   a. Habitat destruction
   b. Photosynthesis
   c. Evolution
   d. Natural selection
4. **What is the main goal of conservation biology?**
   a. To increase human population
   b. To protect and preserve biodiversity
   c. To develop new species
   d. To eliminate all predators
5. **Which of the following is an example of a renewable resource?**
   a. Fossil fuels
   b. Minerals
   c. Solar energy
   d. Coal

Short Answer Questions

1. What is an ecosystem, and what are its main components?
2. Describe the flow of energy through an ecosystem.
3. What is a food chain, and how does it differ from a food web?
4. Explain the importance of biodiversity in ecosystems.
5. What are biogeochemical cycles? Name and briefly describe one.
6. What are some major threats to biodiversity and ecosystems?

Answers to Critical Thinking Questions

1. (b); 2. (b); 3. (a); 4. (b); 5. (c)

REFERENCES

1. Ackerly, D. D., Alex, K., Barnosky, A. D., et al. (2017). Merging paleobiology with conservation biology to guide the future of terrestrial ecosystems. https://core.ac.uk/download/286035957.pdf (accessed on 13th June 2025).
2. Adams, B., Berkes, F., de Athayde, S. F., et al. (2009). The intersections of biological diversity and cultural diversity: Towards integration. https://core.ac.uk/download/9064509.pdf (accessed on 13th June 2025).
3. Adler, P. B., Akçakaya, H. R., Alvarez-Filip, L., Austin, M., Boitani, L., Brooks, T. M., Burgman, M. A., et al. (2015). The IUCN Red List of Ecosystems: Motivations, challenges, and applications. https://core.ac.uk/download/pdf/30671718.pdf (accessed on 13th June 2025).
4. Alkemade, R., Ashmore, M., Bobbink, R., et al. (2009). Global assessment of nitrogen deposition effects on terrestrial plant diversity: A synthesis. https://core.ac.uk/download/49720.pdf (accessed on 13th June 2025).
5. Armitage, R., Jay, G. M. M., & Morad, M. (2004). Managing soil biodiversity: The New Zealand experience. https://core.ac.uk/download/29195192.pdf (accessed on 13th June 2025).
6. Aslaksen, I., & Spash, C. L. (2012). Re-establishing an ecological discourse in the debate over the value of ecosystems and biodiversity. https://core.ac.uk/download/11007889.pdf (accessed on 13th June 2025).
7. Attruk, M. F. H., Nisa, L. C., Putra, R. P. B., et al. (2024). Pemberdayaan masyarakat melalui pendidikan biologi untuk menjaga keanekaragaman hayati dan keseimbangan ekosistem. https://core.ac.uk/download/636367393.pdf (accessed on 13th June 2025).
8. Auge, H., Baessler, C., Bennett, E. M., et al. (2020). Resilience trinity: Safeguarding ecosystem functioning and services across three different time horizons and decision contexts. https://core.ac.uk/download/288113967.pdf (accessed on 13th June 2025).
9. Bailey, C., Bisson, B., Dabulewicz, A., et al. (2023). Citizen and community science approaches to understanding changes in coastal habitats using Anecdata.org. https://core.ac.uk/download/604795638.pdf (accessed on 13th June 2025).
10. Bashir, S., Becker, M. S., Beudels Jamar, R. C., et al. (2015). Developing fencing policies in dryland ecosystems. https://core.ac.uk/download/54248171.pdf (accessed on 13th June 2025).
11. Beisner, B. E., Gessner, M. O., Messier, C., et al. (2022). Pathways for cross-boundary effects of biodiversity on ecosystem functioning. https://core.ac.uk/download/534017789.pdf (accessed on 13th June 2025).
12. Beninde, J., Veith, M., & Hochkirch, A. (2015). Biodiversity in cities needs space: A meta-analysis of factors determining intra-urban biodiversity variation. Ecology Letters, 18, 581–592. https://doi.org/10.1111/ele.12427 (accessed on 13th June 2025).
13. Beyene, B. (2016). Ecological and economic roles of agrobiodiversity. https://core.ac.uk/download/234696321.pdf (accesse*d on 13th June 2025*).
14. Blasi, C., Capotorti, G., Copiz, R., et al. (2018). Urban and rural green infrastructure: Two projects for the metropolitan city of Rome. https://core.ac.uk/download/188823681.pdf (accessed on 13th June 2025).

15. Boitani, L., Mace, G. M., & Rondinini, C. (2014). Challenging the scientific foundations for an IUCN Red List of Ecosystems. https://core.ac.uk/download/54503791.pdf (accessed on 13th June 2025).

16. Bouchard, V., Feunteun, E., Laffaille, P., et al. (2003). Biodiversity in salt marshes: From the patrimonial value to the ecosystem functioning. The case study of the Mont Saint-Michel Bay. https://core.ac.uk/download/12043523.pdf (accessed on 13th June 2025).

17. Bremerich, V., Jähnig, S. C., Martens, K., et al. (2019). The Freshwater Information Platform: A global online network providing data, tools and resources for science and policy support. https://core.ac.uk/download/225124729.pdf (accessed on 13th June 2025).

18. Britchenko, I., Krupskyi, O., Lavrentieva, O., et al. (2020). Issues of shaping the students' professional and terminological competence in science area of expertise in the sustainable development era. https://core.ac.uk/download/305120659.pdf (accessed on 13th June 2025).

19. Charles Stewart Mott Foundation. (2001). Charles Stewart Mott Foundation–2000 Annual Report. https://www.mott.org/wp-content/uploads/2001/06/AR2000.pdf(accessed on 13th June 2025).

20. Costa-Dias, S., Sousa, R., Lobón-Cerviá, J., Laffaille, P., (SPAIN), M. N. d. C. N., (PORTUGAL), U. d. P. -. U., . . . Biodiversité fonctionnelle et gestion des territoires. (2009). The decline of diadromous fish in Western Europe inland waters: Mains causes and consequence. Nova Science Publishers. (accessed on 13th June 2025).

21. Dahale, S., Kumar, A., Miranda, I., et al. (2025). Identifying and forecasting potential biophysical risk areas within a tropical mangrove ecosystem using multi-sensor data. https://core.ac.uk/download/pdf/288485770.pdf (accessed on 13th June 2025).

22. del Castillo, R. F., Echeverría, C., Geneletti, D., et al. (2012). Forest landscape restoration in the drylands of Latin America. https://core.ac.uk/download/19792927.pdf (accessed on 13th June 2025).

23. Elliott, M., Mander, L., Mazik, K., et al. (2016). Ecoengineering with ecohydrology: Successes and failures in estuarine restoration. https://core.ac.uk/download/151161494.pdf (accessed on 13th June 2025).

24. Esler, K. J., Fisher, J. L., Gaertner, M., et al. (2012). Insights into invasion and restoration ecology: Time to collaborate towards a holistic approach to tackle biological invasions. https://core.ac.uk/download/18325893.pdf (accessed on 13th June 2025).

25. Gillespie, A. (2009). From the Galápagos to Tongariro: Recognizing and saving the most important places in the world. https://core.ac.uk/download/29196164.pdf (accessed on 13th June 2025).

26. Goldberger, A., Herman, L., & Fiske, L. (2013). Securing the livelihoods and nutritional needs of fish-dependent communities. https://core.ac.uk/download/80510826.pdf (accessed on 13th June 2025).

27. Greggor, A. L., Clayton, N. S., Phalan, B., & Thornton, A. (2014). Comparative cognition for conservationists. Trends in Ecology & Evolution, 29, 489–495. https://doi.org/10.1016/j.tree.2014.06.004 (accessed on 13th June 2025).

28. Gupta, M.V.; Bartley, D.M.; Acosta, B.O. (eds.). (2004). Use of genetically improved and alien species for aquaculture and conservation of aquatic biodiversity in Africa. WorldFish Center Conf. Proc. (68): 107 p.(accessed on 13th June 2025).

29. *Hassan*in, A., Vogler, A., Smets, E., Delsuc, F., Petersen, G., Bininda-Emonds, O. R. P., Seberg, O., et al. (2009). Assembling the Tree of Life in Europe (AToLE). https://core.ac.uk/download/pdf/288307.pdf (accessed on 13th June 2025).

30. Jay, G. M. M., & Morad, M. (1998). The conservation of New Zealand's biological heritage and the role of environmental planners. https://core.ac.uk/download/29195489.pdf (accessed on 13th June 2025).

31. Jones, K., Schneider, D. C., & Snelgrove, P. (2007). Marine protected areas in Canada with a particular emphasis on Newfoundland: Science, policy and implementation at multiple institutional levels. https://core.ac.uk/download/11700601.pdf (accessed on 13th June 2025).

32. Kowarik, I., Planchuelo, G., & von der Lippe, M. (2020). Endangered plants in novel urban ecosystems are filtered by strategy type and dispersal syndrome, not by spatial dependence on natural remnants. https://core.ac.uk/download/322792243.pdf (accessed on 13th June 2025).

33. McInturff, M. C. A. (2019). Incorporating human beliefs and behaviors into wildlife ecology. https://core.ac.uk/download/287621744.pdf (accessed on 13th June 2025).

34. *Stewart, J. R., 2010. Understanding evoluti*onary processes during past Quaternary climatic cycles: Can it be applied to the future? In: Grant, F., Mergeay, J., Santamaria, L., Young, J. and Watt, A.D., eds. Evolution and Biodiversity: The evolutionary basis of biodiversity and its potential for adaptation to global change. Report of an e. Mallorca, Spain: Centre for Ecol*ogy & Hydrology, INBO, IMEDA, CSIC-UIB.(accessed on 13th J*une 2025).

35. van Apeldoorn, R**. C., de Blust**, G., et al. (2009). Information and communication on the designation and management of Natura 2000 sites. Main Report 3: Towards integrated management. https://core.ac.uk/download/56183.pdf (accessed on *13th June 2025).*

36. *van der Putten, W. H., Mudgal, S., Turbé, A.*, de Toni, A., Lavelle, P., Benito, P., & Ruiz, N. (2010). Soil biodiversity: functions, threats and tools for policy makers. Bio Intelligence Service. http://ec.europa.eu/environment/archives/soil/pdf/biodiversity_report.pdf(accessed on 13th June 2025).

37. van Klink, R., van der Plas, F., van Noordwijk, C. G., WallisDeVries, M. F., & Olff, H. (2015). Effects of large herbivores on grassland arthropod diversity. Biological reviews of the Cambridge Philosophical Society, 90(2), 347–366. https://doi.org/10.1111/brv.12113Setyawan, A. D. (2010). Review: Biodiversity conservation strategy in a native perspective; Case study of shifting cultivation at the Dayaks of Kalimantan. https://core.ac.uk/download/12345898.pdf (accessed on 13th June 2025).

38. Wells, J., D. Roberts, P. Lee, R, Cheng and M. Darveau. 2010. A Forest of Blue—Canada's Boreal Forest: The World's Waterkeeper. International Boreal Conservation Campaign, Seattle. 74 pp.(accessed on 13th June 2025).

39. Zerbi, A., Aliaume, C., Lasserre, G., Garcia, S. M., & Do Chi, T. (2003). The ecosystem approach to fisheries: Issues, terminology, principles, institutional foundations, implementation and outlook. https://core.ac.uk/download/75777389.pdf (accessed on 13th June 2025).

INDEX

C

D

E

F

N

O

P

Q

R

S